Lecture Notes in Computer Science   12544

More information about this subseries at http://www.springer.com/series/7412

Zeynep Akata · Andreas Geiger ·
Torsten Sattler (Eds.)

Pattern Recognition

42nd DAGM German Conference, DAGM GCPR 2020
Tübingen, Germany, September 28 – October 1, 2020
Proceedings

 Springer

Editors
Zeynep Akata (iD)
University of Tübingen
Tübingen, Germany

Andreas Geiger
University of Tübingen
Tübingen, Germany

Torsten Sattler (iD)
Czech Technical University in Prague
Prague, Czech Republic

ISSN 0302-9743   ISSN 1611-3349 (electronic)
Lecture Notes in Computer Science
ISBN 978-3-030-71277-8   ISBN 978-3-030-71278-5 (eBook)
https://doi.org/10.1007/978-3-030-71278-5

LNCS Sublibrary: SL6 – Image Processing, Computer Vision, Pattern Recognition, and Graphics

This Springer imprint is published by the registered company Springer Nature Switzerland AG
The registered company address is: Gewerbestrasse 11, 6330 Cham, Switzerland

Editors
Zeynep Akata 🆔
University of Tübingen
Tübingen, Germany

Andreas Geiger
University of Tübingen
Tübingen, Germany

Torsten Sattler 🆔
Czech Technical University in Prague
Prague, Czech Republic

ISSN 0302-9743   ISSN 1611-3349   (electronic)
Lecture Notes in Computer Science
ISBN 978-3-030-71277-8   ISBN 978-3-030-71278-5   (eBook)
https://doi.org/10.1007/978-3-030-71278-5

LNCS Sublibrary: SL6 – Image Processing, Computer Vision, Pattern Recognition, and Graphics

This Springer imprint is published by the registered company Springer Nature Switzerland AG
The registered company address is: Gewerbestrasse 11, 6330 Cham, Switzerland

Zeynep Akata · Andreas Geiger ·
Torsten Sattler (Eds.)

Pattern Recognition

42nd DAGM German Conference, DAGM GCPR 2020
Tübingen, Germany, September 28 – October 1, 2020
Proceedings

 Springer

Preface

It was our honor and pleasure to organize the 42nd German Conference on Pattern Recognition (DAGM GCPR 2020), held virtually between September 28th and October 1st, 2020. For the first time, DAGM GCPR was held in parallel with the 25th International Symposium on Vision, Modeling, and Visualization (VMV 2020) and the 10th Eurographics Workshop on Visual Computing for Biology and Medicine (VCBM 2020). All three meetings shared organizational support. DAGM GCPR 2020 had 289 participants from 21 countries.

Originally, DAGM GCPR 2020 was planned to take place in Tübingen, located in central Baden-Württemberg in southwest Germany and home to one of Europe's oldest universities. However, holding the conference in person was not possible due to the COVID-19 pandemic, which placed strict restrictions on travel and meetings. Like many other conferences in 2020, DAGM GCPR was thus held fully virtually.

The call for papers for DAGM GCPR 2020 resulted in 89 submissions from 22 countries. As in previous years, DAGM GCPR 2020 offered special tracks on the topics of Computer vision systems and applications (chaired by Bodo Rosenhahn and Carsten Steger), Pattern recognition in the life- and natural sciences (chaired by Joachim Denzler and Xiaoyi Jiang), and Photogrammetry and remote sensing (chaired by Helmut Mayer and Uwe Sörgel). Each paper was subject to a double-blind review process and was reviewed by three reviewers. One of the reviewers acted as a meta-reviewer for the paper, led the discussion of the paper once all reviews were available, and made a publish/reject recommendation to the Program Chairs. For the special track papers, the track chairs acted as meta reviewers.

As in previous years, DAGM GCPR 2020 also welcomed submissions to the Young Researcher Forum (YRF). The YRF is meant to promote promising young researchers, i.e., students who recently finished their Master, and to provide visibility to them. The requirement for the YRF submissions was that the submission had to be based on a Master thesis, with the Master student being the first author of the submission.

Out of the 89 submissions, 20 papers were submitted to the special tracks (14 submissions for the Computer vision systems and applications track, 5 submissions for the Pattern recognition in the life- and natural sciences track, and 1 submission for the Photogrammetry and remote sensing track) and 7 were submitted to the Young Researcher Forum. The Program Chairs decided to reject 5 submissions before review due to violation of the double-blind review process, missing files, etc. Of the remaining submissions, 35 high-quality papers were selected for publication (39% acceptance rate), with one paper being withdrawn later by the authors due to issues with their funding source. Among these 35 accepted papers, 5 were YRF submissions. 4 papers were chosen for oral sessions held jointly with the other two meetings, 10 were selected as DAGM GCPR orals, and 21 were selected for spotlight presentations. All accepted papers were presented live via talks given by one of their authors. These talks were live-streamed on YouTube and remain publicly accessible there. Discussions took place after the sessions through the Discord platform.

Overall, the accepted papers covered a wide spectrum of topics from the areas of pattern recognition, machine learning, image processing, and computer vision. Among the accepted non-YRF papers, 5 papers were nominated for the GCPR best paper award. The papers were selected based on the scores provided by the reviewers and meta-reviewers. A committee consisting of two Program Chairs and four meta-reviewers of the nominated papers selected the best paper and two honorable mentions among the 5 nominees.

Besides the accepted papers, which were presented in a single-track program, DAGM GCPR 2020 featured a day of invited talks and three keynotes, the latter of which were shared with the other two meetings. We are thankful to the seven internationally renowned researchers who accepted our invitations to give invited talks: Matthias Bethge (University of Tübingen, Germany), Sabine Süsstrunk (EPFL, Switzerland), Vittorio Ferrari (Google, Switzerland), Bernt Schiele (MPI, Germany), Siyu Tang (ETH Zurich, Switzerland), Christoph Lampert (IST Austria, Austria), and Davide Scaramuzza (University of Zurich, Switzerland). The keynote talks were given by Vladlen Koltun (Intel), Jan Kautz (NVIDIA), and Hans-Christian Hege (Zuse Institute Berlin). In addition, DAGM GCPR 2020 provided two industry talks, which were delivered by Michael Hirsch (Amazon) and Alexey Dosovitskiy (Google AI Brain).

The success of DAGM GCPR 2020 would have been impossible without the efforts and support of many people and institutions. We thank all the authors for their submissions to DAGM GCPR 2020 and all the reviewers for their commitment and quality of work. We also like to thank our sponsors Amazon (Gold Sponsor), Google (Bronze Sponsor), KAUST (Academic Sponsor), Daimler (Best Paper/Award Sponsor), MVTec Software GmbH (Best Paper/Award Sponsor), and COGNEX (Best Paper/Award Sponsor). We are very grateful for the support from our partners, Eberhard Karls Universität Tübingen, Informatik Forum Stuttgart, Deutsche Arbeitsgemeinschaft für Mustererkennung e.V. (DAGM), and Gesellschaft für Informatik. Special thanks go to all the organizers and the technical team supporting the three meetings. All credit for making DAGM GCPR a successful virtual conference on short notice goes to them. Additionally, we are grateful to Springer for giving us the opportunity to continue publishing the DAGM GCPR proceedings as part of their LNCS series and for a special issue of IJCV dedicated to the best papers from the conference.

As a reader, we hope you will enjoy the proceedings of DAGM GCPR 2020. We hope to see you again at the next DAGM GCPR in Bonn.

October 2020   Zeynep Akata
Andreas Geiger
Torsten Sattler

live-streamed on YouTube and remain publicly accessible there. Discussions took place after the sessions through the Discord platform.

Overall, the accepted papers covered a wide spectrum of topics from the areas of pattern recognition, machine learning, image processing, and computer vision. Among the accepted non-YRF papers, 5 papers were nominated for the GCPR best paper award. The papers were selected based on the scores provided by the reviewers and meta-reviewers. A committee consisting of two Program Chairs and four meta-reviewers of the nominated papers selected the best paper and two honorable mentions among the 5 nominees.

Besides the accepted papers, which were presented in a single-track program, DAGM GCPR 2020 featured a day of invited talks and three keynotes, the latter of which were shared with the other two meetings. We are thankful to the seven internationally renowned researchers who accepted our invitations to give invited talks: Matthias Bethge (University of Tübingen, Germany), Sabine Süsstrunk (EPFL, Switzerland), Vittorio Ferrari (Google, Switzerland), Bernt Schiele (MPI, Germany), Siyu Tang (ETH Zurich, Switzerland), Christoph Lampert (IST Austria, Austria), and Davide Scaramuzza (University of Zurich, Switzerland). The keynote talks were given by Vladlen Koltun (Intel), Jan Kautz (NVIDIA), and Hans-Christian Hege (Zuse Institute Berlin). In addition, DAGM GCPR 2020 provided two industry talks, which were delivered by Michael Hirsch (Amazon) and Alexey Dosovitskiy (Google AI Brain).

The success of DAGM GCPR 2020 would have been impossible without the efforts and support of many people and institutions. We thank all the authors for their submissions to DAGM GCPR 2020 and all the reviewers for their commitment and quality of work. We also like to thank our sponsors Amazon (Gold Sponsor), Google (Bronze Sponsor), KAUST (Academic Sponsor), Daimler (Best Paper/Award Sponsor), MVTec Software GmbH (Best Paper/Award Sponsor), and COGNEX (Best Paper/Award Sponsor). We are very grateful for the support from our partners, Eberhard Karls Universität Tübingen, Informatik Forum Stuttgart, Deutsche Arbeitsgemeinschaft für Mustererkennung e.V. (DAGM), and Gesellschaft für Informatik. Special thanks go to all the organizers and the technical team supporting the three meetings. All credit for making DAGM GCPR a successful virtual conference on short notice goes to them. Additionally, we are grateful to Springer for giving us the opportunity to continue publishing the DAGM GCPR proceedings as part of their LNCS series and for a special issue of IJCV dedicated to the best papers from the conference.

As a reader, we hope you will enjoy the proceedings of DAGM GCPR 2020. We hope to see you again at the next DAGM GCPR in Bonn.

October 2020   Zeynep Akata
                                                               Andreas Geiger
                                                               Torsten Sattler

Preface

It was our honor and pleasure to organize the 42nd German Conference on Pattern Recognition (DAGM GCPR 2020), held virtually between September 28th and October 1st, 2020. For the first time, DAGM GCPR was held in parallel with the 25th International Symposium on Vision, Modeling, and Visualization (VMV 2020) and the 10th Eurographics Workshop on Visual Computing for Biology and Medicine (VCBM 2020). All three meetings shared organizational support. DAGM GCPR 2020 had 289 participants from 21 countries.

Originally, DAGM GCPR 2020 was planned to take place in Tübingen, located in central Baden-Württemberg in southwest Germany and home to one of Europe's oldest universities. However, holding the conference in person was not possible due to the COVID-19 pandemic, which placed strict restrictions on travel and meetings. Like many other conferences in 2020, DAGM GCPR was thus held fully virtually.

The call for papers for DAGM GCPR 2020 resulted in 89 submissions from 22 countries. As in previous years, DAGM GCPR 2020 offered special tracks on the topics of Computer vision systems and applications (chaired by Bodo Rosenhahn and Carsten Steger), Pattern recognition in the life- and natural sciences (chaired by Joachim Denzler and Xiaoyi Jiang), and Photogrammetry and remote sensing (chaired by Helmut Mayer and Uwe Sörgel). Each paper was subject to a double-blind review process and was reviewed by three reviewers. One of the reviewers acted as a meta-reviewer for the paper, led the discussion of the paper once all reviews were available, and made a publish/reject recommendation to the Program Chairs. For the special track papers, the track chairs acted as meta reviewers.

As in previous years, DAGM GCPR 2020 also welcomed submissions to the Young Researcher Forum (YRF). The YRF is meant to promote promising young researchers, i.e., students who recently finished their Master, and to provide visibility to them. The requirement for the YRF submissions was that the submission had to be based on a Master thesis, with the Master student being the first author of the submission.

Out of the 89 submissions, 20 papers were submitted to the special tracks (14 submissions for the Computer vision systems and applications track, 5 submissions for the Pattern recognition in the life- and natural sciences track, and 1 submission for the Photogrammetry and remote sensing track) and 7 were submitted to the Young Researcher Forum. The Program Chairs decided to reject 5 submissions before review due to violation of the double-blind review process, missing files, etc. Of the remaining submissions, 35 high-quality papers were selected for publication (39% acceptance rate), with one paper being withdrawn later by the authors due to issues with their funding source. Among these 35 accepted papers, 5 were YRF submissions. 4 papers were chosen for oral sessions held jointly with the other two meetings, 10 were selected as DAGM GCPR orals, and 21 were selected for spotlight presentations. All accepted papers were presented live via talks given by one of their authors. These talks were

Organization

General Chair

Andreas Geiger — University of Tübingen, Germany; Max Planck Institute for Intelligent Systems, Germany

Program Committee Chairs

Zeynep Akata — University of Tübingen, Germany
Andreas Geiger — University of Tübingen, Germany; Max Planck Institute for Intelligent Systems, Germany
Torsten Sattler — Czech Technical University in Prague, Czech Republic

Special Track Chairs

Joachim Denzler — Friedrich Schiller University Jena, Germany
Xiaoyi Jiang — University of Münster, Germany
Helmut Mayer — Bundeswehr University Munich, Germany
Bodo Rosenhahn — Leibniz University Hannover, Germany
Uwe Sörgel — Universität Stuttgart, Germany
Carsten Steger — MVTec Software GmbH, Germany

Program Committee

Stephan Alaniz — MPI Informatics, Germany
Björn Andres — TU Dresden, Germany
Misha Andriluka — Google, USA
Christian Bauckhage — Fraunhofer IAIS, Germany
Rodrigo Benenson — Google, Switzerland
Horst Bischof — Graz University of Technology, Austria
Andreas Bulling — University of Stuttgart, Germany
Daniel Cremers — TU Munich, Germany
Angela Dai — TU Munich, Germany
Joachim Denzler — Friedrich Schiller University Jena, Germany
Gernot Fink — TU Dortmund, Germany
Boris Flach — Czech Technical University in Prague, Czech Republic
Wolfgang Förstner — University of Bonn, Germany
Uwe Franke — Daimler R&D, Germany
Mario Fritz — CISPA Helmholtz Center for Information Security, Germany
Thomas Fuchs — Memorial Sloan Kettering Cancer Center, USA
Juergen Gall — University of Bonn, Germany
Guillermo Gallego — TU Berlin, Germany

Margrit Gelautz	Vienna University of Technology, Austria
Bastian Goldlücke	University of Konstanz, Germany
Matthias Hein	University of Tübingen, Germany
Christian Heipke	Leibniz Universität Hannover, Germany
Olaf Hellwich	TU Berlin, Germany
Otmar Hilliges	ETH Zurich, Switzerland
Vaclav Hlavac	Czech Technical University in Prague, Czech Republic
Xiaoyi Jiang	University of Münster, Germany
Margret Keuper	University of Mannheim, Germany
Anna Khoreva	Bosch Center for AI, Germany
Reinhard Koch	Kiel University, Germany
Ullrich Köthe	University of Heidelberg, Germany
Walter Kropatsch	TU Vienna, Austria
Arjan Kuijper	Fraunhofer Institute for Computer Graphics Research IGD, Germany; TU Darmstadt, Germany
Christoph Lampert	IST Austria, Austria
Laura Leal-Taixé	TU Munich, Germany
Bastian Leibe	RWTH Aachen University, Germany
Andreas Maier	Friedrich-Alexander University Erlangen-Nuremberg, Germany
Massimiliano Mancini	Sapienza University of Rome, Italy
Helmut Mayer	Bundeswehr University Munich, Germany
Bjoern Menze	TU Munich, Germany
Peter Ochs	Saarland University, Germany
Björn Ommer	Heidelberg University, Germany
Josef Pauli	University of Duisburg-Essen, Germany
Dietrich Paulus	University of Koblenz-Landau, Germany
Thomas Pock	Graz University of Technology, Austria
Gerard Pons-Moll	MPI Informatics, Germany
Christian Riess	Friedrich-Alexander University Erlangen-Nuremberg, Germany
Gerhard Rigoll	TU Munich, Germany
Bodo Rosenhahn	Leibniz University Hannover, Germany
Volker Roth	University of Basel, Switzerland
Carsten Rother	University of Heidelberg, Germany
Hanno Scharr	Forschungszentrum Jülich, Germany
Daniel Scharstein	Middlebury College, USA
Bernt Schiele	MPI Informatics, Germany
Konrad Schindler	ETH Zurich, Switzerland
Katja Schwarz	Max Planck Institute for Intelligent Systems, Germany
Monika Sester	Leibniz Universität Hannover, Germany
Uwe Sörgel	Universität Stuttgart, Germany
Carsten Steger	MVTec Software GmbH, Germany
Rainer Stiefelhagen	Karlsruhe Institute of Technology, Germany
Joerg Stueckler	Max-Planck-Institute for Intelligent Systems, Germany
Paul Swoboda	MPI Informatics, Germany

Siyu Tang	ETH Zurich, Switzerland
Martin Welk	UMIT Hall/Tyrol, Austria
Wenjia Xu	Institute of Electronics, Chinese Academy of Sciences, China

Technical Team

Andreas Engelhardt
Aditya Prakash
Jules Kreuer
Alexander Phi. Goetz
Tim Beckmann

Awards

GCPR Paper Awards

GCPR Best Paper Award

Bias Detection and Prediction of Mapping Errors in Camera Calibration

Annika Hagemann	Robert Bosch GmbH, Germany; Karlsruhe Institute of Technology, Germany
Moritz Knorr	Robert Bosch GmbH, Germany
Holger Janssen	Robert Bosch GmbH, Germany
Christoph Stiller	Karlsruhe Institute of Technology, Germany

GCPR Honorable Mention

Learning to Identify Physical Parameters from Video Using Differentiable Physics

Rama Krishna Kandukuri	Max Planck Institute for Intelligent Systems, Germany; University of Siegen, Germany
Jan Achterhold	Max Planck Institute for Intelligent Systems, Germany
Michael Moeller	University of Siegen, Germany
Joerg Stueckler	Max Planck Institute for Intelligent Systems, Germany

Characterizing the Role of a Single Coupling Layer in Affine Normalizing Flow

Felix Draxler	Heidelberg University, Germany
Jonathan Schwarz	Heidelberg University, Germany
Christoph Schnörr	Heidelberg University, Germany
Ullrich Köthe	Heidelberg University, Germany

DAGM Awards

DAGM German Pattern Recognition Award 2020

Matthias Nießner, TU Munich, for his pioneering research on tracking, reconstructing and visualizing photorealistic 3D face models from video with machine learning and AI.

DAGM MVTec Dissertation Award 2020

Robust Methods for Dense Monocular Non-Rigid 3D Reconstruction and Alignment of Point Clouds

Vladislav Golyanik   Technical University of Kaiserslautern, Germany

DAGM YRF Best Master's Thesis Award 2020

Synthesizing Human Pose from Captions

Yifei Zhang   RWTH Aachen University, Germany

Contents

Characterizing the Role of a Single Coupling Layer in Affine Normalizing Flows

Felix Draxler[1,2,3(✉)] ⓘ, Jonathan Schwarz[1,3] ⓘ, Christoph Schnörr[1,3] ⓘ, and Ullrich Köthe[1,2] ⓘ

[1] Heidelberg Collaboratory for Image Processing, Heidelberg University, Heidelberg, Germany
felix.draxler@iwr.uni-heidelberg.de
[2] Visual Learning Lab, Heidelberg University, Heidelberg, Germany
[3] Image and Pattern Analysis Group, Heidelberg University, Heidelberg, Germany

Abstract. Deep Affine Normalizing Flows are efficient and powerful models for high-dimensional density estimation and sample generation. Yet little is known about how they succeed in approximating complex distributions, given the seemingly limited expressiveness of individual affine layers. In this work, we take a first step towards theoretical understanding by analyzing the behaviour of a *single* affine coupling layer under maximum likelihood loss. We show that such a layer estimates and normalizes conditional moments of the data distribution, and derive a tight lower bound on the loss depending on the orthogonal transformation of the data before the affine coupling. This bound can be used to identify the optimal orthogonal transform, yielding a layer-wise training algorithm for deep affine flows. Toy examples confirm our findings and stimulate further research by highlighting the remaining gap between layer-wise and end-to-end training of deep affine flows.

1 Introduction

Affine Normalizing Flows such as RealNVP [4] are widespread and successful tools for density estimation. They have seen recent success in generative modeling [3,4,9], solving inverse problems [1], lossless compression [6], out-of-distribution detection [12], better understanding adversarial examples [7] and sampling from Boltzmann distributions [13].

These flows approximate arbitrary data distributions $\mu(\mathbf{x})$ by learning an invertible mapping $T(\mathbf{x})$ such that given samples are mapped to normally distributed latent codes $\mathbf{z} := T(\mathbf{x})$. In other words, they reshape the data density μ to form a normal distribution.

While being simple to implement and fast to evaluate, affine flows appear not very expressive at first glance. They consist of invertible layers called coupling

Electronic supplementary material The online version of this chapter (https:// doi.org/10.1007/978-3-030-71278-5_1) contains supplementary material, which is available to authorized users.

© Springer Nature Switzerland AG 2021
Z. Akata et al. (Eds.): DAGM GCPR 2020, LNCS 12544, pp. 1–14, 2021.
https://doi.org/10.1007/978-3-030-71278-5_1

blocks. Each block leaves half of the dimensions untouched and subjects the other half to just parameterized translations and scalings.

Explaining the gap between theory and applications remains an unsolved challenge. Taking the problem apart, a single layer consists of a rotation and an affine nonlinearity. It is often hand-wavingly argued that the deep model's expressivity comes from the rotations between the couplings by allowing different dimensions to influence one another [4].

In this work, we open a rigorous branch of explanation by characterizing the normalizing flow generated by a single affine layer. More precisely, we contribute:

- A single affine layer under maximum likelihood (ML) loss learns first- and second-order moments of the conditional distribution of the changed (active) dimensions given the unchanged (passive) dimensions (Sect. 3.2).
- From this insight, we derive a tight lower bound on how much the affine non-linearity can reduce the loss for a given rotation (Sect. 3.3). This is visualized in Fig. 1 where the bound is evaluated for different rotations of the data.
- We formulate a layer-wise training algorithm that determines rotations using the lower bound and nonlinearities using gradient descent in turn (Sect. 3.4).
- We show that such a single affine layer under ML loss makes the active independent of the passive dimensions if they are generated by a certain rule (Sect. 3.5).

Finally, we show empirically in Sect. 4 that while improving the training of shallow flows, the above new findings do not yet explain the success of deep affine flows and stimulate further research.

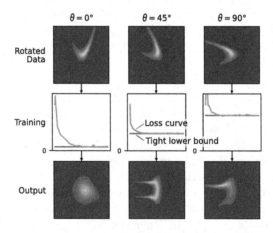

Fig. 1. An affine coupling layer pushes the input density towards standard normal. Its success depends on the rotation of the input *(top row)*. We derive a lower bound for the error that is actually attained empirically *(center row, blue and orange curves)*. The solution with lowest error is clearly closest to standard normal *(bottom row, left)*.

2   Related Work

The connection between affine transformations and the first two moments of a distribution is well-known in the Optimal Transport literature. When the function space of an Optimal Transport (OT) problem with quadratic ground cost is reduced to affine maps, the best possible transport matches mean and covariance of the involved distributions [17]. In the case of conditional distributions, affine maps become conditional affine maps [16]. We show such maps to have the same minimizer under maximum likelihood loss (KL divergence) as under OT costs.

It has been argued before that a single coupling or autoregressive block [14] can capture the moments of conditional distributions. This is one of the motivations for the SOS flow [8], based on a classical result on degree-3 polynomials by [5]. However, they do not make this connection explicit. We are able to give a direct correspondence between the function learnt by an affine coupling and the first two moments of the distribution to be approximated.

Rotations in affine flows are typically chosen at random at initialization and left fixed during training [3,4]. Others have tried training them via some parameterization like a series of Householder reflections [15]. The stream of work most closely related to ours explores the idea to perform layer-wise training. This allows an informed choice of the rotation based on the current estimate of the latent normal distribution. Most of these works propose to choose the least Gaussian dimensions as the active subspace [2,11]. We argue that this is inapplicable to affine flows due to their limited expressivity when the passive dimensions are not informative. To the best of our knowledge, our approach is the first to take the specific structure of the coupling layer into account and derive a tight lower bound on the loss as a function of the rotation.

3   Single Affine Coupling Layer

3.1   Architecture

Normalizing flows approximate data distributions μ available through samples $\mathbf{x} \in \mathbb{R}^D \sim \mu$ by learning an invertible function $T(\mathbf{x})$ such the latent codes $\mathbf{z} := T(\mathbf{x})$ follow an isotropic normal distribution $\mathbf{z} \in \mathbb{R}^D \sim \mathcal{N}(0, \mathbf{1})$. When such a function is found, the data distribution $\mu(\mathbf{x})$ can be approximated using the change-of-variables formula:

$$\mu(\mathbf{x}) = \mathcal{N}(T(\mathbf{x}))|\det \mathbf{J}| =: (T_\sharp^{-1}\mathcal{N})(\mathbf{x}), \tag{1}$$

where $\mathbf{J} = \nabla T(\mathbf{x})$ is the Jacobian of the invertible function, and "$\cdot_\sharp$" is the push-forward operator. New samples $\mathbf{x} \sim \mu$ can be easily generated by drawing $\mathbf{z}$ from the latent Gaussian and transporting them backward through the invertible function:

$$\mathbf{z} \sim \mathcal{N}(0, \mathbf{1}) \quad \Longleftrightarrow \quad \mathbf{x} =: T^{-1}(\mathbf{z}) \sim \mu(\mathbf{x}). \tag{2}$$

Affine Normalizing Flows are a particularly efficient way to parameterize such an invertible function T: They are simple to implement and fast to evaluate in both directions $T(\mathbf{x})$ and $T^{-1}(\mathbf{z})$, along with the Jacobian determinant $\det \mathbf{J}$ [1]. Like most normalizing flow models, they consist of the composition of several invertible layers $T(\mathbf{x}) = (T_L \circ \cdots \circ T_1)(\mathbf{x})$. The layers are called coupling blocks and modify the distribution sequentially. We recursively define the push-forward of the first l blocks as

$$\mu_l = (T_l)_\sharp \mu_{l-1}, \quad \mu_0 = \mu. \tag{3}$$

Each block $T_l, l = 1, \ldots, L$ contains a rotation $\mathbf{Q}_l \in SO(D)$ and a nonlinear transformation τ_l:

$$\mathbf{x}_l = T_l(\mathbf{x}_{l-1}) = (\tau_l \circ \mathbf{Q}_l)(\mathbf{x}_{l-1}), \quad \mathbf{x}_0 = \mathbf{x}. \tag{4}$$

The nonlinear transformation τ_l is given by:

$$\tau_l(\mathbf{y}) = \tau_l\left(\begin{bmatrix} \mathbf{p} \\ \mathbf{a} \end{bmatrix}\right) = \begin{bmatrix} \mathbf{p} \\ \mathbf{a} \odot e^{s_l(\mathbf{p})} + t_l(\mathbf{p}) \end{bmatrix} =: \begin{bmatrix} \mathbf{p} \\ \mathbf{a}' \end{bmatrix} = \mathbf{y}'. \tag{5}$$

Here, $\mathbf{y} = \mathbf{Q}_l \mathbf{x}_{l-1} \sim (\mathbf{Q}_l)_\sharp \mu_{l-1}$ is the rotated input to the nonlinearity (dropping the index l on $\mathbf{y}$ for simplicity) and $\odot$ is element-wise multiplication. An affine nonlinearity first splits its input into *passive* and *active* dimensions $\mathbf{p} \in \mathbb{R}^{D_P}$ and $\mathbf{a} \in \mathbb{R}^{D_A}$. The passive subspace is copied without modification to the output of the coupling. The active subspace is scaled and shifted as a function of the passive subspace, where s_l and $t_l : \mathbb{R}^{D_P} \to \mathbb{R}^{D_A}$ are represented by a single generic feed forward neural network [9] and need not be invertible themselves. The affine coupling design makes inversion trivial by transposing $\mathbf{Q}_l$ and rearranging terms in τ_l.

Normalizing Flows, and affine flows in particular, are typically trained using the Maximum Likelihood (ML) loss [3]. It is equivalent to the Kullback-Leibler (KL) divergence between the push-forward of the data distribution μ and the latent normal distribution [10]:

$$\mathcal{D}_{KL}(T_\sharp \mu \| \mathcal{N}) = -H[\mu] + \frac{D}{2} \log(2\pi) + \mathbb{E}_{\mathbf{x} \sim \mu}\left[\frac{1}{2}\|T(\mathbf{x})\|^2 - \log|\det \mathbf{J}(\mathbf{x})|\right] \tag{6}$$

$$= -H[\mu] + \frac{D}{2} \log(2\pi) + \mathrm{ML}(T_\sharp \mu \| \mathcal{N}), \tag{7}$$

The two differ only by terms independent of the trained model (the typically unknown entropy $H[\mu]$ and the normalization of the normal distribution).

It is unknown whether affine normalizing flows can push arbitrarily complex distributions to a normal distribution [14]. In the remainder of the section, we shed light on this by considering an affine flow that consists of just a single coupling as defined in Eq. (5). Since we only consider one layer, we're dropping the layer index l for the remainder of the section. In Sect. 4, we will discuss how these insights on isolated affine layers transfer to deep flows.

3.2   KL Divergence Minimizer

We first derive the exact form of the ML loss in Eq. (6) for an isolated affine coupling with a fixed rotation $\mathbf{Q}$ as in Eq. (4).

The Jacobian for this coupling has a very simple structure: It is a triangular matrix whose diagonal elements are $\mathbf{J}_{ii} = 1$ if i is a passive dimension and $\mathbf{J}_{ii} = \exp(s_i(\mathbf{p}))$ if i is active. Its determinant is the product of the diagonal elements, so that $\det \mathbf{J}(\mathbf{x}) > 0$ and $\log \det \mathbf{J}(\mathbf{x}) = \sum_{i=1}^{D_A} s_i(\mathbf{p})$. The ML loss thus reads:

$$\text{ML}(T_\sharp \mu \| \mathcal{N}) = \mathbb{E}_{\mathbf{p}, \mathbf{a} \sim \mathbf{Q}_\sharp \mu} \left[\frac{1}{2} \|\mathbf{p}\|^2 + \frac{1}{2} \left\| \mathbf{a} \odot e^{s(\mathbf{p})} + t(\mathbf{p}) \right\|^2 - \sum_{i=1}^{D_A} s_i(\mathbf{p}) \right]. \quad (8)$$

We now derive the minimizer of this loss:

Lemma 1 (Optimal single affine coupling). *Given a distribution μ and a single affine coupling layer T with a fixed rotation $\mathbf{Q}$. Like in Eq. (5), call $(\mathbf{a}, \mathbf{p}) = \mathbf{Q}\mathbf{x}$ the rotated versions of $\mathbf{x} \sim \mu$. Then, at the unique minimum of the ML loss (Eq. (8)), the functions $s, t : \mathbb{R}^{D_P} \to \mathbb{R}^{D_A}$ as in Eq. (4) take the following value:*

$$e^{s_i(\mathbf{p})} = \frac{1}{\sqrt{\text{Var}_{a_i|\mathbf{p}}[a_i]}} = \sigma_{A_i|\mathbf{p}}^{-1}, \quad (9)$$

$$t_i(\mathbf{p}) = -\mathbb{E}_{a_i|\mathbf{p}}[a_i] e^{s_i(\mathbf{p})} = -\frac{m_{A_i|\mathbf{p}}}{\sigma_{A_i|\mathbf{p}}}. \quad (10)$$

We derive this by optimizing for $s(\mathbf{p}), t(\mathbf{p})$ in Eq. (8) for each value of $\mathbf{p}$ separately. The full proof can be found in Appendix A.1.

We insert the optimal s and t to find the active part of the globally optimal affine nonlinearity:

$$\tau(\mathbf{a}|\mathbf{p}) = \mathbf{a} \odot e^{s(\mathbf{p})} + t(\mathbf{p}) = \frac{1}{\sigma_{A|\mathbf{p}}} \odot (\mathbf{a} - \mathbf{m}_{A|\mathbf{p}}). \quad (11)$$

It normalizes $\mathbf{a}$ for each $\mathbf{p}$ by shifting the mean of $\mu(\mathbf{a}|\mathbf{p})$ to zero and rescaling the individual standard deviations to one.

Example 1. Consider a distribution where the first variable p is uniformly distributed on the interval $[-2, 2]$. The distribution of the second variable a is normal, but its mean $m(p)$ and standard deviation $\sigma(p)$ are varying depending on p:

$$\mu(p) = \mathcal{U}([-2, 2]), \quad \mu(a|p) = \mathcal{N}(m(p), \sigma(p)). \quad (12)$$

$$m(p) = \frac{1}{2}\cos(\pi p), \quad \sigma(p) = \frac{1}{8}(3 - \cos(8\pi/3\,p)). \quad (13)$$

We call this distribution "W density". It is shown in Fig. 2a.

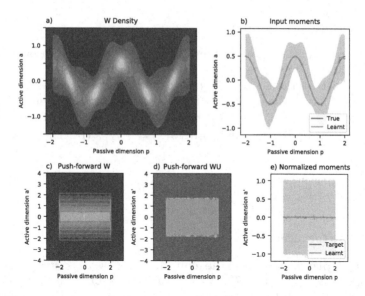

Fig. 2. *(a)* W density contours. *(b)* The conditional moments are well approximated by a single affine layer. *(c, d)* The learnt push-forwards of the W (Example 1) and WU (Example 2) densities remain normal respectively uniform distributions. *(e)* The moments of the transported distributions are close to zero mean and unit variance, shown for the layer trained on the W density.

We now train a single affine nonlinearity τ by minimizing the ML loss, setting $\mathbf{Q} = \mathbf{1}$. As hyperparameters, we choose a subnet for s, t with one hidden layer and a width of 256, a learning rate of 10^{-1}, a learning rate decay with factor 0.9 every 100 epochs, and a weight decay of 0. We train for 4096 epochs with 4096 i.i.d. samples from μ each using the Adam optimizer.

We solve s, t in Lemma 1 for the estimated mean $\hat{m}(p)$ and standard deviation $\hat{\sigma}(p)$ as predicted by the learnt $\hat{s}$ and $\hat{t}$. Upon convergence of the model, they closely follow their true counterparts $m(p)$ and $\sigma(p)$ as shown in Fig. 2b.

Example 2. This example modifies the previous to illustrate that the learnt conditional density $\tau_{\sharp}\mu(\mathbf{a}|\mathbf{p})$ is not necessarily Gaussian at the minimum of the loss.

The W density from above is transformed to the "WU density" by replacing the conditional normal distribution by a conditional uniform distribution with the same conditional mean $m(p)$ and standard deviation $\sigma(p)$ as before.

$$\mu(p) = \mathcal{U}([-2, 2]), \tag{14}$$

$$\mu(a|p) = \mathcal{U}([m(p) - \sqrt{3}\sigma(p), m(p) + \sqrt{3}\sigma(p)]). \tag{15}$$

One might wrongly believe that the KL divergence favours building a distribution that is *marginally* normal while ignoring the conditionals, i.e. $\tau_{\sharp}\mu(p) = \mathcal{N}$. Lemma 1 predicts the correct result, resulting in the following uniform push-forward density depicted in Fig. 2d:

$$T_\sharp \mu(p) = \mu(p) = \mathcal{U}([-2, 2]), \tag{16}$$

$$T_\sharp \mu(a|p) = \mathcal{U}([-\sqrt{3}, \sqrt{3}]). \tag{17}$$

Note how $\tau_\sharp \mu(a|p)$ does not depend on p, which we later generalize in Lemma 2.

3.3   Tight Bound on Loss

Knowing that a single affine layer learns the mean and standard deviation of $\mu(a_i|\mathbf{p})$ for each $\mathbf{p}$, we can insert this minimizer into the KL divergence. This yields a tight lower bound on the loss after training. Even more, it allows us to compute a tight upper bound on the loss improvement by the layer, which we denote $\Delta \geq 0$. This loss reduction can be approximated using samples without training.

Theorem 1 (Improvement by single affine layer). *Given a distribution μ and a single affine coupling layer T with a fixed rotation $\mathbf{Q}$. Like in Eq. (5), call $(\mathbf{a}, \mathbf{p}) = \mathbf{Q}\mathbf{x}$ the rotated versions of $\mathbf{x} \sim \mu$. Then, the KL divergence has the following minimal value:*

$$\mathcal{D}_{KL}(T_\sharp \mu || \mathcal{N}) = \mathcal{D}_{KL}(\mu_P || \mathcal{N}) + \mathbb{E}_\mathbf{p} \left[\sum_{i=1}^{D_A} H[\mathcal{N}(0, \sigma_{A_i|\mathbf{p}})] - H[\mu(\mathbf{a}|\mathbf{p})] \right] \tag{18}$$

$$= \mathcal{D}_{KL}(\mu || \mathcal{N}) - \Delta. \tag{19}$$

The loss improvement by the optimal affine coupling as in Lemma 1 is:

$$\Delta = \frac{1}{2} \sum_{i=1}^{D_A} \mathbb{E}_\mathbf{p}[m_{A_i|\mathbf{p}}^2 + \sigma_{A_i|\mathbf{p}}^2 - 1 - \log \sigma_{A_i|\mathbf{p}}^2]. \tag{20}$$

To proof, insert the minimizer s, t from Lemma 1 into Eq. (8). Then evaluate $\Delta = \mathcal{D}_{KL}(\mu || \mathcal{N}) - \mathcal{D}_{KL}(T_\sharp \mu || \mathcal{N})$ to obtain the statement. The detailed proof can be found in Appendix A.2.

The loss reduction by a single affine layer depends solely on the moments of the distribution of the active dimensions conditioned on the passive subspace. Higher order moments are ignored by this coupling design. Together with Lemma 1, this paints the following picture of an affine coupling layer: It fits a Gaussian distribution to each conditional $\mu(a_i|\mathbf{p})$ and normalizes this Gaussian's moments. The gap in entropy between the fit Gaussian and the true conditional distribution cannot be reduced by the affine transformation. This makes up the remaining KL divergence in Eq. (18).

We now make the connection explicit that a single affine layer can only achieve zero loss on the active subspace iff the conditional distribution is Gaussian with diagonal covariance:

Corollary 1. *If and only if* $(\mathbf{Q}_\sharp\mu)(\mathbf{a}|\mathbf{p})$ *is normally distributed for all p with diagonal covariance, that is:*

$$\mu(\mathbf{a}|\mathbf{p}) = \prod_{i=1}^{D_A} \mathcal{N}(a_i|m_{A_i|\mathbf{p}}, \sigma_{A_i|\mathbf{p}}), \tag{21}$$

a single affine block can reduce the KL divergence on the active subspace to zero:

$$\mathcal{D}_{KL}((T_\sharp\mu)(\mathbf{a}|\mathbf{p})||\mathcal{N}) = 0. \tag{22}$$

The proof can be found in Appendix A.3.

Example 3. We revisit the Examples 1 and 2 and confirm that the minimal loss achieved by a single affine coupling layer on the W-shaped densities matches the predicted lower bound. This is the case for both densities. Figure 3 shows the contribution of the conditional part of the KL divergence $\mathcal{D}_{\mathrm{KL}}((T_\sharp\mu)(a|p)||\mathcal{N})$ as a function of p:

For the W density, the conditional $\mu(a|p)$ is normally distributed. This is the situation of Corollary 1 and the remaining conditional KL divergence is zero. The remaining loss for the WU density is the negentropy of a uniform distribution with unit variance.

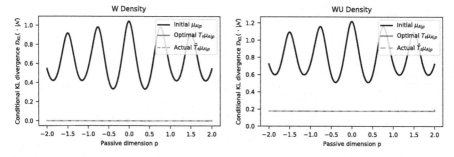

Fig. 3. Conditional KL divergence before *(gray)* and after *(orange)* training for W-shaped densities confirms lower bound *(blue, coincides with orange)*. The plots show the W density from Example 1 *(left)* and the WU density from Example 2 *(right)*. (Color figure online)

3.4   Determining the Optimal Rotation

The rotation $\mathbf{Q}$ of the isolated coupling layer determines the splitting into active and passive dimensions and the axes of the active dimensions (the rotation within the passive subspace only rotates the input into s, t and is irrelevant). The bounds in Theorem 1 heavily depend on these choices and thus depend on the chosen rotation $\mathbf{Q}$. This makes it natural to consider the loss improvement as a function of the rotation: $\Delta(\mathbf{Q})$. When aiming to maximally reduce the loss with a single affine layer, one should choose the subspace maximizing this tight upper bound in Eq. (20):

$$\arg \max_{\mathbf{Q} \in SO(D)} \Delta(\mathbf{Q}). \tag{23}$$

We propose to approximate this maximization by evaluating the loss improvement for a finite set of candidate rotations in Algorithm 1 "Optimal Affine Subspace (OAS)". Note that Step 5 requires approximating Δ from samples. In the regime of low D_P, one can discretize this by binning samples by their passive coordinate $\mathbf{p}$. Then, one computes mean and variance empirically for each bin. We leave the general solution of Eq. (23) for future work.

Algorithm 1. Optimal Affine Subspace (OAS).

1: **Input:** $\mathcal{Q} = \{\mathbf{Q}_1, \ldots, \mathbf{Q}_C\} \subset SO(D)$, $(\mathbf{x}_j)_{j=1}^N$ i.i.d. samples from μ.
2: **for** candidate $\mathbf{Q}_c \in \mathcal{Q}$ **do**
3:    Rotate samples: $\mathbf{y}_j = \mathbf{Q}_c \mathbf{x}_j$.
4:    **for** each active dimension $i = 1, \ldots, D_A$ **do**
5:       Use $(\mathbf{y})_{j=1}^N$ to estimate the conditional mean $m_{A_i|\mathbf{p}}$ and variance $\sigma_{A_i|\mathbf{p}}$ as a function of $\mathbf{p}$. {Example implementation in Example 4}
6:    **end for**
7:    Compute $\Delta_c := \frac{1}{2} \sum_{j=1}^N \sum_{i=1}^{D_A} (m_{A_i|\mathbf{p}_j}^2 + \sigma_{A_i|\mathbf{p}_j}^2 - 1 - \log \sigma_{A_i|\mathbf{p}_j}^2)$ {Equation (20)}.
8: **end for**
9: **Return:** $\arg \max_{\mathbf{Q}_c \in \mathcal{Q}} \Delta_c$.

Example 4. Consider the following two-component 2D Gaussian Mixture Model:

$$\mu = \frac{1}{2} \big(\mathcal{N}([-\delta; 0], \sigma) + \mathcal{N}([\delta; 0], \sigma) \big). \tag{24}$$

We choose $\delta = 0.95, \sigma = \sqrt{1 - \delta^2} = 0.3122\ldots$ so that the mean is zero and the standard deviation along the first axis is one. We now evaluate the loss improvement $\Delta(\theta)$ in Eq. (20) as a function of the angle θ with which we rotate the above distribution:

$$\mu(\theta) := \mathbf{Q}(\theta)_\sharp \mu, \quad [p, a] = \mathbf{Q}(\theta)\mathbf{x} \sim \mu(\theta). \tag{25}$$

Analytically, this can be done pointwise for a given p and then integrated numerically. This will not be possible for applications where only samples are available. As a proof of concept, we employ the previously mentioned binning approach. It groups N samples from μ by their p value into B bins. Then, we compute $m_{A|p_b}$ and $\sigma_{A|p_b}$ using the samples in each bin $b = 1, \ldots, B$.

Figure 4 shows the upper bound as a function of the rotation angle, as obtained from the two approaches. Here, we used $B = 32$ bins and a maximum of $N = 2^{13} = 8192$ samples. Around $N \approx 256$ samples are sufficient for a good agreement between the analytic and empiric bound on the loss improvement and the corresponding angle at the maximum.

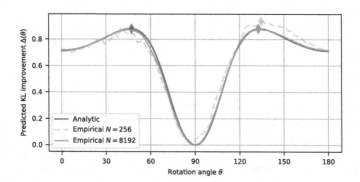

Fig. 4. Tight upper bound given by Eq. (20) for two-component Gaussian mixture as a function of rotation angle θ, determined analytically *(blue)* and empirically *(orange)* for different numbers of samples. The diamonds mark the equivalent outputs of the OAS Algorithm 1. (Color figure online)

Note: For getting a good density estimate using a single coupling, it is crucial to identify the right rotation. If we naively or by chance decide for $\theta = 90°$, the distribution is left unchanged.

3.5 Independent Outputs

An important step towards pushing a multivariate distribution to a normal distribution is making the dimensions independent of one another. Then, the residual to a global latent normal distribution can be solved with one sufficiently expressive 1D flow per dimension, pushing each distribution independently to a normal distribution. The following lemma shows for which data sets a single affine layer can make the active and passive dimensions independent.

Lemma 2. *Given a distribution μ and a single affine coupling layer T with a fixed rotation $\mathbf{Q}$. Like in Eq. 5, call $(\mathbf{a}, \mathbf{p}) = \mathbf{Q}\mathbf{x}$ the rotated versions of $\mathbf{x} \sim \mu$. Then, the following are equivalent:*

1. $\mathbf{a}' := \tau(\mathbf{a}|\mathbf{p}) \perp \mathbf{p}$ *for $\tau(\mathbf{a}|\mathbf{p})$ minimizing the ML loss in Eq. (8),*
2. *There exists $\mathbf{n} \perp \mathbf{p}$ such that $\mathbf{a} = f(\mathbf{p}) + \mathbf{n} \odot g(\mathbf{p})$, where $f, g : \mathbb{R}^{D_P} \to \mathbb{R}^{D_A}$.*

The proof can be found in Appendix A.4.

This results shows what our theory can explain about deep affine flows: It is easy to see that $D - 1$ coupling blocks with $D_A = 1, D_P = D - 1$ can make all variables independent if the data set can be written in the form of $x_i = f(\mathbf{x}_{\neq i}) + x_i g(\mathbf{x}_{\neq i})$. Then, only the aforementioned independent 1D flows are necessary for a push-forward to the normal distribution.

Example 5. Consider again the W-shaped densities from the previous Examples 1 and 2. After optimizing the single affine layer, the two variables p, a' are independent (compare Fig. 2c, d):

$$\text{Example 1: } a' \sim \mathcal{N}(0,1) \perp p, \tag{26}$$

$$\text{Example 2: } a' \sim \mathcal{U}([-\sqrt{3}, \sqrt{3}]) \perp p, \tag{27}$$

4 Layer-Wise Learning

Do the above single-layer results explain the expressivity of deep affine flows? To answer this question, we construct a deep flow layer by layer using the optimal affine subspace (OAS) algorithm Algorithm 1. Each layer l being added to the flow is trained to minimize the residuum between the current push-forward μ_{l-1} and the latent $\mathcal{N}$. The corresponding rotation $\mathbf{Q}_l$ is chosen by maximizing $\Delta(\mathbf{Q}_l)$ and the nonlinearities τ_l are trained by gradient descent, see Algorithm 2.

Algorithm 2. Iterative Affine Flow Construction.

1: Initialize $T^{(0)} = \text{id}$.
2: **repeat**
3:    Compute $\mathbf{Q}_l$ via OAS (Algorithm 1), using samples from $T^{(l-1)}_\sharp \mu$.
4:    Train τ_l on samples $\mathbf{y} = \mathbf{Q}_l \cdot T^{(l-1)}(\mathbf{x})$ for $\mathbf{x} \sim \mu$.
5:    Set $T_l = \tau_l \circ \mathbf{Q}_l$.
6:    Compose $T^{(l)} = T_l \circ T^{(l-1)}$.
7: **until** convergence, e.g. loss or improvement threshold, max. number of layers.
8: **return** Final transport $T^{(L)}$.

Can this ansatz reach the quality of end-to-end affine flows? An analytic answer is out of the scope of this work, and we consider toy examples.

Example 6. We consider a uniform 2D distribution $\mu = \mathcal{U}([-1,1]^2)$. Figure 5 compares the flow learnt layer-wise using Algorithm 2 to flows learnt layer-wise and end-to-end, but with fixed random rotations. Our proposed layer-wise algorithm performs on-par with end-to-end training despite optimizing only the respective last layer in each iteration, and beats layer-wise random subspaces.

Example 7. We now provide more examples on a set of toy distributions. As before, we train layer-wise using OAS and randomly selected rotations, and end-to-end. Additionally, we train a mixed variant of OAS and end-to-end: New layers are still added one by one, but Algorithm 2 is modified such that iteration l optimizes *all* layers 1 through l in an end-to-end fashion. We call this training "progressive" as layers are progressively activated and never turned off again.

We obtain the following results: Optimal rotations always outperform random rotations in layer-wise training. With only a few layers, they also outperform end-to-end training, but are eventually overtaken as the network depth increases. Progressive training continues to be competitive also for deep networks.

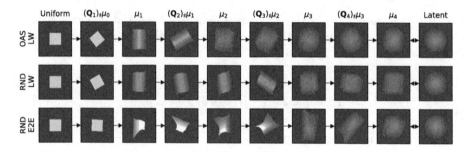

Fig. 5. Affine flow trained layer-wise "LW", using optimal affine subspaces "OAS" *(top)* and random subspaces "RND" *(middle)*. After a lucky start, the random subspaces do not yield a good split and the flow approaches the latent normal distribution significantly slower. End-to-end training "E2E" *(bottom)* chooses a substantially different mapping, yielding a similar quality to layer-wise training with optimal subspaces.

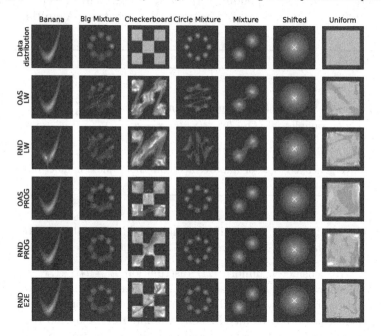

Fig. 6. Affine flows trained on different toy problems *(top row)*. The following rows depic different training methods: layer-wise "LW" *(rows 2 and 3)*, progressively "PROG" *(rows 4-5)* and end-to-end "E2E" *(last row)*. Rotations are "OAS" when determined by Algorithm 1 *(row 2 and 4)* or randomly selected "RND" *(rows 3, 5 and 6)*.

Figure 6 shows the density estimates after twelve layers. At this point, none of the methods show a significant improvement by adding layers. Hyperparameters were optimized for each training configuration to obtain a fair comparison. Densities obtained by layer-wise training exhibit significant spurious structure for both optimal and random rotations, with an advantage for optimally chosen subspaces.

5 Conclusion

In this work, we showed that an isolated affine coupling learns the first two moments of the conditioned data distribution $\mu(\mathbf{a}|\mathbf{p})$. Using this result, we derived a tight upper bound on the loss reduction that can be achieved by such a layer. We then used this to choose the best rotation of the coupling.

We regard our results as a first step towards a better understanding of deep affine flows. We provided sufficient conditions for a data set that can be exactly solved with layer-wise trained affine couplings and a single layer of D independent 1D flows.

Our results can be seen analogously to the classification layer at the end of a multi-layer classification network: The results from Sect. 3 directly apply to the last coupling in a deep normalizing flow. This raises a key question for future work: How do the first $L-1$ layers prepare the distribution μ_{L-1} such that the final layer can perfectly push the data to a Gaussian?

Acknowledgement. This work is supported by Deutsche Forschungsgemeinschaft (DFG, German Research Foundation) under Germany's Excellence Strategy EXC-2181/1 – 390900948 (the Heidelberg STRUCTURES Cluster of Excellence).

Furthermore, we thank our colleagues Lynton Ardizzone, Jakob Kruse, Jens Müller, and Peter Sorrenson for their help, support and fruitful discussions.

References

1. Ardizzone, L., Kruse, J., Rother, C., Köthe, U.: Analyzing inverse problems with invertible neural networks. In: International Conference on Learning Representations (2018)
2. Bigoni, D., Zahm, O., Spantini, A., Marzouk, Y.: Greedy inference with layers of lazy maps. arXiv preprint arXiv:1906.00031 (2019)
3. Dinh, L., Krueger, D., Bengio, Y.: NICE: non-linear independent components estimation. arXiv preprint arXiv:1410.8516 (2014)
4. Dinh, L., Sohl-Dickstein, J., Bengio, S.: Density estimation using real nvp. arXiv preprint arXiv:1605.08803 (2016)
5. Fleishman, A.I.: A method for simulating non-normal distributions. Psychometrika **43**(4), 521–532 (1978)
6. Hoogeboom, E., Peters, J., van den Berg, R., Welling, M.: Integer discrete flows and lossless compression. In: Advances in Neural Information Processing Systems, pp. 12134–12144 (2019)
7. Jacobsen, J.H., Behrmann, J., Zemel, R., Bethge, M.: Excessive invariance causes adversarial vulnerability. arXiv preprint arXiv:1811.00401 (2018)
8. Jaini, P., Selby, K.A., Yu, Y.: Sum-of-squares polynomial flow. arXiv preprint arXiv:1905.02325 (2019)
9. Kingma, D.P., Dhariwal, P.: Glow: generative flow with invertible 1x1 convolutions. In: Advances in Neural Information Processing Systems, pp. 10215–10224 (2018)

10. Marzouk, Y., Moselhy, T., Parno, M., Spantini, A.: Sampling via measure transport: an introduction. In: Ghanem, R., Higdon, D., Owhadi, H. (eds.) Handbook of Uncertainty Quantification, pp. 1–41. Springer, Cham (2016). https://doi.org/10.1007/978-3-319-11259-6_23-1
11. Meng, C., Ke, Y., Zhang, J., Zhang, M., Zhong, W., Ma, P.: Large-scale optimal transport map estimation using projection pursuit. In: Wallach, H., Larochelle, H., Beygelzimer, A., d'Alché-Buc, F., Fox, E., Garnett, R. (eds.) Advances in Neural Information Processing Systems, vol. 32, pp. 8116–8127. Curran Associates, Inc. (2019)
12. Nalisnick, E., Matsukawa, A., Teh, Y.W., Lakshminarayanan, B.: Detecting out-of-distribution inputs to deep generative models using a test for typicality. arXiv preprint arXiv:1906.02994 (2019)
13. Noé, F., Olsson, S., Köhler, J., Wu, H.: Boltzmann generators: sampling equilibrium states of many-body systems with deep learning. Science **365**(6457), eaaw1147 (2019)
14. Papamakarios, G., Nalisnick, E., Rezende, D.J., Mohamed, S., Lakshminarayanan, B.: Normalizing flows for probabilistic modeling and inference. arXiv preprint arXiv:1912.02762 (2019)
15. Putzky, P., Welling, M.: Invert to learn to invert. In: Advances in Neural Information Processing Systems, pp. 446–456 (2019)
16. Tabak, E.G., Trigila, G.: Conditional expectation estimation through attributable components. Inf. Infer. J. IMA **7**(4), 727–754 (2018)
17. Trigila, G., Tabak, E.G.: Data-driven optimal transport. Commun. Pure Appl. Math. **69**(4), 613–648 (2016)

Semantic Bottlenecks: Quantifying and Improving Inspectability of Deep Representations

Max Maria Losch[1]([⊠]), Mario Fritz[2], and Bernt Schiele[1]

[1] Max Planck Institute for Informatics, Saarland Informatics Campus,
Saarbrücken, Germany
{max.losch,bernt.schiele}@mpi-inf.mpg.de
[2] CISPA Helmholtz Center for Information Security, Saarland Informatics Campus,
Saarbrücken, Germany
mario.fritz@cispa.saarland

Abstract. Today's deep learning systems deliver high performance based on end-to-end training but are notoriously hard to inspect. We argue that there are at least two reasons making inspectability challenging: (i) representations are distributed across hundreds of channels and (ii) a unifying metric quantifying inspectability is lacking. In this paper, we address both issues by proposing Semantic Bottlenecks (SB), integrated into pretrained networks, to align channel outputs with individual visual concepts and introduce the model agnostic AUiC metric to measure the alignment. We present a case study on semantic segmentation to demonstrate that SBs improve the AUiC up to four-fold over regular network outputs. We explore two types of SB-layers in this work: while concept-supervised SB-layers (SSB) offer the greatest inspectability, we show that the second type, unsupervised SBs (USB), can match the SSBs by producing one-hot encodings. Importantly, for both SB types, we can recover state of the art segmentation performance despite a drastic dimensionality reduction from 1000s of non aligned channels to 10s of semantics-aligned channels that all downstream results are based on.

1 Introduction

While end-to-end training is key to top performance of deep learning – learned intermediate representations remain opaque to humans with typical training methods. Furthermore, assessing inspectability has remained a fairly elusive concept since its framing has mostly been qualitative (e.g. saliency maps). Given the increasing interest in using deep learning in real world applications, inspectability and a quantification of such is critically missing.

Goals for Inspectability. To address this, prior work on inspectability has proposed to improve the spatial coherency of activation maps [36] or to cluster

Electronic supplementary material The online version of this chapter (https://doi.org/10.1007/978-3-030-71278-5_2) contains supplementary material, which is available to authorized users.

© Springer Nature Switzerland AG 2021
Z. Akata et al. (Eds.): DAGM GCPR 2020, LNCS 12544, pp. 15–29, 2021.
https://doi.org/10.1007/978-3-030-71278-5_2

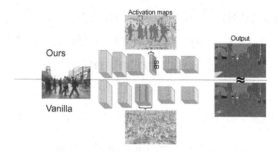

Fig. 1. Semantic Bottleneck (SB) layers can be integrated into any model – aligning channels with semantic concepts (top) while regular deep representations are highly distributed and hard to inspect (bottom). SBs act as basis for all subsequent layers.

representations to acquire outputs of low dimensionality either with supervision [6] or without [8,33]. In contrast, we demand information in each channel to be represented by a single semantic (sub-)concept. This is derived from a simple observation: distributed representations do not lend themselves to trivial interpretation (see bottom of Fig. 1). In order to reduce distributedness we propose to adapt deep networks via three criteria. (i) Reduce the number of channels to a minimum, (ii) associate them with semantic (sub-)concepts, and, at the same time, (iii) aim to lose as little overall performance as possible. In our view such semantics based inspectability can be seen as a way towards achieving true interpretability of deep networks.

Our Contributions are three-fold. Firstly, we introduce two network layers we term Semantic Bottlenecks (SB) based on linear layers to improve alignment with semantic concepts by (i) supervision to visual concepts and (ii) regularizing the output to be one-hot encoded to restrict distributedness. Secondly, we show SBs can be integrated into a state-of-the-art model without impairing performance, even when reducing the number of channels from 4096 to 30. Finally, we introduce the novel AUiC metric to quantify alignment between channel outputs and visual concepts for any model and show our SBs improve the baselines up to four-fold.

2   Related Work

As argued in prior work [23], interpretability can be largely approached in two ways. The first being post-hoc interpretation, for which we take an already trained and well performing model and dissect its decisions a-posteriori to identify important input features via attribution [2,18,27,28,30,40] or attempt to assign meaning to groups of features [3,9,17,29,34,35]. The second approach involves constructing inherently interpretable models – supervised or unsupervised.

Inspectability Without Supervision. Similar to our USBs, [1,21,26,33] and [8] embed an interpretable layer into the network using unsupervised metrics. [1,12,26] base their method on a reconstruction loss that regularizes a latent code, that is used to reconstruct the input, to be more interpretable. Such approaches are based on the successes of VAE [19] based frameworks and their recent successes on learning visual concepts from simple datasets [7,14,16] but have the issue that reconstruction is more challenging than classification itself. Furthermore, they have not yet been shown to work on large scale datasets.

Inspectability with Supervision. Literature on supervised improvements has focused on embedding predefined semantic concepts [6,20,22,25] enabling layers with improved transparency. While [6] proposes a model based on natural language features, [25] performs pretraining on semantic (sub-)concepts and train a secondary model on top for a related task. Our approach constructs a semantically inspectable layer by mapping original representations of a well performing network to a semantically aligned space. This renders our method modular and allows integration into pretrained models.

Quantification of Concept Alignment. In order to investigate the inspectability of deep networks, Bau et al. proposed NetDissect – a method counting number of channels assignable to single visual (sub-)concepts [3]. Here, edges of a bipartite graph, connecting channels and concepts, are weighted by measuring overlap between activation map and pixel annotations. Two additional variants enable (sub-)concept association by combining channels [4,13]. Our AUiC metric leverages the ideas of NetDissect and extends it to satisfy three criteria we deem important for measuring inspectability – which NetDissect does not satisfy.

In contrast to existing literature, we propose semantic bottlenecks which are easy to integrate in any architecture and offer inspectable outputs while retaining performance on a dense prediction task. Additionally, we introduce the model agnostic AUiC metric enabling benchmarking of inspectability.

3   Semantic Bottlenecks

To approach more inspectable intermediate representations we demand information in each channel to be represented by a single semantic (sub-)concept[1]. We propose two variants to achieve this goal: (i) supervise single channels to represent unique concepts and (ii) enforce one-hot outputs to encourage concept-aligned channels and inhibit distributed representations. We construct both variants as layers that can be integrated into pretrained models, mapping non-inspectable representations to an inspectable semantic space. We name these supervised and unsupervised Semantic Bottlenecks (SB).

Case Study. To show the utility of SBs, we choose street scene segmentation on the Cityscapes dataset [11] since it is a difficult task that traditionally requires very large models and has a practical application that has direct benefits from inspectability: autonomous driving. Cityscapes consists of 19 different classes, 2,975 training images and 500 validation images, both densely labeled. We use PSPNet [37] based on ResNet-101 [15], due to its strong performance and because residual networks are abundantly used in computer vision tasks today.

---

[1] For brevity we call all types of concepts simply: concept.

3.1   Supervised Semantic Bottlenecks (SSBs)

Variant (i) supervises each SB channel to represent a single semantic concept using additional concept annotations. Ideally, we possess pixel-annotations for an exhaustive set of subordinate concepts like colors, textures and parts to decide which are required to recover performance on particular tasks. Yet, we show that an encouragingly small task-specific selection is sufficient to satisfy both desiderata of performance and inspectability. We follow Fig. 2a which shows the structure of an SSB. One (or multiple) linear layers (blue trapezoids) receive the distributed input from a pretrained model and are supervised by an auxiliary loss to map them to target concepts (colored boxes). Given the dense prediction task of our case study, we use 1×1-conv modules to retain the spatial resolution.

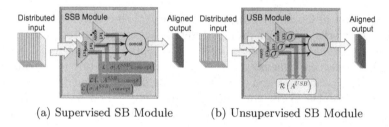

(a) Supervised SB Module   (b) Unsupervised SB Module

Fig. 2. Schematics for the integration of both Semantic Bottleneck (SB) layer types into an existing architecture. To handle independent semantic spaces (e.g. objects, materials), SBs can include parallel bottlenecks. (Color figure online)

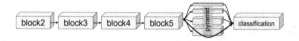

Fig. 3. Simplified PSPNet architecture indicating residual blocks. A detailed overview is presented in the supplement.

Choosing Concepts for Cityscapes. For our supervised SB-layer we choose concepts based on task relevancy for Cityscapes. *Broden+* [31] is a recent collection of datasets which serves as a starting point for the concept annotations we require for the SSBs. It offers thousands of images for objects, parts, materials and textures for which the first three types come with pixel level annotations. Here, a pixel can have multi-label annotations. Based on the 377 part and material concepts available (351 parts sourced from ADE [39] and Pascal-Part [10] and 26 materials sourced from OpenSurfaces [5]), we compile a subset of 70 Cityscapes-relevant concepts (see detailed list in the supplement).

Implementation Details. Since the Broden concepts are not defined on the Cityscapes domain, we train the SSB in a separate step. First, we train the model without bottleneck on Cityscapes. Secondly, the SSB is integrated into the model as additional layer and is trained on Broden while all other parameters are kept fix. To ensure matching dimensionality between SSB and the subsequent layer, we replace it with a new layer of same the type. Finally, all layers after the SSBs are finetuned until convergence following the training in [37], with small adjustments due to restrictions of our hardware resources (stated in supplement). We embed two concept classifier bottlenecks (parts and material) in each SSB, trained independently with cross entropy loss. Given the residual architecture of PSPNet, we integrate SSBs after residual blocks to avoid treatment of skip connections. To cover early as well as late layers we place SSBs after three different locations: block3, block4 and the pyramid module (see Fig. 3). We skip block5, as its output is already integral part of the pyramid module output.

Recovering Performance Using SSBs. As one of our 3 goals for inspectable deep networks, we strive to lose as little performance as possible. We test our SSB-augmented PSPNets on the original Cityscapes task and compare mIoUs (see Table 1). We denote an SSB after *blockX* as *SSB@blockX*. Given our PSPNet baseline mIoU of 76.2%, SSB@block4 is able to recover the full performance, while the applica-

Table 1. Segmentation results on Cityscapes validation set for different placements of SSB. PSPNet‡ is our reference model trained with smaller batchsize.

Configuration	#concepts (materials, parts)	mIoU
PSPNet‡	N/A	76.2
SSB@block3 (512 input feat.)	70 (11, 59)	73.1
SSB@block4 (1024 input feat.)	70 (11, 59)	76.2
SSB@pyramid (4096 input feat.)	70 (11, 59)	72.8

tions to block3 and the pyramid layer result in a slight decrease (73.1% and 72.6% respectively). Our quantification of inspectability (Sect. 4 and 5) will enable reasoning on these performance drops. Regardless, the reduction in the number of channels is substantial (e.g. 1024 reduced to 70 for block4), indicating room to render complex networks more inspectable. This addresses point 1 of our 3 goals (channel reduction). We additionally train SSBs with fewer concepts and a selection of task-irrelevant concepts and find that our choice of relevant concepts outperforms choices containing less relevant concepts. We refer the interested reader to the supplement.

3.2 Unsupervised Semantic Bottlenecks (USBs)

Clearly, the requirement for additional annotation and uncertainty regarding concept choice is a limitation of SSBs. To address this, we investigate the use of annotation free methods to (i) reduce number of channels, (ii) increase semantic association and (iii) lose as little performance as possible. Similar to SSBs, we address point (i) by integrating layers with low dimensionality. To address (ii) we propose to enforce *non*-distributed representations by approaching one-hot encodings, which we implement using softmax and appropriate regularization.

Construction of USBs. As for SSBs, we integrate the USB into a pretrained model. While the SSBs have no activation function, we use softmax for the USBs (see Fig. 2b). We regularize its output-entropy to approach near one-hot outputs during training. We identify two different approaches for regularization. Firstly, the entropy of the softmax output can be minimized via an additional loss. Secondly, parameterizing softmax with the temperature parameter T, allows approaching $\arg\max$ in the limit when $T = 0$. Since one-hot encodings are severely limiting information throughput, we explore the use of N parallel bottlenecks to acquire N-hot encodings. **Softmax application.** Assume we have N parallel bottlenecks, each having K channels. During inference, we want each spatial location (or pixel) to only have N active output values in total – one per bottleneck during inference. Consequently, we apply softmax for each bottleneck and pixel independently. **One-hot regularization.** We investigate two approaches: adding an additional **entropy loss** or utilize a parameterization with temperature T. For the first, we calculate the average entropy of softmax probabilities per bottleneck and pixel and optimize it as additional loss jointly with the classification loss. For the second, we utilize the softmax **temperature parameter**, which we anneal during training. Starting with a high temperature, e.g. $T_0 = 1$ it is reduced quickly in τ training iterations to approach $\arg\max$. We define T at timestep t with polynomial decay: $T_t = T_0 + (T_0 - T_\tau) \cdot \left(1 - \frac{t}{\tau}\right)^\gamma$, where γ specifies how quickly T is decaying.

Implementation Details. In contrast to training SSBs, we finetune the USB parameters jointly with the downstream network while keeping all layers before fixed. Entropy regularization is scaled with factor 0.1 while T is kept at 1.0. For annealing we set $T_0 = 1.0$, $T_\tau = 0.01$ and $\gamma = 10.0$ for rapid decay. During inference, we compute $\arg\max$ **instead of softmax** to acquire one-hot encodings.

Recovering Performance Using USBs. Here we show that introducing USBs result in little to no performance impact while drastically reducing number of channels. Additionally, we show that one-hot encodings can be achieved with appropriate regularization. We report regularization technique, dimensions of the bottlenecks, average entropy across channels as well as active

Table 2. USB performance for two different one-hot regularizations. Annealing during training enables the use of $\arg\max$ on inference to acquire one-hot outputs without performance loss.

loc	One-hot regularization	#channels $N \times K$	Σ: active channels	mIoU soft	mIoU hard
Block3	Entropy loss	2×87	19	73.0	0.1
		5×87	57	74.0	36.7
	Temperature annealing	4×50	60	69.5	69.5
		2×50	33	67.3	67.3
Block4	Entropy loss	2×87	93	75.7	32.9
	Temperature annealing	2×10	18	75.1	75.1
		2×50	97	75.8	75.8
Pyramid	Temperature annealing	2×10	20	71.5	71.5
		2×50	96	75.5	75.5

channels for USB applications to layer block3, block4 and the pyramid layer in Table 2. The *active channels* (Σ) column is counting the channels which are active at least once on the Cityscapes validation set. Given the two evaluated regularizations, we anticipate to find a method retaining mIoU performance when replacing softmax with $\arg\max$ during inference. We observe in comparing the two rightmost columns, that only temperature annealing satisfies this goal. Concluding these results, we identify annealing as the best method. Disadvantageously, it does not recover the mIoU on block3 fully. It appears to be more difficult to learn non-distributed representations early on. Due to $\arg\max$, we

both block4 and pyramid making a big leap forward towards inspectable representations. It is particularly noteworthy, that even a naive choice of task related concepts for supervision can be delineated from distributed, seemingly random representations. We can subsequently conclude that it is feasible to make the information in a deep net much more inspectable by introducing a supervised mapping onto a semantic space. Without supervision, a simple bottleneck having similar mIoU with only 25 channels offers an AUiC of only 0.2 for Cityscapes-Parts on block4 (SSB: 0.3) and 0.04 for Broden (SSB improves three fold: 0.12). Note that SSBs have with 70 channels nearly 3 times as many, such that reducing the channels alone is not sufficient to substantially improve on inspectability.

USBs Align with Cityscapes-Classes. In comparison to SSBs, USBs align much better with Cityscapes-Classes than subordinate concepts. We see here the greatest increase in AUiC, e.g. from 0.05 to 0.4 on block4. On block4 we also report greater inspectability than for regular bottlenecks (0.4 vs 0.2).

Table 4. Averaged stabilities over datasets.

Stability S	Block3	Block4	Pyramid
Original	0.034	0.077	0.099
Bottleneck	0.004	0.069	0.082
SSB	0.047	0.099	0.099
USB	0.938	0.945	0.947

USBs Offer High Stability. Shown in Table 4, we see USBs offering a clear advantage since their outputs are one-hot encoded: alignments are very stable. SSBs on the other hand report only slight stability improvements over baselines. To answer, whether softmax enables greater stability by default (SSBs have no non-linearity), we measure AUiC and S for SSB with softmax. Measuring with softmax $T = 1$, we find a 2-fold increase of stability to 0.20 but a 3-fold decrease in AUiC to 0.07. While softmax alone increases stability, it does not improve AUiC by default. As noted in Sect. 4, a channel is stable if it responds consistently to the same concept no matter the activation value (argmax USBs have only two states). This is not the case for a regular block4 and SSB channel, for which the same channel may be active for multiple concepts albeit with low activation. By our definition, this can be inspectable but is not stable. We conclude that the linear SSB-layer is sufficient to align with semantic concepts yet unable to increase stability by a large margin by default. Note that simple bottlenecks show consistently reduced stability (e.g. 0.069 vs 0.077 for bottleneck vs original on block4).

Representations at Block3 are Difficult to Align. Comparing the AUiC scores between block3 and other locations, it becomes evident that only SSBs improve inspectability. This indicates an intrinsic difficulty in aligning individual channels with semantics that early and could imply a necessity for distributed representations. We leave this as a challenge for future work.

Conclusion. Both SSBs and USBs offer clear advantages over baselines. SSBs are semantically supervised and thus can offer the greatest improvements in AUiC. USBs do not require concept supervision, yet form channels that are well aligned with Cityscapes classes offering a different dimension of inspectability.

5.3   Qualitative Improvements with SBs

To support our quantitative results we compare visualizations of SB-layers and baselines. We show that SB outputs offer substantially improved spatial coherency and consistency. **Top-20 Channels.** To enable comparison between 1000s and 10s of channels, we utilize the mIoU scoring of our AUiC to rank channels. We show the top-20 channels, assigning each a unique color and plotting the arg max per location. Based on our discussion of inspectable channels, this will result in coherent activations for unique concepts *if a channel is aligned*. Visualizations are presented for two images in Fig. 6 for all tested layer locations. 4 additional images are shown in Fig. 5 for the pyramid only.

PSPNet outputs in the first row (*Vanilla*) show the difficulty in interpreting them, since they are highly distributed across channels (also indicated by [13]).

SSB and USB Outputs. Attending to the first image on the left half of Fig. 6, we see spatial coherency greatly improved for SSB and USB outputs over baseline. In particular, note the responses for SSB@block4 which

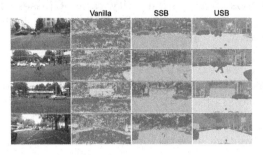

Fig. 5. Top-20 Broden aligned channels only for the pyramid layer from SSB-, USB- and vanilla PSPNet outputs.

show a distinction into wheels (blue color), car windows (dark orange color) and person-legs (light gray color). In relation, the USBs appear to form representations that are early aligned with the output classes, which is especially evident for USB@pyramid. Since it is unsupervised, the USBs offer easy access into what concepts have been learned automatically.

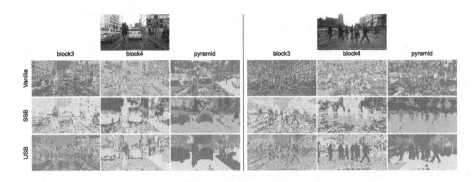

Fig. 6. Top-20 Broden aligned channels from SSB-, USB- and vanilla PSPNet outputs. Each color is mapped to a single output channel. (Color figure online)

6   Conclusion

We proposed supervised and unsupervised Semantic Bottlenecks (SBs) to render deep representations more inspectable by semantic alignment. Additionally, we introduced the AUiC metric quantifying such alignment to enable model agnostic benchmarking and showed that SBs improve baseline scores up to four fold while retaining performance. We identified SSBs offering best semantic alignment while USBs offer greatest alignment stability while requiring no supervision. Our SBs demonstrate that simultaneous performance and inspectability are not a chimera.

Acknowledgements. This research was supported by the Bosch Computer Vision Research Lab Hildesheim, Germany. We thank Dimitrios Bariamis and Oliver Lange for the insightful discussions.

References

1. Al-Shedivat, M., Dubey, A., Xing, E.P.: Contextual explanation networks. arXiv:1705.10301 (2017)
2. Bach, S., Binder, A., Montavon, G., Klauschen, F., Müller, K.R., Samek, W.: On pixel-wise explanations for non-linear classifier decisions by layer-wise relevance propagation. PLoS ONE **10**(7), e0130140 (2015)
3. Bau, D., Zhou, B., Khosla, A., Oliva, A., Torralba, A.: Network dissection: quantifying interpretability of deep visual representations. In: CVPR (2017)
4. Bau, D., et al.: Gan dissection: visualizing and understanding generative adversarial networks. In: ICLR (2019)
5. Bell, S., Upchurch, P., Snavely, N., Bala, K.: Opensurfaces: a richly annotated catalog of surface appearance. ACM Trans. Graph. (TOG) **32**(4), 111 (2013)
6. Bucher, M., Herbin, S., Jurie, F.: Semantic bottleneck for computer vision tasks. In: Jawahar, C.V., Li, H., Mori, G., Schindler, K. (eds.) ACCV 2018. LNCS, vol. 11362, pp. 695–712. Springer, Cham (2019). https://doi.org/10.1007/978-3-030-20890-5_44
7. Burgess, C.P., et al.: Monet: unsupervised scene decomposition and representation. arXiv preprint arXiv:1901.11390 (2019)
8. Chen, C., Li, O., Tao, D., Barnett, A., Rudin, C., Su, J.K.: This looks like that: deep learning for interpretable image recognition. In: NeurIPS, pp. 8930–8941 (2019)
9. Chen, R., Chen, H., Ren, J., Huang, G., Zhang, Q.: Explaining neural networks semantically and quantitatively. In: ICCV, pp. 9187–9196 (2019)
10. Chen, X., Mottaghi, R., Liu, X., Fidler, S., Urtasun, R., Yuille, A.: Detect what you can: detecting and representing objects using holistic models and body parts. In: CVPR (2014)
11. Cordts, M., et al.: The cityscapes dataset for semantic urban scene understanding. In: CVPR (2016)
12. Esser, P., Rombach, R., Ommer, B.: A disentangling invertible interpretation network for explaining latent representations. In: CVPR, pp. 9223–9232 (2020)
13. Fong, R., Vedaldi, A.: Net2vec: quantifying and explaining how concepts are encoded by filters in deep neural networks. In: CVPR, pp. 8730–8738 (2018)
14. Greff, K., et al.: Multi-object representation learning with iterative variational inference. arXiv preprint arXiv:1903.00450 (2019)

15. He, K., Zhang, X., Ren, S., Sun, J.: Deep residual learning for image recognition. In: CVPR (2016)
16. Higgins, I., Matthey, L., Pal, A., Burgess, C., Glorot, X., Botvinick, M., Mohamed, S., Lerchner, A.: beta-vae: learning basic visual concepts with a constrained variational framework. ICLR **2**(5), 6 (2017)
17. Kim, B., Wattenberg, M., Gilmer, J., Cai, C., Wexler, J., Viegas, F., et al.: Interpretability beyond feature attribution: quantitative testing with concept activation vectors (TCAV). In: ICML (2018)
18. Kindermans, P.J., et al.: The (un) reliability of saliency methods. arXiv:1711.00867 (2017)
19. Kingma, D.P., Welling, M.: Auto-encoding variational bayes. arXiv preprint arXiv:1312.6114 (2013)
20. Li, L.J., Su, H., Fei-Fei, L., Xing, E.P.: Object bank: a high-level image representation for scene classification & semantic feature sparsification. In: NeurIPS (2010)
21. Li, O., Liu, H., Chen, C., Rudin, C.: Deep learning for case-based reasoning through prototypes: a neural network that explains its predictions. In: AAAI (2018)
22. Lin, D., Shen, X., Lu, C., Jia, J.: Deep lac: deep localization, alignment and classification for fine-grained recognition. In: CVPR, pp. 1666–1674 (2015)
23. Lipton, Z.C.: The mythos of model interpretability. Queue **16**(3), 30 (2018)
24. Liu, H., Simonyan, K., Yang, Y.: DARTS: differentiable architecture search. In: ICLR (2019)
25. Marcos, D., Lobry, S., Tuia, D.: Semantically interpretable activation maps: whatwhere-how explanations within CNNs. arXiv preprint arXiv:1909.08442 (2019)
26. Melis, D.A., Jaakkola, T.: Towards robust interpretability with self-explaining neural networks. In: NeurIPS (2018)
27. Selvaraju, R.R., Cogswell, M., Das, A., Vedantam, R., Parikh, D., Batra, D., et al.: Grad-cam: visual explanations from deep networks via gradient-based localization. In: ICCV, pp. 618–626 (2017)
28. Shrikumar, A., Greenside, P., Kundaje, A.: Learning important features through propagating activation differences. In: ICML (2017)
29. Simonyan, K., Vedaldi, A., Zisserman, A.: Deep inside convolutional networks: visualising image classification models and saliency maps. arXiv:1312.6034 (2013)
30. Sundararajan, M., Taly, A., Yan, Q.: Axiomatic attribution for deep networks. In: ICML (2017)
31. Xiao, T., Liu, Y., Zhou, B., Jiang, Y., Sun, J.: Unified perceptual parsing for scene understanding. In: ECCV (2018)
32. Xie, S., Zheng, H., Liu, C., Lin, L.: SNAS: stochastic neural architecture search. In: ICLR (2019)
33. Yeh, C.K., Kim, B., Arik, S.O., Li, C.L., Ravikumar, P., Pfister, T.: On conceptbased explanations in deep neural networks. arXiv preprint arXiv:1910.07969 (2019)
34. Yosinski, J., Clune, J., Nguyen, A., Fuchs, T., Lipson, H.: Understanding neural networks through deep visualization. arXiv:1506.06579 (2015)
35. Zeiler, M.D., Fergus, R.: Visualizing and understanding convolutional networks. In: Fleet, D., Pajdla, T., Schiele, B., Tuytelaars, T. (eds.) ECCV 2014. LNCS, vol. 8689, pp. 818–833. Springer, Cham (2014). https://doi.org/10.1007/978-3-319-10590-1_53
36. Zhang, Q., Nian Wu, Y., Zhu, S.C.: Interpretable convolutional neural networks. In: CVPR, pp. 8827–8836 (2018)

37. Zhao, H., Shi, J., Qi, X., Wang, X., Jia, J.: Pyramid scene parsing network. In: CVPR (2017)
38. Zhou, B., Bau, D., Oliva, A., Torralba, A.: Interpreting Deep Visual Representations via Network Dissection. arXiv e-prints arXiv:1711.05611, November 2017
39. Zhou, B., Zhao, H., Puig, X., Fidler, S., Barriuso, A., Torralba, A.: Scene parsing through ade20k dataset. In: CVPR (2017)
40. Zintgraf, L.M., Cohen, T.S., Adel, T., Welling, M.: Visualizing deep neural network decisions: Prediction difference analysis. arXiv:1702.04595 (2017)

Bias Detection and Prediction
of Mapping Errors in Camera Calibration

Annika Hagemann[1,2(✉)], Moritz Knorr[1], Holger Janssen[1],
and Christoph Stiller[2]

[1] Robert Bosch GmbH, Corporate Research, Computer Vision Research Lab,
Hildesheim, Germany
annika.hagemann@de.bosch.com
[2] Institute of Measurement and Control, Karlsruhe Institute of Technology,
Karlsruhe, Germany

Abstract. Camera calibration is a prerequisite for many computer
vision applications. While a good calibration can turn a camera into
a measurement device, it can also deteriorate a system's performance if
not done correctly. In the recent past, there have been great efforts to
simplify the calibration process. Yet, inspection and evaluation of calibration results typically still requires expert knowledge.

In this work, we introduce two novel methods to capture the fundamental error sources in camera calibration: systematic errors (biases)
and remaining uncertainty (variance). Importantly, the proposed methods do not require capturing additional images and are independent of
the camera model. We evaluate the methods on simulated and real data
and demonstrate how a state-of-the-art system for guided calibration can
be improved. In combination, the methods allow novice users to perform
camera calibration and verify both the accuracy and precision.

1 Introduction

In 2000 Zhang published a paper [19] which allowed novice users to perform
monocular camera calibration using only readily available components. Several works, including systems for guided calibration, improved upon the original
idea [10,12,13]. However, we believe that a central building block is still missing: a generic way to evaluate the quality of a calibration result. More precisely,
a way to reliably quantify the remaining biases and uncertainties of a given
calibration. This is of critical importance, as errors and uncertainties in calibration parameters propagate to applications such as visual SLAM [9], ego-motion
estimation [3,17,20] and SfM [1,4]. Despite this importance, typical calibration
procedures rely on relatively simple metrics to evaluate the calibration, such

Electronic supplementary material The online version of this chapter (https://
doi.org/10.1007/978-3-030-71278-5_3) contains supplementary material, which is available to authorized users.

Z. Akata et al. (Eds.): DAGM GCPR 2020, LNCS 12544, pp. 30–43, 2021.
https://doi.org/10.1007/978-3-030-71278-5_3

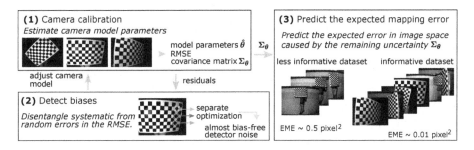

Fig. 1. Proposed camera calibration procedure, including the detection of systematic errors (biases) and the prediction of the expected mapping error.

as the root mean squared error (RMSE) on the calibration dataset. Furthermore, many frequently used metrics lack comparability across camera models and interpretability for non-expert users.

In general, the error sources of camera calibration can be divided into underfit (bias) and overfit (high variance). An underfit can be caused by a camera projection model not being able to reflect the true geometric camera characteristics, an uncompensated rolling shutter, or non-planarity of the calibration target. An overfit, on the other hand, describes that the model parameters cannot be estimated reliably, i.e. a high variance remains. A common cause is a lack of images used for calibration, bad coverage in the image, or a non-diversity in calibration target poses. In this paper, we address the challenge of quantifying both types of errors in target-based camera calibration, and provide three main contributions:

- A method to detect systematic errors (underfit) in a calibration. The method is based on estimating the variance of the corner detector and thereby disentangles random from systematic errors in the calibration residual (Fig. 1).
- A method to predict the expected mapping error (EME) in image space, which quantifies the remaining uncertainty (variance) in model parameters in a model-independent way. It provides an upper bound for the precision that can be achieved with a given dataset (Fig. 1).
- The application of our uncertainty metric EME in calibration guidance, which guides users to poses that lead to a maximum reduction in uncertainty. Extending a recently published framework [10], we show that our metric leads to further improvement of suggested poses.

In combination, these methods allow novice users to perform camera calibration and verify both the accuracy and precision of the result. Importantly, the work presented here explicitly abstracts from the underlying camera model and is therefore applicable in a wide range of scenarios. We evaluate the proposed methods with both simulations and real cameras.

2   Fundamentals

Camera Projection Modeling. From a purely geometric point of view, cameras project points in the 3D world to a 2D image [6]. This projection can be expressed by a function $p : \mathbb{R}^3 \to \mathbb{R}^2$ that maps a 3D point $x = (x, y, z)^T$ from a world coordinate system to a point $\bar{u} = (\bar{u}, \bar{v})^T$ in the image coordinate system. The projection can be decomposed into a coordinate transformation from the world coordinate system to the camera coordinate system $x \to x_c$ and the projection from the camera coordinate system to the image $p_C : x_c \to \bar{u}$:

$$\bar{u} = p(x, \theta, \Pi) = p_C(x_c, \theta) = p_C(Rx + t, \theta), \tag{1}$$

where θ are the *intrinsic* camera parameters and Π and are the *extrinsic* parameters describing the rotation R and translation t. For a plain pinhole camera, the intrinsic parameters are the focal length f and the principal point (ppx, ppy), i.e. $\theta = (f, \text{ppx}, \text{ppy})$. For this case, the projection $p_C(x_c, \theta)$ is given by $u = f/z_c \cdot x_c + \text{ppx}$, $v = f/z_c \cdot y_c + \text{ppy}$. In the following, we will consider more complex camera models, specifically, a standard pinhole camera model (S) with radial distortion $\theta_S = (f_x, f_y, \text{ppx}, \text{ppy}, r_1, r_2)$, and the OpenCV fisheye model (F) $\theta_F = (f_x, f_y, \text{ppx}, \text{ppy}, r_1, r_2, r_3, r_4)$ [8].

Calibration Framework. We base our methods on target-based camera calibration, in which planar targets are imaged in different poses relative to the camera. Without loss of generality, we assume a single chessboard-style calibration target and a single camera in the following. The calibration dataset is a set of images $\mathcal{F} = \{\text{frame}_i\}_{i=1}^{N_{\mathcal{F}}}$. The chessboard calibration target contains a set of corners $\mathcal{C} = \{\text{corner}_i\}_{i=1}^{N_C}$. The geometry of the target is well-defined, thus the 3D coordinates of chessboard-corner i in the world coordinate system are known as $x_i = (x_i, y_i, z_i)^T$. The image coordinates $u_i = (u_i, v_i)^T$ of chessboard-corners are determined by a corner-detector with noise σ_d. Thus, the observed coordinates u_i are assumed to deviate from the true image points $\bar{u}_i$ by an independent identically distributed (i.i.d.) error $\epsilon_d \sim \mathcal{N}(0, \sigma_d)$. Estimation is performed by minimizing a calibration cost function, typically defined by the quadratic sum over reprojection errors

$$\epsilon_{\text{res}}^2 = \sum_{j \in \mathcal{F}} \sum_{i \in \mathcal{C}} ||u_{ij} - p(x_{ij}, \theta, \Pi_j)||^2. \tag{2}$$

For the sake of simplicity, we present formulas for non-robust optimization here. Generally, we advise robustification, e.g. using a Cauchy kernel. Optimization is performed by a non-linear least-squares algorithm, which yields parameter estimates $(\hat{\theta}, \hat{\Pi}) = \text{argmin}(\epsilon_{\text{res}}^2)$.

A common metric to evaluate the calibration is the root mean squared error (RMSE) over all N individual corners coordinates (observations) in the calibration dataset $\mathcal{F}$ [6, p. 133]:

$$\text{RMSE} = \sqrt{\frac{1}{N} \sum_{j \in \mathcal{F}} \sum_{i \in \mathcal{C}} ||u_{ij} - p(x_{ij}, \hat{\theta}, \hat{\Pi}_j)||^2}, \tag{3}$$

The remaining uncertainty in estimated model parameters $\hat{\boldsymbol{\theta}}, \hat{\boldsymbol{\Pi}}$ is given by the parameter's covariance matrix $\boldsymbol{\Sigma}$. The covariance matrix can be computed by backpropagation of the variance of the corner detector σ_d^2:

$$\boldsymbol{\Sigma} = (\boldsymbol{J}_{\text{calib}}^T \boldsymbol{\Sigma}_d^{-1} \boldsymbol{J}_{\text{calib}})^{-1} = \sigma_d^2 (\boldsymbol{J}_{\text{calib}}^T \boldsymbol{J}_{\text{calib}})^{-1}, \tag{4}$$

where $\boldsymbol{\Sigma}_d = \sigma_d^2 \boldsymbol{I}$ is the covariance matrix of the corner detector and $\boldsymbol{J}_{\text{calib}}$ is the Jacobian of calibration residuals [6]. The covariance matrix of intrinsic parameters $\boldsymbol{\Sigma}_\theta$ can be extracted as a submatrix of the full covariance matrix.

3   Related Work

Approaches to evaluating camera calibration can be divided into detecting systematic errors and quantifying the remaining uncertainty in estimated model parameters. Typical choices of uncertainty metrics are the trace of the covariance matrix [10], or the maximum index of dispersion [13]. However, given the variety of camera models, from a simple pinhole model with only three parameters, up to local camera models, with around 10^5 parameters [2,15], parameter variances are difficult to interpret and not comparable across camera models. To address this issue, the parameter's influence on the mapping can be considered. The metric *maxERE* [12] quantifies uncertainty by propagating the parameter covariance into pixel space by means of a Monte Carlo simulation. The value of maxERE is then defined by the variance of the most uncertain image point of a grid of projected 3D points. The *observability* metric [16] weights the uncertainty in estimated parameters (here defined by the calibration cost function's Hessian) with the parameters' influence on a model cost function. Importantly, this model cost function takes into account a potential compensation of differences in the intrinsics by adjusting the extrinsics. The observability metric is then defined by the minimum eigenvalue of the weighted Hessian.

While both of these metrics provide valuable information about the remaining uncertainty, there are some shortcomings in terms of *how* uncertainty is quantified. The *observability* metric does not consider the whole uncertainty, but only the most uncertain parameter direction. Furthermore, it quantifies uncertainty in terms of an increase in the calibration cost, which can be difficult to interpret. *maxERE* quantifies uncertainty in pixel space and is thus easily interpretable. However, it relies on a Monte Carlo Simulation instead of an analytical approach and it does not incorporate potential compensations of differences in the intrinsics by adjusting the extrinsics.

The second type evaluation metrics aims at finding *systematic* errors. As camera characteristics have to be inferred indirectly through observations, there is a high risk of introducing systematic errors in the calibration process by choosing an inadequate projection model, neglecting rolling-shutter effects, or using an out-of-spec calibration target, to give a few examples. If left undetected, these errors will inevitably introduce biases into the application.

Historically, one way to detect systematic errors is to compare the resulting RMSE or reconstruction result against expected values obtained from earlier calibrations or textbooks [7]. However, these values vary for different cameras, lenses, calibration targets, and marker detectors and, hence, only allow capturing gross errors in general. Professional photogrammetry often makes use of highly accurate and precisely manufactured 3D calibration bodies [11]. Images captured from predefined viewpoints are then used to perform a 3D reconstruction of the calibration body. Different length ratios and their deviation from the ground truth are then computed to assess the quality of the calibration by comparing against empirical data. While these methods represent the gold standard due to the accuracy of the calibration body and repeatability, they are often not feasible or too expensive for typical research and laboratory settings and require empirical data for the camera under test. The methods presented in the following relax these requirements but can also be seen as a complement to this standard.

4   Detecting Systematic Errors

In the following, we derive the bias ratio (BR), a novel metric for quantifying the fraction of systematic error contribution to the mean squared reprojection error (MSE). Following the assumptions made in Sect. 2 one finds that asymptotically (by augmentation of [6, p. 136])

$$\text{MSE}_{\text{calib}} = \underbrace{\sigma_d^2(1 - \frac{N_{\mathcal{P}}}{N}) +}_{\text{random error}} \underbrace{\epsilon_{\text{bias}}^2}_{\text{systematic error contribution}}, \tag{5}$$

where $N_{\mathcal{P}}$ is the total number of free intrinsic and extrinsic parameters and ϵ_{bias} denotes the bias introduced through systematic errors. The variance σ_d^2 is generally camera dependent and not known a priori. To disentangle stochastic and systematic error contributions to the MSE, we need a way to determine σ_d^2 independently: The rationale behind many calibration approaches, and in particular guided calibration, is to find most informative camera-target configurations (cf. Fig. 1). For bias estimation, we propose the opposite. We explicitly use configurations which are less informative for calibration but at the same time also less likely to be impacted by systematic errors. More specifically, we decompose the calibration target virtually into several smaller calibration targets $\mathcal{V} = \{\text{target}_i\}_{i=1}^{N_{\mathcal{V}}}$, usually consisting of exclusive sets of the four corners of a checker board tile (cf. Fig. 2a). The poses of each virtual calibration target in each image are then estimated individually while keeping the camera intrinsic parameters fixed. Pose estimation is overdetermined with a redundancy of two (four tile corners and six pose parameters). From the resulting MSE values, MSE_v with $v \in \mathcal{V}$, we compute estimates of σ_d^2 via (5) assuming the bias is negligible within these local image regions

$$\hat{\sigma}_{d_v}^2 = \frac{\text{MSE}_v}{1 - \frac{6}{8}} = 4\,\text{MSE}_v. \tag{6}$$

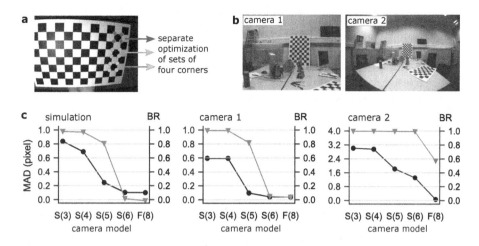

Fig. 2. Detecting systematic errors. **a** Illustration of the virtual decomposition of the calibration target into smaller targets used to estimate the corner detector variance. **b** Exemplary image of the same scene with the two test cameras. **c** Results of the bias ratio (BR) and the robust estimate of the RMSE (MAD) for one simulated and the two real cameras, using models of different complexities. For details see Sect. 7.

To obtain an overall estimate of $\widehat{\sigma}_d^2$, we compute the MSE in (6) across the residuals of all virtual targets, using the MAD as a robust estimator[1]. Finally, we use $\widehat{\sigma}_d^2$ to determine ϵ_{bias}^2 using (5) and compute the *bias ratio* as

$$\text{BR} = \frac{\epsilon_{\text{bias}}^2}{\text{MSE}_{\text{calib}}}. \tag{7}$$

The bias ratio is zero for unbiased calibration and close to one if the results are dominated by systematic errors. The bias ratio is an intuitive metric that quantifies the fraction of bias introduced by systematic errors. A bias ratio below a certain threshold τ_{BR} is a necessary condition for a successful calibration and a precondition for uncertainty estimation.[2] Generally, this kind of analysis can be performed for any separable[3] calibration target.

Practical Implementation. Computation of the bias ratio for target-based calibration procedures:

---

[1] Here, we assume the underlying distribution is Gaussian but might be subject to sporadic outliers. The MAD multiplied by a factor of 1.4826 gives a robust estimate for the standard deviation [14].

[2] To choose a threshold, it can be used that $\frac{1}{1-\text{BR}}$ is approximately F-distributed, representing the ratio of the residual sum of squares (SSE) of the calibration over the SSE of the virtual targets, weighted by their respective degrees of freedom. However, this only holds approximately, as the datapoints are not independent. We therefore use an empirical threshold of $\tau_{BR} = 0.2$, allowing for small biases.

[3] The decomposition of the target must lead to an overdetermined estimation problem.

1. Perform robust camera calibration and extract a robust estimate of $\text{MSE}_{\text{calib}}$ and the optimal parameters $\hat{\boldsymbol{\theta}}$ and $\hat{\boldsymbol{\Pi}}$.
2. Compute the residuals for all $v \in \mathcal{V}$:
   - Decompose the calibration targets found in each image into a total of $N_{\mathcal{V}}$ exclusive virtual calibration targets.
   - Optimize their pose independently leaving $\hat{\boldsymbol{\theta}}$ unchanged.
3. Compute a robust estimate of the MSE over all residuals and determine $\hat{\sigma}_d^2$ using (6).
4. Use $\hat{\sigma}_d^2$ to determine the bias contribution ϵ_{bias}^2 via (5).
5. Finally, compute the bias ratio as $\text{BR} = \epsilon_{\text{bias}}^2 / \text{MSE}_{\text{calib}}$ and test the result against the threshold τ_{BR}.

5   The Expected Mapping Error (EME)

The second type of error source, in addition to biases, is a high remaining uncertainty in estimated model parameters. We will now derive a novel uncertainty metric, the *expected mapping error* (EME), which is interpretable and comparable across camera models. It quantifies the expected difference between the mapping of a calibration result $\boldsymbol{p}_C(\boldsymbol{x}; \hat{\boldsymbol{\theta}})$ and the true (unknown) model $\boldsymbol{p}_C(\boldsymbol{x}; \bar{\boldsymbol{\theta}})$.

Inspired by previous works [5,12], we quantify the mapping difference in *image space*, as pixel differences are easily interpretable: we define a set of points in image space $\mathcal{G} = \{\boldsymbol{u}_i\}_{i=1}^{N_{\mathcal{G}}}$, which are projected to space via the inverse projection $\boldsymbol{p}_C^{-1}(\boldsymbol{u}_i; \bar{\boldsymbol{\theta}})$ using one set of model parameters and then back to the image using the other set of model parameters [2]. The mapping error is then defined as the average distance between original image coordinates $\boldsymbol{u}_i$ and back-projected image points $\boldsymbol{p}_C(\boldsymbol{x}_i; \hat{\boldsymbol{\theta}})$ (see Fig. 3):

$$\tilde{K}(\hat{\boldsymbol{\theta}}, \bar{\boldsymbol{\theta}}) = \frac{1}{N} \sum_{i \in \mathcal{G}} ||\boldsymbol{u}_i - \boldsymbol{p}_C(\boldsymbol{p}_C^{-1}(\boldsymbol{u}_i; \bar{\boldsymbol{\theta}}); \hat{\boldsymbol{\theta}})||^2, \tag{8}$$

where $N = 2N_{\mathcal{G}}$ is the total number of image coordinates. Since small deviations in intrinsic parameters can oftentimes be compensated by a change in extrinsic parameters [16], we allow for a virtual compensating rotation $\boldsymbol{R}$ of the viewing rays. Thus, we formulate the *effective* mapping error as follows:

$$K(\hat{\boldsymbol{\theta}}, \bar{\boldsymbol{\theta}}) = \min_{\boldsymbol{R}} \frac{1}{N} \sum_{i \in \mathcal{G}} ||\boldsymbol{u}_i - \boldsymbol{p}_C(\boldsymbol{R}\, \boldsymbol{p}_C^{-1}(\boldsymbol{u}_i; \bar{\boldsymbol{\theta}}); \hat{\boldsymbol{\theta}})||^2. \tag{9}$$

We now show that for an ideal, bias-free calibration, the effective mapping error $K(\hat{\boldsymbol{\theta}}, \bar{\boldsymbol{\theta}})$ can be *predicted* by propagating parameter uncertainties. Note that the following derivation is independent of the particular choice of K, provided that we can approximate K with a Taylor expansion around $\hat{\boldsymbol{\theta}} = \bar{\boldsymbol{\theta}}$ up to second order:

$$K(\hat{\boldsymbol{\theta}}, \bar{\boldsymbol{\theta}}) \approx K(\bar{\boldsymbol{\theta}}, \bar{\boldsymbol{\theta}}) + \text{grad}(K)\boldsymbol{\Delta\theta} + \frac{1}{2}\boldsymbol{\Delta\theta}^T \boldsymbol{H}_K \boldsymbol{\Delta\theta}$$

$$\approx \frac{1}{N}\boldsymbol{\Delta\theta}^T (\boldsymbol{J}_{\text{res}}{}^T \boldsymbol{J}_{\text{res}})\boldsymbol{\Delta\theta} \tag{10}$$

$$\approx \boldsymbol{\Delta\theta}^T \boldsymbol{H} \boldsymbol{\Delta\theta},$$

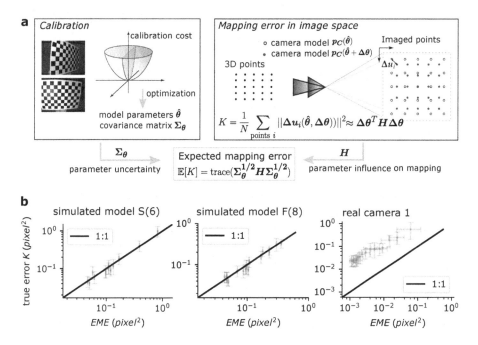

Fig. 3. Predicting the mapping error based on parameter uncertainties. **a** Schematic of the derived uncertainty metric EME = trace($\Sigma_\theta^{1/2} H \Sigma_\theta^{1/2}$). **b** Evaluation in simulation and experiments. The simulation results validate the derived relation (5). For real cameras, the EME is a lower bound to the error, as non-ideal behavior can lead to higher absolute errors. Error bars are 95% bootstrap confidence intervals.

where $\Delta\theta = \bar{\theta} - \hat{\theta}$ is the difference between true and estimated intrinsic parameters, $\mathbf{res}_i(\hat{\theta}, \bar{\theta}) = u_i - p_C(R\, p_C^{-1}(u_i; \bar{\theta}); \hat{\theta})$ are the mapping residuals and $J_{\mathbf{res}} = d\mathbf{res}/d\Delta\theta$ is the Jacobian of the residuals. Furthermore, we defined the *model matrix* $H := \frac{1}{N} J_{\mathbf{res}}{}^T J_{\mathbf{res}}$. For a more detailed derivation of the second step in (10), see Supplementary.

Estimated model parameters $\hat{\theta}$ obtained from a least squares optimization are a random vector, asymptotically following a multivariate Gaussian with mean $\mu_\theta = \bar{\theta}$ and covariance Σ_θ [18, p. 8]. Likewise, the parameter error $\Delta\theta = \bar{\theta} - \hat{\theta}$ follows a multivariate Gaussian, with mean $\mu_{\Delta\theta} = 0$ and covariance $\Sigma_{\Delta\theta} = \Sigma_\theta$. We propagate the distribution of the parameter error $\Delta\theta$ to find the distribution of the mapping error $K(\hat{\theta}, \bar{\theta})$. In short, we find that the mapping error $K(\hat{\theta}, \bar{\theta})$ can be expressed as a linear combination of χ^2 random variables:

$$K(\hat{\theta}, \bar{\theta}) = \Delta\theta^T H \Delta\theta$$

$$= \sum_{i=1}^{N_\theta} \lambda_i Q_i, \quad \text{with} \quad Q_i \sim \chi^2(1). \tag{11}$$

The coefficients λ_i are the eigenvalues of the matrix product $\boldsymbol{\Sigma_\theta}^{1/2}\boldsymbol{H}\boldsymbol{\Sigma_\theta}^{1/2}$ and N_θ is the number of eigenvalues which equals the number of parameters $\boldsymbol{\theta}$. The full derivation of relation (11) is shown in the Supplementary. Importantly, based on expression (11), we can derive the expected value of $K(\hat{\boldsymbol{\theta}}, \bar{\boldsymbol{\theta}})$:

$$\mathbb{E}[K(\hat{\boldsymbol{\theta}}, \bar{\boldsymbol{\theta}})] = \mathbb{E}[\sum_{i=1}^{N_\theta} \lambda_i Q_i] = \sum_{i=1}^{N_\theta} \lambda_i \mathbb{E}[Q_i] = \sum_{i=1}^{N_\theta} \lambda_i$$
$$= \text{trace}(\boldsymbol{\Sigma_\theta}^{1/2}\boldsymbol{H}\boldsymbol{\Sigma_\theta}^{1/2}),$$

where we used that the χ^2-distribution with n degrees of freedom $\chi^2(n)$ has expectation value $\mathbb{E}[\chi^2(n)] = n$. We therefore propose the *expected mapping error* EME $= \text{trace}(\boldsymbol{\Sigma_\theta}^{1/2}\boldsymbol{H}\boldsymbol{\Sigma_\theta}^{1/2})$ as a model-independent measure for the remaining uncertainty.

Practical Implementation. The expected mapping error EME can be determined for any given bundle-adjustment calibration:

1. Run the calibration and extract the RMSE, the optimal parameters $\hat{\boldsymbol{\theta}}$ and the Jacobian $\boldsymbol{J}_{\text{calib}}$ of the calibration cost function.
2. Compute the parameter covariance matrix $\boldsymbol{\Sigma} = \sigma_d^2(\boldsymbol{J}_{\text{calib}}^T\boldsymbol{J}_{\text{calib}})^{-1}$ and extract the intrinsic part $\boldsymbol{\Sigma_\theta}$.
3. Determine the model matrix $\boldsymbol{H}$:
   - Implement the mapping error (Eq. (9)) as a function of the parameter estimate $\hat{\boldsymbol{\theta}}$ and a parameter difference $\boldsymbol{\Delta\theta}$.
   - Numerically compute the Jacobian $\boldsymbol{J}_{\text{res}} = d\text{res}/d\boldsymbol{\Delta\theta}$ at the estimated parameters $\hat{\boldsymbol{\theta}}$ and compute $\boldsymbol{H} = \frac{1}{N}\boldsymbol{J}_{\text{res}}^T\boldsymbol{J}_{\text{res}}$.
4. Compute EME $= \text{trace}(\boldsymbol{\Sigma_\theta}^{1/2}\boldsymbol{H}\boldsymbol{\Sigma_\theta}^{1/2})$.

6   Experimental Evaluation

Simulations. We simulated 3D world coordinates of a single planar calibration target in different poses relative to the camera (random rotations $\varphi_x, \varphi_y, \varphi_z \in [-\frac{\pi}{4}, \frac{\pi}{4}]$, translations $t_z \in [0.5 \text{ m}, 2.5 \text{ m}]$, $t_x, t_y \in [-0.5 \text{ m}, 0.5 \text{ m}]$). We then computed the resulting image coordinates using different camera models. To simulate the detector noise, we added Gaussian noise with $\sigma_d = 0.1$ px to all image coordinates. To validate the bias ratio, we simulated a pinhole camera with two radial distortion parameters, but ran calibrations with different models, including insufficiently complex models (underfit). To validate the uncertainty measure EME $= \text{trace}(\boldsymbol{\Sigma_\theta}^{1/2}\boldsymbol{H}\boldsymbol{\Sigma_\theta}^{1/2})$, we ran calibrations with different numbers of simulated frames ($N_\mathcal{F} \in [3, 20]$) and $n_r = 50$ noise realizations for each set of frames. After each calibration, we computed the true mapping error K with respect to the known ground-truth (Eq. 9) and the EME.

Evaluation with Real Cameras. We tested the metrics for two different real cameras (see Fig. 2b). For each camera, we collected a total of $n = 500$ images of a planar calibration target. As reference, we performed a calibration with all 500 images. To test the bias metric, we ran calibrations with camera models of different complexities (Fig. 2c). To test the uncertainty metric EME, we ran calibrations with different numbers of randomly selected frames ($N_\mathcal{F} \in [3, 20]$, 50 randomly selected datasets for each $N_\mathcal{F}$). For each calibration, we computed both the true mapping error K with respect to the reference and the EME.

7   Results

Validating the Bias Ratio. Figure 2c shows the robust estimate of the RMSE (median absolute deviation, MAD) and the bias ratio for the calibrations of three cameras (one simulated camera and the two real cameras shown in Fig. 2b) for varying numbers of non-zero intrinsic calibration parameters, representing different camera models. In detail, the individual parameter sets are $\boldsymbol{\theta}_{S(3)} = (f, \text{ppx}, \text{ppy})$, $\boldsymbol{\theta}_{S(4)} = (f_x, f_y, \text{ppx}, \text{ppy})$, $\boldsymbol{\theta}_{S(5)} = (f_x, f_y, \text{ppx}, \text{ppy}, r_1)$, $\boldsymbol{\theta}_{S(6)} = (f_x, f_y, \text{ppx}, \text{ppy}, r_1, r_2)$, and $\boldsymbol{\theta}_{F(8)} = (f_x, f_y, \text{ppx}, \text{ppy}, r_1, r_2, r_3, r_4)$ (cf. Sect. 2).

For all cameras, the MAD and BR can be reduced by using a more complex camera model which is to be expected, since the projections are not rectilinear and thus necessitate some kind of (nonlinear) distortion modeling. For the simulated camera and camera 1, a bias ratio below $\tau_{BR} = 0.2$ is reached using the standard camera model (S) with two radial distortion parameters. For camera 2, a low bias ratio cannot be reached even when using OpenCV's fisheye camera model with 8 parameters. This highlights the advantage of the bias ratio over the RMSE: the low RMSE could wrongfully be interpreted as low bias – the bias ratio of BR ≈ 0.6, however, demonstrates that some sort of systematic error remains and a more complex model should be tried.

Validating the Uncertainty Metric. To validate the uncertainty metric EME in simulations, we ran calibrations with different numbers of images using a pinhole with radial distortion $S(6)$ and a fisheye camera $F(8)$. Figure 3b shows the uncertainty metric EME = $\text{trace}(\boldsymbol{\Sigma_\theta}^{1/2} \boldsymbol{H} \boldsymbol{\Sigma_\theta}^{1/2})$ and the real average mapping error. Consistent with Eq. (5), the EME predicts the average mapping error. For the *real* camera, the EME is highly correlated with the true mapping error, however the absolute values of the real errors are higher, which is to be expected in practice. It reflects that (i) the ground-truth is only approximated by the reference calibration, (ii) deviations from the ideal assumptions underlying the covariance matrix (Eq. (4)), and (iii) deviations from the i.i.d. Gaussian error assumption. This limitation affects all metrics that are based on the covariance matrix computed via Eq. (4). The EME therefore provides an *upper bound* to the precision that can be achieved for a given dataset.

Comparison with State-of-the-Art. We compare the EME with the other state-of-the-art uncertainty metrics introduced in Sect. 3. We focus on $\text{trace}(\boldsymbol{\Sigma_\theta})$,

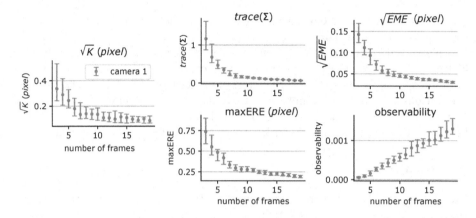

Fig. 4. Comparison of state-of-the-art metrics for real camera 1. On average, the true error K decreases with the number frames. For comparability with maxERE, we show $\sqrt{K}$ and $\sqrt{EME}$ in units of pixels. All metrics are correlated with the true error, but absolute values and the scaling differ. Values are medians across 50 random samples, error bars are 95% bootstrap confidence intervals.

maxERE [12] and *observability* [16], as these are the metrics closest to ours (Fig. 4). All metrics provide information about the remaining uncertainty and are correlated with the true error. However, the metrics quantify uncertainty in very different ways: trace(Σ_θ) quantifies the uncertainty in model parameters, and thus inherently differs depending on the camera model. The *observability* metric accounts for the parameter's effect of the mapping and for compensations via different extrinsics. However, it does not incorporate the full uncertainty, but just the least observable direction. Furthermore, the absolute values are comparatively difficult to interpret, as they measure an increase in the calibration cost. maxERE quantifies the *maximum* expected reprojection error in image space and is therefore easily interpretable. Similar to maxERE, the EME predicts the expected error in image space and is therefore easily interpretable. Instead of a *maximum* error, the EME reflects the *average* error. In contrast to maxERE, the EME does not require a Monte Carlo simulation. Furthermore, the EME can account for a compensation via different extrinsics, which we consider a reasonable assumption in many scenarios.

8 Application in Calibration Guidance

To demonstrate the practical use of the EME, we apply it in calibration guidance. Calibration guidance refers to systems that predict most informative next observations to reduce the remaining uncertainty and then guide users towards these measurements. We choose an exitisting framework, called calibration wizard [10] and extend it with our metric. Calibration wizard predicts the next best pose by minimizing the trace of the intrinsic parameter's covariance matrix

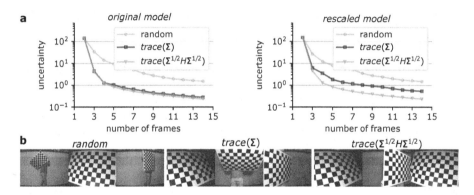

Fig. 5. Application of EME for calibration guidance. **a** For the original model, both metrics lead to a similarly fast reduction in uncertainty. Rescaling the model to a different unit of the focal length results in a reduced performance of trace(Σ_θ), while trace($\Sigma_\theta^{1/2} H \Sigma_\theta^{1/2}$) remains unaffectd. Uncertainty is quantified by the average of the *uncertainty map* proposed by calibration wizard [10]. **b** Examples of suggested poses.

trace(Σ_θ). However, depending on the camera model, parameters will affect the image in very different ways. High variance in a given parameter will not necessarily result in a proportionally high uncertainty in the image. To avoid such an imbalance, we suggest to minimize the uncertainty in image space, instead of parameters, i.e. to replace trace(Σ_θ) with trace($\Sigma_\theta^{1/2} H \Sigma_\theta^{1/2}$).

To compare the methods, we use images of camera 1 (see Fig. 2b). Starting with two random images, the system successively selectes the most informative next image with (i) the original metric trace(Σ_θ), (ii) our metric trace($\Sigma_\theta^{1/2} H \Sigma_\theta^{1/2}$) and (iii) randomly. Using the pinhole model with radial distortion, the poses suggested by trace(Σ_θ) and trace($\Sigma_\theta^{1/2} H \Sigma_\theta^{1/2}$) are similarly well suited, both leading to a significantly faster convergence than random images (Fig. 5). However, when changing the camera model, e.g. by parameterizing the focal length in millimeters instead of pixels, simulated here by a division by 100 ($f \rightarrow 0.01 \cdot f$), the methods differ: the poses proposed by trace($\Sigma_\theta^{1/2} H \Sigma_\theta^{1/2}$) reduce uncertainty significantly faster than trace(Σ_θ). This can be explained by the fact that when minimizing trace(Σ_θ), the variance of less significant parameters will be reduced just as much as the variance of parameters with large effect on the mapping. This example shows that the performance of trace(Σ_θ) can be affected by the choice of the model, while trace($\Sigma_\theta^{1/2} H \Sigma_\theta^{1/2}$) remains unaffected.

9   Conclusion and Future Research

In this paper, we proposed two metrics to evaluate systematic errors and the remaining uncertainty in camera calibration. We have shown that the *bias ratio* (BR) reliably captures underfits, which can result from an insufficiently complex model. Furthermore, we have shown that it is possible to predict the *expected mapping error* (EME) in image space, which provides an upper bound for the precision that can be achieved with a given dataset. Both metrics are model-independent and therefore widely applicable. Finally, we have shown that the EME can be applied for calibration guidance, resulting in a faster reduction in mapping uncertainty than the existing parameter-based approach.

In future, we will extend the metrics to multi-camera systems and extrinsic calibration. Furthermore, we would like to incorporate an analysis of the coverage of the camera field of view into our evaluation scheme.

References

1. Abraham, S., Förstner, W.: Calibration errors in structure from motion. In: Levi, P., Schanz, M., Ahlers, R.J., May, F. (eds.) Mustererkennung 1998, pp. 117–124. Springer, Heidelberg (1998). https://doi.org/10.1007/978-3-642-72282-0_11
2. Beck, J., Stiller, C.: Generalized B-spline camera model. In: 2018 IEEE Intelligent Vehicles Symposium (IV), pp. 2137–2142. IEEE (2018). https://ieeexplore.ieee.org/abstract/document/8500466/
3. Cheong, L.F., Peh, C.H.: Depth distortion under calibration uncertainty. Comput. Vis. Image Underst. **93**(3), 221–244 (2004). https://doi.org/10.1016/j.cviu.2003.09.003. https://linkinghub.elsevier.com/retrieve/pii/S1077314203001437
4. Cheong, L.F., Xiang, X.: Behaviour of SFM algorithms with erroneous calibration. Comput. Vis. Image Underst. **115**(1), 16–30 (2011). https://doi.org/10.1016/j.cviu.2010.08.004. https://linkinghub.elsevier.com/retrieve/pii/S1077314210001852
5. Cramariuc, A., Petrov, A., Suri, R., Mittal, M., Siegwart, R., Cadena, C.: Learning camera miscalibration detection. arXiv preprint arXiv:2005.11711 (2020)
6. Hartley, R., Zisserman, A.: Multiple view geometry in computer vision (2004). https://doi.org/10.1017/CBO9780511811685. oCLC: 171123855
7. Luhmann, T., Robson, S., Kyle, S., Boehm, J.: Close-range Photogrammetry and 3D Imaging. De Gruyter textbook, De Gruyter (2013). https://books.google.de/books?id=TAuBngEACAAJ
8. OpenCV: OpenCV Fisheye Camera Model. https://docs.opencv.org/master/db/d58/group__calib3d__fisheye.html
9. Ozog, P., Eustice, R.M.: On the importance of modeling camera calibration uncertainty in visual SLAM. In: 2013 IEEE International Conference on Robotics and Automation, pp. 3777–3784. IEEE (2013). https://ieeexplore.ieee.org/abstract/document/6631108/
10. Peng, S., Sturm, P.: Calibration wizard: a guidance system for camera calibration based on modelling geometric and corner uncertainty. In: Proceedings of the IEEE International Conference on Computer Vision, pp. 1497–1505 (2019)
11. Rautenberg, U., Wiggenhagen, M.: Abnahme und ueberwachung photogrammetrischer messsysteme nach vdi 2634, blatt 1. PFG 2/2002, S.117-124 (2002). https://www.ipi.uni-hannover.de/fileadmin/ipi/publications/VDI2634_1e.pdf
12. Richardson, A., Strom, J., Olson, E.: AprilCal: assisted and repeatable camera calibration. In: 2013 IEEE/RSJ International Conference on Intelligent Robots and Systems, pp. 1814–1821. IEEE (2013). https://ieeexplore.ieee.org/abstract/document/6696595/

13. Rojtberg, P., Kuijper, A.: Efficient pose selection for interactive camera calibration. In: 2018 IEEE International Symposium on Mixed and Augmented Reality (ISMAR), pp. 31–36. IEEE (2018). https://ieeexplore.ieee.org/abstract/document/8613748/
14. Rousseeuw, P.J., Croux, C.: Alternatives to the median absolute deviation. J. Am. Stat. Assoc. **88**(424), 1273–1283 (1993)
15. Schoeps, T., Larsson, V., Pollefeys, M., Sattler, T.: Why Having 10,000 Parameters in Your Camera Model is Better Than Twelve. arXiv preprint arXiv:1912.02908 (2019). https://arxiv.org/abs/1912.02908
16. Strauss, T.: Kalibrierung von Multi-Kamera-Systemen. KIT Scientific Publishing (2015). https://d-nb.info/1082294497/34
17. Svoboda, T., Sturm, P.: What can be done with a badly calibrated Camera in Ego-Motion Estimation? (1996). https://hal.inria.fr/inria-00525701/
18. Triggs, B., McLauchlan, P.F., Hartley, R.I., Fitzgibbon, A.W.: Bundle adjustment — a modern synthesis. In: Triggs, B., Zisserman, A., Szeliski, R. (eds.) IWVA 1999. LNCS, vol. 1883, pp. 298–372. Springer, Heidelberg (2000). https://doi.org/10.1007/3-540-44480-7_21
19. Zhang, Z.: A flexible new technique for camera calibration. IEEE Trans. Pattern Anal. Mach. Intell. **22**(11), 1330–1334 (2000). https://ieeexplore.ieee.org/abstract/document/888718/
20. Zucchelli, M., Kosecka, J.: Motion bias and structure distortion induced by calibration errors, pp. 68.1–68.10. British Machine Vision Association (2001). https://doi.org/10.5244/C.15.68. http://www.bmva.org/bmvc/2001/papers/51/index.html

Learning to Identify Physical Parameters from Video Using Differentiable Physics

Rama Kandukuri[1,2(✉)] ⓘ, Jan Achterhold[1] ⓘ, Michael Moeller[2] ⓘ,
and Joerg Stueckler[1] ⓘ

[1] Max Planck Institute for Intelligent Systems, Tübingen, Germany
`rama.kandukuri@tuebingen.mpg.de`
[2] University of Siegen, Siegen, Germany

Abstract. Video representation learning has recently attracted attention in computer vision due to its applications for activity and scene forecasting or vision-based planning and control. Video prediction models often learn a latent representation of video which is encoded from input frames and decoded back into images. Even when conditioned on actions, purely deep learning based architectures typically lack a physically interpretable latent space. In this study, we use a differentiable physics engine within an action-conditional video representation network to learn a physical latent representation. We propose supervised and self-supervised learning methods to train our network and identify physical properties. The latter uses spatial transformers to decode physical states back into images. The simulation scenarios in our experiments comprise pushing, sliding and colliding objects, for which we also analyze the observability of the physical properties. In experiments we demonstrate that our network can learn to encode images and identify physical properties like mass and friction from videos and action sequences in the simulated scenarios. We evaluate the accuracy of our supervised and self-supervised methods and compare it with a system identification baseline which directly learns from state trajectories. We also demonstrate the ability of our method to predict future video frames from input images and actions.

1 Introduction

Video representation learning is a challenging task in computer vision which has applications in scene understanding and prediction [19, 20] or vision-based control and planning [9, 10, 12]. Such approaches can be distinguished into supervised or self-supervised methods, the latter typically based on recurrent autoencoder models which are trained for video prediction.

Typical architectures of video prediction models first encode the image in a low dimensional latent scene representation. This latent state is predicted forward eventually based on actions and finally decoded into future frames. Neural

---

Electronic supplementary material The online version of this chapter (https://doi.org/10.1007/978-3-030-71278-5_4) contains supplementary material, which is available to authorized users.

© Springer Nature Switzerland AG 2021
Z. Akata et al. (Eds.): DAGM GCPR 2020, LNCS 12544, pp. 44–57, 2021.
https://doi.org/10.1007/978-3-030-71278-5_4

network based video prediction models like [4,10,23] perform these steps implicitly and typically learn a latent representation which cannot be directly interpreted for physical quantities such as mass, friction, position and velocity. This can limit explainability and generalization for new tasks and scenarios. Analytical models like [3,8,17] in contrast structure the latent space as an interpretable physical parameterization and use analytical physical models to forward the latent state.

In this paper we study supervised and self-supervised learning approaches for identifying physical parameters of objects from video. Our approach encodes images into physical states and uses a differentiable physics engine [3] to forward the physical scene state based on latent physical scene parameters. For self-supervised learning, we apply spatial transformers [14] to decode the predicted physical scene states into images based on known object models. We evaluate our approach in various simulation scenarios such as pushing, sliding and collision of objects and analyze the observability of physical parameters in these scenarios. In our experiments, we demonstrate that physical scene encodings can be learned from video and interactions through supervised and self-supervised training. Our method allows for identifying the observable physical parameters of the objects from the videos. In summary, we make the following contributions in this work

- We propose supervised and self-supervised learning approaches to learn to encode scenes into physical scene representations of objects. Our novel architecture integrates a differentiable physics engine as a forward model. It uses spatial transformers to decode the states back into images for self-supervised learning.
- We analyse the observability of physical parameters in pushing, sliding and collision scenarios. Our approach simultaneously identifies the observable physical parameters during training while learning the network parameters of the encoder.
- We evaluate our approach on simulated scenes and analyse its accuracy in recovering object pose and physical parameters.

1.1   Related Work

Neural Video Prediction. Neural video prediction models learn an embedding of video frames into a latent representation using successive neural network operations such as convolutions, non-linearities and recurrent units. Srivastava et al. [23] embed images into a latent representation recurrently using long short term memory (LSTM [13]) cells. The latent representation is decoded back using a convolutional decoder. Video prediction is achieved by propagating the latent representation of the LSTM forward using predicted frames as inputs. Finn et al. [10] also encode images into a latent representation using successive LSTM convolutions [22]. The decoder predicts motion kernels (5×5 pixels) and composition masks for the motion layers which are used to propagate the input images.

A typical problem of such architectures is that they cannot capture multimodal distributions on future frames well, for example, in the case of uncertain interactions of objects, which leads to blurry predictions. Babaeizadeh et al. [4] introduce a stochastic latent variable which is inferred from the full sequence at training time and sampled from a fixed prior at test time. Visual interaction networks explicitly model object interactions using graph neural networks in a recurrent video prediction architecture [25]. However, these approaches do not learn a physically interpretable latent representation and cannot be used to infer physical parameters. To address this shortcomings, Ye et al. [26] train a variational autoencoder based architecture in a conditional way by presenting training data with variation in each single specific physical property while holding all but a few latent variables fixed. This way, the autoencoder is encouraged to represent this property in the corresponding part of the latent vector. The approach is demonstrated on videos of synthetic 3D scenes with colliding shape primitives. Zhu et al. [27] combine disentangled representation learning based on total correlation [5] with partial supervision of physical properties. These purely deep learning based techniques still suffer from sample efficiency and require significant amounts of training data.

Physics-Based Prediction Models. Several works have investigated differentiable formulations of physics engines which could be embedded as layers in deep neural networks. In [8] an impulse-based velocity stepping physics engine is implemented in a deep learning framework. Collisions are restricted to sphere shapes and sphere-plane interactions to allow for automatic differentiation. The method is used to tune a deep-learning based robot controller but neither demonstrated for parameter identification nor video prediction.

Belbute-Peres et al. [3] propose an end-to-end differentiable physics engine that models frictions and collisions between arbitrary shapes. Gradients are computed analytically at the solution of the resulting linear complementarity problem (LCP) [1]. They demonstrate the method for including a differentiable physics layer in a video prediction network for modelling a 2D bouncing balls scenario with 3 color-coded circular objects. Input to the network are the color segmented images and optical flow estimated from pairs of frames. The network is trained in a supervised way using ground-truth positions of the objects. We propose to use spatial transformers in the decoder such that the network can learn a video representation in a self-supervised way. We investigate 3D scenarios that include pushing, sliding, and collisions of objects and analyze observability of physical parameters using vision and known forces applied to the objects. A different way of formulating rigid body dynamics has been investigated in [11] using energy conservation laws. The method is demonstrated for parameter identification, angle estimation and video prediction for a 2D pendulum environment using an autoencoder network. Similar to our approach, [15] also uses spatial transformers for the decoder. However, differently the physics engine only models gravitational forces between objects and does not investigate full 3D rigid body physics with collision and friction modelling and parameter identification.

Recently, Runia et al. [21] demonstrated an approach for estimating physical parameters of deforming cloth in real-world scenes. The approach minimizes distance in a contrastively learned embedding space which encodes videos of the observed scene and rendered scenes generated with a physical model based on the estimated parameters. In our approach, we train a video embedding network with the physical model as network layer and identify the physical parameters of observed rigid objects during training.

2 Background

2.1 Unconstrained and Constrained Dynamics

The governing equation of unconstrained rigid body dynamics in 3D can be written as

$$\mathbf{f} = \mathbf{M}\dot{\boldsymbol{\xi}} + \text{Coriolis forces} \tag{1}$$

where $\mathbf{f} : [0, \infty[\ \rightarrow \ \mathbb{R}^6$ is the time-dependent torque-force vector, $\mathbf{M} \in \mathbb{R}^{6 \times 6}$ is the mass-inertia matrix and $\dot{\boldsymbol{\xi}} : [0, \infty[\ \rightarrow \ \mathbb{R}^6$ is the time-derivative of the twist vector $\boldsymbol{\xi} = \left(\boldsymbol{\omega}^\top, \mathbf{v}^\top\right)^\top$ stacking rotational and linear velocities $\boldsymbol{\omega}, \mathbf{v} : [0, \infty[\rightarrow \mathbb{R}^3$ [7]. In our experiments we do not consider rotations between two or more frames of reference, therefore we do not have any Coriolis forces. Most of the real world rigid body motions are constrained. To simulate those behaviors we need to constrain the motion with joint, contact and frictional constraints [7].

The force-acceleration based dynamics which we use in Eq. (1) does not work well for collisions since there is a sudden change in the direction of velocity in infinitesimal time [7]. Therefore we use impulse-velocity based dynamics, where even the friction is well-behaved [7], i.e., equations have a solution at all configurations. We discretize the acceleration using the forward Euler method as $\dot{\boldsymbol{\xi}} = (\boldsymbol{\xi}_{t+h} - \boldsymbol{\xi}_t)/h$, where $\dot{\boldsymbol{\xi}}_{t+h}$ and $\dot{\boldsymbol{\xi}}_t$ are the velocities in successive time steps at times $t + h$ and t, and h is the time-step size. Equation (1) now becomes

$$\mathbf{M}\boldsymbol{\xi}_{t+h} = \mathbf{M}\boldsymbol{\xi}_t + \mathbf{f} \cdot h. \tag{2}$$

Constrained Dynamics: The joint constraints are equality constraints and they restrict degrees of freedom of a rigid body. Mathematically this can be written as $\mathbf{J}_e \boldsymbol{\xi}_{t+h} = 0$ where $\mathbf{J}_e$ is the equality Jacobian which gives the directions in which the motion is restricted. The joint constraints exert constraint forces which are solved using Euler-Lagrange equations by solving for the joint force multiplier $\boldsymbol{\lambda}_e$.

The contact constraints are inequality constraints which prevent bodies from interpenetration. This ensures that the minimum distance between two bodies is always greater than or equal to zero. The constraint equations can be written using Newton's impact model [7] as $\mathbf{J}_c \boldsymbol{\xi}_{t+h} \geq -k \mathbf{J}_c \boldsymbol{\xi}_t$. The term $k \mathbf{J}_c \boldsymbol{\xi}_t$ can be replaced with $\mathbf{c}$ which gives $\mathbf{J}_c \boldsymbol{\xi}_{t+h} \geq -\mathbf{c}$, where k is the coefficient of restitution, $\mathbf{J}_c$ is the Jacobian of the contact constraint function at the current state of the system and $\boldsymbol{\lambda}_c$ is the contact force multiplier. Since it is an inequality constraint we introduce slack variables $\mathbf{a}$, which also gives us complementarity constraints [18].

The friction is modeled using a maximum dissipation energy principle since friction damps the energy of the system. In this case we get two inequality constraints since frictional force depends on normal force [2,24]. They can be written as $\mathbf{J}_f\boldsymbol{\lambda}_f + \boldsymbol{\gamma} \geq 0$ and $\mu\boldsymbol{\lambda}_c \geq \mathbf{E}\boldsymbol{\lambda}_f$ where μ is the friction coefficient, $\mathbf{J}_f$ is the Jacobian of the friction constraint function at the current state of the system, $\mathbf{E}$ is a binary matrix which ensures linear independence between equations at multiple contacts, and $\boldsymbol{\lambda}_f$ and $\boldsymbol{\gamma}$ are frictional force multipliers. Since we have two inequality constraints we have two slack variables σ, ζ and two complementarity constraints.

In summary, all the constraints that describe the dynamic behavior of the objects we consider in our scene can be written as the following linear complementarity problem (LCP),

$$
\begin{pmatrix} 0 \\ 0 \\ \mathbf{a} \\ \sigma \\ \zeta \end{pmatrix} - \begin{pmatrix} \mathbf{M} & -\mathbf{J}_{eq}^T & -\mathbf{J}_c^T & -\mathbf{J}_f^T & 0 \\ \mathbf{J}_{eq} & 0 & 0 & 0 & 0 \\ \mathbf{J}_c & 0 & 0 & 0 & 0 \\ \mathbf{J}_f & 0 & 0 & 0 & \mathbf{E} \\ 0 & 0 & \mu & -\mathbf{E}^T & 0 \end{pmatrix} \begin{pmatrix} \boldsymbol{\xi}_{t+h} \\ \boldsymbol{\lambda}_{eq} \\ \boldsymbol{\lambda}_c \\ \boldsymbol{\lambda}_f \\ \boldsymbol{\gamma} \end{pmatrix} = \begin{pmatrix} -\mathbf{M}\boldsymbol{\xi}_t - h\mathbf{f}_{ext} \\ 0 \\ \mathbf{c} \\ 0 \\ 0 \end{pmatrix}, \quad (3)
$$

$$
\text{subject to} \quad \begin{pmatrix} \mathbf{a} \\ \sigma \\ \zeta \end{pmatrix} \geq 0, \quad \begin{pmatrix} \boldsymbol{\lambda}_c \\ \boldsymbol{\lambda}_f \\ \boldsymbol{\gamma} \end{pmatrix} \geq 0, \quad \begin{pmatrix} \mathbf{a} \\ \sigma \\ \zeta \end{pmatrix}^T \begin{pmatrix} \boldsymbol{\lambda}_c \\ \boldsymbol{\lambda}_f \\ \boldsymbol{\gamma} \end{pmatrix} = 0.
$$

The above LCP is solved using a primal-dual algorithm as described in [18]. It is embedded in our deep neural network architecture in a similar way as in [1] and [3], which facilitates backpropagation of gradients at its solution.

3 Method

We develop a deep neural network architecture which encodes images into physical states $\mathbf{s}_i = \left(\mathbf{x}_i^\top, \boldsymbol{\xi}^\top\right)^\top$ where $\mathbf{x}_i = \left(\mathbf{q}_i^\top, \mathbf{p}_i^\top\right)^\top$ with orientation $\mathbf{q}_i \in \mathbb{S}^3$ as unit quaternion and position $\mathbf{p}_i \in \mathbb{R}^3$ of object i. We propagate the state using the differentiable physics engine which is integrated as layer on the encoding in the deep neural network. For self-supervised learning, a differentiable decoder subnetwork generates images from the integrated state representation of the objects.

We aim to learn the system's dynamics by regressing the state trajectories and learning the physical parameters of the objects. These parameters can be the masses of the bodies and the coefficient of friction between two bodies. We initialize the objects at certain locations in the scene with some velocity and start the simulation by applying forces. In the following, we will detail our network architecture and training losses.

3.1   Network Architecture

Encoder. For supervised learning experiments, we use convolutional layers followed by fully connected layers with exponential linear units (ELU) [6] to encode poses from images. The encoder receives the image I_t and is encoded as pose $\mathbf{x}_t$. We need at least two images to infer velocities from images. For self-supervised learning experiments, we use a variational encoder [16] with the same base architecture as in the supervised case. Here, the encoder receives the image I_t and outputs the mean and log variance of a Gaussian hidden state distribution $p(\mathbf{z}_t \mid I_t)$. A latent space sample is obtained using the reparameterization trick [16], and the pose $\mathbf{x}_t$ is regressed from it using a fully connected layer.

We use three images so that we can average out the velocities in case of collisions when the two frames are collected just before and after collision. We use the difference in poses to estimate velocity instead of directly training the network to output velocities. This gives us the average velocity, not the final velocity. For example in 1D, when a block of mass m is acting under an effective force f_{eff} between times t_0 and t_1, the velocity at time t_1 is given by

$$v(t_1) = \underbrace{\frac{p(t_1) - p(t_0)}{t_1 - t_0}}_{\text{average velocity}} + \frac{1}{2}\frac{f_{\text{eff}}}{m}(t_1 - t_0) \tag{4}$$

If we would let the network learn the velocities, it would require to implicitly learn the physics which we want to avoid by the use of the differentiable physics engine. The encoded states are provided as input to the differentiable physics engine.

Trajectory Integration. We integrate a trajectory of poses from the initial pose estimated by the encoder and the velocity estimates by the differentiable physics engine. In each time step, we calculate the new pose of each object $\mathbf{x} = \left(\mathbf{q}^\top, \mathbf{p}^\top\right)^\top$ where $\mathbf{q} \in \mathbb{S}^3$ is a unit quaternion representing rotation and $\mathbf{p} \in \mathbb{R}^3$ is the position from the resulting velocities of the LCP $\boldsymbol{\xi}_t = \left(\boldsymbol{\omega}_t^\top, \mathbf{v}_t^\top\right)^\top$ by

$$\mathbf{p}_t = \mathbf{p}_t + \mathbf{v}_t \cdot h$$
$$\mathbf{q}_t = \mathbf{q}_t \times \text{quat}(e^{0.5\boldsymbol{\omega}_t h}) \tag{5}$$

where $\text{quat}(\cdot)$ is an operator which converts a rotation matrix into a quaternion.

Decoder. We use a spatial transformer network layer [14] to decode object poses into images. These transformations provide structure to the latent space and thus allow the network to train in a self-supervised way. The poses estimated by the physics engine or inferred by the encoder are given as inputs to the network along with content patches of the objects and the background. The content patches and the background are assumed known and extracted from training images using ground-truth masks. The spatial transformer renders these objects at appropriate locations on the background assuming the camera intrinsics, its view pose in the scene and parameterization of the plane are known.

3.2 Training Losses

Supervised Learning. For supervised learning, we initialize the physics engine with inferred poses $\mathbf{x}_{1:N,i}^{enc}$ for each object i from the encoder where N is the (video) sequence length. Estimated poses $\hat{\mathbf{x}}_{1:N,i}$ by the physics engine as well as the inferred poses by the encoder are compared with ground truth poses $\mathbf{x}_{1:N,i}^{gt}$ to infer physical parameters,

$$L_{\text{supervised}} = \sum_i e(\mathbf{x}_{1:N,i}^{gt}, \mathbf{x}_{1:N,i}^{enc}) + \alpha e(\mathbf{x}_{1:N,i}^{gt}, \hat{\mathbf{x}}_{1:N,i}),$$

$$e(\mathbf{x}_1, \mathbf{x}_2) := \frac{1}{N} \sum_{t=1}^{N} \left\| \ln(\mathbf{q}_{2,t}^{-1} \mathbf{q}_{1,t}) \right\|_2^2 + \left\| \mathbf{p}_{2,t} - \mathbf{p}_{1,t} \right\|_2^2, \tag{6}$$

where α is a weighting constant $\alpha = 0.1$ in our experiments), t is the time step and i indexes objects. We use the quaternion geodesic norm to measure differences in rotations.

Self-supervised Learning. For self-supervised learning, we initialize the physics engine with inferred poses $\mathbf{x}_{1:N}^{enc}$ from the encoder. Both estimated poses by the physics engine and the inferred poses are reconstructed into images $\hat{I}_{1:N}^{rec}$ and $I_{1:N}^{rec}$ using our decoder, respectively. The images are compared to input frames $I_{1:N}^{gt}$ to identify the physical parameters and train the network. We impose a KL divergence loss between the inferred encoder distribution $p(\mathbf{z}_t \mid I_t)$ and the standard normal distribution prior $p(\mathbf{z}_t) = \mathcal{N}(\mathbf{0}, \mathbf{I})$,

$$L_{\text{self-supervised}} = \frac{1}{N} \left\| I_{1:N}^{gt} - I_{1:N}^{rec} \right\|_2^2 + \frac{\alpha}{N} \left\| I_{1:N}^{gt} - \hat{I}_{1:N}^{rec} \right\|_2^2 \tag{7}$$

$$+ \sum_{t=1}^{N} KL\left(p(\mathbf{z}_t \mid I_t) \| p(\mathbf{z}_t)\right).$$

System Identification. For reference, we also directly optimize for the physical parameters based on the ground-truth trajectories $\mathbf{p}_{1:N}^{gt}$ without the image encoder. For this we use the first state as an input to the differentiable physics engine. In this case, the loss function is $L_{\text{sys-id}} = \sum_i e(\mathbf{x}_i^{gt}, \hat{\mathbf{x}}_i)$.

4 Experiments

We evaluate our approach in 3D simulated scenarios including pushing, sliding and collision of objects (see Fig. 1).

4.1 Simulated Scenarios and Observability Analysis

In this section, we discuss and analyze the different scenarios for the observability of physical parameters. To this end, we simplify the scenarios into 1D or 2D scenarios where dynamics equations are simpler to write.

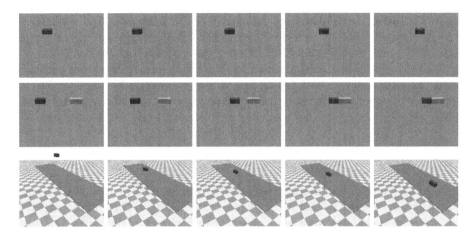

Fig. 1. 3D visualization of the simulated scenes. Top: block pushed on a flat plane. Middle: block colliding with another block. Bottom: block falling and sliding down on an inclined plane.

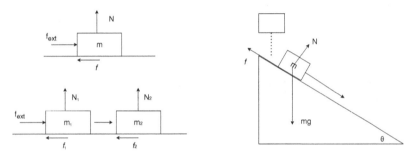

Fig. 2. 1D/2D sketches of scenarios. Top left: block pushed on a flat plane. Bottom left: block colliding with another block. Right: block sliding down on an inclined plane.

Block Pushed on a Flat Plane. In this scenario, a block of mass m, lying on a flat plane is pushed with a force $\mathbf{f}_{ext}$ at the center of mass as shown in Fig. 2 (top left). In this 1D example, since we only have a frictional constraint we can use Eq. (2) in combination with the frictional force $f = \mu N$ to describe the system, where μ is the coefficient of friction, $g = 9.81\,\mathrm{m/s^2}$ is the acceleration due to gravity and $N = mg$ is the normal force since the body has no vertical motion. The velocity v_{t+h} in the next time step hence is

$$v_{t+h} = v_t + \frac{f_{ext}}{m}h - \mu g h \tag{8}$$

We observe that only either one of mass or friction can be inferred at a time. Thus, in our experiments we fix one of the parameters and learn the other.

Block Colliding with Another Block. To learn both mass and coefficient of friction simultaneously, we introduce a second block with known mass (m_2) made of the same material like the first one. This ensures that the coefficient of friction (μ) between the plane and the two blocks is same. Since we are pushing the blocks, after collision, both blocks move together. In the 1D example in Fig. 2 (bottom left), when applied an external force (f_{ext}), the equation to calculate the linear velocities $v_{1/2,t+h}$ of both objects in the next time step becomes

$$v_{1,t+h} = v_{1_t} + \frac{f_{ext}}{m_1}h - \mu gh, \quad v_{2,t+h} = v_{2_t} + \frac{f'}{m_2}h - \mu gh, \qquad (9)$$

where μgm_1 and μgm_2 are frictional forces acting on each block and f' is the equivalent force on the second body when moving together. Now, in our experiments we can learn both mass and coefficient of friction together given the rest of the parameters in the equation.

Block Freefall and Sliding Down on an Inclined Plane. In this scenario the block slides down the inclined plane after experiencing a freefall as shown in Fig. 2 (right). In the 1D example, since the freefall is unconstrained (ignoring air resistance), the velocity update is given by $v_{t+h} = v_t + gh$. For block sliding down on an inclined plane, the equation to calculate velocity in the next time is $v_{t+h} = v_t + g(\sin\theta - \mu\cos\theta)\,h$, where θ is the plane inclination. We can see that we can only infer the coefficient of friction μ and due to the free fall we do not need to apply additional forces.

4.2 Results

We simulated the scenarios in 3D using the bullet physics engine using PyBullet[1]. Note that the bullet physics engine is different to the LCP physics engine in our network and can yield qualitatively and numerically different results. The bodies are initialized at random locations to cover the whole workspace. Random forces between 1–10 N are applied at each time step. These forces are applied in $+x$, $-x$, $+y$ and $-y$ directions which are chosen at random but kept constant for a single trajectory while the magnitude of the forces randomly varies in each time step. In total, 1000 different trajectories are created with 300 time steps each for each scenario. We render top-down views at 128 × 128 resolution. Training and test data are split with ratio 9:1. For evaluation we show the evolution of the physical parameters during the training. We also give the average relative position error by the encoder which is the average of the difference between ground truth positions and estimated poses divided by object size.

System Identification Results. As a baseline result, system identification (see Sect. 3.2) can be achieved within 200 epochs with an average position error

---

[1] https://pybullet.org.

Table 1. Supervised learning results for the 3 scenarios. The physical parameters are well identified (blue lines) close to the ground truth values (red lines).

Inference	Block Pushed On a Flat Plane	Block Sliding Down the Inclined Plane	Block Colliding With Another Block
Mass	position inference error: 4%	Not feasible	position inference error: 8%
Coefficient of friction	position inference error: 2%	position inference error: 5%	rotation inference error: 8°

for all the scenarios between 0.7–1.2%. The physical parameters reach nominal values with high accuracy. Detailed results are given in the supplementary material.

Supervised Learning Results. We train our network using the supervised loss in Sect. 3.2. We warm up the encoder by pre-training with ground truth poses so that when optimizing for physics parameters the training of the encoder is stable. We then continue training the encoder on the full supervised loss. From Table 1, we observe that all the learned physical parameters (in blue) slightly oscillate around the ground truth values (in red). The average inferred position error for all the scenarios is between 2–8% and the average inferred rotation error for the collision scenario is 8°. The parameter learning seems to be robust to this degree of accuracy in the estimated initial states.

Self-supervised Learning Results. Now, we train the network in a self-supervised way (see Sect. 3.2). In this experiment, we generate sequences where the objects start at random locations with zero initial velocity, since the initial velocity estimate is ambiguous for our self-supervised learning approach. We obtain average velocities from the estimated poses (Eq. (4)). Since the pose estimation error is high in self-supervised experiments, the accuracy in velocity especially at the beginning of training is not sufficient for self-supervised learning. We pre-train the encoder in an encoder-decoder way so that when optimizing for physics parameters the training is stable. We continue training the encoder on the full self-supervised loss. To provide the network with gradients for localizing

Table 2. Self-supervised learning results for the pushing and collision scenarios. While the encoder error is slightly higher than in the supervised learning case, the physical parameters are identified (blue lines) close to the ground truth values (red lines).

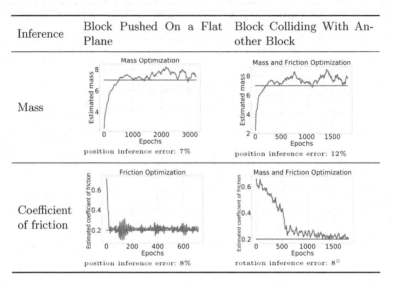

Inference	Block Pushed On a Flat Plane	Block Colliding With Another Block
Mass	Mass Optimization ... position inference error: 7%	Mass and Friction Optimization ... position inference error: 12%
Coefficient of friction	Friction Optimization ... position inference error: 8%	Mass and Friction Optimization ... rotation inference error: 8°

the objects, we use Gaussian smoothing on the input and reconstructed images starting from kernel size 128 and standard deviation 128, and reducing it to kernel size 5 and standard deviation 2 by the end of training. From Table 2, we observe that our approach can still recover the physical parameters at good accuracy. Expectably, they are less accurate than in the supervised learning experiment. The average inferred position error for all the scenarios is between 7–12% and the average inferred rotation error for the collision scenario is 8°. Through the use of spatial transformers our approach is limited to rendering top-down views and cannot handle 3D translation and rotation in our third scenario.

4.3   Qualitative Video Prediction Results

The learned model in Sect. 4.2 can be used for video prediction. The images in the top row in Figs. 3(a) and 3(b) are the ground truth, the images in the middle row are the reconstructions from the predicted trajectories by our network and the images in the bottom row are the difference images. We roll out a four second trajectory. We can observe that the positions of the objects are well predicted by our approach, while the approach yields small inaccuracies in predicting rotations which occur after the collision of the objects. Further video prediction results are included in the supplementary material.

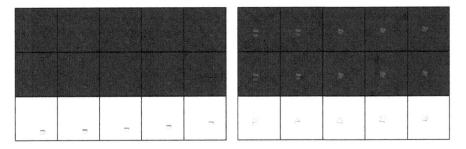

Fig. 3. Qualitative video prediction results for block pushing (left) and collision scenarios (right) with our method. Top: simulated images (from left to right frames 0, 30, 60, 120, 180). Middle: predicted images by our approach. Bottom: difference images.

4.4 Discussion and Limitations

Our approach achieves good results for supervised and self-supervised learning in the evaluated scenarios. We have studied observability and feasibility of learning physical parameters and video embedding by our approach. At its current stage, our architecture makes several assumptions on the scenes which could be addressed in future research. Our approach for using 2D spatial transformers for image generation restricts the self-supervised learning approach to known object shape and appearance and top down views. For real scenes our methods needs information about the applied forces which can be obtained from a known dynamics model (e.g. of a robot) or force sensors. For self-supervised learning, methods for bridging the sim-to-real domain gap have to be investigated.

5 Conclusion

In this paper we study supervised and self-supervised learning approaches to learn image encodings and identify physical parameters. Our deep neural network architecture integrates differentiable physics with a spatial transformer network layer to learn a physical latent representation of video and applied forces. For supervised learning, an encoder regresses the initial object state from images. Self-supervised learning is achieved through the implementation of a spatial transformer which decodes the predicted positions by the encoder and the physics engine back into images. This way, the model can also be used for video prediction with known actions by letting the physics engine predict positions and velocities conditioned on the actions. We evaluate our approach in scenarios which include pushing, sliding and collision of objects. We analyze the observability of physical parameters and assess the quality of the reconstruction of these parameters using our learning approaches. In future work we plan to investigate further scenarios including learning the restitution parameter and extend our self-supervised approach to real scenes and full 3D motion of objects.

Acknowledgements. We acknowledge support from Cyber Valley, the Max Planck Society, and the German Federal Ministry of Education and Research (BMBF) through the Tuebingen AI Center (FKZ: 01IS18039B). The authors thank the International Max Planck Research School for Intelligent Systems (IMPRS-IS) for supporting Jan Achterhold.

References

1. Amos, B., Kolter, J.Z.: Optnet: differentiable optimization as a layer in neural networks. In: International Conference on Machine Learning, pp. 136–145 (2017)
2. Anitescu, M., Potra, F.A.: Formulating dynamic multi-rigid-body contact problems with friction as solvable linear complementarity problems. Nonlinear Dyn. **14**, 231–247 (1997)
3. de Avila Belbute-Peres, F., Smith, K., Allen, K., Tenenbaum, J., Kolter, J.Z.: End-to-end differentiable physics for learning and control. In: Advances in Neural Information Processing Systems, pp. 7178–7189 (2018)
4. Babaeizadeh, M., Finn, C., Erhan, D., Campbell, R., Levine, S.: Stochastic variational video prediction. In: Proceedings of the International Conference on Learning Representations (2018)
5. Chen, R.T.Q., Li, X., Grosse, R.B., Duvenaud, D.K.: Isolating sources of disentanglement in variational autoencoders. In: Advances in Neural Information Processing Systems, pp. 2610–2620 (2018)
6. Clevert, D.A., Unterthiner, T., Hochreiter, S.: Fast and accurate deep network learning by exponential linear units (ELUs). In: Proceedings of the International Conference on Learning Representations (2016)
7. Cline, M.B.: Rigid body simulation with contact and constraints. Ph.D. thesis (2002). https://doi.org/10.14288/1.0051676. https://open.library.ubc.ca/collections/ubctheses/831/items/1.0051676
8. Degrave, J., Hermans, M., Dambre, J., Wyffels, F.: A differentiable physics engine for deep learning in robotics. Front. Neurorobotics **13** (2016). https://doi.org/10.3389/fnbot.2019.00006
9. Finn, C., Levine, S.: Deep visual foresight for planning robot motion. In: International Conference on Robotics and Automation, pp. 2786–2793 (2017)
10. Finn, C., Goodfellow, I.J., Levine, S.: Unsupervised learning for physical interaction through video prediction. In: Advances in Neural Information Processing Systems, pp. 64–72 (2016)
11. Greydanus, S., Dzamba, M., Yosinski, J.: Hamiltonian neural networks. In: Advances in Neural Information Processing Systems, pp. 15379–15389 (2019)
12. Hafner, D., et al.: Learning latent dynamics for planning from pixels. In: International Conference on Machine Learning, pp. 2555–2565 (2019)
13. Hochreiter, S., Schmidhuber, J.: Long short-term memory. Neural Comput. **9**, 1735–80 (1997)
14. Jaderberg, M., Simonyan, K., Zisserman, A., Kavukcuoglu, K.: Spatial transformer networks. In: Advances in Neural Information Processing Systems, pp. 2017–2025 (2015)
15. Jaques, M., Burke, M., Hospedales, T.M.: Physics-as-inverse-graphics: joint unsupervised learning of objects and physics from video. In: Proceedings of the International Conference on Learning Representations (2020)
16. Kingma, D.P., Welling, M.: Auto-encoding variational bayes. In: Proceedings of the International Conference on Learning Representations (2014)

17. Kloss, A., Schaal, S., Bohg, J.: Combining learned and analytical models for predicting action effects. CoRR abs/1710.04102 (2017)
18. Mattingley, J., Boyd, S.: CVXGEN: a code generator for embedded convex optimization. Optim. Eng. **13** (2012). https://doi.org/10.1007/s11081-011-9176-9
19. Mottaghi, R., Bagherinezhad, H., Rastegari, M., Farhadi, A.: Newtonian scene understanding: unfolding the dynamics of objects in static images. In: Proceedings of the IEEE Conference on Computer Vision and Pattern Recognition (2016)
20. Mottaghi, R., Rastegari, M., Gupta, A., Farhadi, A.: "What happens if..." learning to predict the effect of forces in images. In: European Conference on Computer Vision (2016)
21. Runia, T.F.H., Gavrilyuk, K., Snoek, C.G.M., Smeulders, A.W.M.: Cloth in the wind: a case study of estimating physical measurement through simulation. In: Proceedings of the IEEE Conference on Computer Vision and Pattern Recognition (2020)
22. Shi, X., Chen, Z., Wang, H., Yeung, D.Y., Wong, W.K., Woo, W.C.: Convolutional LSTM network: a machine learning approach for precipitation nowcasting. In: Advances in Neural Information Processing Systems, pp. 802–810 (2015)
23. Srivastava, N., Mansimov, E., Salakhutdinov, R.: Unsupervised learning of video representations using lstms. In: International Conference on Machine Learning (2015)
24. Stewart, D.: Rigid-body dynamics with friction and impact. SIAM Rev. **42**, 3–39 (2000). https://doi.org/10.1137/S0036144599360110
25. Watters, N., Zoran, D., Weber, T., Battaglia, P., Pascanu, R., Tacchetti, A.: Visual interaction networks: learning a physics simulator from video. In: Advances in Neural Information Processing Systems (2017)
26. Ye, T., Wang, X., Davidson, J., Gupta, A.: Interpretable intuitive physics model. In: European Conference on Computer Vision (2018)
27. Zhu, D., Munderloh, M., Rosenhahn, B., Stückler, J.: Learning to disentangle latent physical factors for video prediction. In: German Conference on Pattern Recognition (2019)

Assignment Flow for Order-Constrained OCT Segmentation

Dmitrij Sitenko$^{(\boxtimes)}$, Bastian Boll, and Christoph Schnörr

Image and Pattern Analysis Group (IPA) and Heidelberg Collaboratory for Image Processing (HCI), Heidelberg University, Heidelberg, Germany
dmitrij.sitenko@iwr.uni-heidelberg.de

Abstract. At the present time Optical Coherence Tomography (OCT) is among the most commonly used non-invasive imaging methods for the acquisition of large volumetric scans of human retinal tissues and vasculature. Due to tissue-dependent speckle noise, the elaboration of automated segmentation models has become an important task in the field of medical image processing.

We propose a novel, purely data driven *geometric approach to order-constrained* 3D *OCT retinal cell layer segmentation* which takes as input data in any metric space. This makes it unbiased and therefore amenable for the detection of local anatomical changes of retinal tissue structure. To demonstrate robustness of the proposed approach we compare four different choices of features on a data set of manually annotated 3D OCT volumes of healthy human retina. The quality of computed segmentations is compared to the state of the art in terms of mean absolute error and Dice similarity coefficient.

1 Introduction

Overview. Optical Coherence Tomography (OCT) is a non-invasive imaging technique which measures the intensity response of back scattered light from millimeter penetration depth. We focus specifically on the application of OCT in ophthalmology for aquisition of high-resolution volume scans of the human retina. This provides information about retinal tissue structure in vivo to understand human eye functionalities. OCT devices record multiple two-dimensional B-scans in rapid succession and combine them into a single volume in a subsequent alignment step. Taking an OCT scan only takes minutes and can help detect symptoms of pathological conditions such as glaucoma, diabetes, multiple sclerosis or age-related macular degeneration.

The relative ease of data aquisition also enables to use multiple OCT volume scans of a single patient over time to track the progression of a pathology or quantify the success of therapeutic treatment. To better leverage the availability of raw OCT data in both clinical settings and empirical studies, much work has focused on automatic extraction of relevant information, in particular automatic cell layer segmentation, detection of fluid and reconstruction of vascular structures. The difficulty of these tasks lies in challenging signal-to-noise

Z. Akata et al. (Eds.): DAGM GCPR 2020, LNCS 12544, pp. 58–71, 2021.
https://doi.org/10.1007/978-3-030-71278-5_5

ratio which is influenced by multiple factors including mechanical eye movement during registration and the presence of speckle.

Related Work. Effective segmentation of OCT volumes is a very active area of research. Several methods for segmenting human retina were proposed in [11,12,16] and [2] which rely on graphical models. To increase robustness, the retina segmentation approaches proposed in [23] and [9] employ shape priors using soft constraints. In [19] Rathke first introduced a parallelizable segmentation method based on probabilistic graphical models with global low-rank shape prior. Variational approaches given in [8,26] and [18] model retina layers by zero level sets with properly chosen functionals including soft constraints. Much recent work has focused on the use of deep learning to address the task of cell layer segmentation in a purely data driven way. Methods presented in [17,21] rely on the U-net architecture [20] which yields good predictive performance in settings with limited availability of training data. To enforce global order of cell layers along a spatial axis as well as additional regularization, local predictions have been tied together through graph-based methods [10] or through a second machine learning component [14]. However, if global context is already used in feature extraction, the risk of overfitting remains and unseen pathologies may result in unpredictable behavior.

Approach. Our segmentation approach is a smooth image labeling algorithm based on geometric numerical integration on an elementary statistical manifold. It can work with input data from any metric space, making it agnostic to the choice of feature extraction and suitable as plug-in replacement in diverse pipelines. In addition to respecting the natural order of cell layers, our segmentation process has a high amount of built-in parallelism such that modern graphics acceleration hardware can easily be leveraged. We evaluate the effectiveness of our novel approach for a selection of input features ranging from traditional covariance descriptors to convolutional neural networks.

Contribution. We propose a geometric assignment approach that extends the approach introduced by [4] to retinal layer segmentation with the following novel characteristics:

(i) By leveraging a continuous characterization of layer ordering, our method is able to simultaneously perform local regularization and to incorporate the global topological ordering constraint in a *single smooth* labeling process. The segmentation is computed from a distance matrix containing pairwise distances between data for each voxel and prototypical data for each layer in some feature space. This highlights the ability to extract features from raw OCT data in a variety of different ways and to use the proposed segmentation as a plug-in replacement for other graph-based methods.

(ii) Computationally fast and high-quality cell layer segmentations of OCT volumes are obtained by using only local features for each voxel. This is in contrast to competing deep learning approaches which commonly use information from an entire B-scan as input. In addition, the exclusive use of local features combats bias introduced through limited data availability in training and enables the incorporation of three-dimensional information without compromising runtime scalability.

(iii) The highly parallelizable segmentation approach with global voxel interaction enables robust cell layer segmentation of entire OCT volumes without using any prior knowledge other than local regularity and order of cell layers. In particular, no global shape prior is used as opposed to segmentation methods relying on graphical models like, e.g., [19]. Figure 1 shows a typical result obtained with our novel approach after segmenting healthy retina tissues with labels specified in (Fig. 2).

Organization. Our paper is organized as follows. The assignment flow approach is briefly summarized in Sect. 2 and extended in Sect. 4 in order to take into account the order of layers as a global constraint. In Sect. 3, we consider the Riemannian manifold $\mathcal{P}_d$ of positive definite matrices as a suitable feature space for local OCT data descriptors. The resulting features are subsequently compared to local features extracted by a convolutional network in Sect. 5. The evaluation of performance measures for OCT segmentation of our novel approach are reported in Sect. 5 and compared to the state-of-the-art method given in [16].

Fig. 1. From left to right: 3D OCT volume scan dimension $512 \times 512 \times 256$ of healthy human retina with ambiguous locations of layer boundaries. The resulting segmentation of 11 layers expressing the order preserving labeling of the proposed approach. Illustration of boundary surfaces between different segmented cell layers.

2   Assignment Flow

We summarize the assignment flow approach introduced by [4] and refer to the recent survey [22] for more background and a review of recent related work.

Assignment Manifold. Let $(\mathcal{F}, d_{\mathcal{F}})$ be a metric space and $\mathcal{F}_n = \{f_i \in \mathcal{F} : i \in \mathcal{I}\}$, $|\mathcal{I}| = n$ given data. Assume that a predefined set of prototypes $\mathcal{F}_* = \{f_j^* \in \mathcal{F} : j \in \mathcal{J}\}$, $|\mathcal{J}| = c$ is given. *Data labeling* denotes assignments $j \to i$, $f_j^* \to f_i$ to be determined in a spatially regularized fashion. The assignments at each pixel $i \in \mathcal{I}$ are encoded by *assignment vectors* $W_i = (W_{i1}, \dots, W_{ic})^\top \in \mathcal{S}$ in the relative interior $\mathcal{S} = \mathrm{rint}\,\Delta_c$ of the probability simplex, that becomes a Riemannian manifold $(\mathcal{S}, g)$ endowed with the Fisher-Rao metric g from information geometry. The *assignment manifold* $(\mathcal{W}, g)$, $\mathcal{W} = \mathcal{S} \times \cdots \times \mathcal{S}$ ($n = |\mathcal{I}|$ factors) is the product manifold whose points encode the label assignments at all pixels.

Assignment Flow. Based on the given data and prototypes, the distance vector field $D_{\mathcal{F};i} = \big(d_{\mathcal{F}}(f_i, f_1^*), \ldots, d_{\mathcal{F}}(f_i, f_c^*)\big)^\top$, $i \in \mathcal{I}$ is well defined. This data representation is lifted to the assignment manifold by the *likelihood map* and the *likelihood vectors*, respectively,

$$L_i \colon \mathcal{S} \to \mathcal{S}, \quad L_i(W_i) = \frac{W_i e^{-\frac{1}{\rho} D_{\mathcal{F};i}}}{\langle W_i, e^{-\frac{1}{\rho} D_{\mathcal{F};i}} \rangle}, \quad i \in \mathcal{I}. \tag{2.1}$$

This map is based on the affine e-connection of information geometry and the scaling parameter $\rho > 0$ is used for normalizing the a-prior unknown scale of the components of $D_{\mathcal{F};i}$ that depends on the specific application at hand. The likelihood vectors are spatially regularized by the *similarity map* and the *similarity vectors*, respectively,

$$S_i \colon \mathcal{W} \to \mathcal{S}, \quad S_i(W) = \mathrm{Exp}_{W_i}\Big(\sum_{k \in \mathcal{N}_i} w_{ik}\, \mathrm{Exp}_{W_i}^{-1}\big(L_k(W_k)\big)\Big), \quad i \in \mathcal{I}, \tag{2.2}$$

where $\mathrm{Exp}_p(v) = \frac{pe^{v/p}}{\langle p, e^{v/p}\rangle}$ is the exponential map corresponding to the e-connection and positive weights w_{ik}, $k \in \mathcal{N}_i$, that sum up to 1 on every patch around pixel i indexed by $\mathcal{N}_i$, determine the regularization properties.

The *assignment flow* is induced on the assignment manifold $\mathcal{W}$ by the locally coupled system of nonlinear ODEs

$$\dot{W}_i = R_{W_i} S_i(W), \qquad W_i(0) = \mathbb{1}_{\mathcal{S}}, \quad i \in \mathcal{I}, \tag{2.3}$$

where the map $R_p = \mathrm{Diag}(p) - pp^\top$, $p \in \mathcal{S}$ turns the right-hand side into a tangent vector field and $\mathbb{1}_{\mathcal{W}} \in \mathcal{W}$ denotes the barycenter of the assignment manifold $\mathcal{W}$. The solution $W(t) \in \mathcal{W}$ is numerically computed by geometric integration [27] and determines a labeling $W(T)$ for sufficiently large T after a trivial rounding operation. Convergence and stability of the assignment flow have been studied by [28].

3   OCT Data Representation by Covariance Descriptors

In this section, we briefly sketch the basic geometric notation for representation of OCT data in terms of covariance descriptors $f_i \in \mathcal{F}_n$ [25] and identify the metric data space $(\mathcal{F}, d_{\mathcal{F}})$ underlying (2.1).

The Manifold $\mathcal{P}_d$. The Riemannian manifold $(\mathcal{P}_d, g)$ of positive definite matrices of dimension $\frac{(d+1)(d)}{2}$ and the Riemannian metric are given by

$$\mathcal{P}_d = \{S \in \mathbb{R}^{d \times d} \colon S = S^\top,\ S \text{ is positive definite}\} \tag{3.1a}$$

$$g_S(U, V) = \mathrm{tr}(S^{-1} U S^{-1} V), \qquad U, V \in T_S \mathcal{P}_d = \{S \in \mathbb{R}^{d \times d} \colon S^\top = S\}. \tag{3.1b}$$

The Riemannian distance is given by

$$d_{\mathcal{P}_d}(S,T) = \left(\sum_{i\in[d]} \left(\log \lambda_i(S,T) \right)^2 \right)^{1/2}, \tag{3.2}$$

whereas the globally defined exponential map reads

$$\exp_S(U) = S^{\frac{1}{2}} \operatorname{expm}(S^{-\frac{1}{2}} U S^{-\frac{1}{2}}) S^{\frac{1}{2}}, \tag{3.3}$$

with $\operatorname{expm}(\cdot)$ denoting the matrix exponential. Given a smooth objective function $J\colon \mathcal{P}_d \to \mathbb{R}$, the Riemannian gradient is given by

$$\operatorname{grad} J(S) = S\big(\partial J(S)\big)S \in T_S\mathcal{P}_d, \tag{3.4}$$

where the symmetric matrix $\partial J(S)$ is the Euclidean gradient of J at S.

Region Covariance Descriptors. To apply the introduced geometric framework, we model each OCT volume by a mapping $I : \mathcal{D} \to \mathbb{R}_+$ where $\mathcal{D} \subset \mathbb{R}^3$ is an underlying spatial domain.

To each voxel $v \in \mathcal{D}$, we associate the local feature vector $f\colon \mathcal{D} \to \mathbb{R}^{10}$,

$$f(v) := (I(v), \nabla_x I(v), \nabla_y I(v), \nabla_z I(v), \sqrt{2}\nabla_{xy} I(v), \ldots, \nabla_{zz} I(v))^\top \tag{3.5}$$

assembled from the raw intensity value $I(v)$ as well as first- and second-order responses of derivatives filters capturing information from larger scales following [13]. By introducing a suitable geometric graph spanning $\mathcal{D}$, we can associate a neighborhood $\mathcal{N}(i)$ of fixed size with each voxel $i \in \mathcal{I}$ as in (2.2). For each neighborhood, we define the regularized *region covariance descriptor* S_i as

$$S_i := \sum_{j\in\mathcal{N}(i)} \theta_{ij}(f_j - \overline{f_i})(f_j - \overline{f_i})^T + \epsilon I, \quad \overline{f_i} = \sum_{k\in\mathcal{N}(i)} \theta_{ik} f_k, \tag{3.6}$$

as a weighted empirical covariance matrix with respect to feature vectors f_j. The small value $1 \gg \epsilon > 0$ acts as a regularization parameter enforcing positive definiteness of S_i. In the following, we use the shorthand notation $[n] = \{1, \ldots, n\}$ for natural numbers n.

Computing Prototypical Covariance Descriptors. Given a set of covariance descriptors

$$\mathcal{S}_N = \{(S_1, \omega_1), \ldots, (S_N, \omega_N)\} \subset \mathcal{P}_d \tag{3.7}$$

together with positive weights ω_i, we next focus on the solution of the problem

$$\overline{S} = \arg \min_{S\in\mathcal{P}_d} J(S; \mathcal{S}_N), \quad J(S; \mathcal{S}_N) = \sum_{i\in[N]} \omega_i d_{\mathcal{P}_d}^2(S, S_i), \tag{3.8}$$

with the distance $d_{\mathcal{P}_d}$ given by (3.2). From (3.3), we deduce

$$U = \exp_S^{-1} \circ \exp_S(U) = S^{\frac{1}{2}} \operatorname{logm}\left(S^{-\frac{1}{2}} \exp_S(U) S^{-\frac{1}{2}}\right) S^{\frac{1}{2}} \tag{3.9}$$

with the matrix logarithm logm = expm^{-1} [15, Section 11]. The efficient mean retrieval of (3.8) regarding the evaluation of (3.2) requires a nontrivial matrix decomposition that has to be applied multiple times to every voxel (vertex) of a 3D gridgraph. This results in an overall quite expensive approach in particular for a large data set. Therefore we reduce the computational costs by relying on an approximation of the Riemannian mean by employing surrogate metrics and distances introduced below.

Log-Euclidean Distance and Means. A computationally cheap approach was proposed by [3] (among several other ones). Based on the operations

$$S_1 \odot S_2 = \text{expm}\left(\text{logm}(S_1) + \text{logm}(S_2)\right), \tag{3.10a}$$

$$\lambda \cdot S = \text{expm}\left(\lambda \, \text{logm}(S)\right), \tag{3.10b}$$

the set $(\mathcal{P}_s, \odot, \cdot)$ becomes isomorphic to the vector space where $\odot$ plays the role of addition. Consequently, the mean of the data $\mathcal{S}_N$ given by (3.7) is defined analogous to the arithmetic mean by

$$\overline{S} = \text{expm}\left(\sum_{i \in [N]} \omega_i \, \text{logm}(S_i)\right). \tag{3.11}$$

While computing the mean is considerably cheaper than integrating the flow induced by (3.4) with respect to objective (3.8), the geometry (curved structure) of the manifold $\mathcal{P}_d$ is ignored. Therefore, in the next section, we additionally consider another approximation of the Riemannian mean that better respects the underlying geometry but can still be evaluated efficiently.

S-Divergence and Means. For an approximation of the objective function (3.8), we replace the Riemannian $d_g^2(p, q)$ distance by the *Stein divergence* proposed by Sra [24]

$$D_s(S_1, S_2) = \log \det\left(\frac{S_1 + S_2}{2}\right) - \frac{1}{2}\log \det(S_1 S_2), \qquad S, S_1, S_2 \in \mathcal{P}_d, \tag{3.12}$$

and avoid involved generalized eigenvalue problem for evaluation of (3.2) by replacing (3.8) with

$$\overline{S} = \arg\min_{S \in \mathcal{P}_d} J_s(S; \mathcal{S}_N), \qquad J_s(S; \mathcal{S}_N) = \sum_{i \in [N]} \omega_i D_s(S, S_i). \tag{3.13}$$

We refer to, e.g., [5,6] for a more complete exposition of divergence functions. The Riemannian gradient flow for this specific problem reads

$$\dot{S} = -\text{grad}\, J_s(S; \mathcal{S}_N) \overset{(3.4)}{=} -S\partial J(S; \mathcal{S}_N)S \tag{3.14a}$$

$$= -\frac{1}{2}\left(SR(S; \mathcal{S}_N)S - S\right), \qquad R(S; \mathcal{S}_N) = \sum_{i \in [N]} \omega_i \left(\frac{S + S_i}{2}\right)^{-1}. \tag{3.14b}$$

Discretizing the flow using the geometric explicit Euler scheme with step size h yields,

$$S_{(t+1)} = \exp_{S_{(t)}} \left(-h \operatorname{grad} J_s(S_{(t)}; \mathcal{S}_N) \right) \tag{3.15a}$$

$$\overset{(3.3)}{=} S_{(t)}^{\frac{1}{2}} \operatorname{expm} \left(\frac{h}{2} \left(I - S_{(t)}^{\frac{1}{2}} R(S_{(t)}; \mathcal{S}_N) S_{(t)}^{\frac{1}{2}} \right) \right) S_{(t)}^{\frac{1}{2}}. \tag{3.15b}$$

Using as initial point $S_{(0)}$ the log-Euclidean mean (3.11) defines the following algorithm that we use for mean retrieval throughout the present paper.

Algorithm 1: Geometric Matrix Mean Based on the S-divergence.

Initialization, ϵ (termination threshold)
$t = 0, \quad S_{(0)}$ solves (3.11)
$\epsilon_0 > \epsilon$ (any value ϵ_0)
while $\epsilon_t > \epsilon$ **do**

    $LL^\top = S_{(t)}$
    $L_i L_i^\top = \frac{S_{(t)} + S_i}{2}$ for $i \in [N]$
    $U = I - S_{(t)}^{\frac{1}{2}} \left(\sum_{i \in [N]} \omega_i (L_i L_i^\top)^{-1} \right) S_{(t)}^{\frac{1}{2}}$
    $S_{(t+1)} = S_{(t)}^{\frac{1}{2}} \operatorname{expm}(\frac{h}{2} U) S_{(t)}^{\frac{1}{2}}$
    $\epsilon_{t+1} := \|U\|_F, \quad t \leftarrow t + 1$

---

4   Ordered Layer Segmentation

In this section, we work out an extension of the assignment flow (Sect. 2) which is able to respect the order of cell layers as a global constraint while remaining in the same smooth geometric setting. In particular, existing schemes for numerical integration still apply to the novel variant.

4.1   Ordering Constraint

With regard to segmenting OCT data volumes, the order of cell layers is crucial prior knowledge. Figure 2 illustrates for a schematic OCT volume acquisition of 11 retina layers and 3 separating membranes (ILM, ELM, BM) and typical scan notations used throughout the paper. To incorporate this knowledge into the geometric setting of Sect. 2, we require a smooth notion of ordering which allows to compare two probability distributions. In the following, we assume prototypes $f_j^* \in \mathcal{F}$, $j \in [n]$ in some feature space $\mathcal{F}$ to be indexed such that ascending label indices reflect the physiological order of cell layers.

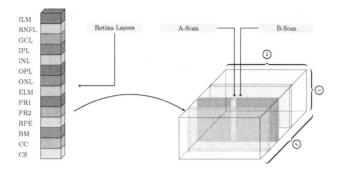

Fig. 2. OCT volume acquisition: ① is the A-scan axis (single A-scan is marked yellow). Multiple A-scans taken in rapid succession along axis ② form a two-dimensional B-scan (single B-scan is marked blue). The complete OCT volume is formed by repeating this procedure along axis ③. A list of retina layers and membranes we expect to find in every A-scan is shown on the left. (Color figure online)

Definition 1 (Ordered Assignment Vectors). *A pair of voxel assignments* $(w_i, w_j) \in \mathcal{S}^2$, $i < j$ *within a single A-scan is called* ordered, *if* $w_j - w_i \in K = \{By\colon y \in \mathbb{R}_+^c\}$ *which is equivalent to* $Q(w_j - w_i) \in \mathbb{R}_+$ *with the matrices*

$$B_{ij} := \begin{cases} -1 & \text{if } i = j \\ 1 & \text{if } i - j = 1 \,, \\ 0 & \text{else} \end{cases} \quad Q_{i,j} = \begin{cases} 1 & \text{if } i \geq j \\ 0 & \text{else} \end{cases} \quad (4.1)$$

4.2 Ordered Assignment Flow

Likelihoods as defined in (2.1) emerge by lifting $-\frac{1}{\rho}D_{\mathcal{F}}$ regarded as Euclidean gradient of $-\frac{1}{\rho}\langle D_{\mathcal{F}}, W \rangle$ to the assignment manifold. It is our goal to encode order preservation into a generalized likelihood matrix $L_{\mathrm{ord}}(W)$. To this end, consider the assignment matrix $W \in \mathcal{S}^N$ for a single A-scan consisting of N voxels. We define the related matrix $Y(W) \in \mathbb{R}^{N(N-1)\times c}$ with rows indexed by pairs $(i, j) \in [N]^2$, $i \neq j$ in fixed but arbitrary order. Let the rows of Y be given by

$$Y_{(i,j)}(W) = \begin{cases} Q(w_j - w_i) & \text{if } i > j \\ Q(w_i - w_j) & \text{if } i < j \end{cases} \,. \quad (4.2)$$

By construction, an A-scan assignment W is ordered exactly if all entries of the corresponding $Y(W)$ are nonnegative. This enables to express the ordering constraint on a single A-scan in terms of the energy objective

$$E_{\mathrm{ord}}(W) = \sum_{(i,j)\in[N]^2, \, i\neq j} \phi(Y_{(i,j)}(W)) \,. \quad (4.3)$$

where $\phi\colon \mathbb{R}^c \to \mathbb{R}$ denotes a smooth approximation of $\delta_{\mathbb{R}^c_+}$. In our numerical experiments, we choose

$$\phi(y) = \left\langle \gamma \exp\left(-\frac{1}{\gamma}y\right), \mathbb{1} \right\rangle \tag{4.4}$$

with a constant $\gamma > 0$. Suppose a full OCT volume assignment matrix $W \in \mathcal{W}$ is given and denote the set of submatrices for each A-scan by $C(W)$. Then order preserving assignments consistent with given distance data $D_{\mathcal{F}}$ in the feature space $\mathcal{F}$ are found by minimizing the energy objective

$$E(W) = \langle D_{\mathcal{F}}, W \rangle + \sum_{W_A \in C(W)} E_{\mathrm{ord}}(W_A) . \tag{4.5}$$

We consequently define the generalized likelihood map

$$L_{\mathrm{ord}}(W) = \exp_W\left(-\nabla E(W)\right) = \exp_W\left(-\frac{1}{\rho}D_{\mathcal{F}} - \sum_{W_A \in C(W)} \nabla E_{\mathrm{ord}}(W_A)\right) \tag{4.6}$$

and specify a corresponding assignment flow variant.

Definition 2 (Ordered Assignment Flow). *The dynamical system*

$$\dot{W} = R_W S(L_{ord}(W)), \qquad W(0) = \mathbb{1}_{\mathcal{W}} \tag{4.7}$$

evolving on $\mathcal{W}$ is called the ordered assignment flow.

By applying known numerical schemes [27] for approximately integrating the flow (4.7), we find a class of discrete-time image labeling algorithms which respect the physiological cell layer ordering in OCT data. In Sect. 5, we benchmark the simplest instance of this class, emerging from the choice of geometric Euler integration.

5   Experimental Results and Discussion

OCT-Data. In the following sections, we describe experiments performed on a set of volumes with annotated OCT B-scans extracted by a spectral domain OCT device (Heidelberg Engineering, Germany). Further, we always assume an OCT volume in question to consist of N_B B-scans, each comprising N_A A-scans with N voxels.

While raw OCT volume data has become relatively plentiful in clinical settings, large volume datasets with high-quality gold-standard segmentation are not widely available at the time of writing. By extracting features which represent a given OCT scan *locally* as opposed to incorporating global context at every stage, it is our hypothesis that superior generalization can be achieved in the face of limited data availability. This is most expected for pathological cases in which global shape of cell layers may deviate drastically from seen examples in the training data. Our approach consequently differs from common deep learning methods which explicitly aim to incorporate global context into the feature extraction process. Utilization of shape prior limits the methods ability to generalize to unseen data if large deviation from the expected global shape seen in training is present.

Prototypes on $\mathcal{P}^d$. For applying the framework introduced in Sect. 2, we interpret covariance features (3.6) as data points $f_i \in \mathcal{F}_n$ evolving on the natural metric space (3.1a) and model each retina tissue indexed by $l \in \{1, \ldots, C\}$ with a random variable S_l taking values $\{S_l^k\}_{k=1}^{N_l}$. To generalize the retina layer detection to multiple OCT data sets instead of just using a single prototype (3.13), we partition the samples $\{S_l^k\}_{k=1}^{N_l}$ into K_l disjoint sets $\{S_1^l, \ldots, S_{K_l}^l\}$ with representatives $\{\tilde{S}_l^1, \ldots, \tilde{S}_l^{K_l}\}$. These are serving as prototypes f_j^*, $j \in \mathcal{J}$ which are determined offline for each $l \in \{1, \ldots, 14\}$ as the minimal expected loss measured by the Stein divergence (3.12) according to K-means like functional

$$E_{p_l}(S^l) = \sum_{j=1}^{K_l} p(j) \sum_{S_i \in S_j} \frac{p(i|j)}{p(j)} D_S(S_l^i, \tilde{S}_l^j), \quad p(i,j) = \frac{1}{N_l}, \; p(j) = \frac{N_j}{N_l}, \quad (5.1)$$

with marginals $p_l(j) = \sum_{i=1}^{N_j} p_l(j|S_l^i)$ and using Algorithm 1 for mean retrieval. The experimental results discussed next illustrate the relative influence of the covariance descriptors and regularization property of the ordered assignment flow, respectively. Throughout, we fixed the grid connectivity $\mathcal{N}_i$ for each voxel $i \in \mathcal{I}$ to $3 \times 5 \times 5$. Figure 3, second row, illustrates a typical result of nearest neighbor assignment and the volume segmentation *without* ordering constraints. As the second raw shows, the high texture similarity between the choroid and GCL layer yields wrong predictions resulting in violation of biological retina ordering through the whole volume which cannot be resolved with the based assignment flow approach given in Sect. 2. In third row of Fig. 3, we plot the *ordered* volume segmentation by stepwise increasing the parameter γ defined in (4.4), which controls the ordering regularization by means of the novel generalized likelihood matrix (4.6). The direct comparison with the ground truth remarkably shows how the ordered labelings evolve on the assignment manifold while simultaneously giving accurate data-driven detection of RNFL, OPL, INL and the ONL layer. For the remaining critical inner layers, the local prototypes extracted by (5.1) fail to segment the retina properly, due to the presence of vertical shadow regions originating from the scanning process of the OCT-data.

CNN Features. In addition to the covariance features in Sect. 3, we compare a second approach to local feature extraction based on a convolutional neural network architecture. For each node $i \in [n]$, we trained the network to directly predict the correct class in $[c]$ using raw intensity values in $\mathcal{N}_i$ as input. As output, we find a score for each layer which can directly be transformed into a distance vector suitable as input to the ordered assignment flow (4.7) via (4.6). The specific network used in our experiments has a ResNet architecture comprising four residually connected blocks of 3D convolutions and ReLU activation.

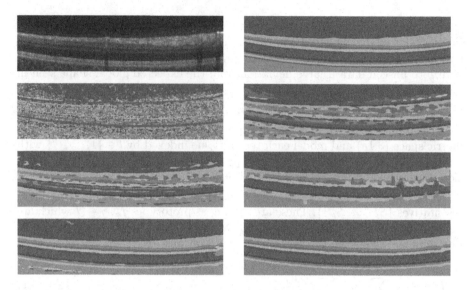

Fig. 3. From top to bottom. 1st row: One B-scan from OCT-volume showing the shadow effects with annotated ground truth on the right. **2nd row:** Nearest neighbor assignment based on prototypes computed with Stein divergence and result of the segmentation returned by the basic assignment flow (Sect. 2) on the right. **3rd row:** Illustration of the proposed *layer-ordered* volume segmentation based on covariance descriptors with ordered volume segmentation for different $\gamma = 0.5$ on left and $\gamma = 0.1$ on the right (cf. Eq. (4.4)). **4th row:** Illustration of local rounding result extracted from Res-Net and the result of ordered flow on the right.

Model size was hand-tuned for different sizes of input neighborhoods, adjusting the number of convolutions per block as well as corresponding channel dimensions. In particular, labeling accuracy is increased for detection of RPE and PR2 layers, as illustrated in the last raw of Fig. 3.

Evaluation. To assess the segmentation performance of our proposed approach, we compared to the state of the art graph-based retina segmentation method of 10 intra-retinal layers developed by the Retinal Image Analysis Laboratory at the Iowa Institute for Biomedical Imaging [1,11,16], also referred to as the IOWA Reference Algorithm. We quantify the region agreement with manual segmentation regarded as gold standard. Specifically, we calculate the DICE similarity coefficient [7] and the mean absolute error for segmented cell layer within the pixel size of 3.87 μm compared to human grader on an OCT volume consisting of 61 B-scans reported in Table 1. To allow a direct comparison to the proposed segmentation method, the evaluation was performed on layers summarized in Table 1. We point out that in general our method is not limited to any number of segmented layers if ground truth is available and further performance evaluations though additional comparison with the method proposed in [19] will be included in the complete report of the proposed approach which is beyond scope of this paper. The OCT volumes were imported into OCTExplorer 3.8.0 and segmented using the predefined Macular-OCT IOWA software.

Both methods detect the RNFL layer with high accuracy whereas for the underlying retina tissues the automated segmentation with ordered assignment flow indicates the smallest mean absolute error and the highest Dice similarity index underpinning the superior performance of order preserving labeling in view of accuracy.

Fig. 4. Visualization of segmented intraretinal surfaces: **Left:** IOWA layer detection of 10 boundaries, **Middle:** Proposed labeling result based on local features extraction, **Right:** Ground truth. For a quantitative comparison: see Table 1.

Table 1. Mean absolute error measures are given in pixels (1 pixel − 3.87 µm). **Left:** Dice and mean absolute error of the proposed approach. **Right:** Resulting metrics achieved by IOWA reference algorithm. All numbers demonstrate the superior performance of our novel order-preserving labeling approach.

OAF	DICE index	Mean absolute error	IOWA	DICE index	Mean absolute error
RNFL	0.9962	3.5920	RNFL	0.9906	2.8290
GCL	0.8390	1.3091	GCL	0.7933	2.1063
IPL	0.8552	4.0340	IPL	0.7148	5.0753
INL	0.8714	6.0180	INL	0.7696	6.0090
OPL	0.8886	4.5345	OPL	0.8510	5.4852
ONL+ELM+PR1	0.9070	1.9550	ONL+ELM+PR1	0.8374	7.0928
PR2+RPE	0.9784	2.6511	PR2+RPE	0.9006	12.4891

6   Conclusion

In this paper we presented a novel, fully automated and purely data driven approach for retina segmentation in OCT-volumes. Compared to methods [9, 16] and [19] that have proven to be particularly effective on tissue classification with a priory known retina shape orientation, our ansatz merely relies on local features and yields ordered labelings which are directly enforced through the underlying geometry of statistical manifold. Consequently, by building on the feasible concept of spatially regularized assignment [22], the ordered flow (Definition 2) possesses the potential to be extended towards the detection of pathological retina changes and vascular vessel structure, which is the objective of our current research.

Acknowledgement. We thank Dr. Stefan Schmidt and Julian Weichsel for sharing with us their expertise on OCT sensors, data acquisition and processing. In addition, we thank Prof. Fred Hamprecht and Alberto Bailoni for their guidance in training deep networks for feature extraction from 3D data.

References

1. Abràmoff, M.D., Garvin, M.K., Sonka, M.: Retinal imaging and image analysis. IEEE Rev. Biomed. Eng. **3**, 169–208 (2010)
2. Antony, B., et al.: Automated 3-D segmentation of intraretinal layers from optic nerve head optical coherence tomography images. In: Progress in Biomedical Optics and Imaging - Proceedings of SPIE, vol. 7626, pp. 249–260 (2010)
3. Arsigny, V., Fillard, P., Pennec, X., Ayache, N.: Geometric means in a novel vector space structure on symmetric positive definite matrices. SIAM J. Matrix Anal. Appl. **29**(1), 328–347 (2007)
4. Åström, F., Petra, S., Schmitzer, B., Schnörr, C.: Image labeling by assignment. J. Math. Imaging Vis. **58**(2), 211–238 (2017)
5. Bauschke, H.H., Borwein, J.M.: Legendre functions and the method of random Bregman projections. J. Convex Anal. **4**(1), 27–67 (1997)
6. Censor, Y.A., Zenios, S.A.: Parallel Optimization: Theory, Algorithms, and Applications. Oxford University Press, New York (1997)
7. Dice, L.R.: Measures of the amount of ecologic association between species. Ecology **26**(3), 297–302 (1945)
8. Duan, J., Tench, C., Gottlob, I., Proudlock, F., Bai, L.: New variational image decomposition model for simultaneously denoising and segmenting optical coherence tomography images. Phys. Med. Biol. **60**, 8901–8922 (2015)
9. Dufour, P.A., et al.: Graph-based multi-surface segmentation of OCT data using trained hard and soft constraints. IEEE Trans. Med. Imaging **32**(3), 531–543 (2013)
10. Fang, L., Cunefare, D., Wang, C., Guymer, R., Li, S., Farsiu, S.: Automatic segmentation of nine retinal layer boundaries in OCT images of non-exudative AMD patients using deep learning and graph search. Biomed. Opt. Expr. **8**(5), 2732–2744 (2017)
11. Garvin, M.K., Abramoff, M.D., Wu, X., Russell, S.R., Burns, T.L., Sonka, M.: Automated 3-D intraretinal layer segmentation of macular spectral-domain optical coherence tomography images. IEEE Trans. Med. Imaging **9**, 1436–1447 (2009)
12. Nicholson, B., Nielsen, P., Saebo, J., Sahay, S.: Exploring tensions of global public good platforms for development: the case of DHIS2. In: Nielsen, P., Kimaro, H.C. (eds.) ICT4D 2019. IAICT, vol. 551, pp. 207–217. Springer, Cham (2019). https://doi.org/10.1007/978-3-030-18400-1_17
13. Hashimoto, M., Sklansky, J.: Multiple-order derivatives for detecting local image characteristics. Comput. Vis. Graph. Image Process. **39**(1), 28–55 (1987)
14. He, Y., et al.: Deep learning based topology guaranteed surface and MME segmentation of multiple sclerosis subjects from retinal OCT. Biomed. Opt. Expr. **10**(10), 5042–5058 (2019)
15. Higham, N.: Functions of Matrices: Theory and Computation. SIAM (2008)

16. Kang, L., Xiaodong, W., Chen, D.Z., Sonka, M.: Optimal surface segmentation in volumetric images-a graph-theoretic approach. IEEE Trans. Pattern Anal. Mach. Intell. **28**(1), 119–134 (2006)
17. Liu, X., et al.: Semi-supervised automatic segmentation of layer and fluid region in retinal optical coherence tomography images using adversarial learning. IEEE Access **7**, 3046–3061 (2019)
18. Novosel, J., Vermeer, K.A., de Jong, J.H., Wang, Z., van Vliet, L.J.: Joint segmentation of retinal layers and focal lesions in 3-D OCT data of topologically disrupted retinas. IEEE Trans. Med. Imaging **36**(6), 1276–1286 (2017)
19. Rathke, F., Schmidt, S., Schnörr, C.: Probabilistic intra-retinal layer segmentation in 3-D OCT images using global shape regularization. Med. Image Anal. **18**(5), 781–794 (2014)
20. Ronneberger, O., Fischer, P., Brox, T.: U-Net: convolutional networks for biomedical image segmentation. In: Navab, N., Hornegger, J., Wells, W.M., Frangi, A.F. (eds.) MICCAI 2015. LNCS, vol. 9351, pp. 234–241. Springer, Cham (2015). https://doi.org/10.1007/978-3-319-24574-4_28
21. Roy, A., et al.: ReLayNet: retinal layer and fluid segmentation of macular optical coherence tomography using fully convolutional networks. Biomed. Opt. Expr. **8**(8), 3627–3642 (2017)
22. Schnörr, C.: Assignment flows. In: Grohs, P., Holler, M., Weinmann, A. (eds.) Handbook of Variational Methods for Nonlinear Geometric Data, pp. 235–260. Springer, Cham (2020). https://doi.org/10.1007/978-3-030-31351-7_8
23. Song, Q., Bai, J., Garvin, M.K., Sonka, M., Buatti, J.M., Wu, X.: Optimal multiple surface segmentation with shape and context priors. IEEE Trans. Med. Imaging **32**(2), 376–386 (2013)
24. Sra, S.: Positive definite matrices and the S-divergence. Proc. Am. Math. Soc. **144**(7), 2787–2797 (2016)
25. Tuzel, O., Porikli, F., Meer, P.: Region covariance: a fast descriptor for detection and classification. In: Leonardis, A., Bischof, H., Pinz, A. (eds.) ECCV 2006. LNCS, vol. 3952, pp. 589–600. Springer, Heidelberg (2006). https://doi.org/10.1007/11744047_45
26. Yazdanpanah, A., Hamarneh, G., Smith, B.R., Sarunic, M.V.: Segmentation of intra-retinal layers from optical coherence tomography images using an active contour approach. IEEE Trans. Med. Imaging **30**(2), 484–496 (2011)
27. Zeilmann, A., Savarino, F., Petra, S., Schnörr, C.: Geometric numerical integration of the assignment flow. Inverse Probl. **36**(3), 034004 (33pp) (2020)
28. Zern, A., Zeilmann, A., Schnörr, C.: Assignment flows for data labeling on graphs: convergence and stability. CoRR abs/2002.11571 (2020)

Boosting Generalization in Bio-signal Classification by Learning the Phase-Amplitude Coupling

Abdelhak Lemkhenter[✉][ID] and Paolo Favaro[ID]

Department of Computer Science, University of Bern, Bern, Switzerland
{abdelhak.lemkhenter,paolo.favaro}@inf.unibe.ch

Abstract. Various hand-crafted feature representations of bio-signals rely primarily on the amplitude or power of the signal in specific frequency bands. The phase component is often discarded as it is more sample specific, and thus more sensitive to noise, than the amplitude. However, in general, the phase component also carries information relevant to the underlying biological processes. In fact, in this paper we show the benefits of learning the coupling of both phase and amplitude components of a bio-signal. We do so by introducing a novel self-supervised learning task, which we call *phase-swap*, that detects if bio-signals have been obtained by merging the amplitude and phase from different sources. We show in our evaluation that neural networks trained on this task generalize better across subjects and recording sessions than their fully supervised counterpart.

1 Introduction

Bio-signals, such as Electroencephalograms and Electrocardiograms, are multi-variate time-series generated by biological processes that can be used to assess seizures, sleep disorders, head injuries, memory problems, heart diseases, just to name a few [19]. Although clinicians can successfully learn to correctly interpret such bio-signals, their protocols cannot be directly converted into a set of numerical rules yielding a comparable assessment performance. Currently, the most effective way to transfer this expertise into an automated system is to gather a large number of examples of bio-signals with the corresponding labeling provided by a clinician, and to use them to train a deep neural network. However, collecting such labeling is expensive and time-consuming. In contrast, bio-signals without labeling are more readily available in large numbers.

Recently, self-supervised learning (SelfSL) techniques have been proposed to limit the amount of required labeled data. These techniques define a so-called *pretext* task that can be used to train a neural network in a supervised manner on data without manual labeling. The pretext task is an artificial problem, where a model is trained to output what transformation was applied to the data. For instance, a model could be trained to output the probability that a time-series had been time-reversed [25]. This step is often called pre-training and it can be

© Springer Nature Switzerland AG 2021
Z. Akata et al. (Eds.): DAGM GCPR 2020, LNCS 12544, pp. 72–85, 2021.
https://doi.org/10.1007/978-3-030-71278-5_6

carried out on large data sets as no manual labeling is required. The training of the pre-trained neural network then continues with a small learning rate on the small target data set, where labels are available. This second step is called *fine-tuning*, and it yields a substantial boost in performance [21]. Thus, SelfSL can be used to automatically learn physiologically relevant features from unlabelled bio-signals and improve classification performance.

SelfSL is most effective if the pretext task focuses on features that are relevant to the target task. Typical features work with the amplitude or the power of the bio-signals, but as shown in the literature, the phase carries information about the underlining biological processes [2, 15, 20]. Thus, in this paper, we propose a pretext task to learn the coupling between the amplitude and the phase of the bio-signals, which we call *phase swap* (PS). The objective is to predict whether the phase of the Fourier transform of a multivariate physiological time-series segment was swapped with the phase of another segment.

We show that features learned through this task help classification tasks generalize better, regardless of the neural network architecture.

Our contributions are summarized as follows

- We introduce phase swap, a novel self-supervised learning task to detect the coupling between the phase and the magnitude of physiological time-series;
- With phase swap, we demonstrate experimentally the importance of incorporating the phase in bio-signal classification;
- We show that the learned representation generalizes better than current state of the art methods to new subjects and to new recording sessions;
- We evaluate the method on four different data sets and analyze the effect of various hyper-parameters and of the amount of available labeled data on the learned representations.

2 Related Work

Self-supervised Learning. Self-supervised learning refers to the practice of pre-training deep learning architectures on user-defined pretext tasks. This can be done on large volumes of unlabeled data since the annotations can be automatically generated for these tasks. This is a common practice in the Natural Language Processing literature. Examples of such works include Word2Vec [17], where the task is to predict a word from its context, and BERT [3], where the model is pretrained as a masked language model and on the task of detecting consecutive sentences. The self-supervision framework has also been gaining popularity in Computer Vision. Pretext tasks such as solving a jigsaw puzzle [21], predicting image rotations [5] and detecting local inpainting [11] have been shown to be able to learn useful data representations for downstream tasks. Recent work explores the potential of self-supervised learning for EEG signals [1] and time series in general [10]. In [1], the focus is on long-term/global tasks such as determining whether two given windows are nearby temporally or not.

Deep Learning for Bio-signals. Bio-signals include a variety of physiological measures across time such as: Electroencephalogram (EEG), Electrocardiogram (ECG)

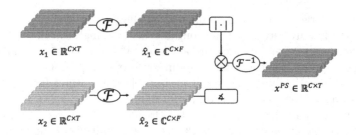

Fig. 1. Illustration of the phase-swap operator Φ. The operator takes two signals as input and then combines the amplitude of the first signal with the phase of the second signal in the output.

(ECG), Electromyogram (EMG), Electrooculography (EOG), etc. These signals are used by clinicians in various applications, such as sleep scoring [18] or seizure detection [23]. Similarly to many other fields, bio-signals analysis has also seen the rise in popularity of deep learning methods for both classification [7] and representation learning [1]. The literature review [22] showcases the application of deep learning methods to various EEG classification problems such as brain computer interfaces, emotion recognition and seizure detection. The work by Banville et al. [1] leverages self-supervised tasks based on the relative temporal positioning of pairs/triplets of EEG segments to learn a useful representation for a downstream sleep staging application.

Phase Analysis. The phase component of bio-signals has been analyzed before. Busch et al. [2] show a link between the phase of the EEG oscillations, in the alpha (8–12 Hz) and theta (4–8 Hz) frequency bands, and the subjects' ability to perceive the flash of a light. The phase of the EEG signal is also shown to be more discriminative for determining firing patterns of neurons in response to certain types of stimuli [20]. More recent work, such as [15], highlights the potential link between the phase of the different EEG frequency bands and cognition during proactive control of task switching.

3   Learning to Detect the Phase-Amplitude Coupling

In this section, we define the *phase swap* operator and the corresponding SelfSL task, and present the losses used for pre-training and fine-tuning.

Let $D_{i,j}^W = \{(x^{i,j,k}, y^{i,j,k})\}_{k=1}^N$ be the set of samples associated with the i-th subject during the j-th recording session. Each sample $x^{i,j,k} \in \mathbf{R}^{C \times W}$ is a multivariate physiological time-series window where C and W are the number of channels and the window size respectively. $y^{i,j,k}$ is the class of the k-th sample. Let $\mathcal{F}$ and $\mathcal{F}^{-1}$ be the Discrete Fourier Transform operator and its inverse, respectively. These operators will be applied to a given vector x extracted from the bio-signals. In the case of multivariate signals, we apply these operators channel-wise.

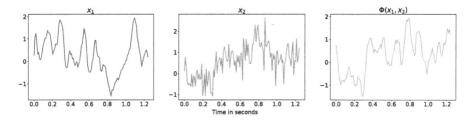

Fig. 2. Illustration of the PS operator on a pair of 1.25 s segments taken from the Fpz-Cz channel in the SC data set [18]. The original signals are x_1 and x_2.

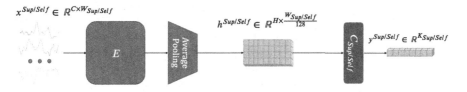

Fig. 3. Training with either the self-supervised or the supervised learning task.

For the sake of clarity, we provide the definitions of the absolute value and the phase element-wise operators. Let $z \in \mathbf{C}$, where $\mathbf{C}$ denotes the set of complex numbers. Then, the absolute value, or *magnitude*, of z is denoted $|z|$ and the phase of z is denoted $\angle z$. With such definitions, we have the trivial identity $z = |z|\angle z$.

Given two samples $x^{i,j,k}$, $x^{i,j,k'} \in D_{i,j}^W$, the *phase swap* (PS) operator Φ is

$$\Phi\left(x^{i,j,k}, x^{i,j,k'}\right) \doteq \mathcal{F}^{-1}\left[\left|\mathcal{F}\left(x^{i,j,k}\right)\right| \odot \angle\mathcal{F}\left(x^{i,j,k'}\right)\right] = x_{swap}^{i,j,k}, \tag{1}$$

where $\odot$ is the element-wise multiplication (see Fig. 1). Note that the energy per frequency is the same for both $x_{swap}^{i,k}$ and $x^{i,k}$ and that only the phase, *i.e.*, the synchronization between the different frequencies, changes. Examples of phase swapping between different pairs of signals are shown in Fig. 2.

Notice how the shape of the oscillations change drastically when the PS operator is applied and no trivial shared patterns seem to emerge.

The PS pretext task is defined as a binary classification problem. A sample belongs to the positive class if it is transformed using the PS operator, otherwise it belongs to the negative class. In all our experiments, both inputs to the PS operator are sampled from the same patient during the same recording session. Because the phase is decoupled from the amplitude of white noise, our model has no incentive to detect noise patterns. On the contrary, it will be encouraged to focus on the structural patterns in the signal in order to detect whether the phase and magnitude of the segment are coupled or not.

We use the FCN architecture proposed by Wang et al. [24] as our core neural network model $E : \mathbf{R}^{C \times W} \rightarrow \mathbf{R}^{H \times W/128}$. It consists of 3 convolutions blocks using a Batch Normalization layer [8] and a ReLU activation followed by a pooling layer. The output of E is then flattened and fed to two Softmax layers C_{Self} and C_{Sup}, which are trained on the self-supervised and supervised tasks respectively.

Instead of a global pooling layer, we use an average pooling layer with a stride of 128. This allows us to keep the number of weights of the supervised network $C_{Sup} \circ E$ constant when the self-supervised task is defined on a different window size. The overall framework is illustrated in Fig. 3. Note that the encoder network E is the same for both tasks.

The loss function for training on the SelfSL task is the cross-entropy

$$\mathcal{L}_{Self}\left(y^{Self}, E, C_{Self}\right) = -\frac{1}{N} \sum_{i=1}^{N} \sum_{k=1}^{K_{Self}} y_{i,k}^{Self} \log\left(C_{Self} \circ E(x_i)\right)_k, \quad (2)$$

where $y_{i,k}^{Self}$ and $(C_{Self} \circ E(x_i))_k$ are the one-hot representations of the true SelfSL pretext label and the predicted probability vector respectively. We optimize Eq. (2) with respect to the parameters of both E and C_{Self}. Similarly, we define the loss function for the (supervised) fine-tuning as the cross-entropy

$$\mathcal{L}_{Sup}\left(y^{Sup}, E, C_{Sup}\right) = -\frac{1}{N} \sum_{i=1}^{N} \sum_{k=1}^{K_{Sup}} y_{i,k}^{Sup} \log\left(C_{Sup} \circ E(x_i)\right)_k, \quad (3)$$

where $y_{i,k}^{Sup}$ denotes the label for the target task. The $y_{i,k}^{Sup/Self}$ vectors are in $\mathbf{R}^{N \times K_{Sup/Self}}$, where N and $K_{Sup/Self}$ are the number of samples and classes respectively. In the fine-tuning, E is initialized with the parameters obtained from the optimization of Eq. (2) and C_{Sup} with random weights, and then they are both updated to optimize Eq. (3), but with a small learning rate.

4 Experiments

4.1 Data Sets

In our experiments, we use the Expanded SleepEDF [6,12,18], the CHB-MIT [23] and ISRUC-Sleep [13] data sets as they contain recordings from multiple patients. This allows us to study the generalization capabilities of the learned feature representation to new recording sessions and new patients. The Expanded SleepEDF database contains two different sleep scoring data sets

- Sleep Cassette Study (SC) [18]: Collected between 1987 and 1991 in order to study the effect of age on sleep. It includes 78 patients with 2 recording sessions each (3 recording sessions were lost due to hardware failure).
- Sleep Telemetry Study (ST) [12]: Collected in 1994 as part of a study of the effect of Temazepam on sleep in 22 different patients with 2 recordings sessions each.

Both data sets define sleep scoring as a 5-way classification problem. The 5 classes in question are the sleep stages: Wake, NREM 1, NREM 2, NREM 3/4, REM. The NREM 3 and 4 are merged into one class due to their small number of samples (these two classes are often combined together in sleep studies).

The third data set we use in our experiments is the CHB-MIT data set [23] recorded at the Children's Hospital Boston from pediatric patients with intractable seizures. It includes multiples recording files across 22 different patients. We retain the 18 EEG channels that are common to all recording files. The sampling rate for all channels 256 Hz. The target task defined on this data set is predicting whether a given segment is a seizure event or not, *i.e.*, a binary classification problem. For all the data sets, the international 10–20 system [16] was adopted for the choice of the positioning of the EEG electrodes.

The last data set we use is ISRUC-Sleep [13], for sleep scoring as a 4-way classification problem. We use the 14 channels extracted in the Matlab version of the data set. This data set consists of three subgroups: subgroups I and II contain respectively recordings from 100 and 8 subjects with sleep disorders, whereas subgroup III contains recordings from 10 healthy subjects. This allows us to test the generalization from diagnosed subjects to healthy ones.

For the SC, ST and ISRUC-sleep data sets we resample the signals to 102.4 Hz. This resampling allows us to simplify the neural network architectures we use, because in this case most window sizes can be represented by a power of 2, *e.g.*, a window of 2.5 s corresponds to 256 samples. We normalize each channel per recording file in all data sets to have zero mean and a standard deviation of one.

4.2 Training Procedures and Models

In the supervised baseline (respectively, self-supervised pre-training), we train the randomly initialized model $C_{Sup} \circ E$ (respectively, $C_{Self} \circ E$) on the labeled data set for 10 (respectively, 5) epochs using the Adam optimizer [14] with a learning rate of 10^{-3} and $\beta = (0.9, 0.999)$. We balance the classes present in the data set using resampling (no need to balance classes in the self-supervised learning task). In fine-tuning, we initialize E's weights with those obtained from the SelfSL training and then train $C_{Sup} \circ E$ on the labeled data set for 10 epochs using the Adam optimizer [14], but with a learning rate of 10^{-4} and $\beta = (0.9, 0.999)$. As in the fully supervised training, we also balance the classes using re-sampling. In all training cases, we use a default batch size of 128.

We evaluate our self-supervised framework using the following models

- **PhaseSwap**: The model is pre-trained on the self-supervised task and fine-tuned on the labeled data;
- **Supervised**: The model is trained solely in a supervised fashion;
- **Random**: C_{Sup} is trained on top of a frozen randomly initialized E;
- **PSFrozen**: We train C_{Sup} on top of the frozen weights of the model E pre-trained on the self-supervised task.

4.3 Evaluation Procedures

We evaluate our models on train/validation/test splits in our experiments. In total we use at most 4 sets, which we refer to as the training set, the Validation

Set, the Test set A and the Test set B. The validation set and test set A and the training set share the same patient identities, while B contains recordings from other patients. The validation set and test set A use distinct recording sessions. Validation Set and the training set share the same patient identities and recording sessions with a 75% (for the training set) and 25% (Validation Set) split. We use each test set for the following purposes

- **Validation Set**: this set serves as a validation set;
- **Test set A**: this set allows us to evaluate the generalization error on new recording sessions for patients observed during training;
- **Test set B**: this set allows us to evaluate the generalization error on new recording sessions for patients not observed during training.

We use the same set of recordings and patients for both the training of the self-supervised and supervised tasks. For the ST, SC and ISRUC data sets we use class re-balancing only during the supervised fine-tuning. However, for the CHB-MIT data set, the class imbalance is much more extreme: The data set consists of less than 0.4% positive samples. Because of that, we under-sample the majority class both during the self-supervised and supervised training. This prevents the self-supervised features from completely ignoring the positive class. Unless specified otherwise, we use $W_{Self} = 5\,\mathrm{s}$ and $W_{Sup} = 30\,\mathrm{s}$ for the ISRUC, ST and SC data sets, $W_{Self} = 2\,\mathrm{s}$ and $W_{Sup} = 10\,\mathrm{s}$ for the CHB-MIT data set, where W_{Self} and W_{Sup} are the window size for the self-supervised and supervised training respectively. For the ISRUC, ST and SC data sets, the choice of W_{Sup} corresponds to the granularity of the provided labels. For the CHB-MIT data set, although labels are provided at a rate 1 Hz, the literature in neuroscience usually defines a minimal duration of around 10 s for an epileptic event in humans [4], which motivates our choice of $W_{Sup} = 10\,\mathrm{s}$.

Evaluation Metric. As an evaluation metric, we use the balanced accuracy

$$Acc^{Balanced}(y, \hat{y}) = \frac{1}{K} \sum_{k=1}^{K} \frac{\sum_{i=1}^{N} \hat{y}_{i,k} y_{i,k}}{\sum_{i=1}^{N} y_{i,k}}, \tag{4}$$

which is defined as the average of the recall values per class, where K, N, y and $\hat{y}$ are respectively the number of classes, the number of samples, the one-hot representation of true labels and the predicted labels.

4.4 Generalization on the Sleep Cassette Data Set

We explore the generalization of the self-supervised trained model by varying the number of different patients used in the training set for the SC data set. The r_{train} is the percentage of patient identities used for training, in Validation Set and in Test set A. In Table 1, we report the balanced accuracy on all test sets for various values of r_{train}. The self-supervised training was done using a window size of $W_{Self} = 5\,\mathrm{s}$. We observe that the **PhaseSwap** model performs the best for all values of r_{train}. We also observe that the performance gap between the

Table 1. Comparison of the performance of the **PhaseSwap** model on the SC data set for various values of r_{train}. For $r_{train} = 100\%^*$ we use all available recordings for the training and the Validation sets. Results with different r_{train} are not comparable.

r_{train}	Experiment	Validation set	Test set A	Test set B
20%	**PhaseSwap**	**84.3%**	**72.0%**	**69.6%**
20%	**Supervised**	79.4%	67.9%	66.0%
50%	**PhaseSwap**	**84.9%**	**75.1%**	**73.3%**
50%	**Supervised**	81.9%	71.7%	69.4%
75%	**PhaseSwap**	**84.9%**	**77.6%**	**76.1%**
75%	**Supervised**	81.6%	73.7%	72.8%
100%*	**PhaseSwap**	**84.3%**	–	–
100%*	**Supervised**	83.5%	–	–

PhaseSwap and **Supervised** models is narrower for larger values for r_{train}. This is to be expected since including more identities in the training set allows the **Supervised** model to generalize better. For $r_{train} = 100\%^*$ we use all recording sessions across all identities for the training set and in the Validation Set (since all identities and sessions are used, the Test sets A and B are empty). The results obtained for this setting show that there is still a slight benefit with the **PhaseSwap** pre-training even when labels are available for most of the data.

4.5 Generalization on the ISRUC-Sleep Data Set

Using the ISRUC-Sleep data set [13], we aim to evaluate the performance of the PhaseSwap model on healthy subjects when it was trained on subjects with sleep disorders. For the self-supervised training, we use $W_{Self} = 5$ s. The results are reported in Table 2. Note that we combined the recordings of subgroup II and the ones not used for the training from subgroup I into a single test set since they are from subjects with sleep disorders. We observe that for both experiments, $r_{train} = 25\%$ and $r_{train} = 50\%$, the PhaseSwap model outperforms the supervised baseline for both test sets. Notably, the performance gap on subgroup III is larger than 10%. This can be explained by the fact that sleep disorders can drastically change the sleep structure of the affected subjects, which in turn leads the supervised baseline to learn features that are specific to the disorders/subjects present in the training set.

4.6 Comparison to the Relative Positioning Task

The Relative Positioning (RP) task was introduced by Banville et al. [1] as a self-supervised learning method for EEG signals, which we briefly recall here. Given x_t and $x_{t'}$, two samples with a window size W and starting points t and t' respectively, the RP task defines the following labels

Table 2. Comparison of the performance of the **PhaseSwap** model on the ISRUC-Sleep data set for various values of r_{train}.

r_{train}	Model	Validation set	Test set B (subgroup I + II)	Test set B (subgroup III)
25%	PhaseSwap	75.8%	**67.3%**	**62.8%**
25%	Supervised	**75.9%**	63.1%	47.9%
50%	PhaseSwap	**76.3%**	68.2%	**67.1%**
50%	Supervised	75.5%	**68.3%**	57.3%

Table 3. Comparison between the PS and RP pre-training on the SC data set.

Pre-training	Validation set	Test set A	Test set B	SelfSL validation accuracy
Supervised	79.4%	67.9%	66.0%	–
PS	**84.3%**	**72.0%**	**69.6%**	86.9%
RP	80.3%	66.2%	65.4%	56.9%

$C_{Self}(|h_t - h_{t'}|) = \mathbb{1}\,(|t - t'| \le \tau_{pos}) - \mathbb{1}\,(|t - t'| > \tau_{neg})$, where $h_t = E(x_t)$, $h_{t'} = E(x_{t'})$, $\mathbb{1}(\cdot)$ is the indicator function, and τ_{pos} and τ_{neg} are predefined quantities. Pairs that yield $C_{Self} = 0$ are discarded. $|\cdot|$ denotes the element-wise absolute value operator.

Next, we compare our self-supervised task to the RP task [1]. For both settings, we use $W_{Self} = 5\,\mathrm{s}$ and $r_{train} = 20\%$. For the RP task we choose $\tau_{pos} = \tau_{neg} = 12 \times W_{Self}$. We report the balanced accuracy for all test sets on the SC data set in Table 3. We observe that our self-supervised task outperforms the RP task. This means that the features learned through the PS task allow the model to perform better on unseen data.

4.7   Results on the Sleep Telemetry and CHB-MIT Data Sets

In this section, we evaluate our framework on the ST and CHB-MIT data sets. For the ST data set, we use $W_{Self} = 1.25\,\mathrm{s}$, $W_{Sup} = 30\,\mathrm{s}$ and $r_{train} = 50\%$. For the CHB-MIT data set, we use $W_{Self} = 2\,\mathrm{s}$, $W_{Sup} = 10\,\mathrm{s}$, $r_{train} = 25\%$ and 30 epochs for the supervised fine-tuning/training. As shown in Table 4, we observe that for the ST data set, the features learned through the PS task produce a significant improvement, especially on Test sets A and B. For the CHB-MIT data set, the PS fails to provide the performance gains as observed for the previous data sets. We believe that this is due to the fact that the PS task is too easy on this particular data set: Notice how the validation accuracy is above 99%. With a trivial task, self-supervised pre-training fails to learn any meaningful feature representations.

In order to make the task more challenging, we introduce a new variant, which we call **PS + Masking**, where we randomly zero out all but 6 randomly selected channels for each sample during the self-supervised pre-training. The

Table 4. Evaluation of the **PhaseSwap** model on the ST and CHB-MIT datasets.

Dataset	Experiment	Val. set	Test set A	Test set B	SelfSL val. accuracy
ST	**Supervised**	69.2%	52.3%	46.7%	–
ST	**PhaseSwap**	**74.9%**	**60.4%**	**52.3%**	71.3%
CHB-MIT	**Supervised**	**92.6%**	89.5%	58.0%	–
CHB-MIT	**PhaseSwap**	92.2%	86.8%	55.1%	99.8%
CHB-MIT	**PS+Masking**	91.7%	**90.6%**	**59.8%**	88.1%

Table 5. Comparison of the performance of the **PhaseSwap** model on the SC data set for various values of the window size W_{Self}.

W_{Self}	Experiment	Validation set	Test set A	Test set B
1.25 s	**PhaseSwap**	84.3%	72.0%	69.6%
2.5 s	**PhaseSwap**	**84.6%**	71.9%	70.0%
5 s	**PhaseSwap**	83.4%	**72.5%**	**70.9%**
10 s	**PhaseSwap**	83.6%	71.6%	69.9%
30 s	**PhaseSwap**	83.9%	71.0%	69.2%
–	**Supervised**	79.4%	68.1%	66.1%

model obtained through this scheme performs the best on both sets A and B and is comparable to the **Supervised** baseline on the validation set. As for the reason why the PS training was trivial on this particular data set, we hypothesize that this is due to the high spatial correlation in the CHB-MIT data set samples. This data set contains a high number of homogeneous channels (all of them are EEG channels), which in turn result in a high spatial resolution of the brain activity. At such a spatial resolution, the oscillations due to the brain activity show a correlation both in space and time [9]. However, our PS operator ignores the spatial aspect of the oscillations. When applied, it often corrupts the spatial coherence of the signal, which is then easier to detect than the temporal phase-amplitude incoherence. This hypothesis is supported by the fact that the random channel masking, which in turn reduces the spatial resolution during the self-supervised training, yields a lower training accuracy, *i.e.*, it is a non-trivial task.

4.8   Impact of the Window Size

In this section, we analyze the effect of the window size W_{Self} used for the self-supervised training on the final performance. We report the balanced accuracy on all our test sets for the SC data set in Table 5. For all these experiments, we use 20% of the identities in the training set. The capacity of the **Supervised** model $C_{Sup} \circ E$ is independent of W_{Self} (see Sect. 3), and thus so is its performance. We observe that the best performing models are the ones using $W_{Self} = 2.5$ s for the Validation Set and $W_{Self} = 5$ s for sets A and B. We argue that the features

Table 6. Balanced accuracy reported on the SC data set for the four training variants.

Experiment	Validation set	Test set A	Test set B
Supervised	79.4%	67.9%	66.0%
PhaseSwap	**84.3%**	**72.0%**	**69.6%**
PSFrozen	75.2%	68.1%	67.1%
Random	70.1%	62.1%	63.9%

learned by the self-supervised model are less specific for larger window sizes. The PS operator drastically changes structured parts of the time series, but barely affects pure noise segments. As discussed in Sect. 3, white noise is invariant with respect to the PS operator. With smaller window sizes, most of the segments are either noise or structured patterns, but as the window size grows, its content becomes a combination of the two.

4.9   Frozen vs Fine-Tuned Encoder

In Table 6, we analyze the effect of freezing the weights of E during the supervised fine-tuning. We compare the performance of the four variants described in Sect. 4.2 on the SC data set. All variants use $W_{Self} = 5$ s, $W_{Sup} = 30$ s and $r_{train} = 20\%$. As expected, we observe that the **PhaseSwap** variant is the most performant one since it is less restricted in terms of training procedure than **PSFrozen** and **Random**. Moreover, the **PSFrozen** outperforms the **Random** variant on all test sets and is on par with the **Supervised** baseline on the Test set B. This confirms that the features learned during pre-training are useful for the downstream classification even when the encoder model E is frozen during the fine-tuning. The last variant, **Random**, allows us to disentangle the contribution of the self-supervised task from the prior imposed by the architecture choice for E. As we can see in Table 6, the performance of the **PhaseSwap** variant is significantly higher than the latter variant, confirming that the self-supervised task chosen here is the main factor behind the performance gap.

4.10   Architecture

Most of the experiments in this paper use the FCN architecture [24]. In this section, we illustrate that the performance boost of the PhaseSwap method does not depend on the neural network architecture. To do so, we also analyze the performance of a deeper architecture in the form of the Residual Network (ResNet) proposed by Humayun et al. [7]. We report in Table 7 the balanced accuracy computed using the SC data set for two choices of $W_{Self} \in \{2.5\,\text{s}, 30\,\text{s}\}$ and two choices of $r_{train} \in \{20\%, 100\%^*\}$. The table also contains the performance of the FCN model trained using the PS task as a reference. We do not report the results for the RP experiment using $W_{Self} = 30$ s as we did not manage to make the self-supervised pre-training converge. All ResNet models were trained

Table 7. Evaluation of the **PhaseSwap** model using the ResNet architecture on the SC data set. Values denoted with a * are averages across two runs.

r_{train}	W_{Self}	Architecture	Experiment	Val. set	Test set A	Test set B
20%	5 s	FCN	FCN + PS	84.3%	72.0%	69.6%
20%	5 s	ResNet	phase swap	82.1%	**72.5%**	69.6%
20%	5 s	ResNet	RP	72.3%	67.4%	65.9%
20%	–	ResNet	supervised	79.1%*	70.0%*	66.5%*
20%	30 s	ResNet	phase swap	**83.6%**	70.7%	**69.3%**
100%*	5 s	FCN	FCN + PS	84.3%	–	–
100%*	5 s	ResNet	phase swap	81.2%	–	–
100%*	5 s	ResNet	RP	79.1%	–	–
100%*	–	ResNet	supervised	**84.2%***	–	–
100%*	30 s	ResNet	phase swap	**84.2%**	–	–

for 15 epochs for the supervised fine-tuning. For $r_{train} = 20\%$, we observe that pre-training the ResNet on the PS task outperforms both the supervised and RP pre-training. We also observe that for this setting, the model pre-trained with $W_{Self} = 30$ s performs better on both the validation set and test set B compared to the one pre-trained using $W_{Self} = 5$ s. Nonetheless, the model using the simpler architecture still performs the best on those sets and is comparable to the best performing one on set A. We believe that the lower capacity of the FCN architecture prevents the learning of feature representations that are too specific to the pretext task compared the ones learned with the more powerful ResNet. For the setting $r_{train} = 100\%^*$, the supervised ResNet is on par with a model pre-trained on the PS task with $W_{Self} = 30$ s. Recall that $r_{train} = 100\%^*$ refers to the setting where all recording session and patients are used for the training set. Based on these results, we can conclude that there is a point of diminishing returns in terms of available data beyond which the self-supervised pre-training might even deteriorate the performance of the downstream classification tasks.

5   Conclusions

We have introduced the phase swap pretext task, a novel self-supervised learning approach suitable for bio-signals. This task aims to detect when bio-signals have mismatching phase and amplitude components. Since the phase and amplitude of white noise are uncorrelated, features learned with the phase swap task do not focus on noise patterns. Moreover, these features exploit signal patterns present both in the amplitude and phase domains. We have demonstrated the benefits of learning features from the phase component of bio-signals in several experiments and comparisons with competing methods. Most importantly, we find that pre-training a neural network with limited capacity on the phase swap task builds features with a strong generalization capability across subjects and observed sessions. One possible future extension of this work, as suggested by the results on the CHB-MIT data set [23], is to incorporate spatial correlations in the PS operator through the use of a spatio-temporal Fourier transformation.

Acknowledgement. This research is supported by the Interfaculty Research Cooperation "Decoding Sleep: From Neurons to Health & Mind" of the University of Bern.

References

1. Banville, H., Moffat, G., Albuquerque, I., Engemann, D.A., Hyvärinen, A., Gramfort, A.: Self-supervised representation learning from electroencephalography signals. In: 2019 IEEE 29th International Workshop on Machine Learning for Signal Processing (MLSP), pp. 1–6. IEEE (2019)
2. Busch, N.A., Dubois, J., VanRullen, R.: The phase of ongoing EEG oscillations predicts visual perception. J. Neurosci. **29**(24), 7869–7876 (2009)
3. Devlin, J., Chang, M.W., Lee, K., Toutanova, K.: Bert: pre-training of deep bidirectional transformers for language understanding. arXiv preprint arXiv:1810.04805 (2018)
4. Fisher, R.S., Scharfman, H.E., deCurtis, M.: How can we identify ictal and interictal abnormal activity? In: Scharfman, H.E., Buckmaster, P.S. (eds.) Issues in Clinical Epileptology: A View from the Bench. AEMB, vol. 813, pp. 3–23. Springer, Dordrecht (2014). https://doi.org/10.1007/978-94-017-8914-1_1
5. Gidaris, S., Singh, P., Komodakis, N.: Unsupervised representation learning by predicting image rotations. In: 6th International Conference on Learning Representations, ICLR 2018, Vancouver, BC, Canada, 30 April–3 May 2018, Conference Track Proceedings (2018)
6. Goldberger, A.L., et al.: Physiobank, physiotoolkit, and physionet: components of a new research resource for complex physiologic signals. Circulation **101**(23), e215–e220 (2000)
7. Humayun, A.I., Sushmit, A.S., Hasan, T., Bhuiyan, M.I.H.: End-to-end sleep staging with raw single channel EEG using deep residual convnets. In: 2019 IEEE EMBS International Conference on Biomedical & Health Informatics (BHI), pp. 1–5. IEEE (2019)
8. Ioffe, S., Szegedy, C.: Batch normalization: accelerating deep network training by reducing internal covariate shift. In: Proceedings of the 32nd International Conference on International Conference on Machine Learning, ICML 2015, vol. 37, pp. 448–456. JMLR.org (2015)
9. Ito, J., Nikolaev, A.R., Van Leeuwen, C.: Spatial and temporal structure of phase synchronization of spontaneous alpha EEG activity. Biol. Cybern. **92**(1), 54–60 (2005)
10. Jawed, S., Grabocka, J., Schmidt-Thieme, L.: Self-supervised learning for semi-supervised time series classification. In: Lauw, H.W., Wong, R.C.-W., Ntoulas, A., Lim, E.-P., Ng, S.-K., Pan, S.J. (eds.) PAKDD 2020. LNCS (LNAI), vol. 12084, pp. 499–511. Springer, Cham (2020). https://doi.org/10.1007/978-3-030-47426-3_39
11. Jenni, S., Jin, H., Favaro, P.: Steering self-supervised feature learning beyond local pixel statistics. In: Proceedings of the IEEE/CVF Conference on Computer Vision and Pattern Recognition, pp. 6408–6417 (2020)
12. Kemp, B., Zwinderman, A.H., Tuk, B., Kamphuisen, H.A., Oberye, J.J.: Analysis of a sleep-dependent neuronal feedback loop: the slow-wave microcontinuity of the EEG. IEEE Trans. Biomed. Eng. **47**(9), 1185–1194 (2000)

13. Khalighi, S., Sousa, T., Santos, J.M., Nunes, U.: ISRUC-sleep: a comprehensive public dataset for sleep researchers. Comput. Methods Programs Biomed. **124**, 180–192 (2016)
14. Kingma, D.P., Ba, J.: Adam: a method for stochastic optimization. In: Bengio, Y., LeCun, Y. (eds.) 3rd International Conference on Learning Representations, ICLR 2015, San Diego, CA, USA, 7–9 May 2015, Conference Track Proceedings (2015)
15. López, M.E., Pusil, S., Pereda, E., Maestú, F., Barceló, F.: Dynamic low frequency EEG phase synchronization patterns during proactive control of task switching. Neuroimage **186**, 70–82 (2019)
16. Malmivuo, J., Plonsey, R., et al.: Bioelectromagnetism: Principles and Applications of Bioelectric and Biomagnetic Fields. Oxford University Press, Oxford (1995)
17. Mikolov, T., Chen, K., Corrado, G., Dean, J.: Efficient estimation of word representations in vector space. arXiv preprint arXiv:1301.3781 (2013)
18. Mourtazaev, M., Kemp, B., Zwinderman, A., Kamphuisen, H.: Age and gender affect different characteristics of slow waves in the sleep eeg. Sleep **18**(7), 557–564 (1995)
19. Naït-Ali, A.: Advanced Biosignal Processing. Springer, Heidelberg (2009). https://doi.org/10.1007/978-3-540-89506-0
20. Ng, B.S.W., Logothetis, N.K., Kayser, C.: EEG phase patterns reflect the selectivity of neural firing. Cereb. Cortex **23**(2), 389–398 (2013)
21. Noroozi, M., Favaro, P.: Unsupervised learning of visual representations by solving jigsaw puzzles. In: Leibe, B., Matas, J., Sebe, N., Welling, M. (eds.) ECCV 2016. LNCS, vol. 9910, pp. 69–84. Springer, Cham (2016). https://doi.org/10.1007/978-3-319-46466-4_5
22. Roy, Y., Banville, H., Albuquerque, I., Gramfort, A., Falk, T.H., Faubert, J.: Deep learning-based electroencephalography analysis: a systematic review. J. Neural Eng. **16**(5), 051001 (2019)
23. Shoeb, A.H.: Application of machine learning to epileptic seizure onset detection and treatment. Ph.D. thesis, Massachusetts Institute of Technology (2009)
24. Wang, Z., Yan, W., Oates, T.: Time series classification from scratch with deep neural networks: a strong baseline. In: 2017 International Joint Conference on Neural Networks (IJCNN), pp. 1578–1585. IEEE (2017)
25. Wei, D., Lim, J.J., Zisserman, A., Freeman, W.T.: Learning and using the arrow of time. In: Proceedings of the IEEE Conference on Computer Vision and Pattern Recognition, pp. 8052–8060 (2018)

Long-Tailed Recognition
Using Class-Balanced Experts

Saurabh Sharma[1(✉)], Ning Yu[1,2], Mario Fritz[3], and Bernt Schiele[1]

[1] Max Planck Institute for Informatics, Saarland Informatics Campus,
Saarbrücken, Germany
{ssharma,ningyu,schiele}@mpi-inf.mpg.de
[2] University of Maryland, College Park, USA
[3] CISPA Helmholtz Center for Information Security, Saarland Informatics Campus,
Saarbrücken, Germany
fritz@cispa.saarland

Abstract. Deep learning enables impressive performance in image recognition using large-scale artificially-balanced datasets. However, real-world datasets exhibit highly class-imbalanced distributions, yielding two main challenges: relative imbalance amongst the classes and data scarcity for mediumshot or fewshot classes. In this work, we address the problem of long-tailed recognition wherein the training set is highly imbalanced and the test set is kept balanced. Differently from existing paradigms relying on data-resampling, cost-sensitive learning, online hard example mining, loss objective reshaping, and/or memory-based modeling, we propose an ensemble of class-balanced experts that combines the strength of diverse classifiers. Our ensemble of class-balanced experts reaches results close to state-of-the-art and an extended ensemble establishes a new state-of-the-art on two benchmarks for long-tailed recognition. We conduct extensive experiments to analyse the performance of the ensembles, and discover that in modern large-scale datasets, relative imbalance is a harder problem than data scarcity. The training and evaluation code is available at https://github.com/ssfootball04/class-balanced-experts.

1 Introduction

In the past decades, deep learning has boosted success in image recognition to a new level [14]. The availability of large-scale datasets with thousands of images in each class [4,47] has been a major factor in this revolution. However, these datasets are manually curated and artificially balanced, as opposed to real-world datasets that exhibit a highly skewed and class-imbalanced distribution in a long-tailed shape: a few common classes and many more rare classes. To address this practical challenge, in this work, we focus on the problem of long-tailed recognition, wherein datasets exhibit a natural power-law distribution [32], allowing us to assess model performance on four folds: *Manyshot* classes ($\geq$100 samples), *Mediumshot* classes (20–100 samples), *Fewshot* classes (<20 samples), and *All* classes. Training data follows a highly class-imbalanced distribution, and testing data is balanced so that equally good performance over all classes is crucial [24].

© Springer Nature Switzerland AG 2021
Z. Akata et al. (Eds.): DAGM GCPR 2020, LNCS 12544, pp. 86–100, 2021.
https://doi.org/10.1007/978-3-030-71278-5_7

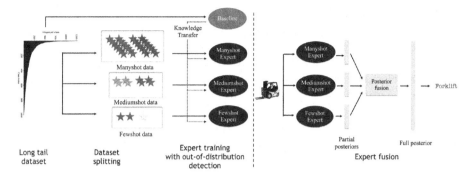

Fig. 1. Our pipeline for long-tailed recognition: an ensemble of experts trained on class-balanced subsets of *Manyshot, Mediumshot,* and *Fewshot* data. We *transfer knowledge* from *Manyshot* to *Mediumshot* and *Fewshot* classes by initialising experts with a *Baseline* model trained on all the data. Expert models classify samples outside their subset as out-of-distribution and output partial posteriors that are fused into a full posterior to obtain the final prediction.

The two main challenges for a long-tailed classification model are *relative imbalance* amongst the classes, and *data scarcity* or unobservable data modes [13]. Existing techniques for imbalanced classification have focused on data re-sampling [6, 13] and cost-sensitive learning [3, 23] to re-weigh the loss objective or counter *relative imbalance*, while techniques for fewshot learning have employed data augmentation [7, 36, 40, 41], classifier weight prediction [9, 28, 29], or prototype-based non-parametric methods [24, 30, 33] to address *data scarcity*.

Unlike the aforementioned paradigms, we instead revisit the classic approach of ensemble of experts [17, 19, 44] and adapt it to long-tailed recognition. We first decompose the imbalanced classification problem into balanced classification problems by splitting the long-tailed training classes into balanced subsets. Then we train an expert on each balanced subset, so-called *Manyshot, Mediumshot,* or *Fewshot* data, with out-of-distribution detection for samples outside an expert's class-balanced subset. This explicitly tackles the issue of *relative imbalance*, and prevents competition between *Manyshot* and *Fewshot* classes during training.

Further, to use all available data for learning feature representations and to *transfer knowledge* from *Manyshot* to *Mediumshot* and *Fewshot* classes, we initialise the feature extractor of each expert using a *Baseline* model trained on the entire dataset. This simple and effective approach reaches close to state-of-the-art results without involving more complex models or sophisticated loss objectives. Moreover, the decomposition into class-balanced subsets allows us to analyse the upper bound on performance in each data regime. Specifically, our experiments with an *Oracle* upper bound allow us to bring *Fewshot* and *Mediumshot* accuracy on par with *Manyshot* accuracy, revealing that in modern large-scale datasets the data scarcity for *Mediumshot* and *Fewshot* classes can be effectively handled using knowledge transfer from *Manyshot* classes. Therefore, relative imbalance is a more severe problem.

We also leverage the flexibility and modularity of the ensemble framework to create larger and more diverse ensembles using existing solutions for long-tailed recognition. In particular, we involve the following methods in the solution space: (1) a *Baseline* model without any bells or whistles; (2) feature learning followed by classifier finetuning with uniform class sampling [31,41]; (3) data augmentation using feature generation networks [7,36,41]; and (4) knowledge transfer through prototype-based memory representation [24,30]. The extended ensemble consisting of all these models outperforms the current state-of-the-art on two benchmark datasets by a significant margin.

Our **contributions** in this work can be summarised as follows:

(1) We propose an effective and modular ensemble of experts framework for long-tailed recognition that decomposes the imbalanced classification problem into multiple balanced classification problems. Our framework utilises all available data for learning feature representations and transfers this knowledge from *Manyshot* to *Mediumshot* and *Fewshot* classes. The results of our ensemble of class-balanced experts are close to the state-of-the-art performance on two long-tailed benchmark datasets, ImageNet-LT and Places-LT [24].
(2) We enrich our ensemble with a diverse set of existing solutions for long-tailed recognition, namely data re-sampling, data augmentation using synthesised features, and prototype-based classification, and establish a new state-of-the-art for long-tailed recognition.
(3) We analyse the upper bound performance of our approach in the following manner: we assume Oracle access to the experts containing the ground truth classes of the test samples in their class-balanced subsets. We discover that *data scarcity* for rare classes is not a severe issue in modern large-scale datasets. Rather, *relative imbalance* is the main bottleneck.

2   Related Work

Imbalanced Classification and Long-Tailed Recognition. There is a long history of research in imbalanced classification [1,13,32], in binary and more generally multi-class classification problems. Classic problems that naturally encounter class imbalance are face attribute detection [18,26], object detection [23,48], and image defect detection [43]. Prior work on image classification [37,38] deals with long-tailed datasets, but only recently a benchmark for the problem on the ImageNet and Places dataset was proposed by [24]. They also propose splits for open-world classification, but in this work we only consider long-tailed recognition and we report the performance of our methods on the proposed ImageNet-LT and Places-LT. We summarise below the existing solutions for imbalanced classification and long-tailed recognition.

Data Re-sampling Heuristics and Cost-Sensitive Learning. These are classic ways to tackle long-tailed recognition. A more balanced data distribution is achieved by randomly over-sampling fewshot classes or randomly under-sampling of manyshot classes [6,13]. However, over-sampling suffers from over-fitting on fewshot classes while under-sampling cannot take full benefit of available data for generalization on manyshot classes. Other work has focused on hard example mining [5] or cost-sensitive learning [3,23] reasoned from class frequencies. Instead, to augment our ensemble of class-balanced experts, we use a uniform class sampling procedure in mini-batch training for finetuning the classifier after a representation learning phase, which has the advantage that all data is used to learn representations while decision boundary learning takes class imbalance into account. This has also been employed before in related zero-shot learning [41] and fewshot learning [31] work.

Synthetic Data Augmentation. This is a classic technique that synthesises features for minority classes based on feature space similarities [2,12]. More recently, generative models have been employed in zero-shot [7,40,41] and fewshot learning [36] literature to automatically generate images or feature embeddings for data-starved classes. In this work, we use the f-VAEGAN-D2 model from [41] that generates feature embeddings conditioned on available class embeddings using a VAE-GAN model, and integrate it into our ensemble of experts framework.

Prototype-Based Models and Knowledge Transfer. Prototype-based networks [30,33] maintain a memory module for all the classes such that each class is equally represented regardless of sample frequency. In particular, Liu et al. [24] learn prototype-based features on-the-fly to effectively transfer knowledge from manyshot classes to fewshot classes. We integrate their model into our ensemble due to its ability to perform consistently well across the entire class spectrum. Transfer learning [27] addresses data imbalance by transferring abundant features of manyshot classes to those of fewshot classes. Recent work includes transferring the intra-class variance [42] and transferring semantic deep features [24,46]. We instead transfer knowledge across the dataset by initialising our expert models with a baseline model pre-trained on the entire dataset.

Ensemble Learning. Ensemble methods are a well-studied topic in machine learning literature. In particular, a variety of ensemble-based methods using boosting [11,34], bagging [8,20], stacking [35], and evolutionary selection of classifiers [21] have been employed for imbalanced datasets. However, they all consider ensembles with the same kind of model and task. Our approach is related to the work of Hinton et al. [17] who train an ensemble of experts over disjoint semantically-close subsets of classes, thereby each expert deals with a different classification task. We instead train our experts on subsets of classes that are intrinsically balanced to counter relative imbalance and prevent competition between manyshot and fewshot classes during training. Moreover, we integrate a diverse set of models for long-tailed recognition into our ensemble of experts.

Out-of-Distribution Detection and Confidence Calibration. Modern neural networks can function both as classification models and detectors for out-of-distribution examples [15]. Recent works focus on adding small perturbations in input space and applying temperature scaling [22], and adding loss terms to push out-of-distribution examples towards uniform confidence [16]. Related work on confidence calibration tries to fix overconfident predictions on in-distribution data using temperature scaling [10]. We instead focus on learning an ensemble of class-balanced experts for long-tailed recognition, where the problem of out-of-distribution detection arises when dealing with samples from outside an expert's subset, and jointly calibrate experts' confidences to fuse their posteriors.

3   Method

We propose an ensemble of experts for solving the problem of long-tailed recognition. We split the long-tailed dataset into (approximately) class-balanced subsets, and a separate classification model, or expert, is trained for each subset. Expert models identify samples belonging to classes outside their subset as out-of-distribution; therefore we train them to produce low confidence predictions on these samples. During inference, each classification model yields a partial posterior distribution for test samples, the ensemble of which is fused to form a complete posterior distribution. Our entire pipeline is depicted in Fig. 1. The modularity of our framework allows us to explictly address the problem of *relative imbalance*, and moreover analyse the upper bounds for performance in each data regime using Oracle access to experts containing ground truth classes of test samples in their class-balanced subsets.

3.1   Long-Tailed Recognition Using Class-Balanced Experts

The task of long-tailed visual recognition is as follows: given class-imbalanced training set $\mathcal{D}_{Train} = \{(x_i, y_i)\}_{i=1}^{n}$ and class-balanced validation set $\mathcal{D}_{Val}$ and class-balanced test set $\mathcal{D}_{Test}$, the objective is to maximise test accuracy on four folds, *Manyshot* classes ($\geq$100 samples), *Mediumshot* classes (20–100 samples), *Fewshot* classes ($<$20 samples), and *All* classes. This is a hard problem, since any high performing model must deal with the two problems of relative imbalance and data scarcity.

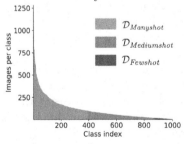

Fig. 2. Dataset splitting: We decompose ImageNet-LT into (relatively) class-balanced *Manyshot*, *Mediumshot*, and *Fewshot* data subsets.

Relative imbalance leads to biased classification boundaries wherein accuracy on fewshot samples is compromised in favor of manyshot samples that dominate the training objective. Data scarcity leads to representations that do not model unobserved data modes and is more severe. To tackle both these issues, we sort the class-imbalanced training set $\mathcal{D}_{Train}$ according to class frequencies and partition it into contiguous class-balanced subsets $\mathcal{D}_{Manyshot}$, $\mathcal{D}_{Mediumshot}$ and $\mathcal{D}_{Fewshot}$. This is visualised in Fig. 2. For each subset, we train separate classification models or experts, that are initialised using a model pre-trained on the entire dataset. Consequently we obtain the expert models $\mathcal{E}_{Manyshot}$, $\mathcal{E}_{Mediumshot}$ and $\mathcal{E}_{Fewshot}$ corresponding to each class-balanced subset. The feature extractor part of each expert model $\mathcal{E}_-$ is initialised using the *Baseline* model pre-trained on the entire training set $\mathcal{D}_{Train}$. This enables knowledge transfer from *Manyshot* to *Mediumshot* and *Fewshot* classes. In this work, the expert models $\mathcal{E}_-$ and the *Baseline* model are deep fully convolutional neural networks with softmax classifiers.

3.2 Out-of-Distribution Detection for Experts

The expert models identify samples from classes outside their class-balanced subset as out-of-distribution or OOD for short, therefore we train them using an out-of-distribution detection strategy. Observe that this is a hard problem, since here OOD examples come from within the same distribution albeit from extra classes within the dataset, as opposed to standard out-of-distribution detection wherein OOD samples come from an entirely different dataset.

Training with Reject Class. We add a reject class to the softmax classifier of each expert. For instance, $\mathcal{E}_{Manyshot}$ treats samples from $\mathcal{D}_{Mediumshot} \cup \mathcal{D}_{Fewshot}$ as a single reject class. This introduces imbalance since the reject class has far more samples than any other class, therefore we undersample reject class samples appropriately during training. We correct for the statistical bias by incrementing its logit score by the log of the undersampling ratio. We note that samples in the reject class have very high variance and are therefore hard to fit.

3.3 Fusing Expert Posteriors

We consider various baseline strategies and propose a novel joint calibration module to fuse expert posteriors $\mathcal{E}_-(x)$ into a complete posterior distribution. The final prediction and confidence scores are taken from this posterior, denoted as $q(x)$, using the argmax operation.

KL-Divergence Minimisation. We find the full posterior distribution for each sample, by minimising its KL-divergence with all the partial posterior distributions predicted by the experts [17], that is,

$$\min_{q(x)} \sum_{\mathcal{E}_-} KL(\mathcal{E}_-(x) \| q(x))$$

where $q(x)$ is parameterised using logits z and a softmax function as $q(x) = softmax(z)$. Note that probabilities corresponding to out-of-distribution classes for the expert $\mathcal{E}_-$ are summed up into one probability score in $q(x)$ to align the two distributions.

Soft-Voting. We find the full posterior by summing up the partial posteriors directly and normalising the sum to 1,

$$q(x) = \frac{\sum_{\mathcal{E}_-} g(\mathcal{E}_-(x))}{\sum_{\mathcal{E}_-} \mathbb{1}}$$

Here $g(.)$ is a function that converts an expert's partial posterior into a full posterior. Since experts are trained with a reject class, $g(.)$ averages reject class probability score across out-of-distribution classes corresponding to expert $\mathcal{E}_-$.

Expert Selection. We train a 3-way classifier on the validation set, taking the partial posterior vectors $\mathcal{E}_-(x)$ of each expert $\mathcal{E}_-$ as input, to predict for a sample x the expert model $\mathcal{E}_-$ that contain's the sample's ground truth class in its class-balanced subset. Thus, for instance, the classifier learns to predict that a manyshot sample lies in the class-balanced subset of the manyshot expert $\mathcal{E}_{Manyshot}$. The full posterior $q(x)$ is then given by $g(\mathcal{E}_-(x))$ for the predicted expert $\mathcal{E}_-$, where $g(.)$ is defined similarly as before.

Model Stacking. We train a single layer linear softmax classifier to predict the full posterior q(x) from the partial posterior vectors $\mathcal{E}_-(x)$ of each expert $\mathcal{E}_-$. The vectors $\mathcal{E}_-(x)$ are concatenated to form a feature embedding for the softmax classifier which is trained by optimising the cross entropy loss on the validation set. This is a standard way for ensemble fusion known as model stacking [39].

Joint Calibration. We calibrate the partial posteriors $\mathcal{E}_-(x)$ by learning scaling and shift parameters before adding up the posteriors similarly to soft-voting,

$$q(x) = \frac{\sum_{\mathcal{E}_-} g(\sigma_{SM}(w_{\mathcal{E}_-} \odot z_{\mathcal{E}_-}(x) + b_{\mathcal{E}_-}))}{\mathbb{Z}}$$

where σ_{SM} denotes the softmax operation, $w_{\mathcal{E}_-}$ and $b_{\mathcal{E}_-}$ are scale and shift parameters respectively, $z_{\mathcal{E}_-}(x)$ denotes the logit scores of expert $\mathcal{E}_-$ for sample x, $\odot$ denotes elementwise multiplication of two vectors, $\mathbb{Z}$ is a normalisation factor, and $g(.)$ is defined as before. We learn scale and shift parameters by minimising the cross entropy loss on the validation set. This module effectively learns the right alignment for experts' partial posteriors before performing soft-voting.

4   Experiments

Datasets. We use the object-centric ImageNet-LT and scene-centric Places-LT datasets for long-tailed recognition, released by Liu et al. [24]. The training set statistics are depicted in Table 1. ImageNet-LT has an imbalanced training set with 115,846 images for 1,000 classes from ImageNet-1K [4].

Table 1. Statistics for training sets in ImageNet-LT and Places-LT.

Datasets	Attributes	Many	Medium	Few	All
ImageNet-LT	Classes	391	473	136	1,000
	Samples	89,293	24,910	1,643	115,846
Places-LT	Classes	132	162	71	365
	Samples	52,862	8,834	804	62,500

The class frequencies follow a natural power-law distribution [32] with a maximum number of 1,280 images per class and a minimum number of 5 images per class. The validation and testing sets are balanced and contain 20 and 50 images per class respectively. Places-LT has an imbalanced training set with 62,500 images for 365 classes from Places-2 [47]. The class frequencies follow a natural power-law distribution [32] with a maximum number of 4,980 images per class and a minimum number of 5 images per class. The validation and testing sets are balanced and contain 20 and 100 images per class respectively.

Evaluation Metrics. We report average top-1 accuracy across the four folds, *Manyshot* classes (≥ 100 samples), *Mediumshot* classes (20–100 samples), *Fewshot* classes (<20 samples), and *All* classes. Since the test set is balanced across all classes, the average accuracy and mean precision coincide. These four metrics are important for fine-grained evaluation since high accuracy on *All* classes does not imply high accuracy on *Fewshot* classes or *Mediumshot* classes.

Implementation Details. For the *Baseline* model, we take a Resnet-10 backbone for ImageNet-LT, following [24]. We initialise the model with Gaussian weights, use an initial learning rate of 0.2, and train for 100 epochs with a cosine learning rate schedule [25]. For Places-LT, we start with an ImageNet pre-trained Resnet-152 model, and finetune it with 0.01 learning rate for the first 30 epochs followed by 0.1 exponential decay in every 10 epochs. To train expert models, we initialise the feature extractor of each expert $\mathcal{E}_-$ from the *Baseline* model, and finetune it on its class-balanced subset. For $\mathcal{E}_{Mediumshot}$ and $\mathcal{E}_{Fewshot}$, we freeze the lower layers of the feature extractor and only learn the top few layers. The number of learnable layers is a hyperparameter that is fixed by measuring performance on the validation set. To train experts with the reject class, we fix the undersampling ratio for samples from the reject class by measuring performance on the validation set. Note that the classifier for each expert $\mathcal{E}_-$ is smaller than the *Baseline* model; it equals the number of classes in the expert's class-balanced subset, plus an additional reject class.

4.1   Oracle Performance

To estimate the upper bound of our approach, we consider the performance with *Oracle* access to expert selection information, that is, with apriori knowledge of the expert $\mathcal{E}_-$ that contains the ground-truth class of a test sample in its class-balanced subset. The results are depicted in Table 2 and Table 3. The *Oracle* outperforms the *Baseline* by a significant margin on *Mediumshot*, *Fewshot* and *All* accuracy. Moreover, it is significantly interesting to note that the Oracle accuracies on *Mediumshot* and *Fewshot* classes are on par with *Manyshot* accuracy. This illustrates that performance drops on *Mediumshot* and *Fewshot* classes result from *relative*

Table 2. Performance of Oracle vs Baseline on ImageNet-LT.

Method	Many	Medium	Few	All
Baseline	**54.3**	26.2	5.8	34.4
Experts (Oracle)	54.2	**43.3**	**45.7**	**47.9**

Table 3. Performance of Oracle vs Baseline on Places-LT.

Method	Many	Medium	Few	All
Baseline	45.4	25.6	9.0	29.5
Experts (Oracle)	**47.3**	**46.1**	**46.5**	**46.6**

imbalance rather than *data scarcity*. Therefore, in principle, it is possible for a classification model to match *Fewshot* and *Mediumshot* accuracy with *Manyshot* accuracy in modern large-scale datasets. It is also interesting to see that the *Manyshot* accuracy does not improve much by using an *Oracle*, suggesting that *Manyshot* accuracy is already saturated in the *Baseline* model.

4.2   Effect of Joint Calibration Module

We apply the methods outlined in Sect. 3.3 for fusing expert posteriors and compare their performance on ImageNet-LT and Places-LT. The results are depicted in Table 4 and Table 5. KL-div minimisation and Soft-voting yield the highest *Fewshot* accuracy, however *All* accuracy is much lower than the other methods. Expert selection and Stacking are better than KL-div minimisation and Soft-voting on *Manyshot*, *Mediumshot* and *All* accuracy, but worse on *Fewshot* accuracy. The Joint-calibration module obtains the best *Manyshot*, *Mediumshot* and *All* accuracy, even though *Fewshot* accuracy suffers.

Table 4. Effect of joint calibration module for ImageNet-LT.

Module	Many	Medium	Few	All
KL-div min	25.3	20.5	**39.1**	21.9
Soft-voting	26.3	21.3	38.9	25.6
Expert selection	38.3	32.6	17.2	32.8
Stacking	28.1	27.5	33.8	28.6
Joint calibration	**43.2**	**34.3**	18.9	**35.7**

Table 5. Effect of joint calibration module for Places-LT.

Module	Many	Medium	Few	All
KL-div min	30.2	31.7	**28.9**	30.4
Soft-voting	30.0	31.8	28.9	30.6
Expert selection	32.6	31.8	24.5	30.7
Stacking	28.2	36.0	26.2	31.3
Joint calibration	**37.2**	**35.3**	26.3	**34.2**

4.3   Diverse Ensembles with Experts

In this section, we extend our ensemble using existing long-tailed recognition solutions and analyse the performance of various combinations of models in the ensemble. We experiment with the following models: (i) The *Baseline* model, (ii) The three expert models, $\mathcal{E}_{Manyshot}$, $\mathcal{E}_{Mediumshot}$ and $\mathcal{E}_{Fewshot}$ fused using Soft-voting, collectively referred to as *Experts*, (iii) Classifier finetuning with uniform class sampling,wherein we freeze the feature extractor of the *Baseline* model and

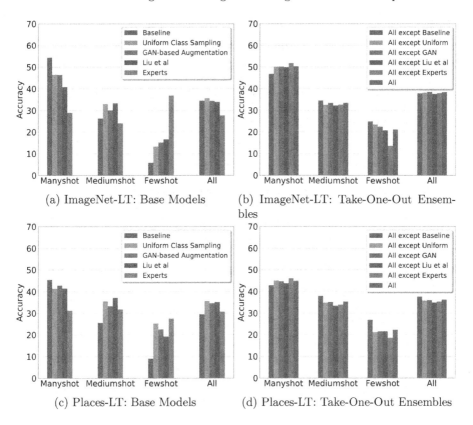

(a) ImageNet-LT: Base Models   (b) ImageNet-LT: Take-One-Out Ensembles

(c) Places-LT: Base Models   (d) Places-LT: Take-One-Out Ensembles

Fig. 3. From L-R: Performance of - Base Models, and Take-One-Out ensembles. All results are evaluated on the testing set. Top and bottom rows correspond to ImageNet-LT and Places-LT respectively. Best viewed in color with zoom.

finetune the classifier with uniform class sampling. This is referred to as *Uniform class sampling* or *Uniform*, (iv) Data augmentation for *Mediumshot* and *Fewshot* classes using a conditional generative model from class embeddings to feature embeddings, denoted as *GAN based augmentation* or simply *GAN*, (v) Knowledge transfer from *Manyshot* to *Fewshot* classes using a learned convex combination of class prototypes from [24], denoted as *Liu et al.*. The performances of these base models are depicted in Fig. 3a and Fig. 3c. Notice how the performance of the *Baseline* model degrades from *Manyshot* to *Mediumshot* to *Fewshot* accuracy. The *Expert* models give the highest accuracy on the *Fewshot* classes, but are worse on *Manyshot* accuracy.

We combine all these models into a single ensemble, take one model out and see the effect on the performance. To keep the analysis simple, we use Soft-voting for fusing posteriors from all the models, since it doesn't involve learning additional parameters. This ablation is depicted in Fig. 3b and Fig. 3d. As expected, the diverse ensembles give higher *All* accuracy than the base models. Taking

Experts out causes performance drop on *Mediumshot, Fewshot* and *All* accuracy, and increase in accuracy on *Manyshot* classes. This suggests that the *Experts* are important in the ensemble for high *Mediumshot* and *Fewshot* accuracy. On the other hand, taking the *Baseline* model out of the ensemble causes an increase in *Fewshot* accuracy while *Manyshot* accuracy drops. The ablation also reveals the inherent trade-off between *Manyshot* and *Fewshot* accuracy; an appropriate combination of models can tilt accuracy in favor of *Manyshot* or *Fewshot* classes.

4.4 Comparison to the State-of-the-Art

We now compare our ensemble of class-balanced experts and the diverse ensemble described in the previous section to the state-of-the-art on the test set of ImageNet-LT and Places-LT. All ensemble combinations use the joint calibration module to fuse model posteriors as it gives us the highest average accuracy. The results are depicted in Table 6 and Table 7. We observe that Ours (Experts) gives us close to state-of-the-art results, and Ours (All) establishes a new state-of-the-art on both the benchmark datasets. This validates our hypothesis that an ensemble of class-balanced expert models is a simple and effective strategy for dealing with long-tailed datasets.

Table 6. Results on ImageNet-LT, using backbone Resnet-10. *Results obtained from the author's code. ‡Results taken directly from [24].

Methods	Many	Medium	Few	All
Lifted Loss‡ [26]	35.8	30.4	17.9	30.8
Focal Loss‡ [23]	36.4	29.9	16	30.5
Range Loss‡ [45]	35.8	30.3	17.6	30.7
FSLwF‡ [9]	40.9	22.1	15	28.4
Liu et al.‡ [24]	43.2	35.1	18.5	35.6
Baseline	**54.3**	26.2	5.7	34.4
Uniform	46.5	33.0	13.3	35.6
GAN	46.4	30.0	15.2	34.4
Liu et al.* [24]	40.8	33.3	16.6	33.9
Ours (*Experts*)	43.2	34.3	18.9	35.7
Ours (*All*)	48.2	**37.0**	**21.5**	**39.2**

Table 7. Results on Places-LT, using backbone Resnet-152. *Results obtained from the author's code. ‡Results taken directly from [24].

Methods	Many	Medium	Fews	All
Lifted Loss‡ [26]	41.1	35.4	24.0	35.2
Focal Loss‡ [23]	41.1	34.8	22.4	34.6
Range Loss‡ [45]	41.1	35.4	23.2	35.1
FSLwF‡ [9]	43.9	29.9	**29.5**	34.9
Liu et al.‡ [24]	44.7	37.0	25.3	35.9
Baseline	**45.4**	25.6	9.0	29.5
Uniform	41.3	35.5	25.2	35.6
GAN	42.7	33.3	22.5	34.6
Liu et al.* [24]	41.4	37.1	19.2	35.2
Ours (*Experts*)	37.2	35.3	26.3	34.2
Ours (*All*)	43.6	**39.9**	27.7	**38.9**

4.5 Discussion

There is significant difference between the results depicted in Table 2 and Table 3, and Table 6 and Table 7. This shows that the various strategies used for fusing expert posteriors are sub-optimal. To analyse the underlying cause, we take our ensemble of class-balanced experts and plot a confusion matrix, each entry showing the percentage of samples from dataset $\mathcal{D}_-$ that are classified by expert model

$\mathcal{E}_-$. For the preliminary analysis we use Soft-voting for fusing expert posteriors. Figure 4a shows the result for Places-LT. The plot shows there is significant confusion amongst experts; experts aren't selected optimally for classes to which a test sample belongs. We term this phenomenon as *Expert collision*.

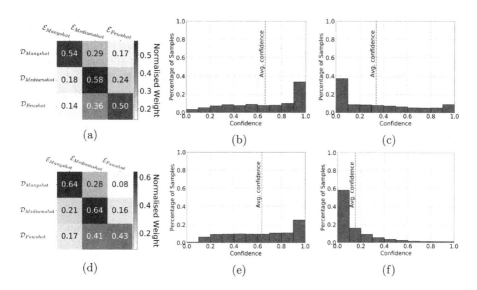

(a)   (b)   (c)

(d)   (e)   (f)

Fig. 4. Top (bottom): Before (after) joint calibration. L-R: Expert confusion matrix, confidence histograms of $\mathcal{E}_{Manyshot}$ for samples it correctly classifies in $\mathcal{E}_{Manyshot}$, and $\mathcal{E}_{Fewshot}$ for the same samples. All results on Places-LT. Joint calibration aligns experts' confidences and decreases *expert collision*.

We further consider each expert's confidence in its predictions. We take the confidence or the maximum softmax probability (MSP) from the expert posteriors and plot confidence histograms. We do this for $\mathcal{E}_{Manyshot}$ on its class-balanced subset $\mathcal{D}_{Manyshot}$, for samples from the test set it correctly classifies, and for $\mathcal{E}_{Fewshot}$ on the same test samples from $\mathcal{D}_{Manyshot}$. This is depicted in Fig. 4b and Fig. 4c. The plots show that $\mathcal{E}_{Manyshot}$ has high confidence predictions while $\mathcal{E}_{Fewshot}$ has low confidence predictions on these samples. However, to avoid *Expert collision* both the confidence histograms should have a reasonable margin in between and not overlap. Figure 4d and Fig. 4e, 4f respectively show the confusion matrix and confidence histograms after joint calibration. It's essential to align confidences of the three experts correctly, and this is precisely what *joint calibration* does by learning scale and shift parameters for each class.

5   Conclusion

This article presented an ensemble of class-balanced experts framework for long-tailed recognition. Our effective and modular strategy explicitly tackles *relative imbalance* without resorting to complex models or sophisticated loss objectives. We decompose the imbalanced classification problem into balanced classification problems that are more tractable, and train separate expert models for *Manyshot, Mediumshot* and *Fewshot* subsets of the data with a reject class for samples lying outside an expert's class-balanced subset. We scale and shift experts' partial posteriors to jointly calibrate experts' predictions, and our ensemble of class-balanced experts reaches close to state-of-the-art performance on two long-tailed benchmarks. We also extend our ensemble with diverse existing solutions for long-tailed recognition and establish a new state-of-the-art on the two benchmark datasets. Moreover, our experiments with an Oracle upper bound reveal that performance drops on *Mediumshot* accuracy and *Fewshot* accuracy are caused by *relative imbalance* and not *data scarcity* for rare classes. Therefore, it is possible to bring *Mediumshot* and *Fewshot* accuracy on par with *Manyshot* accuracy by remedying *relative imbalance* in modern large-scale datasets, which motivates further research in this direction.

References

1. Bengio, S.: The battle against the long tail. In: Workshop on Big Data and Statistical Machine Learning (2015)
2. Chawla, N.V., Bowyer, K.W., Hall, L.O., Kegelmeyer, W.P.: Smote: synthetic minority over-sampling technique. JAIR **16**, 321–357 (2002)
3. Cui, Y., Jia, M., Lin, T.Y., Song, Y., Belongie, S.: Class-balanced loss based on effective number of samples. In: CVPR (2019)
4. Deng, J., Dong, W., Socher, R., Li, L.J., Li, K., Fei-Fei, L.: ImageNet: a large-scale hierarchical image database. In: CVPR (2009)
5. Dong, Q., Gong, S., Zhu, X.: Class rectification hard mining for imbalanced deep learning. In: ICCCV (2017)
6. Estabrooks, A., Jo, T., Japkowicz, N.: A multiple resampling method for learning from imbalanced data sets. Comput. Intell. **20**, 18–36 (2004)
7. Felix, R., Vijay Kumar, B.G., Reid, I., Carneiro, G.: Multi-modal cycle-consistent generalized zero-shot learning. In: Ferrari, V., Hebert, M., Sminchisescu, C., Weiss, Y. (eds.) ECCV 2018. LNCS, vol. 11210, pp. 21–37. Springer, Cham (2018). https://doi.org/10.1007/978-3-030-01231-1_2
8. Galar, M., Fernandez, A., Barrenechea, E., Bustince, H., Herrera, F.: A review on ensembles for the class imbalance problem: bagging-, boosting-, and hybrid-based approaches. IEEE Trans. Syst. Man Cybern. Part C (Appl. Rev.) **42**, 463–484 (2011)
9. Gidaris, S., Komodakis, N.: Dynamic few-shot visual learning without forgetting. In: CVPR (2018)
10. Guo, C., Pleiss, G., Sun, Y., Weinberger, K.Q.: On calibration of modern neural networks. In: Proceedings of the 34th International Conference on Machine Learning, vol. 70, pp. 1321–1330. JMLR. org (2017)
11. Guo, H., Viktor, H.L.: Learning from imbalanced data sets with boosting and data generation: the databoost-im approach. KDD Explor. Newslett. **6**, 30–39 (2004)
12. Han, H., Wang, W.Y., Mao, B.H.: Borderline-smote: a new over-sampling method in imbalanced data sets learning. In: ICIC (2005)
13. He, H., Garcia, E.A.: Learning from imbalanced data. TKDE **21**, 1263–1284 (2009)

14. He, K., Zhang, X., Ren, S., Sun, J.: Deep residual learning for image recognition. In: CVPR (2016)
15. Hendrycks, D., Gimpel, K.: A baseline for detecting misclassified and out-of-distribution examples in neural networks. In: Proceedings of International Conference on Learning Representations (2017)
16. Hendrycks, D., Mazeika, M., Dietterich, T.: Deep anomaly detection with outlier exposure. In: Proceedings of the International Conference on Learning Representations (2019)
17. Hinton, G., Vinyals, O., Dean, J.: Distilling the knowledge in a neural network. arXiv preprint arXiv:1503.02531 (2015)
18. Huang, C., Li, Y., Chen, C.L., Tang, X.: Deep imbalanced learning for face recognition and attribute prediction. TPAMI **42**, 2781–2794 (2019)
19. Jacobs, R.A., Jordan, M.I., Nowlan, S.J., Hinton, G.E.: Adaptive mixtures of local experts. Neural Comput. **3**(1), 79–87 (1991)
20. Khoshgoftaar, T.M., Van Hulse, J., Napolitano, A.: Comparing boosting and bagging techniques with noisy and imbalanced data. IEEE Trans. Syst. Man Cybern. Part A Syst. Hum. **41**, 552–568 (2010)
21. Krawczyk, B., Woźniak, M., Schaefer, G.: Cost-sensitive decision tree ensembles for effective imbalanced classification. Appl. Soft Comput. **14**, 554–562 (2014)
22. Liang, S., Li, Y., Srikant, R.: Enhancing the reliability of out-of-distribution image detection in neural networks. In: 6th International Conference on Learning Representations, ICLR 2018 (2018)
23. Lin, T.Y., Goyal, P., Girshick, R., He, K., Dollár, P.: Focal loss for dense object detection. In: ICCV (2017)
24. Liu, Z., Miao, Z., Zhan, X., Wang, J., Gong, B., Yu, S.X.: Large-scale long-tailed recognition in an open world. In: CVPR (2019)
25. Loshchilov, I., Hutter, F.: SGDR: stochastic gradient descent with warm restarts. arXiv preprint arXiv:1608.03983 (2016)
26. Oh Song, H., Xiang, Y., Jegelka, S., Savarese, S.: Deep metric learning via lifted structured feature embedding. In: CVPR (2016)
27. Oquab, M., Bottou, L., Laptev, I., Sivic, J.: Learning and transferring mid-level image representations using convolutional neural networks. In: Proceedings of the IEEE Conference on Computer Vision and Pattern Recognition, pp. 1717–1724 (2014)
28. Qi, H., Brown, M., Lowe, D.G.: Low-shot learning with imprinted weights. In: Proceedings of the IEEE Conference on Computer Vision and Pattern Recognition, pp. 5822–5830 (2018)
29. Qiao, S., Liu, C., Shen, W., Yuille, A.L.: Few-shot image recognition by predicting parameters from activations. In: Proceedings of the IEEE Conference on Computer Vision and Pattern Recognition, pp. 7229–7238 (2018)
30. Snell, J., Swersky, K., Zemel, R.: Prototypical networks for few-shot learning. In: NeurIPS (2017)
31. Sun, Q., Liu, Y., Chua, T.S., Schiele, B.: Meta-transfer learning for few-shot learning. In: CVPR (2019)
32. Van Horn, G., Perona, P.: The devil is in the tails: fine-grained classification in the wild. arXiv preprint arXiv:1709.01450 (2017)
33. Vinyals, O., Blundell, C., Lillicrap, T., Wierstra, D., et al.: Matching networks for one shot learning. In: NIPS (2016)
34. Wang, B.X., Japkowicz, N.: Boosting support vector machines for imbalanced data sets. Knowl. Inf. Syst. **25**, 1–20 (2010)

35. Wang, S., Yao, X.: Diversity analysis on imbalanced data sets by using ensemble models. In: CIDM (2009)
36. Wang, Y.X., Girshick, R., Hebert, M., Hariharan, B.: Low-shot learning from imaginary data. In: CVPR (2018)
37. Wang, Y.-X., Hebert, M.: Learning to learn: model regression networks for easy small sample learning. In: Leibe, B., Matas, J., Sebe, N., Welling, M. (eds.) ECCV 2016. LNCS, vol. 9910, pp. 616–634. Springer, Cham (2016). https://doi.org/10.1007/978-3-319-46466-4_37
38. Wang, Y.X., Ramanan, D., Hebert, M.: Learning to model the tail. In: NeurIPS (2017)
39. Wolpert, D.H.: Stacked generalization. Neural Netw. **5**, 241–259 (1992)
40. Xian, Y., Lorenz, T., Schiele, B., Akata, Z.: Feature generating networks for zero-shot learning. In: CVPR (2018)
41. Xian, Y., Sharma, S., Schiele, B., Akata, Z.: f-vaegan-d2: a feature generating framework for any-shot learning. In: CVPR (2019)
42. Yin, X., Yu, X., Sohn, K., Liu, X., Chandraker, M.: Feature transfer learning for face recognition with under-represented data. In: CVPR (2019)
43. Yu, N., Shen, X., Lin, Z., Mech, R., Barnes, C.: Learning to detect multiple photographic defects. In: WACV (2018)
44. Yuksel, S.E., Wilson, J.N., Gader, P.D.: Twenty years of mixture of experts. IEEE Trans. Neural Netw. Learn. Syst. **23**(8), 1177–1193 (2012)
45. Zhang, X., Fang, Z., Wen, Y., Li, Z., Qiao, Y.: Range loss for deep face recognition with long-tailed training data. In: ICCV (2017)
46. Zhong, Y., et al.: Unequal-training for deep face recognition with long-tailed noisy data. In: CVPR (2019)
47. Zhou, B., Lapedriza, A., Khosla, A., Oliva, A., Torralba, A.: Places: a 10 million image database for scene recognition. TPAMI **40**, 1452–1464 (2017)
48. Zhu, X., Anguelov, D., Ramanan, D.: Capturing long-tail distributions of object subcategories. In: CVPR (2014)

Analyzing the Dependency of ConvNets on Spatial Information

Yue Fan$^{(\boxtimes)}$, Yongqin Xian$^{(\boxtimes)}$, Max Maria Losch$^{(\boxtimes)}$, and Bernt Schiele$^{(\boxtimes)}$

Max Planck Institute for Informatics, Saarbrücken, Germany
{yfan,yxian,mlosch,schiele}@mpi-inf.mpg.de

Abstract. Intuitively, image classification should profit from using spatial information. Recent work, however, suggests that this might be overrated in standard CNNs. In this paper, we are pushing the envelope and aim to investigate the reliance on spatial information further. We propose to discard spatial information via shuffling locations or average pooling during both training and testing phases to investigate the impact on individual layers. Interestingly, we observe that spatial information can be deleted from later layers with small accuracy drops, which indicates spatial information at later layers is not necessary for good test accuracy. For example, the test accuracy of VGG-16 only drops by 0.03% and 2.66% with spatial information completely removed from the last 30% and 53% layers on CIFAR-100, respectively. Evaluation on several object recognition datasets with a wide range of CNN architectures shows an overall consistent pattern.

1 Introduction

Despite the impressive performances of convolutional neural networks (CNNs) on computer vision tasks [9,10,16,18,25], their inner workings remain mostly obfuscated to us, especially how the information is encoded throughout layers. Generally, the majority of modern CNNs for image classification utilize a collection of filters with local receptive fields to capture hierarchical patterns across all the convolutional layers [10,16,25]. Such design choices are based on the assumption that spatial information remains important at every convolutional layer, and better representations can be attained by gradually enlarging the receptive field to incorporate more contexts. This further leads to lots of approaches that help capture spatial correlations between features in order to improve model performance [1,13,26]. For example, a popular class of those methods is the visual attention mechanism [15,19] which enables more powerful representations by enhancing the most salient region of the image.

However, recent works on restricting the receptive field of CNN architectures for scrambled inputs [2] or using wavelet feature networks of shallow depth [20],

Electronic supplementary material The online version of this chapter (https://doi.org/10.1007/978-3-030-71278-5_8) contains supplementary material, which is available to authorized users.

Z. Akata et al. (Eds.): DAGM GCPR 2020, LNCS 12544, pp. 101–115, 2021.
https://doi.org/10.1007/978-3-030-71278-5_8

have all found it to be possible to acquire competitive performances on the respective tasks. This raises doubts on the necessity of spatial information for classification and whether the network can still maintain the performance when the spatial information is completely removed from the training process.

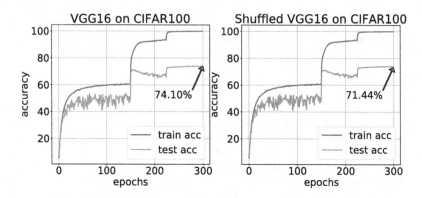

Fig. 1. Shuffling the feature maps from the last 54% layers in VGG-16 randomly and spatially only reduces the final test accuracy by 2.66% (from 74.10% to 71.44%) on CIFAR-100, and the training processes look surprisingly similar, which implies that spatial information may not be necessary for good classification accuracy.

In this work, we re-design the structure of the network to separate the spatial information and channel-wise information independently, with the goal of analyzing the dependency of the network on them. Spatial information refers to the spatial ordering on the feature map. To this end, we propose *channel-wise shuffle* to eliminate channel information, and *spatial shuffle, patch-wise spatial shuffle* and *GAP+FC* to eliminate spatial information. Surprisingly, we find that the spatial information is not necessary at later layers, and the modified CNNs, i.e. without accessing any spatial information at later layers, can still achieve competitive results on several object recognition datasets. As an example, Fig. 1 shows the training processes of a standard VGG-16 and a modified VGG-16 with spatial shuffle on CIFAR-100. In the shuffled VGG-16, feature maps must first go through a random spatial shuffle operation before convolved with the filters from the last 54% layers. Interestingly, the test accuracy only drops 2.66%, and the training process is nearly identical to the standard VGG-16. This observation generalizes to various CNN architectures: removing spatial information from the last 30% layers gives a surprisingly little test accuracy decrease within 1% across architectures and datasets, and the accuracy decrease is still within 7% even if the last 50% layers are manipulated. This indicates that spatial information is overrated for standard CNNs and not necessary to reach competitive performances. Finally, our investigation on the detection task shows that although the unavailability of spatial information at later layers does hinder the CNN to localize objects, the impact is not as fatal as expected; at the same time, the classification ability of the model is not affected.

The main contributions of our work are as follows: we find that spatial information at later layers is not really necessary for good classification test accuracy and that even though the depth of the network plays an important role, later layers do not require spatial integration. As a side effect, GAP+FC leads to a smaller model with fewer parameters with small test accuracy drops.

2   Related Work

Intuitively, object recognition benefits from gradually enlarged receptive field and spatial integration. For that reason extensive efforts have been made to enhance the aggregation of spatial information in the decision-making progress of CNNs. [5,32] have made attempts to generalize the strict spatial sampling of convolutional kernels to allow for globally spread out sampling, and [31] have spurred a range of follow-up work on embedding global context layers with the help of spatial down-sampling. Another emerging interest of augmenting CNNs with self-attention has also made progress in several vision tasks. [27] presents a non-local operation that computes the response at a position as a weighted sum of the features at all positions to capture long-range dependencies and shows that self-attention is an instantiation of their non-local operations. [3] show improvements on image classification and achieve state-of-the-art results on video action recognition tasks with a variant of non-local operations. Even a fully attentional model is verified to be effective for various visual tasks [21].

While all of these works have improved on a related classification metric in some way, it is not entirely evident whether the architectural changes alone can be credited, as there is an increasing number of work on questioning the importance of the extent of spatial information for common CNNs. One of the most recent observations by [2] indicates that the VGG-16 architecture trained on ImageNet is invariant to scrambled images to a large extent. Furthermore, they construct a modified ResNet architecture with a limited receptive field as small as 33×33, similar to the style of the traditional Bag-of-Visual-Words and reach competitive results on ImageNet. In contrast to their work, we make a clear distinction between first and last layers, and we show empirically spatial information at last layers are not necessary for good test accuracy.

[23] assumes that current CNNs do not respect the spatial information due to the pooling operation; CNNs look for features in the image without paying attention to their pose during prediction. This limitation motivates the work of [23] where they make use of dynamic routing among capsules to encode the spatial information. Moreover, the widely used global average pooling in most recently proposed architectures [10,17] implies that collapsing spatial information at the very end does not affect the test accuracy. On a related note, [8] indicates that models trained solely on ImageNet do not learn shape sensitive representations with constructing object-texture mismatched images, which would be expected to require global spatial information. Instead, the models are mostly sensitive to local texture features.

Our work aims to push the envelope further to investigate the necessity of spatial information in the processing pipeline of CNNs. While related work has put the attention mainly on altering the input and does not differentiate between last and first layers, we are interested in taking measures that remove the spatial information at different intermediate layers to shed light on how CNNs process spatial information, evaluating its importance and providing insights for architectural design choices.

3 Methods and Experimental Setup

In this section, we design methods to systematically study the phenomenon found in Fig. 1 that spatial information appears to be neglectable to some extent. We test how information is represented throughout the network's layers by discarding spatial or channel information in different ways in intermediate layers and applying them to well-established architectures. Experiments are conducted on object recognition and detection tasks. Section 3.1 elaborates details on our approaches, and the experimental setup is discussed in Sect. 3.2.

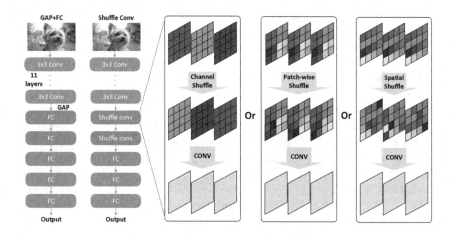

Fig. 2. An example of VGG-16 modified by our methods. The leftmost architecture shows the modification (in red) from *GAP+FC*, where the last two convolutional layers are replaced by fully-connected layers after a GAP layer. The middle architecture shows the modification (in red) from shuffle conv, where the last two convolutional layers are replaced by one of the shuffling methods and an ordinary convolution. *Spatial shuffle* randomly and independently permutes pixels on each feature map at a global scale in the sense that a pixel can end up anywhere on the feature map. *Patch-wise shuffle* first divides the feature map into grids; then it randomly permutes the pixel locations within each grid independently. *Channel shuffle* randomly permutes the order of feature maps, leaving the spatial ordering unchanged. (Color figure online)

3.1   Approaches to Constrain Information

We propose four different methods, namely *channel-wise shuffle*, *spatial shuffle*, *patch-wise spatial shuffle*, and *GAP+FC*, to remove either spatial or channel information from the training. Spatial information here refers to the awareness of the relative spatial position between activations on the same feature map, and channel information stands for the dependency across feature maps. The left part of Fig. 2 illustrates an example of VGG-16 with its last two layers modified by GAP+FC or any of the three shuffle methods.

Spatial Shuffle extends the ordinary convolution operation by prepending a random spatial shuffle operation to permute the input to the convolution. As illustrated in Fig. 2: Given an input tensor of size $c \times h \times w$ with c being the number of feature maps for a convolutional layer, we first take one feature map from the input tensor and flatten it into a 1-d vector with $h \times w$ elements, whose ordering is then permuted randomly. The resulting vector is finally reshaped back into $h \times w$ and substitute the original feature map. This procedure is independently repeated c times for each feature map so that activations from the same location in the previous layer are misaligned, thereby preventing the information from being encoded by the spatial arrangement of the activations. The shuffled output becomes the input of an ordinary convolutional layer in the end. Even though shuffling itself is not differentiable, gradients can still be propagated through in the same way as pooling operations. Therefore it can be embedded into the model directly for end-to-end training. As the indices are recomputed within each forward pass, the shuffled output is also independent across training and testing steps.

Images in the same batch are shuffled in the same way for the sake of simplicity since we find empirically that it does not make a difference whether the images in the same batch are shuffled in different ways.

Patch-Wise Spatial Shuffle is a variant of *spatial shuffle*. In contrast, patch-wise spatial shuffle does not perform on a global scale but a local scale by dividing the feature map into grids. Each patch in the grid is subsequently shuffled independently. Afterwards, an ordinary convolution is performed as usual. Note that the two operations are equivalent when the patch size is the same as the feature map size. Figure 2 demonstrates an example of patch-wise spatial shuffle with a 2×2 patch size, where the random permutation of pixel locations is restricted within each patch.

Channel-Wise Shuffle is used to investigate the importance of channel information which is normally deemed as essential [28–30]. It keeps the spatial ordering of activations and randomly permutes the ordering of feature maps to prevent the model from utilizing channel information. An illustration can be seen in Fig. 2, channel-wise shuffle is also performed independently across training and testing steps.

GAP+FC denotes Global Average Pooling and Fully Connected Layers. *Spatial Shuffle* is an intuitive way of destroying spatial information. However, shuffling introduces undesirable randomness into the model; non-deterministic feature maps from an image lead to fluctuations in the model prediction, so an evaluation needs multiple forward passes to acquire an estimate of the mean of the output. A simple deterministic alternative achieving a similar goal is to deploy Global Average Pooling (GAP) after an intermediate layer, and all the subsequent ones are substituted by fully connected layers. Compared to *Spatial Shuffle* that introduces an extra computational burden at each forward pass, it is a much more efficient way to avoid learning spatial information at intermediate layers because it shrinks the spatial size of all subsequent feature maps to one; therefore, the number of FLOPs and parameters are also reduced.

3.2   Experimental Setup

This section details the experimental setup for the classification and object detection tasks. We test different architectures on three datasets: CIFAR-100, Small-ImageNet-32x32 [4], and Pascal VOC 2007 + 2012. Small-ImageNet-32x32 is a down-sampled version of the original ImageNet (from 256×256 to 32×32). We report top-1 accuracy and mAP [6,7] in classification and detection experiments respectively. We will take an existing architecture and apply the modification to different layers. The rest of the setup and hyper-parameters for modified architectures remain the same as the original architectures.

Classification: For the VGG architecture, the modification is only performed on the convolutional layers, as illustrated in Fig. 2. For the ResNet architecture, one bottleneck sub-module is considered as a single piece, and the modification is applied onto the 3×3 convolutions within the sub-module since they are the only operations with spatial extent. Features that go through the skip connection branch are also shuffled in the shuffle experiments to prevent the model from learning to ignore the information from the residual branch. The rest of the configuration remains the same (see supplemental material for an example of modified ResNet-50 architecture).

For CIFAR-100 and Small-ImageNet-32x32 experiments, the original ResNet architecture down-samples the input image by a factor of 32 and gives 1×1 feature maps at last layers, therefore shuffling is noneffective. To make shuffling non-trivial, we set the first convolution in ResNet to 3×3 with stride 1 and the first max-pooling layer is removed so that the final feature map size is 4×4.

To alleviate the effect of mismatched training details, we first reproduce the reported results for all experiments and then train our modified architectures under the same training setting. All models in the same set of experiments (e.g. VGG-16 on CIFAR-100) use the same set of hyper-parameters, and they share the same initialization from the same random seed. During testing, we make sure to use a different random seed than during training.

Detection: We use the training set and validation set of VOC 2012+2007 as the training data and report mAP on VOC 2007 test set. We shuffle the last layer in the backbone model to test the robustness of localization against the absence of spatial information.

4 Results

We first compare the test accuracy of VGG-16 on CIFAR-100 with spatial or channel information missing from a different number of last layers in Sect. 4.1. An in-depth study of our main observations on CIFAR-100 and Small-ImageNet-32x32 for VGG-16 and ResNet-50 is conducted in Sect. 4.2. In Sect. 4.3, we investigate the model robustness against the loss of spatial information in various degree by controlling the amount of spatial information that passes through the network. Finally, we present the detection results on VOC datasets in Sect. 4.4.

4.1 Spatial and Channel-Wise Shuffle on VGG-16

In this section, we first investigate the invariance of pre-trained models to the absence of the spatial or channel information at test time, then we impose this invariance at training time with methods in Sect. 3.1.

Shuffle the Last 30% Layers Channel-Wise: Our baseline is a VGG-16 trained on CIFAR-100 that achieves 74.10% test accuracy. We first test its robustness against the absence of the channel information at test time by substituting the last 30% convolutional layers with the channel-wise shuffle convolution. As is expected, the test accuracy drops to 1.04% (Table 1), which is the same as the random guessing on CIFAR-100. Following the same training scheme of the baseline, we then train another VGG-16 with channel-wise shuffle added to its last 30% convolutional layers. This model can reach around 67% test accuracy no matter whether channel-wise shuffle is applied at test time. However, it still performs significantly worse than the baseline, which indicates that the expressiveness of the model is much limited without utilizing the ordering of feature maps even though the spatial information is preserved.

Table 1. Top-1 accuracy of VGG-16 on CIFAR-100 with spatial/channel-wise shuffle enabled at either training or test time for the last 30% layers. A model from standard training does not possess robustness against spatial shuffle (23.49%) and channel-wise shuffle (1.04%). However, when imposed in training, the model achieves 74.07% test accuracy for spatial shuffle and 67.56% for channel-wise shuffle, showing impressive robustness to the loss of spatial information.

Train scheme	No shuffle	Channel shuffle	Channel shuffle	No shuffle	Spatial shuffle	Spatial shuffle	No shuffle
Test scheme	No shuffle	Channel shuffle	No shuffle	Channel shuffle	Spatial shuffle	No shuffle	Spatial shuffle
Top-1(%)	74.10	67.56	67.80	1.04	74.07	73.74	23.49

Shuffle the Last 30% Layers Spatially: As a comparison to channel shuffle, we repeat the same experiment on spatial shuffle, and the result is presented in the second half of Table 1. No shuffle → spatial shuffle of the pre-trained VGG-16 gives 23.49% test accuracy, which is similar to the test accuracy of a one-hidden-layer perceptron (with 512 hidden units and ReLU activation) on CIFAR-100 (25.61%) when evaluated with the random spatial shuffle. However, if the spatial shuffle is infused into the model at training time, then the baseline test accuracy can be retained no matter whether random spatial shuffle appears at test time (74.07% for spatial shuffle → spatial shuffle and 73.74% for spatial shuffle → no shuffle).

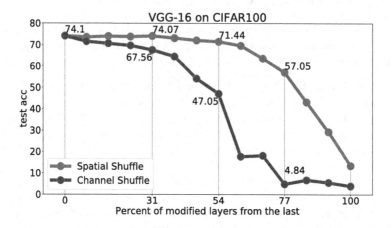

Fig. 3. Classification accuracy of VGG-16 on CIFAR-100 with different shuffle schemes. The very slow decrease of the test accuracy of spatial shuffle implies a far less important role of spatial information for classification. The test accuracy is not much affected, given that the spatial shuffle modifies 31% of its layers. Even with 54% later layers shuffled spatially, the test accuracy only decreases by 2.66%, and the same number of the test accuracy decrease in channel-wise shuffle happens when the last layer is modified.

Shuffle Other Layers: To systematically study the impact of spatial and channel information, we gradually increase the number of modified layers from the last in VGG-16 and report the corresponding test accuracy in Fig. 3. All models are trained with the same setup, and shuffling is performed both at training and test time; the x-axis is the percentage of modified layers counting from the last layer on with 0 referring the baseline.

Besides an overall decreasing trend for both shuffling with the increase of the percent of modified layers, the test accuracy of spatial shuffle drops unexpectedly slowly, e.g. merely 2.66% test accuracy drop when up to 54% of layers from the last are shuffled spatially. Likewise, when spatial information is removed from the last 77% layers, it still has a reasonable test accuracy (57.05%), whereas the test accuracy of channel-wise shuffle is only 4.84%.

Discussion: This indicates that although a standard model makes use of both spatial dimension and channel dimension to encode information, the spatial information plays a surprisingly less pivotal role than the channel information. The model is even able to adapt to the complete absence of spatial information at later layers if spatial information is removed explicitly at training time, which strengthens the claims from [2,23] that CNNs intrinsically possess invariance to the spatial relationship among features to some extent. Moreover, the unsuccessful adaptation to channel-wise shuffle implies that the large model capacity may mainly come from the channel order and shuffling the channel order causes unrecoverable damage to the model.

4.2 Spatial Information at Later Layers is Not Necessary

In this section, we design more experiments to study the reliance of different layers on spatial information: we modify the last convolutional or bottleneck layers of VGG-16 or ResNet-50 by *Spatial Shuffle* (both at training and test time) and *GAP+FC* such that the spatial information is removed in different ways. Our modification on the baseline model always starts from the last layer and is consecutively extended to the first layer. The modified networks are then trained on the training set with the same setup and evaluated on the hold-out validation set.

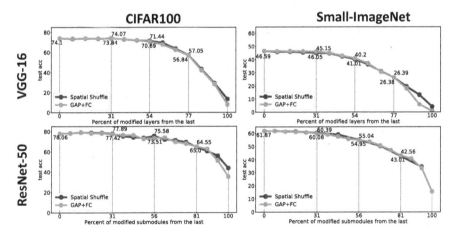

Fig. 4. Classification results of GAP+FC and spatial shuffle for VGG-16 and ResNet-50 on CIFAR-100 and Small-ImageNet-32x32. The x-axis is the percent of modified layers/sub-modules counting from the last one. Models on the same dataset are trained with the same setup. It can be observed consistently across experiments that the baseline test accuracy is preserved for a long time even though spatial information is eliminated from the last several layers by spatial shuffle or GAP+FC, suggesting that spatial information at later layers is not necessary for good test accuracy. The difference between the baseline models and the models whose latter half of the layers are modified by GAP+FC or spatial shuffle is, however, still in a reasonable range between 2.48% (ResNet-50 with spatial shuffle on CIFAR-100) to 6.92% (ResNet-50 with GAP+FC on Small-ImageNet-32x32).

Results on CIFAR-100 and Small-ImageNet-32x32: Results of VGG-16 and ResNet-50 on CIFAR-100 and Small-ImageNet-32x32 are shown in Fig. 4. The x-axis is the percent of modified later layers, and 0 is the baseline model test accuracy without modifying any layer.

As we can see, *Spatial Shuffle* and *GAP+FC* have a similar overall behaviour consistently across architectures and datasets: the baseline test accuracy is retained for a long time before it starts to decrease with the increase of the percent of modified layers. When the last 30% layers are modified by GAP+FC or spatial shuffle, there is no or little test accuracy decrease across experiments (0.17% for ResNet-50 on CIFAR-100 and 1.44% for VGG-16 on Small-ImageNet with spatial shuffle). And the test accuracy decrease is still in a reasonable range (2.48% with spatial shuffle on CIFAR-100 and 6.92% for GAP+FC on Small-ImageNet-32x32 for ResNet-50), even with around half of the last layers modified. At 77% to 81% of the modified later layers, the test accuracy just starts to show a significant difference to the baseline in the range of 8.58% (ResNet-50 with spatial shuffle on CIFAR-100) to 20.21% (VGG-16 with GAP+FC on Smalll-ImageNet-32x32).

Our experiments here clearly show that spatial information can be neglected from a significant number of later layers with no or small test accuracy drop if the invariance is imposed at training, which suggests that *spatial information at last layers is not necessary for good test accuracy*. We should, however, notice that it does not indicate that models whose prediction is based on spatial information can not generalize well. Besides, unlike the common design manner that layers at different depth inside the network are normally treated equally, e.g. the same module is always used throughout the architecture [12,14,24], our observation implies it is beneficial to have different designs for different layers since there is no necessity to encode spatial information in the later layers. As a side effect, GAP+FC can reduce the number of model parameters with little test accuracy drop. For example, GAP+FC achieves nearly identical results (46.05%) to the VGG-16 baseline (46.59%), while reducing the number of parameters from 37.70M to 29.31M on Small-ImageNet-32x32.

4.3   Patch-Wise Spatial Shuffle

In this section, we study the relation between the model test accuracy and the amount of spatial information that propagates throughout a network. The latter is controlled by patch-wise spatial shuffle with different patch sizes. The larger the patch size is, the less the preserved spatial information. Patch-wise spatial shuffle reduces to spatial shuffle when the patch size is the same as the feature map size, in which case no spatial information remains. Our experiments are conducted on CIFAR-100 for VGG-16 and ResNet-50, and we only shuffle a single layer at a time since the model is not able to recover the "damage" caused by shuffling an early layer (see more in the supplemental material).

The result of patch-wise spatial shuffling of different patch sizes is shown in Fig. 5. We can see that the patch size does not make much difference in terms of the test accuracy at later layers, e.g. results of patch size 2, 4 and 8 for ResNet-50 at 8–14 layers are similar. However, the test accuracy has a rapid decrease with the increase of the patch size at first layers, indicating a relatively important role of spatial information at first layers. Nevertheless, this role might not be as much important as what is commonly believed, as the ResNet-50 still has 40.76% test accuracy when the input image is completely shuffled.

4.4   Detection Results on VOC Datasets

Object detection should intuitively suffer more from spatial shuffling than classification since the spatial information should help to localize objects. In this section, we show some initial results on Pascal VOC [6, 7].

We design an analogue to YOLO [22] as our detection model. The architecture consists of a backbone and a detection head; the backbone is a ResNet-50 without the classifier, and the detection head has three bottlenecks and a 3×3 convolutional layer whose outputs is in the same format as [22]. Different to [22], we deploy a 3×3 convolution instead of a fully connected layer in the end to output the final detection results. The latter gives the model potential access to the object feature, which may be exploited by the model to predict its location. In order to prevent the undesirable shortcut, we use a 3×3 convolution so that the prediction of a bounding box at a certain location does not depend on all activation on the feature map.

By using a pre-trained ResNet-50 on ImageNet, we can reach 66% mAP on VOC2007 test set after fine-tuning, which is the same as the number in [22]. To avoid pretraining a spatially shuffled model on ImageNet, we compare a spatially shuffled model and a non spatially shuffled model, both trained from

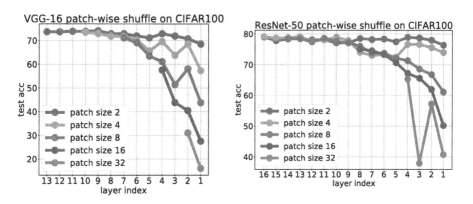

Fig. 5. The result of patch-wise spatial shuffling of VGG-16 and ResNet-50 on CIFAR-100. Only a single layer is shuffled at a time. Layer index 13 and 16 stand for the last layer of VGG-16 and ResNet-50, respectively. With the increase of the patch size, the test accuracy decreases faster at first layers than that at last layers. It is interesting to see that both models' test accuracy do not fall into the random guess (16.02% for VGG-16 and 40.76% for ResNet-50) at layer index one and patch size 32, where the input image is completely shuffled.

Fig. 6. Left: Qualitative detection results on the VOC 2007 test set. Examples are the first 11 images in the test set. The left result is from the baseline, and the right result is from the shuffled model. Right: Detection error analysis of our baseline and the shuffled model shows a doubled localization error in the shuffled model and the rest types of error are in the same level as the baseline.

scratch on VOC. Our models are trained for 500 epochs with exponentially decaying learning rate starting from 0.001. Our baseline model achieves 50% mAP on VOC2007 test set without using an ImageNet pre-trained backbone. The result of the shuffled model, where we apply random shuffle to the last layer of the backbone, is 34%. While this sounds like a large drop, it turns out that the classification performance is essentially preserved and only the localization performance is suffering. To analyze this effect in detail, we use the method and tools proposed in [11]. The diagnosis tool classifies each prediction from the model as either correct prediction or a type of error based on its class label and IoU with the ground truth. More details can be found in [11].

The results in Fig. 6 right show that the misclassification to the wrong class and background are of similar percents for both models, and the localization error doubles for the shuffled model (an increase from 14.2% to 28.4%). Though random shuffling indeed affects the model's localization ability, it is unexpected that the effect is not fatal. Because random shuffling switches features, it is highly likely the model trained with spatial shuffle has to predict the correct bounding box for one object based on some other features. We should also notice that a prediction is counted as a localization error if it has the correct class label and the IoU to the ground truth is less than 0.5. Therefore, classification-wise speaking, the shuffled model got 73.7% (45.3% + 28.4%) of its predictions correct, which is at the same level as the baseline (73.3% = 59.1% + 14.2%).

Qualitative Results: Figure 6 left shows some qualitative results from both models. Those examples are the first 11 images in the VOC2007 test set. We can see that the localization error actually mainly comes from small objects for which the shuffled model tends to predict several bounding boxes on one object, and the bounding box of the relatively big object is not really off, e.g. the shuffled model managed to localize the dining table in the middle right image and the horse in the middle left image while the baseline can not.

5   Conclusion

To conclude, we empirically show that a significant number of later layers of CNNs are robust to the absence of spatial information, which is commonly assumed to be important for object recognition tasks. Modern CNNs can tolerate the loss of spatial information from the last 30% of layers at around 1% accuracy drop; and the test accuracy only decreases by less than 7%, when spatial information is removed from the last half of layers on CIFAR-100 and Small-ImageNet-32x32. Though the depth of the network is essential for good test accuracy, later layers do not require spatial integration.

References

1. Bell, S., Lawrence Zitnick, C., Bala, K., Girshick, R.: Inside-outside net: detecting objects in context with skip pooling and recurrent neural networks. In: Proceedings of the IEEE Conference on Computer Vision and Pattern Recognition, pp. 2874–2883 (2016)
2. Brendel, W., Bethge, M.: Approximating CNNs with bag-of-local-features models works surprisingly well on imageNet. In: International Conference on Learning Representations (2019). https://openreview.net/forum?id=SkfMWhAqYQ
3. Chen, Y., Kalantidis, Y., Li, J., Yan, S., Feng, J.: A 2-nets: double attention networks. In: Advances in Neural Information Processing Systems, pp. 352–361 (2018)
4. Chrabaszcz, P., Loshchilov, I., Hutter, F.: A downsampled variant of imagenet as an alternative to the cifar datasets. arXiv preprint arXiv:1707.08819 (2017)
5. Dai, J., Qi, H., Xiong, Y., Li, Y., Zhang, G., Hu, H., Wei, Y.: Deformable convolutional networks. In: Proceedings of the IEEE International Conference on Computer Vision, pp. 764–773 (2017)
6. Everingham, M., Van Gool, L., Williams, C.K.I., Winn, J., Zisserman, A.: The PASCAL Visual Object Classes Challenge 2007 (VOC 2007) Results. http://www.pascal-network.org/challenges/VOC/voc2007/workshop/index.html
7. Everingham, M., Van Gool, L., Williams, C.K.I., Winn, J., Zisserman, A.: The PASCAL Visual Object Classes Challenge 2012 (VOC 2012) Results. http://www.pascal-network.org/challenges/VOC/voc2012/workshop/index.html
8. Geirhos, R., Rubisch, P., Michaelis, C., Bethge, M., Wichmann, F.A., Brendel, W.: Imagenet-trained CNNs are biased towards texture; increasing shape bias improves accuracy and robustness. In: International Conference on Learning Representations (2019). https://openreview.net/forum?id=Bygh9j09KX
9. Girshick, R., Donahue, J., Darrell, T., Malik, J.: Rich feature hierarchies for accurate object detection and semantic segmentation. In: Proceedings of the IEEE Conference on Computer Vision and Pattern Recognition, pp. 580–587 (2014)
10. He, K., Zhang, X., Ren, S., Sun, J.: Deep residual learning for image recognition. In: CVPR (2016)
11. Hoiem, D., Chodpathumwan, Y., Dai, Q.: Diagnosing error in object detectors. In: Fitzgibbon, A., Lazebnik, S., Perona, P., Sato, Y., Schmid, C. (eds.) ECCV 2012. LNCS, vol. 7574, pp. 340–353. Springer, Heidelberg (2012). https://doi.org/10.1007/978-3-642-33712-3_25
12. Howard, A.G., et al.: MobileNets: efficient convolutional neural networks for mobile vision applications. arXiv e-prints arXiv:1704.04861 (Apr 2017)

13. Hu, J., Shen, L., Sun, G.: Squeeze-and-excitation networks. In: Proceedings of the IEEE Conference on Computer Vision and Pattern Recognition, pp. 7132–7141 (2018)
14. Iandola, F.N., Moskewicz, M.W., Ashraf, K., Han, S., Dally, W.J., Keutzer, K.: SqueezeNet: AlexNet-level accuracy with 50x fewer parameters and <1mb model size. ArXiv abs/1602.07360 (2017)
15. Jaderberg, M., Simonyan, K., Zisserman, A., et al.: Spatial transformer networks. In: Advances in Neural Information Processing Systems, pp. 2017–2025 (2015)
16. Krizhevsky, A., Sutskever, I., Hinton, G.E.: Imagenet classification with deep convolutional neural networks. In: Proceedings of the 25th International Conference on Neural Information Processing Systems - Volume 1, pp. 1097–1105. NIPS 2012, Curran Associates Inc., USA (2012). http://dl.acm.org/citation.cfm?id=2999134.2999257
17. Lin, M., Chen, Q., Yan, S.: Network in network. CoRR abs/1312.4400 (2013)
18. Long, J., Shelhamer, E., Darrell, T.: Fully convolutional networks for semantic segmentation. In: Proceedings of the IEEE Conference on Computer Vision and Pattern Recognition, pp. 3431–3440 (2015)
19. Mnih, V., Heess, N., Graves, A., et al.: Recurrent models of visual attention. In: Advances in Neural Information Processing Systems, pp. 2204–2212 (2014)
20. Oyallon, E., Belilovsky, E., Zagoruyko, S.: Scaling the scattering transform: deep hybrid networks. In: Proceedings of the IEEE International Conference on Computer Vision, pp. 5618–5627 (2017)
21. Ramachandran, P., Parmar, N., Vaswani, A., Bello, I., Levskaya, A., Shlens, J.: Stand-alone self-attention in vision models. arXiv preprint arXiv:1906.05909 (2019)
22. Redmon, J., Divvala, S., Girshick, R., Farhadi, A.: You only look once: unified, real-time object detection. In: Proceedings of the IEEE Conference on Computer Vision and Pattern Recognition, pp. 779–788 (2016)
23. Sabour, S., Frosst, N., Hinton, G.E.: Dynamic routing between capsules. In: Guyon, I., Luxburg, U.V., Bengio, S., Wallach, H., Fergus, R., Vishwanathan, S., Garnett, R. (eds.) Advances in Neural Information Processing Systems, vol. 30, pp. 3856–3866. Curran Associates, Inc. (2017)
24. Sandler, M., Howard, A., Zhu, M., Zhmoginov, A., Chen, L.C.: Mobilenetv2: inverted residuals and linear bottlenecks. In: Proceedings of the IEEE Conference on Computer Vision and Pattern Recognition, pp. 4510–4520 (2018)
25. Simonyan, K., Zisserman, A.: Very deep convolutional networks for large-scale image recognition. arXiv preprint arXiv:1409.1556 (2014)
26. Szegedy, C., et al.: Going deeper with convolutions. In: Proceedings of the IEEE Conference on Computer Vision and Pattern Recognition, pp. 1–9 (2015)
27. Wang, X., Girshick, R., Gupta, A., He, K.: Non-local neural networks. In: Proceedings of the IEEE Conference on Computer Vision and Pattern Recognition, pp. 7794–7803 (2018)
28. Xie, S., Girshick, R., Dollár, P., Tu, Z., He, K.: Aggregated residual transformations for deep neural networks. In: Proceedings of the IEEE Conference on Computer Vision and Pattern Recognition, pp. 1492–1500 (2017)
29. Zhang, T., Qi, G.J., Xiao, B., Wang, J.: Interleaved group convolutions. In: Proceedings of the IEEE International Conference on Computer Vision, pp. 4373–4382 (2017)
30. Zhang, X., Zhou, X., Lin, M., Sun, J.: ShuffleNet: an extremely efficient convolutional neural network for mobile devices. In: Proceedings of the IEEE Conference on Computer Vision and Pattern Recognition, pp. 6848–6856 (2018)

31. Zhao, H., Shi, J., Qi, X., Wang, X., Jia, J.: Pyramid scene parsing network. In: Proceedings of the IEEE Conference on Computer Vision and Pattern Recognition, pp. 2881–2890 (2017)
32. Zhu, X., Hu, H., Lin, S., Dai, J.: Deformable convnets v2: more deformable, better results. In: Proceedings of the IEEE Conference on Computer Vision and Pattern Recognition, pp. 9308–9316 (2019)

Learning Monocular 3D Vehicle Detection Without 3D Bounding Box Labels

Lukas Koestler[1,2(✉)], Nan Yang[1,2], Rui Wang[1,2], and Daniel Cremers[1,2]

[1] Technical University of Munich, Munich, Germany
lukas.koestler@tum.de
[2] Artisense, Munich, Germany
https://lukaskoestler.com/ldwl

Abstract. The training of deep-learning-based 3D object detectors requires large datasets with 3D bounding box labels for supervision that have to be generated by hand-labeling. We propose a network architecture and training procedure for learning monocular 3D object detection without 3D bounding box labels. By representing the objects as triangular meshes and employing differentiable shape rendering, we define loss functions based on depth maps, segmentation masks, and ego- and object-motion, which are generated by pre-trained, off-the-shelf networks. We evaluate the proposed algorithm on the real-world KITTI dataset and achieve promising performance in comparison to state-of-the-art methods requiring 3D bounding box labels for training and superior performance to conventional baseline methods.

Keywords: 3D object detection · Differentiable rendering · Autonomous driving

1 Introduction

Three-dimensional object detection is a crucial component of many autonomous systems because it enables the planning of collision-free trajectories. Deep-learning-based approaches have recently shown remarkable performance [33] but require large datasets for training. More specifically, the detector is supervised with 3D bounding box labels which are obtained by hand-labeling LiDAR point clouds [10]. On the other hand, methods that optimize pose and shape of individual objects utilizing hand-crafted energy functions do not require 3D bounding box labels [8,32]. However, these methods cannot benefit from training data and produce worse predictions in our experiments. To leverage deep learning and overcome the need for hand-labeling, we thus introduce a training scheme for monocular 3D object detection which does not require 3D bounding box labels for training.

We build upon Pseudo-LiDAR [33], a recent supervised 3D object detector that utilizes a pre-trained image-to-depth network to back-project the image into

Electronic supplementary material The online version of this chapter (https://doi.org/10.1007/978-3-030-71278-5_9) contains supplementary material, which is available to authorized users.

Z. Akata et al. (Eds.): DAGM GCPR 2020, LNCS 12544, pp. 116–129, 2021.
https://doi.org/10.1007/978-3-030-71278-5_9

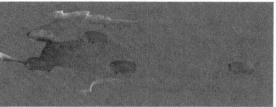

Fig. 1. We propose a monocular 3D vehicle detector that requires no 3D bounding box labels for training. The right image shows that the predicted vehicles (*colored shapes*) fit the ground truth bounding boxes (*red*). Despite the noisy input depth (*lower left*), our method is able to accurately predict the 3D poses of vehicles due to the proposed fully differentiable training scheme. We additionally show the projections of the predicted bounding boxes (*colored boxes, upper left*) (Color figure online).

a point cloud and then applies a 3D neural network. To replace the direct supervision by 3D bounding box labels, our method additionally uses 2D instance segmentation masks, as well as, ego- and object-motion as inputs during training. We show that our method works with off-the-shelf, pre-trained networks: Mask R-CNN [13] for segmentation and struct2depth [4] for motion estimation. Therefore, we introduce no additional labeling requirements for training in comparison to Pseudo-LiDAR. During inference the motion network is not required.

Due to the Pseudo-LiDAR-based architecture, our approach can utilize depth maps from mono-to-depth, or stereo-to-depth methods, which can be self-supervised or supervised. We show experiments for all four combinations. For depth maps generated by a self-supervised mono-to-depth network [11], only Mask R-CNN needs to be trained supervisedly and we use a model pre-trained on the general COCO dataset [22], therefore avoiding any supervision on the KITTI dataset.

1.1 Related Work

Object Detection. Two-dimensional object detection is a fundamental task in computer vision, where two-stage, CNN-based detectors [29] have shown impressive performance. Mask R-CNN [13] extends this approach to include the prediction of instance segmentation masks with high accuracy.

In contrast, image-based 3D object detection is still an open problem because depth information has to be inferred from 2D image data. Approaches based on per-instance optimization minimize a hand-crafted energy function for each object individually; the function encodes prior knowledge about pose and shape and considers input data, e.g., the back-projection of an estimated depth map [8], an image-gradient-based fitness measure [38], or the photometric constraint for stereo images together with 2D segmentation masks [32]. Initial deep-learning-based methods for stereo images [6] and monocular images [5] generate object proposals which are then ranked by a neural network. Subsequent approaches employ geometric constraints to lift 2D detections into 3D [25,27]. Kundu et al. [19] propose to compare the predicted pose and shape of each object to the ground truth depth map and segmentation mask, which yields two additional loss terms during training. They employ rendering to define the loss function and approximate the gradient using finite differences. Their approach relies on 3D bounding box labels for supervision and uses the additional loss terms to improve the final performance. Li et al. [21] propose Stereo-RCNN which combines deep learning and per-instance optimization for object detection from stereo images. Similar to our approach, Stereo-RCNN does not supervise the 3D position using 3D bounding box labels. In contrast to our method, they use the 3D bounding box labels to directly supervise the 3D dimensions, the viewpoint, and the perspective keypoint. Replacing the 3D bounding box labels by estimated 3D dimensions, viewpoints, and perspective keypoints is a non-trivial extension of their work. Furthermore, it is not studied how well their algorithm would handle the inevitable noise in the estimated 3D dimensions, viewpoints, and perspective keypoints if they are not computed from the highly accurate ground truth labels. Moreover, Stereo-RCNN is designed specifically for stereo images, while the proposed method is designed for monocular images and can be easily extended to the stereo setting (cf. Sect. 3). Wang et al. [33] back-project the depth map obtained from an image-to-depth network to a point cloud and then use networks initially designed for LiDAR data [18,26] for detection. Their method, Pseudo-LiDAR, showed that representing depth information in the form of point clouds is advantageous and has inspired our work.

Learning Without Direct Supervision. In the context of autonomous driving, self-supervised learning has been used successfully for depth prediction [11,35], as well as depth and ego-motion prediction [4]. Using only 2D supervision for 3D estimation is common in object reconstruction where the focus lies on estimating pose and shape for a diverse class of objects, but networks are commonly trained and evaluated on artificial datasets without noise. Generally, neural networks are trained to extract the 3D shape of an object from a single image. Initial works [17,34] use multi-view images with known viewpoints to define a loss based on the ground truth segmentation mask in each image and the differentiably rendered shape. Subsequent methods [14,16] overcome the dependence on known poses by including the pose into the prediction pipeline and thus require only 2D supervision.

The aforementioned approaches rely on rendering a 2D image from the 3D representation to define loss functions based on the input. To enable training, the renderer has to be differentiable with respect to the 3D representation. Loper and Black [23] proposed a mesh-based, differentiable renderer called OpenDR, which was extended in [14]. Other methods use approximations to ray casting for voxel volumes [34], differentiable point clouds [16], or differentiable rasterization for triangular meshes [17].

1.2   Contribution

We propose a monocular 3D vehicle detector that is trained without 3D bounding box labels by leveraging differentiable shape rendering. The major inputs

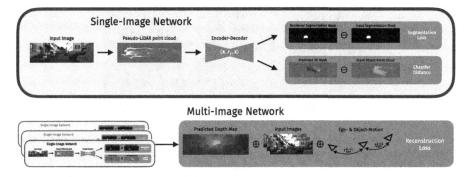

Fig. 2. The proposed model contains a single-image network and a multi-image network extension. The single-image network back-projects the input depth map estimated from the image into a point cloud. A Frustum PointNet encoder predicts the pose and shape of the vehicle which are then decoded into a predicted 3D mesh and segmentation mask through differentiable rendering. The predictions are compared to the input segmentation mask and back-projected point cloud to define two loss terms. The multi-image network architecture takes three temporally consecutive images as the inputs, and the single-image network is applied individually to each image. Our network predicts a depth map for the middle frame based on the vehicle's pose and shape. A pre-trained network predicts ego-motion and object-motion from the images. The reconstruction loss is computed by differentiably warping the images into the middle frame.

for our model are 2D segmentation masks and depth maps, which we obtain from pre-trained, off-the-shelf networks. Therefore, our method does not require 3D bounding box labels for supervision. Two-dimensional ground truth and LiDAR point clouds are only required for training the pre-trained networks. We thus overcome the need for hand-labeled datasets which are cumbersome to obtain and contribute towards the wider applicability of 3D object detection. We train and evaluate the detector on the KITTI object detection dataset [10]. The experiments show that our model achieves comparable results to state-of-the-art supervised monocular 3D detectors despite not using 3D bounding box labels for training. We further show that replacing the input monocular depth with stereo depth yields competitive stereo 3D detection performance, which shows the generality of our 3D detection framework.

2   Learning 3D Vehicle Detection Without 3D Bounding Box Labels

The proposed model consists of a single-image network that can learn from single, monocular images and a multi-image extension that additionally learns from temporally consecutive frames. Figure 2 depicts the proposed architecture. We utilize pre-trained networks to compute depth maps, segmentation masks, and ego- and object-motion, which are used as inputs to the network and for the loss functions during training. During inference only the single-image network and the pre-trained image-to-depth and segmentation networks are required.

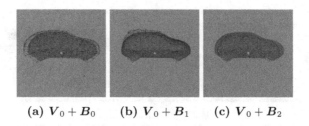

(a) $V_0 + B_0$ (b) $V_0 + B_1$ (c) $V_0 + B_2$

Fig. 3. Shape manifold visualization. The mean shape is shown in red, and the deformed meshes are shown as black wireframes. The resulting shape space can represent longer (3a), higher (3b), and smaller (3c) cars.

2.1 Shape Representation

We use a mesh representation given by a mean mesh together with linear vertex displacements which are obtained from the manifold proposed in [8] by a semi-manual process and are available on the project page. The mean vertex positions are denoted $V_0 \in \mathbb{R}^{N \times 3}$, the K vertex displacement matrices are denoted $B_k \in \mathbb{R}^{N \times 3}$, $k = 1, \ldots, K$, the shape coefficients are denoted $z = (z_1, \ldots, z_K)$ and the deformed vertex positions in the canonical coordinate system are denoted $V_{def} \in \mathbb{R}^{N \times 3}$. The deformed vertex positions are the linear combination

$$V_{def} = V_0 + \sum_{k=1}^{K} z_k \cdot B_k . \tag{1}$$

2.2 Single-Image Network

The input depth map is back-projected into a point cloud, which decouples the architecture from the depth source as in [33]. The point cloud is filtered with the object segmentation mask to obtain the object point cloud. For depth maps from monocular images, the object point cloud frequently has outliers at occlusion boundaries, which are filtered out based on their depth values.

Afterward, a Frustum PointNet encoder [26] predicts the position $x \in \mathbb{R}^3$, orientation $r_y \in [0, 2\pi)$, and shape $z \in \mathbb{R}^K$ of the vehicle. The shape coefficients z are applied in a canonical, object-attached coordinate system to obtain the deformed mesh based on our proposed shape manifold (Subsect. 2.1) using Eq. 1. The deformed mesh is rotated by r_y around the y-axis and translated by x to obtain the mesh in the reference coordinate system.

The deformed mesh in the reference coordinate system is rendered differentiably to obtain a predicted segmentation mask S_{obj} and a predicted depth map D_{obj}. The rendered depth map D_{obj} that incorporates the predicted pose and shape of the vehicle is used only in the multi-image network. For the image areas which do not belong to the vehicle, as defined by the input segmentation mask, we utilize the input depth map as the background depth and render the depth

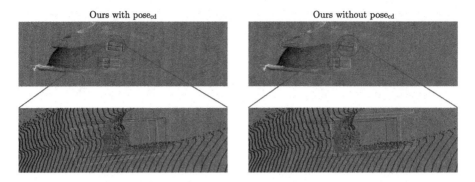

Fig. 4. Qualitative results with and without $pose_{cd}$ (cf. Sect. 2.3). We show the ground truth (*red*) and the predictions (*green*). Without the proposed $pose_{cd}$ the model learns to tightly fit the point cloud which leads to worse results due to errors in the point cloud. With $pose_{cd}$ the segmentation loss can correct the erroneous position of the point cloud and the predicted position is more accurate (Color figure online).

from the deformed mesh otherwise. For rendering the predicted depth map and segmentation mask we utilize a recent implementation [14] of the differentiable renderer proposed in [23]. Additional details are in the supplementary material.

2.3   Loss Functions

In order to train without 3D bounding box labels we use three losses, the segmentation loss $\mathcal{L}_{seg}$, the chamfer distance $\mathcal{L}_{cd}$, and the photometric reconstruction loss $\mathcal{L}_{rec}$. The first two are defined for single images and the photometric reconstruction loss relies on temporal photo-consistency for three consecutive frames (Fig. 2). The total loss is the weighted sum of the single image loss for each frame and the reconstruction loss

$$\mathcal{L}_{tot} = w_{rec} \cdot \mathcal{L}_{rec} + \frac{1}{3} \cdot \sum_{t} \mathcal{L}_{single}^{t}, \tag{2}$$

where the single image loss is the weighted sum of the segmentation loss and chamfer distance

$$\mathcal{L}_{single} = w_{cd} \cdot \mathcal{L}_{cd} + w_{seg} \cdot \mathcal{L}_{seg}. \tag{3}$$

To capture multi-scale information, the segmentation and reconstruction loss are computed for image pyramids [3] with eight levels, which we form by repeatedly applying a 5×5 binomial kernel with stride two. For each pyramid level the loss values are the mean over the pixel-wise loss values which ensures equal weighting for each level.

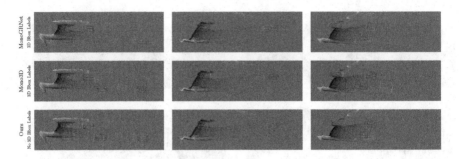

Fig. 5. Qualitative comparison of MonoGRNet [27] (*first row*), Mono3D [5] (*second row*), and our method (*third row*) with depth maps from BTS [20]. We show ground truth bounding boxes for cars (*red*), predicted bounding boxes (*green*), and the back-projected point cloud. In comparison to Mono3D, the prediction accuracy of the proposed approach is increased specifically for further away vehicles. As in the quantitative evaluation (cf. Table 1), the performance of MonoGRNet and our model is comparable (Color figure online).

Segmentation Loss. The segmentation loss penalizes the difference between the input segmentation mask S_{in} and the differentiably rendered segmentation mask S_{obj} using the squared L^2 norm.

$$\mathcal{L}_{seg} = ||S_{in} - S_{obj}||^2 . \tag{4}$$

Chamfer Distance. The chamfer distance for point clouds, which was used in the context of machine learning by [9], penalizes the 3D distance between two point clouds. Its original formulation is symmetric w.r.t. the two point clouds. In contrast, the situation analyzed in this paper does not posses this symmetry. For each point r_i in the input object point cloud, there must exist a corresponding vertex v in the deformed mesh, while due to occlusion or truncation, the reverse is not true. Therefore, we use a non-symmetric version of the chamfer distance

$$\mathcal{L}_{cd} = \frac{1}{M} \sum_i \min_j \rho(||r_i - v_j||). \tag{5}$$

We employ the Huber loss $\rho : \mathbb{R} \to \mathbb{R}_0^+$ to gain robustness against outliers.

For depth maps obtained from monocular image-to-depth networks, we notice weak performance of the chamfer distance (cf. Table 3) due to a high bias in the position of the input object point cloud, which is caused by the global scale ambiguity (cf. Fig. 4). To use the orientation information captured in the object point cloud without deteriorating the position estimate, we introduce pose$_{cd}$. The network outputs an auxiliary position x_{aux}, and the chamfer distance is then calculated using this position

$$\mathcal{L}_{cd} = \mathcal{L}_{cd}(x_{aux}, r_y). \tag{6}$$

The auxiliary position x_{aux} is predicted by a separate network head. We cut the gradient flow between the main network and the additional head to not

Table 1. Result for the proposed KITTI validation set. We report the average precision (AP) in percent for the car category in the bird's-eye view (BEV) and in 3D. The AP is the average over 40 values as introduced in [31]. Our method convincingly outperforms the supervised baseline method Mono3D and shows promising performance in comparison to a state-of-the-art supervised method MonoGRNet.

Method	Input	Without 3D Bbox	$AP_{BEV,0.7}$			$AP_{3D,0.7}$		
			Easy	Mode	Hard	Easy	Mode	Hard
Ours	Mono	✓	19.23	9.60	5.34	6.13	3.10	1.70
MonoGRNet [27]	Mono		**23.07**	**16.37**	**10.05**	**13.88**	**9.01**	**5.67**
Mono3D [5]	Mono		1.92	1.13	0.77	0.40	0.21	0.17

influence the main network, which necessitates the use of another loss term that back-propagates through the predicted position x.

Multi-image Reconstruction Loss. The multi-image network is inspired by the recent success of self-supervised depth prediction from monocular images [4, 11], which relies on differentiably warping temporally consecutive images into a common frame to define the reconstruction loss. The single-image network is applied to three consecutive images I^{t-1}, I^t, I^{t+1} of the same vehicle and the reconstruction loss is defined in the middle frame. The reconstruction loss is formulated as in [4] and we use their pre-trained network to estimate the ego-motion and object motion required for warping.

Hindsight Loss. To overcome the multi-modality of the loss w.r.t. the orientation of the vehicle, we apply the hindsight loss mechanism [12], which has been frequently used in the context of self-supervised object reconstruction [14,16]. The network predicts orientation hypotheses in L bins and the hindsight loss is the minimum of the total loss over the hypotheses.

3   Experiments

We quantitatively compare our method with other state-of-the-art monocular 3D detection methods on the publicly available KITTI 3D object detection dataset [10]. Note that since our method is the first monocular 3D detector trained without 3D bounding box labels, the compared-against methods are supervised methods that are trained with ground truth 3D bounding box labels. We conduct an extensive ablation study on the different loss terms to show the efficacy of each proposed component. Because the accuracy of the input point cloud plays a crucial role for the proposed model, we show experiments with depth maps estimated from different methods. Finally, we compare against methods based on per-instance optimization.

Fig. 6. Qualitative comparison of ShapePriors [8] (first row) and our approach (second row) with depth maps from BTS. We show ground truth bounding boxes for cars (*red*), predicted bounding boxes (*green*), and the back-projected point cloud. ShapePriors is initialized with detections from 3DOP [6] as in the original paper, which leads to false positives (*left column*). For the quantitative evaluation (cf. Subsect. 3.2) we control for this difference and our approach still shows better performance. The comparison shows that learning can produce more robust and accurate prediction than per-instance optimization. Both methods do not require 3D bounding box labels for training. (Color figure online)

KITTI Object Detection. The KITTI dataset consists of sequences that are used for numerous benchmarks, e.g. 3D object detection and depth prediction. This leads to an overlap of the common validation set for object detection [6] and the popular Eigen [7] train set for monocular depth estimation. The overlap was already noted by [33]. Unlike in [33], we use a subset of the validation set that has no sequence-level overlap with the Eigen training set or the KITTI 2015 stereo training set. Following works can integrate pre-trained mono-to-depth and stereo-to-depth networks directly. The split files can be found on the project page. Results on the standard validation set [6] are given in the supplementary material and they unsurprisingly show better performance than on the proposed split.

For the confidence score we estimate the KITTI category (easy, moderate, and hard) from the data. We shift and scale the baseline scores $1 - \mathcal{L}_{single}$ such that objects which are estimated to be easy have a higher score than any object which is estimated to be moderate. The same holds for moderate objects in comparison to hard objects. This gives a slight improvement in average precision and details are in the supplementary material.

Pre-trained Networks. For Mask R-CNN [13] we use the implementation of [1] and their pre-trained weights on the COCO [22] dataset. For ego- and object-motion estimation we utilize the official implementation of struct2depth [4] and their pre-trained weights on the Eigen train split. For depth estimation we use Monodepth 2 [11], BTS [20], SGM [15], and GA-Net [37]. For Monodepth 2 we use the official implementation and their pre-trained weights on Zhou's [39] subset of the Eigen train split; this model is trained with supervision from monocular images of resolution 1024×320 and utilizes pre-training on ImageNet [30]. For BTS we use the official implementation and their pre-trained weights on the Eigen train split. For SGM we use the public implementation provided by

Table 2. Depth source ablation study. The average precision of the proposed model improves when using a supervised instead of an unsupervised image-to-depth method and when using stereo images instead of monocular images. Our more general method delivers the best performs among methods trained without 3D bounding box labels, but worse performance as the stereo-specific Stereo-RCNN which uses partial 3D bounding box information for training. Our approach clearly improves upon the common baseline 3DOP and the recent DirectShape and TLNet. Stereo-RCNN does not directly supervise the 3D position, but directly supervises the 3D bounding box dimensions. Additionally, they compute the viewpoint and perspective keypoint from the ground truth 3D bounding box label and use them for supervision and thus require 3D bounding box labels during training. Replacing the 3D bbox labels by estimated 3D dimensions, viewpoints, and perspective keypoints is a non-trivial extension of their work.

Method	Input	Without 3D Bbox	$AP_{BEV,0.7}$			$AP_{3D,0.7}$		
			Easy	Mode	Hard	Easy	Mode	Hard
Ours (Monodepth)	Mono	✓	10.78	5.43	2.99	4.53	2.16	1.17
Ours (BTS)	Mono	✓	19.23	9.60	5.34	6.13	3.10	1.70
Ours (SGM)	Stereo	✓	31.51	15.78	8.76	8.42	4.08	2.23
Ours (GA-Net)	Stereo	✓	<u>68.16</u>	<u>35.82</u>	<u>20.45</u>	<u>38.45</u>	<u>18.78</u>	<u>10.44</u>
Stereo-RCNN [21]	Stereo	(✓)	**71.51**	**53.81**	**35.56**	**56.68**	**38.30**	**25.45**
TLNet [28]	Stereo		24.92	17.01	11.25	13.74	9.45	6.13
DirectShape [32]	Stereo	✓	24.91	16.03	10.28	12.60	7.36	4.33
3DOP [6]	Stereo		8.72	5.52	3.29	2.68	1.48	1.05

[2] and piecewise linear interpolation in 2D to complete the disparity map. For GA-Net we use the official implementation and their pre-trained weights on Scene Flow [24] and the KITTI 2015 stereo training set. For matching consecutive segmentation masks we use a similar procedure to [4]; however, we first warp the segmentation masks into a common frame using optical flow [36].

Evaluation Results. For monocular object detection, we compare to two supervised monocular 3D detection networks: MonoGRNet [27] is a state-of-the-art monocular detector and Mono3D [5] is a common baseline method. Table 1 shows the evaluation results. Our results are superior to the ones generated by Mono3D in all categories. While MonoGRNet outperforms our method, the performance gap is relatively small. This difference is smaller for the easy category than for the moderate category, which shows that handling distant objects and occlusions when learning without 3d bounding box labels is challenging.

Table 3. Ablation study using depth maps from BTS [20]. Using the chamfer distance without the proposed pose$_{cd}$ reduces the accuracy significantly. Learning pose and shape without 3D bounding box labels is an under-constraint problem and the performance decreases (cf. last row). Without multi-image training the performance in the BEV is similar but the performance in 3D is decreased.

Method	$AP_{BEV,0.7}$			$AP_{3D,0.7}$		
	Easy	Mode	Hard	Easy	Mode	Hard
Full Model	19.23	**9.60**	**5.34**	**6.13**	**3.10**	**1.70**
W/o $\mathcal{L}_{cd}$	9.75	5.21	2.75	3.50	1.73	0.98
W/o pose$_{cd}$	4.53	2.84	1.58	0.94	0.48	0.26
W/o $\mathcal{L}_{seg}$	4.22	2.23	1.16	0.76	0.41	0.18
W/o $\mathcal{L}_{rec}$	**19.60**	9.48	5.30	4.88	2.26	1.20
W/ B_k	16.02	8.12	4.51	5.24	2.59	1.32

3.1 Ablation Study

Input Depth. Table 2 shows that the average precision with BTS [20], a supervised mono-to-depth network, is better than the performance with the self-supervised Monodepth 2 [11], due to the superior depth estimation accuracy. This leads to the question: *Does the performance of the proposed model constantly improve if more accurate depth maps are used as input?* When switching from mono to stereo, better depth maps are estimated, and the AP is dramatically improved, as can be seen in Table 2. Besides, using depth maps from GA-Net [37], a stereo-to-depth network trained in a supervised fashion, outperforms using depth maps from the traditional stereo matching algorithm SGM [15] by a notable margin. In Table 2, we also show the results of state-of-the-art stereo 3D detectors, Stereo-RCNN [21], DirectShape [32], 3DOP [6], and TLNet [28]. The proposed approach ranks first among the methods that do not use 3D bounding box labels for training.

Loss Terms. We demonstrate the significance of using the chamfer distance together with the proposed pose$_{cd}$ in Fig. 4 and Table 3. Simultaneously estimating pose and shape generally resulted in worse performance and training instabilities due to the inherent scale ambiguity. The best results we achieved are obtained with the mean shape – the shape variability of cars within the KITTI dataset is small and thus a fixed shape is a reasonable approximation. More details can be found in the supplementary material. During our experiments, the reconstruction loss in the multi-image setting contributes marginal improvements, which may be due to the noise in the ego-motion and object-motion predictions, which were taken from the self-supervised struct2depth [4]; details are included in the supplementary material.

3.2   Comparison with Non-Learning-Based Methods

We choose ShapePriors [8] for comparison because it uses very similar input data; ShapePriors uses depth maps and initial 3D detections, while our method uses depth maps and 2D segmentation masks during inference. We compare both methods using depth maps generated by GA-Net.

The initial 3D detections were taken from 3DOP in the original paper. To facilitate a fair quantitative comparison, we initialize the position with the median of the object point cloud in the x and z direction and the minimum in the y direction. For the orientation and the 2D bounding box we use the ground truth. Because we require the ground truth label for the orientation initialization and the segmentation mask for the position initialization, we match segmentation masks and labels. Thus, the results presented here are not comparable to the other results within this paper.

Under these conditions, ShapePriors achieves 23.65% $AP_{BEV,0.7,easy}$ and ours 77.47%. For the qualitative comparison (cf. Fig. 6) ShapePriors is initialized with detections from 3DOP [6] as in the original paper. The quantitative and qualitative comparisons show that per-instance optimization delivers less robust and accurate predictions than learning. Similarly, the comparison against Direct-Shape (cf. Table 2) indicates that learning can extract meaningful priors from the training data and ultimately deliver superior performance.

4   Conclusion

We propose the first monocular 3D vehicle detection method for real-world data that can be trained without 3D bounding box labels. By proposing a differentiable-rendering-based architecture we can train our model from unlabeled data using pre-trained networks for instance segmentation, depth estimation, and motion prediction. During inference only the instance segmentation and depth estimation networks are required. Without ground truth labels for training, we decisively outperform a baseline supervised monocular detector and show promising performance in comparison to a state-of-the-art supervised method.

Furthermore, we demonstrate the generality of the proposed framework by using depth maps from a stereo-to-depth network and without further changes achieving state-of-the-art performance for stereo 3D object detection without 3D bounding box labels for training. While this paper demonstrates that monocular 3D object detection without 3D bounding box labels for training is viable, many directions for future research remain, e.g. the explicit integration of stereo images, the extension to pedestrians and cyclists, training on large, unlabelled datasets, or the integration of an occlusion aware segmentation loss.

References

1. Abdulla, W.: Mask R-CNN for object detection and instance segmentation on keras and tensorflow. https://github.com/matterport/Mask_RCNN

2. Audet, S., Kitta, Y., Noto, Y., Sakamoto, R., Takagi, A.: fixstars/libSGM: stereo semi global matching by CUDA. https://github.com/fixstars/libSGM
3. Burt, P., Adelson, E.: The Laplacian pyramid as a compact image code. IEEE Trans. Commun. **31**(4), 532–540 (1983)
4. Casser, V., Pirk, S., Mahjourian, R., Angelova, A.: Depth prediction without the sensors: leveraging structure for unsupervised learning from monocular videos. In: AAAI Conference on Artificial Intelligence, vol. 33, pp. 8001–8008. AAAI Press (2019)
5. Chen, X., Kundu, K., Zhang, Z., Ma, H., Fidler, S., Urtasun, R.: Monocular 3D object detection for autonomous driving. In: CVPR, pp. 2147–2156. IEEE (2016)
6. Chen, X., et al.: 3D object proposals for accurate object class detection. In: Cortes, C., Lawrence, N.D., Lee, D.D., Sugiyama, M., Garnett, R. (eds.) NeurIPS, pp. 424–432. Curran Associates (2015)
7. Eigen, D., Puhrsch, C., Fergus, R.: Depth map prediction from a single image using a multi-scale deep network. In: Ghahramani, Z., Welling, M., Cortes, C., Lawrence, N.D., Weinberger, K.Q. (eds.) NeurIPS, pp. 2366–2374. Curran Associates (2014)
8. Engelmann, F., Stückler, J., Leibe, B.: Joint object pose estimation and shape reconstruction in urban street scenes using 3D shape priors. In: Rosenhahn, B., Andres, B. (eds.) GCPR 2016. LNCS, vol. 9796, pp. 219–230. Springer, Cham (2016). https://doi.org/10.1007/978-3-319-45886-1_18
9. Fan, H., Su, H., Guibas, L.J.: A point set generation network for 3D object reconstruction from a single image. In: CVPR, pp. 2463–2471. IEEE (2017)
10. Geiger, A., Lenz, P., Urtasun, R.: Are we ready for autonomous driving? The KITTI vision benchmark suite. In: CVPR, pp. 3354–3361. IEEE (2012)
11. Godard, C., Aodha, O.M., Firman, M., Brostow, G.J.: Digging into self-supervised monocular depth estimation. In: ICCV, pp. 3827–3837. IEEE (2019)
12. Guzmán-Rivera, A., Batra, D., Kohli, P.: Multiple choice learning: learning to produce multiple structured outputs. In: Bartlett, P.L., Pereira, F.C.N., Burges, C.J.C., Bottou, L., Weinberger, K.Q. (eds.) NeurIPS, pp. 1808–1816. Curran Associates (2012)
13. He, K., Gkioxari, G., Dollár, P., Girshick, R.B.: Mask R-CNN. In: ICCV, pp. 2980–2988. IEEE (2017)
14. Henderson, P., Ferrari, V.: Learning single-image 3D reconstruction by generative modelling of shape, pose and shading. Int. J. Comput. Vis. **128**(4), 835–854 (2020)
15. Hirschmüller, H.: Accurate and efficient stereo processing by semi-global matching and mutual information. In: CVPR, pp. 807–814. IEEE (2005)
16. Insafutdinov, E., Dosovitskiy, A.: Unsupervised learning of shape and pose with differentiable point clouds. In: Bengio, S., Wallach, H.M., Larochelle, H., Grauman, K., Cesa-Bianchi, N., Garnett, R. (eds.) NeurIPS, pp. 2807–2817. Curran Associates (2018)
17. Kato, H., Ushiku, Y., Harada, T.: Neural 3D mesh renderer. In: CVPR, pp. 3907–3916. IEEE (2018)
18. Ku, J., Mozifian, M., Lee, J., Harakeh, A., Waslander, S.L.: Joint 3D proposal generation and object detection from view aggregation. In: IROS, pp. 1–8. IEEE (2018)
19. Kundu, A., Li, Y., Rehg, J.M.: 3D-RCNN: instance-level 3d object reconstruction via render-and-compare. In: CVPR, pp. 3559–3568. IEEE (2018)
20. Lee, J.H., Han, M., Ko, D.W., Suh, I.H.: From big to small: multi-scale local planar guidance for monocular depth estimation (2019). https://arxiv.org/abs/1907.10326v5

21. Li, P., Chen, X., Shen, S.: Stereo R-CNN based 3D object detection for autonomous driving. In: CVPR, pp. 7644–7652. IEEE (2019)
22. Lin, T.-Y., et al.: Microsoft COCO: common objects in context. In: Fleet, D., Pajdla, T., Schiele, B., Tuytelaars, T. (eds.) ECCV 2014. LNCS, vol. 8693, pp. 740–755. Springer, Cham (2014). https://doi.org/10.1007/978-3-319-10602-1_48
23. Loper, M.M., Black, M.J.: OpenDR: an approximate differentiable renderer. In: Fleet, D., Pajdla, T., Schiele, B., Tuytelaars, T. (eds.) ECCV 2014. LNCS, vol. 8695, pp. 154–169. Springer, Cham (2014). https://doi.org/10.1007/978-3-319-10584-0_11
24. Mayer, N., et al.: A large dataset to train convolutional networks for disparity, optical flow, and scene flow estimation. In: CVPR, pp. 4040–4048. IEEE (2016)
25. Mousavian, A., Anguelov, D., Flynn, J., Kosecka, J.: 3D bounding box estimation using deep learning and geometry. In: CVPR, pp. 5632–5640. IEEE (2017)
26. Qi, C.R., Liu, W., Wu, C., Su, H., Guibas, L.J.: Frustum pointnets for 3D object detection from RGB-D data. In: CVPR, pp. 918–927. IEEE (2018)
27. Qin, Z., Wang, J., Lu, Y.: Monogrnet: a geometric reasoning network for monocular 3D object localization. In: AAAI Conference on Artificial Intelligence, pp. 8851–8858. AAAI Press (2019)
28. Qin, Z., Wang, J., Lu, Y.: Triangulation learning network: from monocular to stereo 3D object detection. In: CVPR, pp. 7615–7623. IEEE (2019)
29. Ren, S., He, K., Girshick, R.B., Sun, J.: Faster R-CNN: towards real-time object detection with region proposal networks. In: Cortes, C., Lawrence, N.D., Lee, D.D., Sugiyama, M., Garnett, R. (eds.) NeurIPS, pp. 91–99. Curran Associates (2015)
30. Russakovsky, O., et al.: Imagenet large scale visual recognition challenge. Int. J. Comput. Vis. 115(3), 211–252 (2015)
31. Simonelli, A., Bulò, S.R., Porzi, L., Lopez-Antequera, M., Kontschieder, P.: Disentangling monocular 3D object detection. In: ICCV, pp. 1991–1999. IEEE (2019)
32. Wang, R., Yang, N., Stückler, J., Cremers, D.: Directshape: photometric alignment of shape priors for visual vehicle pose and shape estimation. In: ICRA (2020)
33. Wang, Y., Chao, W., Garg, D., Hariharan, B., Campbell, M.E., Weinberger, K.Q.: Pseudo-LiDar from visual depth estimation: Bridging the gap in 3D object detection for autonomous driving. In: CVPR, pp. 8445–8453. IEEE (2019)
34. Yan, X., Yang, J., Yumer, E., Guo, Y., Lee, H.: Perspective transformer nets: learning single-view 3d object reconstruction without 3D supervision. In: Lee, D.D., Sugiyama, M., von Luxburg, U., Guyon, I., Garnett, R. (eds.) NeurIPS, pp. 1696–1704. Curran Associates (2016)
35. Yang, N., Wang, R., Stückler, J., Cremers, D.: Deep virtual stereo odometry: leveraging deep depth prediction for monocular direct sparse odometry. In: Ferrari, V., Hebert, M., Sminchisescu, C., Weiss, Y. (eds.) ECCV 2018. LNCS, vol. 11212, pp. 835–852. Springer, Cham (2018). https://doi.org/10.1007/978-3-030-01237-3_50
36. Yin, Z., Darrell, T., Yu, F.: Hierarchical discrete distribution decomposition for match density estimation. In: CVPR, pp. 6044–6053. IEEE (2019)
37. Zhang, F., Prisacariu, V.A., Yang, R., Torr, P.H.S.: GA-NET: guided aggregation net for end-to-end stereo matching. In: CVPR, pp. 185–194. IEEE (2019)
38. Zhang, Z., Tan, T., Huang, K., Wang, Y.: Three-dimensional deformable-model-based localization and recognition of road vehicles. IEEE Trans. Image Process. 21(1), 1–13 (2012)
39. Zhou, T., Brown, M., Snavely, N., Lowe, D.G.: Unsupervised learning of depth and ego-motion from video. In: CVPR, pp. 6612–6619. IEEE (2017)

Observer Dependent Lossy Image Compression

Maurice Weber[1(✉)], Cedric Renggli[1], Helmut Grabner[2], and Ce Zhang[1]

[1] Department of Computer Science, ETH Zürich, Switzerland
maurice.weber@inf.ethz.ch
[2] ZHAW School of Engineering, Winterthur, Switzerland

Abstract. Deep neural networks have recently advanced the state-of-the-art in image compression and surpassed many traditional compression algorithms. The training of such networks involves carefully trading off entropy of the latent representation against reconstruction quality. The term quality crucially depends on the observer of the images which, in the vast majority of literature, is assumed to be human. In this paper, we aim to go beyond this notion of compression quality and look at human visual perception and image classification *simultaneously*. To that end, we use a family of loss functions that allows to optimize deep image compression depending on the observer and to interpolate between human perceived visual quality and classification accuracy, enabling a more unified view on image compression. Our extensive experiments show that using perceptual loss functions to train a compression system preserves classification accuracy much better than traditional codecs such as BPG without requiring retraining of classifiers on compressed images. For example, compressing ImageNet to 0.25 bpp reduces Inception-ResNet classification accuracy by only 2%. At the same time, when using a human friendly loss function, the same compression system achieves competitive performance in terms of MS-SSIM. By combining these two objective functions, we show that there is a pronounced trade-off in compression quality between the human visual system and classification accuracy.

1 Introduction

Image compression algorithms aim at finding representations of images that use as little storage—measured in bits—as possible. Opposed to lossless image compression, where the goal is to achieve a high compression rate while requiring perfect reconstruction, lossy image compression enables even higher compression rates by allowing for a loss in reconstruction quality. Recently, image compression based on deep neural networks (DNNs) has achieved remarkable results in both lossless [33] and lossy image compression [2,4,32,35,41,43], outperforming many traditional codecs. One distinct advantage of such methods is their flexibility with regards to the term *reconstruction quality* which crucially depends

Electronic supplementary material The online version of this chapter (https://doi.org/10.1007/978-3-030-71278-5_10) contains supplementary material, which is available to authorized users.

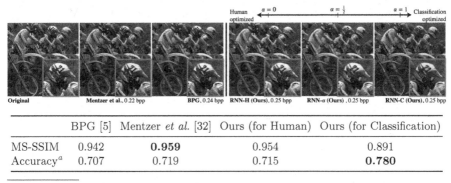

	BPG [5]	Mentzer *et al.* [32]	Ours (for Human)	Ours (for Classification)
MS-SSIM	0.942	**0.959**	0.954	0.891
Accuracy[a]	0.707	0.719	0.715	**0.780**

[a] Accuracy uncompressed: 0.803

Fig. 1. Accuracy evaluated on ImageNet-1K with off-the-shelf Inception-ResNet-V2. MS-SSIM on Kodak. Both datasets are compressed at ∼0.25 bpp with different methods. Our Classification optimized system induces very low loss in classification accuracy at high compression rates, compared to human optimized approaches.

on the observer of the compressed images. Previous research in lossy image compression expressed quality largely in terms of human visual perception and optimized for the human visual system (HVS), using distortion measures such as multiscale structural similarity [47] (MS-SSIM) or mean squared error (MSE) as training objectives. However, due to recent advances in computer vision systems, increasingly more images are observed solely by machines and bypass humans. Consequently, a natural question that arises is whether or not there exists a relation between quality perceived by humans and quality perceived by computer vision systems, and if so, how can we trade off quality between different types of observers? In other words, is a compression system optimized for the human observer also optimal for machines? We investigate these questions by specifically looking at classification of natural images as one of the most well studied tasks in computer vision. The training of modern classifiers is typically a costly and time-consuming undertaking and parameters of the best performing classifiers are often publicly available. With that in mind, we are interested in a compression system that generalizes well in the following sense. *Firstly*, we want to compress images such that no retraining of classifiers on compressed images is required. *Secondly*, the compression system should be agnostic to classifier architectures. *Thirdly*, it should also generalize well to other visual tasks such as fine-grained visual categorization of natural images. Together, these generalization requirements encourage using publicly available, pretrained classifiers on compressed images from the same domain or on related tasks where classifiers were obtained with transfer learning. Our method for classification oriented compression relies on a feature reconstruction loss using deep features extracted from the hidden layers of a convolutional neural network trained for image classification. This type of loss function has been used in the context of super-resolution [7,21,27], style-transfer [15,21] and variational autoencoders [12] with remarkable success.

In order to optimize for human visual perception, we make use of MS-SSIM as a measure of quality perceived by humans, since it has been reported to correlate better with the HVS than MSE. Finally, the convex combination of the two objectives allows to investigate the trade-off between human visual perception and classification in the context of image compression. In summary, the contributions of our work are threefold:

- We show that training deep image compression with a perceptual loss function preserves classification accuracy much better than human optimized compression systems. In addition, our experiments show that *(1) we do not have to retrain classifiers on compressed images in order to preserve accuracy on highly compressed images*, and *(2) using VGG-based feature reconstruction loss generalizes to other models, indicating that deep CNN features are shared between CNN architectures.*
- By looking at the convex combination between human and classification-friendly loss, we present a *simple way to trade off compression quality* in terms of human perception against image classification. Since we only rely on the training objective, our method can be integrated to any learned lossy image compression system.
- Our extensive experimental study indicates that *there exists a pronounced trade-off* between compression quality perceived by the human observer and classification accuracy. We show how improved compression quality for the human observer comes at the cost of degraded classification accuracy, and vice versa.

We emphasize that the contribution of this work is not presenting a new type of loss function nor in a new deep compression architecture. Rather, we make us of existing techniques in order to present a method to trading off compression quality depending on the observer and to show that it is possible to explicitly optimize compression for subsequent classification.

2   Related Work

Deep Image Compression. Image compression using DNNs has recently become an active area of research. The most popular types of architectures used for image compression are based on autoencoders [2,4,32,35,41] and recurrent neural networks [22,42,43] (RNNs). Typically, the networks are trained in an end-to-end manner to minimize a pixel-wise notion of distortion such as MSE, MS-SSIM or L_1-distance between original and decoded image.

Compression for Computer Vision. Image compression in combination with other computer vision tasks has been studied in a number of recent works. Liu *et al.* [29] propose an image compression framework based on JPEG that is favorable to DNN classifiers. Also starting from an engineered codec, Liu *et al.* [30] propose a 3D image compression framework based on JPEG2000 which is tailored to segmentation of 3-D medical images. Both works differ from ours in that we look at learned image compression, rather than modifying an engineered one.

A few examples exist in the literature, where a classifier is learned from features extracted from the encoded representations. Gueguen et al. [17] train a modified ResNet-50 directly on the blockwise discrete cosine transform coefficients from the middle of the JPEG codec. Torfason et al. [44] make use of the compressive autoencoder proposed in [41] and train neural networks for classification and segmentation on the latent (quantized) representations and on the decoded images. These works stand orthogonal to ours in that we do not allow training on compressed versions of images. Rather, we train the compression algorithm such that it maintains information relevant for subsequent classification, keeping the classifiers fixed. We furthermore focus on agnosticity to architectures of inference algorithms. Finally, since compression artifacts typically compromise the performance of classifiers, Dodge and Karam [11] study the effect of JPEG compression on image classification with neural networks.

Feature Reconstruction Loss. This class of similarity functions makes use of deep features extracted from convolutional neural networks. Recent advances in generative modelling have shown that using this type of loss functions, high quality images can be generated and have been applied to a variety of tasks. Gatys et al. [14,15] apply the idea to style transfer and texture synthesis, while Johnson et al. [21] and Bruna et al. [7] achieve remarkable results in super resolution [7,21] and style transfer [21]. Ledig et al. [27] further develop the idea and enhance the CNN feature loss with adversarial training to achieve state-of-the-art results in single image super resolution. In the image compression domain, steps in this direction have also been made. Agustsson et al. [3], Santurkar et al. [37] and Liu et al. [28] enhance pixel-wise distortion and adversarial training with a feature reconstruction loss. Furthermore, Chinen et al. [8] and Zhang et al. [48] both propose a similarity metric based on deep features extracted from VGG-16 trained for image classification. These works have in common that their focus is on the human observer, while we exploit properties of feature reconstruction loss in the context of compression geared towards subsequent image classification. Feature reconstructions loss has also been used in the context of compression artifact removal. Galteri et al. [13] train a generative adversarial network in combination with a VGG-based perceptual loss function to remove compression artifacts in images. It is shown that this can significantly increase the quality of compressed images in terms of MS-SSIM and in terms of object detection accuracy. However, contrary to our work, no clear trade-off between the human observer and image classification is investigated.

3   Method

In this section, we outline our approach to compressing images for human visual perception, classification accuracy and the interpolation between the two. Throughout this paper we adopt the compression architecture proposed by Toderici et al. [43], based on recurrent neural networks. We emphasize that we only focus on the objective functions to account for different types of observers.

Compression Framework. Let $\mathcal{X} \subseteq \mathbb{R}^d$ denote a set of training images, $\mathcal{Z} \subseteq \mathbb{Z}$ the quantization levels and $d: \mathbb{R}^d \times \mathbb{R}^d \to \mathbb{R}$ a notion of distortion between images.

Our goal is to find a compression system consisting of an encoder $E\colon \mathbb{R}^d \to \mathbb{R}^m$ that maps input images $\mathbf{x}$ to their latent representation $\mathbf{z} = E(\mathbf{x})$, a quantizer $q\colon \mathbb{R}^m \to \mathcal{Z}^m$ that discretizes $\mathbf{z}$ to $\hat{\mathbf{z}} = q(\mathbf{z})$, and a decoder $D\colon \mathcal{Z}^m \to \mathbb{R}^d$ that maps the quantized representation back to image space, $\hat{\mathbf{x}} = D(\hat{\mathbf{z}})$. The goal is then to minimize the rate-distortion trade-off over the training set $\mathcal{X}$, i.e. for $\beta \geq 0$, we want to minimize $\sum_{\mathbf{x} \in \mathcal{X}} d(\mathbf{x}, \hat{\mathbf{x}}) + \beta H(\hat{\mathbf{z}})$, where H denotes the entropy. As a compression architecture, we adopt the RNN-based model proposed in [43] with gated recurrent units (GRUs), allowing for variable bitrates. An input image $\mathbf{x}$ is passed through the encoder and quantizer, mapping the latent codes stochastically to $\mathcal{Z}^m = \{-1, +1\}^m$. The quantized representation is subsequently decoded, yielding an estimate of the original image. This is repeated with the residual error fed to the encoder to obtain an estimate at the next bitrate, using information from the hidden states of the previous iterations. Formally, a single iteration at unrolling step $t \geq 1$, can be represented as $\hat{\mathbf{x}}_t = \hat{\mathbf{x}}_{t-1} + D_t(Q(E_t(\mathbf{r}_t)))$ with $\hat{\mathbf{x}}_0 = \mathbf{0}$ and $\mathbf{r}_1 = \mathbf{x}$ and where E_t and D_t are encoder and decoder carrying information from the previous unrolling steps. Finally, we remark that, since $\mathcal{Z}$ contains a finite number of quantization levels, we set $\beta = 0$ in the training objective.

Optimizing for Human Visual Perception. In order to optimize the compression system for the human observer, we choose a measure of distortion that approximately models human visual perception. The multiscale structural similarity index (MS-SSIM) [47] is based on the assumption that the human eye is adapted for extracting structural information from images and incorporates image details at multiple resolutions. It is furthermore reported to correlate better with human visual perception than MSE. Since MS-SSIM is differentiable, we follow [22,32,35] and minimize directly $d_H(\mathbf{x}, \hat{\mathbf{x}}) = 1 - \mathrm{MS\text{-}SSIM}(\mathbf{x}, \hat{\mathbf{x}})$. We refer to compression optimized with d_H as **RNN-H**. An alternative approach would be to use other, human-centric distortion metrics such as LPIPS [48] or the approach proposed in [8]. However, as these approaches are based on CNNs they bear the additional challenge of dealing with checkerboard-like artifacts [34].

Optimizing for Classification. Suppose we are given a CNN classifier f trained on a set of images and labels $(\mathcal{X}', \mathcal{Y}')$ and corresponding training and validation splits $(\mathcal{X}'_{train}, \mathcal{Y}'_{train})$ and $(\mathcal{X}'_{val}, \mathcal{Y}'_{val})$. When we optimize compression for classification accuracy, we are interested in finding an encoder, quantizer and decoder such that the accuracy evaluated on the decoded validation set $D(q(E(\mathcal{X}'_{val})))$ is maintained as well as possible, *without* further retraining the classifier on decoded images. Formally, we wish to maximize $\sum_{\mathbf{x} \in \mathcal{X}'_{val}} \mathbb{1}\{f(\mathbf{x}) = f(\hat{\mathbf{x}})\}$. We are thus not interested in matching decoded and original images on a pixel-wise basis, but rather on preserving features which are relevant for subsequent classification. Image classification is a task which is typically invariant to translations and local deformations (see *e.g.* [6,31]), which motivates the use of an objective function with similar properties. For example, using a pixel-wise distortion, such as MSE, which is not invariant to such deformations would be a suboptimal choice. Furthermore, minimizing MSE encourages the generator to produce images that are pixel-wise averages of plausible solutions [12], resulting in overly smooth images. In other words, high frequency information such as textures will tend to get lost in the compression process. While this is less problematic for the HVS, which is more susceptible to low frequency changes, CNNs are sensitive to any change in frequency [29].

Features learned by convolutional neural networks [26] for image classification provide a promising alternative. The intuition is that, if such features are maintained in the compression process, then the compressed representations are encouraged to encode information relevant to classification rather than to the human observer. Moreover, it is known that CNNs provide stability to small geometric deformations and translations, thanks to rectification and pooling units [6]. This is beneficial for our purpose, since we do not want to put too much emphasis on such deformations as they do not affect classification. Finally, feature reconstruction loss typically leads to high frequency artifacts ([12] and references therein) and checkerboard patterns [34]. While this harms human perceived visual quality, our experiments indicate that this is not the case for classification. These considerations make distortion measures based on CNN features promising candidates for classification oriented image compression.

In order to define a distortion measure that incorporates these properties, we fix a CNN classifier f_L trained on a dataset $(\mathcal{X}'', \mathcal{Y}'')$. Denote by ϕ_i the responses of the i-th convolutional layer after activation and let $\mathcal{I}$ be a set of such layers. Note that $\mathcal{I}$ is not required to include all layers. We then define the distortion measure associated with the loss network f_L and layers $\mathcal{I}$ to be MSE in feature space

$$d_{C,\mathcal{I}}(\mathbf{x}, \hat{\mathbf{x}}) = \sum_{i \in \mathcal{I}} \gamma_i \|\phi_i(\mathbf{x}) - \phi_i(\hat{\mathbf{x}})\|_2^2, \tag{1}$$

where $\gamma_i := (H_i \times W_i \times C_i)^{-1}$ and H_i, W_i, C_i represent the spatial dimensions of the corresponding layer. Note that we do not restrict the loss network to be trained on the same dataset as the compression system or the classifier f, however we do require that $\mathcal{X}'' \cap \mathcal{X}'_{val} = \varnothing$. Furthermore, the classifier f might have a different underlying architecture than the loss network f_L. This formulation allows to investigate the generalizability of the compression system to new datasets and CNN architectures. We refer to compression optimized with $d_{C,\mathcal{I}}$ as **RNN-C**.

From Human Visual Perception to Classification. In a scenario where images are consumed by both humans and classifiers, we would like to be able to trade off reconstruction quality between the two observers. In other words, we want to have a compressed representation of an image that contains features relevant for classification *and* looks visually pleasing for the human observer. At the same time, this enables us to investigate the relation between human visual perception and classification accuracy. For that purpose, we consider the convex combination between distortions d_H and $d_{C,\mathcal{I}}$

$$d_{\alpha,\mathcal{I}}(\mathbf{x}, \hat{\mathbf{x}}) = (1 - \alpha) \cdot \lambda_H \cdot d_H(\mathbf{x}, \hat{\mathbf{x}}) + \alpha \cdot d_{C,\mathcal{I}}(\mathbf{x}, \hat{\mathbf{x}}) \tag{2}$$

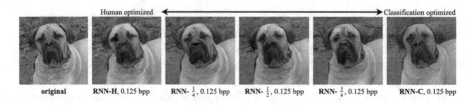

Fig. 2. Sample image from the Stanford Dogs dataset. RNN-H results in smoother and blurrier images, RNN-C on the other hand produces sharp images but suffers from checkerboard-like artifacts stemming from the CNN based loss function.

and control the trade-off with the parameter $\alpha \in [0, 1]$. The parameter λ_H is a scaling parameter which keeps the two losses on the same magnitude and is set to 5,000. We refer to compression optimized with $d_{\alpha, \mathcal{I}}$ as **RNN-α**.

4   Experiments

In this section we experimentally validate our approach to trading off compression quality between human visual perception and classification accuracy, making use of the proposed family of loss functions. All models are implemented in Python using the Tensorflow [1] library.[1]

Image Compression. We use the RNN compression architecture proposed by Toderici *et al.* [43] with GRUs and the additive reconstruction framework. Our implementation differs from the original version in two aspects. Firstly, during training, we feed as input the full resolution images, rather than 32×32 image patches. And secondly, instead of optimizing the L_1-distance in image space, we use the family of loss functions (2) as training objective. Furthermore, we do not use the lossless entropy coding scheme proposed in their original work. While this would likely result in reduced bitrates, and thereby further improve our results, we omit this in order to reduce complexity and focus exclusively on the distortion during training. If not stated otherwise, we train the networks for 8 unrolling steps, yielding rates between 0.125 and 1.0 bpp. As training data $\mathcal{X}$, we use the training split of the ILSVRC-2012 [36] dataset, commonly known as ImageNet-1K. We preprocess the images by resizing such that the smallest side equals 256 pixels and aspects are preserved using bilinear interpolation. During training, we take random crops of size 224×224 and randomly flip them horizontally. During validation, we use the central crop of size 224×224. We follow [32] and normalize with a mean and variance obtained from a subset of the training set. We train all our networks using the Adam optimizer [24] for three epochs with the learning rate set to 4e-4 and minibatches of size four. All models are trained on eight Nvidia Titan X GPUs with 12GB RAM.

---

[1] The source code is available at https://github.com/DS3Lab/odlc.

Measures of Distortion. We train the compression networks using the loss function defined in Eq. (2) and use VGG-16 trained on the ImageNet-1K training split as our loss network f_L. Preliminary experiments have shown that choosing $\mathcal{I} = \{\phi_{1.1}, \phi_{5.1}\}$ in (1), where $\phi_{i.j}$ denotes the j-th convolutional layer after activation in the i-th block of VGG-16, performed best. Including the entire set of convolutions or only the first or last layer did not yield any improvements. We provide detailed plots to compare the different loss compositions with respect to accuracy in the supplementary materials. The weights of the loss network are frozen and left unchanged during training. We experiment with different values for the parameter α, starting the training each time from scratch. Namely, in order to optimize for human visual perception, we set $\alpha = 0$, while for classification oriented compression, we set $\alpha = 1$. To investigate the trade-off between human vision and classification, we train with $\alpha \in \{\frac{1}{4}, \frac{1}{2}, \frac{3}{4}\}$, also starting training from scratch each time.

Comparison with Other Methods. We compare our approach to the traditional compression algorithms JPEG [46], WebP [16] and BPG [5] which achieves state-of-the-art performance in HVS oriented compression. Following [32,35], BPG is used in the non-default 4:4:4 chroma format. Finally, we also compare against the state-of-the-art learned compression method presented in Mentzer *et al.* [32], using their publicly available weights and code. Since the available weights do not compress images below 0.3 bpp, we train two models with the same hyperparameters but different number of bottleneck channels to achieve lower bitrates.

Datasets. We evaluate our approach on four publicly available datasets. For classification we use the ImageNet-1K dataset [36], as well as two datasets used for fine-grained categorization, CUB-200–2011 [45] and Stanford Dogs [23]. In order to evaluate the performance in terms of human visual perception we use the Kodak Photo CD dataset [25] and the ImageNet-1K validation split.

CNN Architectures. On ImageNet-1K, we use DenseNet-121 [20], Inception-ResNet-V2 [39], Inception-V3 [40], MobileNet-V1 [19], ResNet-50 [18], Xception [9] and VGG-16 [38] for inference and use the weights provided by the Keras Library [10]. For fine-grained categorization on CUB-200–2011 and Stanford Dogs, we use Inception-V3, Inception-ResNet-V2, MobileNet-V1, ResNet-50 and VGG-16. To obtain the classifiers, we use ImageNet pre-trained networks and fine-tune all layers on the original uncompressed training split.

Evaluating Classification Accuracy. In order to compare the different compression algorithms with regard to classification accuracy, we evaluate a collection of CNN architectures on datasets compressed with different algorithms and at different bitrates. Note that all classifiers are trained on the uncompressed respective training datasets, *without* retraining on decoded data. The evaluation procedure is as follows. Since generally, the images do not have the same resolution, we resize them such that the smaller side equals S_{comp} and aspects are preserved. We then take the central crop of size $S_{comp} \times S_{comp}$ yielding square images. After this step, given a compression algorithm, we encode the images for a predefined grid of quality parameters and compute the bpp values for each image and quality parameter. For each quality level, we subsequently take the average over the entire validation set, yielding the final bpp values. Finally, we decode and take the central crop of size $S_{inf} \times S_{inf}$ of the decoded image, which is then fed to the classifier. This results in a set of (bpp, accuracy) points for each classifier and compression method. For CNNs that expect inputs of size $S_{inf} = 299$ we set $S_{comp} = 336$ and for those with $S_{inf} = 224$, we set $S_{comp} = 256$.

Evaluating Human Visual Perception. The procedure to compare the compression methods for human perceived visual quality is as follows. To account for the variable resolution on ImageNet-1K, we resize each image with bilinear interpolation such that the smallest side equals 256 pixels and aspects are preserved. We then take the central crop of size 256×256. Since the Kodak images are all of equal resolution, we skip this first resizing step and keep the original resolution. We then compress the images using a predefined grid of quality parameters and compute their bpp values which are averaged over the validation set. Finally, we compute the MS-SSIM scores between decoded and original (resized) image. This yields a set of (bpp, MS-SSIM) points for each compression method.

4.1   Results

We start by investigating the trade-off between human perception and classification accuracy using compression trained with an increasingly more classification friendly loss. We then look at compression in terms of classification accuracy, followed by our results on human perception.

From Human Visual Perception to Classification. In order to investigate the relation between compression quality perceived by humans in terms of MS-SSIM, and by CNN classifiers, we train the compression networks with loss functions that interpolate between human friendly and classification friendly loss, i.e. for values of α in $\{0, \frac{1}{4}, \frac{1}{2}, \frac{3}{4}, 1\}$. This trade-off can be seen qualitatively in Fig. 2. Optimizing for MS-SSIM, results in images that appear smoother and more blurry. Classification optimized compression on the other hand results in sharper images but suffers from checkerboard-like artifacts. This type of degradation is a known issue for feature visualization and super resolution (see *e.g.* [34]) and – in our case – stems from the convolution based loss function which incurs artifacts in gradients. In order to quantitatively investigate the trade-off, we visualize the relation in Fig. 3. We plot MS-SSIM on Kodak (left axis) and ImageNet-1K validation accuracy (right axis) against the tradeoff parameter α corresponding to RNN compression trained with different loss functions. The Figures indicate that we can indeed trade off accuracy against MS-SSIM by optimizing compression with our family of loss functions. Interestingly, we observe that by increasing the trade-off parameter α from 0 to 0.25, we substantially increase accuracy while the reduction in MS-SSIM is relatively small. The same holds for the other direction. Finally, we observe that the trade-off is much more pronounced in the low bitrate regime.

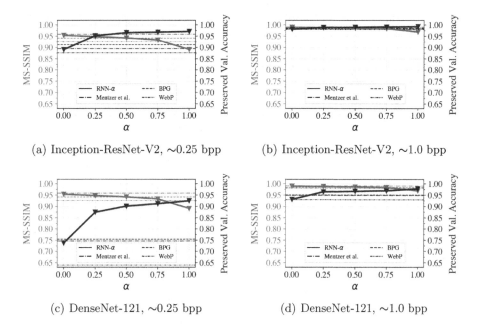

(a) Inception-ResNet-V2, ~0.25 bpp   (b) Inception-ResNet-V2, ~1.0 bpp

(c) DenseNet-121, ~0.25 bpp   (d) DenseNet-121, ~1.0 bpp

Fig. 3. MS-SSIM evaluated on Kodak (left axis, grey), validation accuracy evaluated on ImageNet-1k (right axis, blue). As α increases, MS-SSIM decreases, while validation accuracy increases. The trade-off is especially pronounced for the low bitrates and DenseNet-121 is in general more sensitive to compression than Inception-ResNet-V2 (Color figure online).

ImageNet Classification. Table 1 shows the classification accuracies for a wider collection of CNN architectures. We see that our RNN-C outperforms both the traditional codecs BPG, WebP and JPEG as well as our RNN-H and the deep image compression method proposed in [32], across all architectures and bitrates considered. In the case of the loss network VGG-16 this is to be expected, since we explicitly train the compression network to produce images whose VGG-features – which are fed to the fully connected layers for classification – match the ones from their uncompressed version. Interestingly however, we see that RNN-C generalizes well to architectures different from the loss network and maintains the accuracy remarkably well, indicating that hidden representations are shared among CNN architectures. It is again noticeable that the advantage of RNN-C is much more pronounced for low bitrates.

Fine-Grained Visual Categorization. In order to explore the generalization properties of our compression system to new tasks, we evaluate our method on two well known datasets for fine-grained visual categorization, namely Stanford Dogs and CUB-200–2011. We emphasize that the compression system is trained on the ImageNet-1K training split. Figures 4(a) and 4(b) indicate that RNN-C outperforms both the traditional codecs, RNN-H compression and the approach

Table 1. Validation accuracy on ImageNet-1K. Our RNN-C compression consistently outperforms all other methods across bitrates and architectures.

ImageNet-1K Validation Accuracy								
	224×224 input					299×299 input		
	bpp	DenseNet-121	MobileNet	ResNet-50	VGG-16[a]	bpp	Inception-V3	Xception
Low Bitrates (∼0.13 bpp)								
RNN-C	0.125	**0.5914**	**0.4903**	**0.6092**	**0.6009**	0.125	**0.6691**	**0.6815**
RNN-H	0.125	0.4109	0.3221	0.4309	0.3797	0.125	0.5550	0.5722
Mentzer *et al.* [32]	0.142	0.4744	0.3774	0.4931	0.4387	0.140	0.6069	0.6269
BPG	0.157	0.4661	0.3421	0.4711	0.4448	0.132	0.5750	0.6050
JPEG	0.136	0.0480	0.0493	0.0426	0.0320	0.113	0.2675	0.2166
Medium Bitrates (∼0.65 bpp)								
RNN-C	0.625	**0.7252**	**0.6256**	**0.7246**	**0.6998**	0.625	**0.7678**	**0.7787**
RNN-H	0.625	0.6709	0.5688	0.6744	0.6450	0.625	0.7434	0.7543
Mentzer *et al.* [32]	0.652	0.6842	0.5909	0.6975	0.6670	0.648	0.7567	0.7652
BPG	0.725	0.6857	0.5834	0.6894	0.6634	0.581	0.7377	0.7523
WebP	0.589	0.6306	0.5263	0.6323	0.6247	0.602	0.7268	0.7429
JPEG	0.582	0.6166	0.5111	0.6333	0.6365	0.686	0.7390	0.7476
High Bitrates (∼1.0 bpp)								
RNN-C	1.000	**0.7303**	**0.6347**	**0.7316**	**0.7037**	1.000	**0.773**	**0.7840**
RNN-H	1.000	0.6998	0.5984	0.7018	0.6732	1.000	0.7631	0.7728
Mentzer *et al.* [32]	1.037	0.7076	0.6183	0.7152	0.6841	1.034	0.7667	0.7767
BPG	1.048	0.7085	0.6151	0.7168	0.6841	1.066	0.7618	0.7756
WebP	0.997	0.6930	0.6050	0.7039	0.6829	1.055	0.7589	0.7699
JPEG	1.087	0.6808	0.5865	0.6963	0.6918	0.962	0.7517	0.7622
Original	-	0.7453	0.6590	0.7465	0.7088	-	0.7786	0.7907

[a] Loss network used to train RNN-C compression.

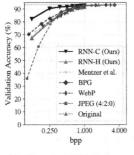
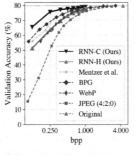
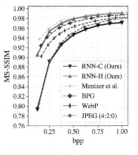

(a) Stanford Dogs Acc.   (b) CUB-200-2011 Acc.   (c) MS-SSIM on Kodak

Fig. 4. Validation accuracy is displayed in (a) and (b) where our RNN-C consistently outperforms BPG, JPEG, WebP, RNN-H and deep compression [32] across bitrates. MS-SSIM on Kodak is shown in (c), indicating that our RNN-H is comptetitive to the state-of-the-art while RNN-C is comparable to JPEG. In each figure, RNN compression is trained from scratch on ImageNet-1K.

presented in [32] in terms of preserved classification accuracy with Inception-ResNet-V2 on both datasets. Similarly to ImageNet-1K classification, we see that the difference is especially pronounced below 0.5 bpp.

Human Visual Perception. Figure 4(c) shows that RNN-H outperforms RNN-C on Kodak across bitrates. In comparison to neural compression from [32], RNN-H performs competitive, although worse for the lowest bitrate. Comparing our method against the traditional codecs, RNN-H slightly outperforms BPG for bitrates above 0.2 bpp. Additionally, RNN-H clearly outperforms WebP and JPEG, while RNN-C performs competitive to JPEG.

5 Discussion

In this paper we investigate the trade-off in learned image compression with RNNs [43] between human visual perception and image classification. To that end, we use a family of loss functions that enables us to either optimize compression for the human observer, or towards subsequent image classification. Our experiments show that when using the human friendly loss, RNN compression performs competitive to a state-of-the-art learned compression method [32] and to the traditional codec BPG [5] in terms of MS-SSIM. JPEG and WebP perform consistently worse than our approach. We use MS-SSIM as a model for image similarity perceived by humans which, although being a widely adopted measure of distortion, is only an approximation to a true model of the human visual system. Our classification friendly loss, based on features extracted from VGG-16, induces a compression system which by a large margin outperforms both the the other approaches in terms of preserved classification accuracy. Our experiments furthermore indicate that our approach is agnostic to the CNN architecture used for classification and does not require the classifiers to be retrained on compressed images. This suggests that we can indeed explicitly optimize image compression for subsequent classification. We observe a clear trade-off between quality perceived by the human visual system and classification accuracy, meaning that, for a fixed bitrate, an increase in accuracy always comes at the cost of degraded quality for the human observer, and vice versa. Across classifiers, this trade-off is much more pronounced for bitrates below 0.5 bpp. Finally, we find that by moving the loss function only marginally towards classification, we can substantially increase the preserved accuracy while incurring only a minor reduction in MS-SSIM. This improves compression in a scenario where images are consumed by humans and classifiers simultaneously and allows a user to trade off reconstruction quality accordingly.

An interesting line of future work could include investigating other types of distortion measures used for the human oriented training loss, for example metrics that are based on CNNs which have been reported to correlate better with human perceptual similarity. Additionally, while classification is one of the most basic computer vision tasks, it would be interesting to explore whether the approach presented here also generalizes to other tasks such as image sementation and object detection.

Acknowledgments. CZ and the DS3Lab gratefully acknowledge the support from the Swiss National Science Foundation (Project Number 200021_184628), Innosuisse/SNF BRIDGE Discovery (Project Number 40B2-0_187132), European Union Horizon 2020 Research and Innovation Programme (DAPHNE, 957407), Botnar Research Centre for Child Health, Swiss Data Science Center, Alibaba, Cisco, eBay, Google Focused Research Awards, Oracle Labs, Swisscom, Zurich Insurance, Chinese Scholarship Council, and the Department of Computer Science at ETH Zurich.

References

1. Abadi, M., et al.: TensorFlow: large-scale machine learning on heterogeneous systems (2015). https://www.tensorflow.org/, software available from tensorflow.org
2. Agustsson, E., et al.: Soft-to-hard vector quantization for end-to-end learning compressible representations. Adv. Neural Inf. Process. Syst. **30**, 1141–1151 (2017)
3. Agustsson, E., Tschannen, M., Mentzer, F., Timofte, R., Van Gool, L.: Generative adversarial networks for extreme learned image compression. arXiv preprint arXiv:1804.02958 (2018)
4. Ballé, J., Laparra, V., Simoncelli, E.P.: End-to-end optimized image compression. In: International Conference on Learning Representations (ICLR) (2017)
5. Bellard, F.: BPG image format (2014). https://bellard.org/bpg/
6. Bruna, J., Mallat, S.: Invariant scattering convolution networks. IEEE Trans. Pattern Anal. Mach. Intell. **35**(8), 1872–1886 (2013)
7. Bruna, J., Sprechmann, P., LeCun, Y.: Super-resolution with deep convolutional sufficient statistics. In: International Conference on Learning Representations (ICLR), May 2016
8. Chinen, T., et al.: Towards a semantic perceptual image metric. In: 2018 25th IEEE International Conference on Image Processing (ICIP), October 2018
9. Chollet, F.: Xception: deep learning with depthwise separable convolutions. In: The IEEE Conference on Computer Vision and Pattern Recognition (CVPR), July 2017
10. Chollet, F., et al.: Keras (2015). https://keras.io
11. Dodge, S., Karam, L.: Understanding how image quality affects deep neural networks. In: 2016 Eighth International Conference on Quality of Multimedia Experience (QoMEX), pp. 1–6 (2016)
12. Dosovitskiy, A., Brox, T.: Generating images with perceptual similarity metrics based on deep networks. Adv. Neural Inf. Process. Syst. **29**, 658–666 (2016)
13. Galteri, L., Seidenari, L., Bertini, M., Del Bimbo, A.: Deep generative adversarial compression artifact removal. In: The IEEE International Conference on Computer Vision (ICCV), October 2017
14. Gatys, L., Ecker, A.S., Bethge, M.: Texture synthesis using convolutional neural networks. In: Advances in Neural Information Processing Systems, pp. 262–270 (2015)
15. Gatys, L.A., Ecker, A.S., Bethge, M.: Image style transfer using convolutional neural networks. In: The IEEE Conference on Computer Vision and Pattern Recognition (CVPR), June 2016
16. Google: Webp image format (2015). https://developers.google.com/speed/webp/. Accessed 17 Mar 2019
17. Gueguen, L., Sergeev, A., Kadlec, B., Liu, R., Yosinski, J.: Faster neural networks straight from jpeg. Adv. Neural Inf. Process. Syst. **31**, 3933–3944 (2018)

18. He, K., Zhang, X., Ren, S., Sun, J.: Deep residual learning for image recognition. In: The IEEE Conference on Computer Vision and Pattern Recognition (CVPR), June 2016
19. Howard, A.G., et al.: Mobilenets: efficient convolutional neural networks for mobile vision applications (2017). arXiv:1704.04861
20. Huang, G., Liu, Z., van der Maaten, L., Weinberger, K.Q.: Densely connected convolutional networks. In: The IEEE Conference on Computer Vision and Pattern Recognition (CVPR), July 2017
21. Johnson, J., Alahi, A., Fei-Fei, L.: Perceptual losses for real-time style transfer and super-resolution. In: Leibe, B., Matas, J., Sebe, N., Welling, M. (eds.) ECCV 2016. LNCS, vol. 9906, pp. 694–711. Springer, Cham (2016). https://doi.org/10.1007/978-3-319-46475-6_43
22. Johnston, N., et al.: Improved lossy image compression with priming and spatially adaptive bit rates for recurrent networks. In: The IEEE Conference on Computer Vision and Pattern Recognition (CVPR), June 2018
23. Khosla, A., Jayadevaprakash, N., Yao, B., Fei-Fei, L.: Novel dataset for fine-grained image categorization. In: First Workshop on Fine-Grained Visual Categorization, IEEE Conference on Computer Vision and Pattern Recognition, June 2011
24. Kingma, D., Ba, J.: Adam: a method for stochastic optimization. In: International Conference on Learning Representations (ICLR), May 2015
25. Kodak, E.: Kodak lossless true color image suite (PhotoCD PCD0992)
26. LeCun, Y., Kavukcuoglu, K., Farabet, C.: Convolutional networks and applications in vision. In: Proceedings of 2010 IEEE International Symposium on Circuits and Systems, pp. 253–256 (2010)
27. Ledig, C., et al.: Photo-realistic single image super-resolution using a generative adversarial network. In: The IEEE Conference on Computer Vision and Pattern Recognition (CVPR), July 2017
28. Liu, H., Chen, T., Shen, Q., Yue, T., Ma, Z.: Deep image compression via end-to-end learning. In: The IEEE Conference on Computer Vision and Pattern Recognition (CVPR) Workshops, June 2018
29. Liu, Z., et al.: DeepN-JPEG: a deep neural network favorable JPEG-based image compression framework. In: 2018 55th ACM/ESDA/IEEE Design Automation Conference (DAC), pp. 1–6 (2018)
30. Liu, Z., et al.: Machine vision guided 3D medical image compression for efficient transmission and accurate segmentation in the clouds. In: The IEEE Conference on Computer Vision and Pattern Recognition (CVPR), June 2019
31. Mallat, S.: Understanding deep convolutional networks. Philos. Trans. Roy. Soc. A **374**(2065) (2016)
32. Mentzer, F., Agustsson, E., Tschannen, M., Timofte, R., Van Gool, L.: Conditional probability models for deep image compression. In: The IEEE Conference on Computer Vision and Pattern Recognition (CVPR), June 2018
33. Mentzer, F., Agustsson, E., Tschannen, M., Timofte, R., Van Gool, L.: Practical full resolution learned lossless image compression. In: The IEEE Conference on Computer Vision and Pattern Recognition (CVPR), June 2019
34. Odena, A., Dumoulin, V., Olah, C.: Deconvolution and checkerboard artifacts. Distill (2016). http://distill.pub/2016/deconv-checkerboard/
35. Rippel, O., Bourdev, L.: Real-time adaptive image compression. In: Proceedings of the 34th International Conference on Machine Learning, vol. 70, pp. 2922–2930, August 2017
36. Russakovsky, O., et al.: Imagenet large scale visual recognition challenge. Int. J. Comput. Vis. **115**(3), 211–252 (2015)

37. Santurkar, S., Budden, D., Shavit, N.: Generative compression. In: 2018 Picture Coding Symposium (PCS), pp. 258–262 (2018)
38. Simonyan, K., Zisserman, A.: Very deep convolutional networks for large-scale image recognition. In: International Conference on Learning Representations (ICLR), May 2015
39. Szegedy, C., Ioffe, S., Vanhoucke, V., Alemi, A.A.: Inception-v4, inception-resnet and the impact of residual connections on learning. In: Proceedings of the Thirty-First AAAI Conference on Artificial Intelligence, AAAI 2017, pp. 4278–4284. AAAI Press (2017)
40. Szegedy, C., Vanhoucke, V., Ioffe, S., Shlens, J., Wojna, Z.: Rethinking the inception architecture for computer vision. In: The IEEE Conference on Computer Vision and Pattern Recognition (CVPR), June 2016
41. Theis, L., Shi, W., Cunningham, A., Huszár, F.: Lossy image compression with compressive autoencoders. In: International Conference on Learning Representations (ICLR) (2017)
42. Toderici, G., et al.: Variable rate image compression with recurrent neural networks. In: International Conference on Learning Representations (ICLR) (2016)
43. Toderici, G., et al.: Full resolution image compression with recurrent neural networks. In: The IEEE Conference on Computer Vision and Pattern Recognition (CVPR), pp. 5435–5443, July 2017
44. Torfason, R., Mentzer, F., Agustsson, E., Tschannen, M., Timofte, R., Gool, L.V.: Towards image understanding from deep compression without decoding. In: International Conference on Learning Representations (ICLR), April 2018
45. Wah, C., Branson, S., Welinder, P., Perona, P., Belongie, S.: The Caltech-UCSD Birds-200-2011 Dataset. Technical report. CNS-TR-2011-001, California Institute of Technology (2011)
46. Wallace, G.K.: The jpeg still picture compression standard. IEEE Trans. Cons. Electron. **38**(1), xviii–xxxiv (1992)
47. Wang, Z., Simoncelli, E.P., Bovik, A.C.: Multiscale structural similarity for image quality assessment. In: The Thirty-Seventh Asilomar Conference on Signals, Systems and Computers, vol. 2, pp. 1398–1402 (2003)
48. Zhang, R., Isola, P., Efros, A.A., Shechtman, E., Wang, O.: The unreasonable effectiveness of deep features as a perceptual metric. In: The IEEE Conference on Computer Vision and Pattern Recognition (CVPR), June 2018

Adversarial Synthesis of Human Pose from Text

Yifei Zhang[1,2], Rania Briq[1(✉)], Julian Tanke[1], and Juergen Gall[1]

[1] Computer Vision Group, University of Bonn, Bonn, Germany
{briq,tanke,gall}@iai.uni-bonn.de
[2] Bonn-Aachen International Center for Information Technology,
RWTH-Aachen University, Bonn, Germany
yifei.zhang@rwth-aachen.de

Abstract. This work focuses on synthesizing human poses from human-level text descriptions. We propose a model that is based on a conditional generative adversarial network. It is designed to generate 2D human poses conditioned on human-written text descriptions. The model is trained and evaluated using the COCO dataset, which consists of images capturing complex everyday scenes with various human poses. We show through qualitative and quantitative results that the model is capable of synthesizing plausible poses matching the given text, indicating that it is possible to generate poses that are consistent with the given semantic features, especially for actions with distinctive poses.

1 Introduction

Given a text description like "A tennis player hitting a tennis ball with a racquet", we can directly imagine a human pose that matches the description. Such ability would be useful for applications like retrieving images with semantically similar poses or animating avatars based on text descriptions. Synthesizing the human pose, however, is very difficult since the articulated body pose is much more complex than rigid or nearly convex shapes like objects or faces. Although previous works on synthesizing images from text describing a scene [16,22,23,25,28,32,33] achieve astonishing results when the images contain objects such as flowers, animals with small pose variations like birds or general scenes such as mountains or playing fields, the synthesized humans in these scenes appear quite unrealistic due to distorted or incorrect poses. This failure is due to the uniqueness of the human pose which is highly articulated and versatile. Conversely, most existing works for modeling humans rely on the pose as part of the intermediate feature representation [12,18]. Synthesizing poses in complex scenes is therefore an essential step towards synthesizing images with realistic human poses.

Electronic supplementary material The online version of this chapter (https://doi.org/10.1007/978-3-030-71278-5_11) contains supplementary material, which is available to authorized users.

Z. Akata et al. (Eds.): DAGM GCPR 2020, LNCS 12544, pp. 145–158, 2021.
https://doi.org/10.1007/978-3-030-71278-5_11

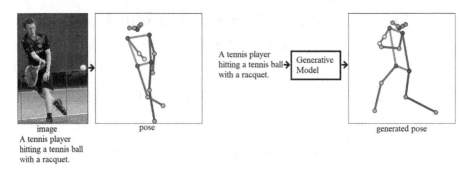

image
A tennis player
hitting a tennis ball
with a racquet.

pose

A tennis player hitting a tennis ball with a racquet.

Generative Model

generated pose

Fig. 1. The image on the left hand side shows an example from the COCO dataset that is annotated by an image caption describing the image and the human pose. In this work, we use only the image caption to generate the human pose.

In this work, we focus on synthesizing versatile human poses from text as shown in Fig. 1. We examine how well the synthesized poses match the text description and whether it is possible to achieve semantic consistency in their feature spaces. To achieve this goal, we design a model based on Generative Adversarial Networks (GANs) [6] to generate a single person pose conditioned on a given human-level text description. In order to condition the network to generate a pose that matches the text, the text is first encoded into an embedding using a pre-trained language model and then fed-forward through a convolutional network. The generated and real poses will then be assessed by a critic network whose objective is to maximize the earth mover's distance between the real and generated samples distributions. Similar to the pose representation in detection-based human pose estimation [29], we represent the pose by a set of heatmaps each corresponding to a body keypoint. Additionally, to resolve the highly unstable nature of GAN training, we experiment with different GAN models and loss functions and thoroughly evaluate their impact on the synthesized poses. We evaluate the approach on the COCO dataset and show that it is possible to generate human poses that are consistent with a given text.

2   Related Work

Generative models are a powerful tool for learning data distributions. Recent advancements in deep network architectures have enabled modeling complex and high-dimensional data such as images [27]. Examples of deep generative models include Deep Belief Networks (DBNs) [10], Variational Autoencoder (VAEs) [11] and the more recent approach of Generative Adversarial Networks (GANs) [6]. In the field of computer vision, GANs have been employed for different tasks for content synthesis, including unconditional image synthesis [6, 24], image synthesis conditioned on text [13, 14, 16, 22, 23, 25, 28, 32, 33], generating text description conditioned on images [5], style transfer between images [4], and transferring a target pose to a given person's pose in an image [19].

Image synthesis conditioned on text has gained traction in computer vision research recently. Motivations for such works include matching features between the semantic and visual space. Reed *et al.* [25] combine a GAN with a deep symmetric structured text-image joint embedding to synthesize plausible images of birds and flowers from human-written text descriptions. Zhang *et al.* [33] propose a GAN composed of two stages and generate hierarchical representations that are transferred between several stacked GANs. Reed *et al.* [25] and Zhang *et al.* [33] also attempt to generalize their models to generate images with multiple types of objects using the COCO dataset. However, their approaches do not directly address the human pose, and the persons in the synthesized images have deformed poses. In a subsequent work, Reed *et al.* [26] generate images based on text and show that using a sparse set of keypoints allows for synthesizing a higher resolution image. Zhou *et al.* [35] alter the pose of a person in a given image based on a text description. In the first stage they generate a pose by predicting a pose from a set of pose clusters created from the training set. However, they use a pedestrian dataset in which the poses are simple and contain mainly small variations of standing or walking persons. Since the approach assumes that all poses can be represented by a small set of clusters, the approach cannot be applied to datasets with versatile and highly articulated poses such as COCO. In a more recent work, Xu *et al.* [32] proposed a more advanced attentional GAN, which is multi-stage and attention-based, such that it can synthesize fine-grained details by paying attention to the relevant words in the text. Their model outperforms the previous works but individuals still appear deformed in the generated images. Li *et al.* [14] propose an object-driven attention module that generates images conditioned on the class label. However, they do not explicitly handle the human case and the humans still look deformed despite improved results. In the fashion domain, Zhu *et al.* [36] manipulate the clothing of a person in a given image based on a text description without altering the pose. Other related works such as [7,9,15,34] deal with searching for or synthesizing plausible human poses that match object affordances in a given scene.

3   Generating Human Poses from Text

The goal of our approach is to generate human poses that match a textual description as illustrated in Fig. 1. To this end, we use a conditional Wasserstein GAN as shown in Fig. 2. The text description is first converted into a vector and used to condition the GAN, which predicts heatmaps for each joint, which are finally converted into a human pose. Before we discuss the network architecture in Sect. 3.2, we discuss the representation of the text and the human pose.

3.1   Feature Representation

We need to define representations for the text description as well as the human pose such that they can be used in a convolutional network. The text is encoded by the mapping $\varphi : \mathbb{T} \to \mathbb{R}^{300}$, which maps a text sequence into a 300 dimensional embedding space. For the text embedding, we use fastText [2,20]. As is

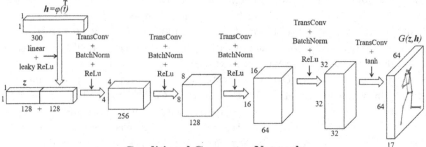

Conditional Generator Network

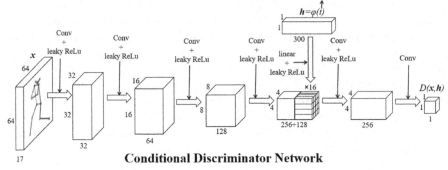

Conditional Discriminator Network

Fig. 2. The architecture of the proposed network. The generator G takes a noise vector $z \in \mathbb{R}^{128}$ and a text encoding vector $h \in \mathbb{R}^{300}$ as input and generates a pose heatmap $G(z, h) \in \mathbb{R}^{J \times 64 \times 64}$ where J is the number of keypoints. The discriminator D takes a real or generated pose heatmap H and a text encoding vector h as input. The discriminator predicts a single value $D(x, h) \in \mathbb{R}$ indicating its confidence about the sample being real or generated. For upsampling, transposed convolution layers are used.

common for human pose estimation [31], we represent the human pose by a heatmap $x \in \mathbb{R}^{m \times n}$ for each joint j. The heatmaps are modeled by a Gaussian distribution centered at the keypoint coordinate. Compared to a skeleton representation based on joint coordinates, heatmaps allow for representing joints that are invisible due to occlusion or truncation by setting the heatmaps to zero and allow an implementation based on a convolutional network rather than a fully connected one. The choice of the heatmap-based representation is also validated by our experiments, in which we compare the proposed representation with a skeleton representation that is regressed by a fully connected network. Given these two representations for the text description and the human pose, we will describe the network architecture that generates heatmaps from the embedded text in the following section.

3.2 Architecture

In order to learn to predict plausible poses from text, we use adversarial training as illustrated in Fig. 2. In our experiments, we show that a vanilla GAN performs poorly. We therefore use a Wasserstein GAN (WGAN), which is a more stable variant for training GANs with their continuous and nearly everywhere differentiable loss functions [1].

The model consists of a conditional generator network G and a conditional discriminator network D. The input to the generator is a concatenation of a noise vector $z \sim N(0, I)$, where N denotes a normal distribution, with the embedded text description $h = \varphi(t)$ reduced by a layer from 300 to 128 dimensions. Given z and h, the network infers J heatmaps with resolution $m \times n$, i.e. $G(z, h) \in \mathbb{R}^{J \times m \times n}$. The discriminator network takes either real or generated heatmaps as input. Since our goal is to generate heatmaps or poses that match the text description, we condition the network on the embedded text $h = \varphi(t)$ as well. Since the heatmaps have a higher dimensionality with $J \times 64 \times 64$ than h, we first apply the inverse transformations of the generator until the resolution is reduced to 4×4. We then concatenate the embedded text, by duplicating it 16 times after a layer that reduces the vector h from 300 to 128 dimensions. Both networks are trained together where D's objective is to maximize the distance between the generated heatmaps $G(z, h)$ and the real heatmaps x sampled from the training dataset $\mathbb{P}_r$. Unlike in the unconditional case, D has to distinguish two types of errors: heatmaps that correspond to unrealistic human poses as well as heatmaps that correspond to realistic poses, but the poses do not match the text description. The two errors are penalized by the following two terms:

$$
\begin{aligned}
L_{D^*} = &- \mathbb{E}_{(x,h) \sim \mathbb{P}_r, z \sim \mathbb{P}_z}[D(x, h) - D(G(z, h), h)] \\
&- \mathbb{E}_{(x,h) \sim \mathbb{P}_r, \hat{h} \sim \mathbb{P}_h}[D(x, h) - D(x, \hat{h})]
\end{aligned}
\tag{1}
$$

where $(x, h) \sim \mathbb{P}_r$ is a pair of a heatmap and the corresponding text encoding from the training set $\mathbb{P}_r$ and $G(z, h)$ is the generated pose for the same text embedding h and a random noise vector z. For the second term, we sample a second text encoding $\hat{h}$ from the training set independently of x, i.e. $\hat{h} \sim \mathbb{P}_h$.

In order to optimize the WGAN using the dual objective of Kantorovich-Rubinstein [30], the discriminator network needs to be Lipschitz continuous, i.e. $|D(x_2) - D(x_1)| \leq |x_2 - x_1|$ for any x_1, x_2. Enforcing the Lipschitz constraint requires to constrain the gradient norm of the discriminator to 1. This can be achieved in two ways. The first approach uses a Lipschitz penality (LP) [21]:

$$
R_{LP} = \mathbb{E}_{(\hat{x},h) \sim \mathbb{P}_{\hat{x},h}}[\max(0, \|\nabla_{\hat{x},h} D(\hat{x}, h)\|_2 - 1)^2].
\tag{2}
$$

The Lipschitz penalty term is one sided and it is only active if the gradient norm is larger than 1. The second approach is termed Gradient Penalty (GP) [8]:

$$
R_{GP} = \mathbb{E}_{(\hat{x},h) \sim \mathbb{P}_{\hat{x},h}}[(\|\nabla_{\hat{x},h} D(\hat{x}, h)\|_2 - 1)^2]
\tag{3}
$$

which prefers that the gradient is one. In both cases, we sample $\hat{x}$ uniformly along straight lines between a real heatmap x and a generated heatmap $G(z, h)$ conditioned on the matching text encoding h, i.e. $\hat{x} = \epsilon x + (1 - \epsilon) \cdot G(z, h)$ where ϵ is uniformly sampled in $[0, 1]$. In our experiments, we evaluate the model when either of these terms is used. The loss function of D is therefore:

$$L_D = L_{D*} + \lambda R \qquad (4)$$

where R is either R_{LP} or R_{GP}, which are denoted by WGAN-LP or WGAN-GP, respectively, and λ is the regularization parameter for the Lipschitz constraint. To improve the training of G, a term with interpolated text encodings is added to the standard loss of G:

$$L_G = - \mathbb{E}_{z \sim \mathbb{P}_z, h \sim \mathbb{P}_h} \left[D(G(z, h), h) \right]$$
$$- \mathbb{E}_{z \sim \mathbb{P}_z, h_1, h_2 \sim \mathbb{P}_h} \left[D\left(G\left(z, \frac{h_1 + h_2}{2}\right), \frac{h_1 + h_2}{2}\right) \right]. \qquad (5)$$

Here, $h, h_1, h_2 \sim \mathbb{P}_h$ are text encodings from the training set, and $\frac{1}{2}h_1 + \frac{1}{2}h_2$ is an interpolated encoding between two training samples. The second term adds many more text encoding samples that lie near the real distribution manifold for G to learn [25].

To obtain poses from the J heatmaps generated by the model, we take the point with the maximum activation in each channel as the location of the corresponding keypoint j if its confidence value is above 0.2, otherwise we omit the keypoint. This means that our model is not limited to generate full body poses, but it can generate full body poses as well as poses of the upper body only as shown in Fig. 3.

4   Dataset and Training

Dataset. We use the COCO (Common Objects in Context) [17] dataset for training and evaluating the model. This dataset contains more than $100k$ annotated images of everyday scenes and every image has five human-written text descriptions describing the scene. Additionally, the persons are annotated by 17 body keypoints. In order to ensure that the text description refers to the person, we only include images which contain a single person and at least 8 visible keypoints.

Training. We first train an unconditional model, i.e. only pose heatmaps are used while the text is excluded. In this way, we pre-train the model on all annotated poses of COCO and we are not limited to the training samples where the text refers to the annotated person, so that the model learns to generate realistic pose. In this setting, the network parameters related to the text encoding are set to zero, while the remaining network parameters are updated. The samples are created by cropping each annotated person using the provided bounding box. In total, there are $116,021$ annotated poses in the training set and $4,812$ poses in the validation set. G is updated after every 5 iterations of updating D. We use $\lambda = 10$ as weight for the regularizer in (4).

After pre-training, we train the conditional model using both the pose heatmaps and the text from the images with a single person. For the second stage, there are in total $17,326$ images with a single annotated person in the training set and 714 images in the validation set. During training, we randomly

Fig. 3. Examples of generated poses from text. The first row shows the ground-truth pose from the validation set. The text on the top is the associated text. The three poses below each ground-truth pose are synthesized by the model from the text on the top with different noise vectors z. It can be seen that some poses such as 'throwing' (third column) are more distinct than others such as 'holding' (second column). For throwing, we can see that the wrist joint is raised. For 'working on the computer' (fifth column), we can see a sitting pose with the wrists extended appearing to be typing.

select one of the annotated captions per image. We apply an affine transformation such that the bounding box is located at the center of the image. At this stage, all network parameters are updated and we increase the weight of λ to 150 due to the small number of training samples. To improve training, we also perform some slight data augmentation on the heatmaps by randomly flipping them horizontally and rotating them between $-10°$ and $+10°$ around the center.

5   Experiments

Qualitative Results. Figure 3 shows some qualitative poses generated by the model, and the ground truth poses as reference. The captions used here are randomly selected from the validation set. We can see that the text encodings are indeed effectively guiding the synthesis of the poses, such that most of the generated poses resemble the real pose and they can reflect the given text, in particular for distinct actions.

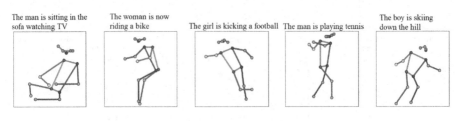

Fig. 4. Poses synthesized from text that is not part of the COCO dataset.

We also evaluated if the model overfits to the text description of the COCO dataset or if it can as well generate plausible poses from other text descriptions as well. Since we do not have any ground-truth poses, we used sentences that relate to activities, such that it is rather clear what the target poses should look like. The results appear in Fig. 4. As can be seen, the generated poses match the input text well.

It is also interesting to see what the model can produce if we only feed it with parts of a sentence. Figure 5 shows the results. It can be seen that specific verbs and nouns like 'playing' and 'tennis' matter more in interpreting the context and guiding the model in generating human poses although verbs such as 'playing' are generic and can map to various poses, unlike 'ski' for example.

Comparison to Regression. To demonstrate the benefit of representing human poses by heatmaps, we also trained a WGAN-LP that uses a fully connected network to regress the keypoint coordinates instead of a convolutional neural network that predicts a heatmap for each keypoint. In addition to the coordinate prediction, the generator predicts a probability value for the keypoint visibility. For this, we use an additional entropy loss based on the ground truth visibility flags of the training data. The regression approach is less intuitive than the heatmap-based approach and it is more difficult to train. Figure 6 shows some qualitative poses generated by the regression model. If we compare the poses with Fig. 3, we clearly see that the regression approach generates less realistic poses than the proposed approach that is based on heatmaps.

Quantitative Evaluation. In order to show that the model learned to generate unseen samples that are close to the real distribution, we calculate the distance of the nearest neighbor (NN) pose in the training set of each generated sample conditioned on the text from the validation set and denote it by $\bar{d}_{nn}^p$. This distance is calculated by generating poses conditioned on the captions from the validation set and then for each such generated pose, we take the distance to its nearest neighbor and finally average the results over all the generated poses. For comparison, in addition to training our algorithm with the Lipshitz-LP term (WGAN-LP), we also train our model using the Lipschitz-GP term (WGAN-GP) and the vanilla GAN.

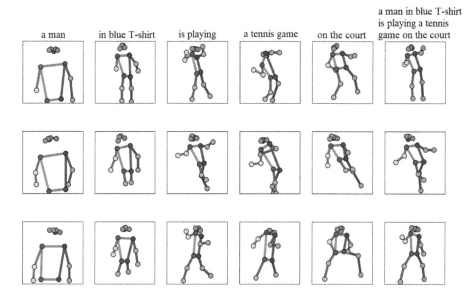

Fig. 5. Poses synthesized from parts of a sentence. The noise input in each row is fixed and varies across the rows. Here what really made the pose unique are the words 'tennis game' since the verb 'playing' can apply to many different poses.

Table 1 shows the results. The vanilla GAN has the largest distance and we observed that a mode collapse occurs such that there were many repetitions and unrealistic poses in the generated results. When the model is trained using WGAN-GP or WGAN-LP, the NN distance is much smaller where WGAN-LP performs slightly better than WGAN-GP. If a regression-based approach is used instead of the heatmaps, the distance is much higher. The nearest neighbor distance, however, measures only if the generated poses are plausible, but it does not indicate if the generated pose matches the input text. Therefore, in order to show that the text is guiding the pose generation, we calculate the distance to the pose corresponding to the nearest training sample based on the caption, which is obtained by the Euclidean distance in the text embedding space. We denote this distance by $\bar{d}^p_{t_{nn}}$. As for the other distance, WGAN-LP performs slightly better than WGAN-GP and the vanilla GAN performs worst. The regression-based approach performs also worse than the proposed method. We also report the average distance to all poses of the training set, which we denote by $\bar{d}^p_{all}$. We provide additional qualitative results for the three approaches in the supplementary material.

To further evaluate the conditional model using the poses in the validation set, we propose the following conditional measure with respect to the validation set. For a text encoding h_i in the validation set, the model synthesizes $k = 10$ poses using k different noise vectors z. We then calculate three distances for each of the k poses: the first, $\bar{d}^p_{nn}$, is the distance to the nearest neighbor among poses

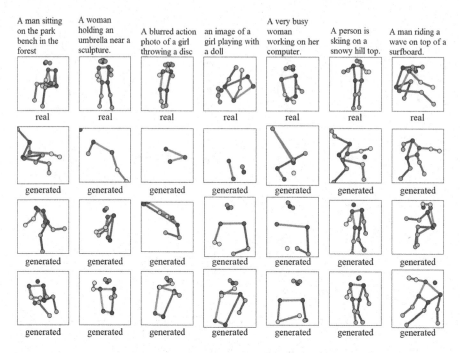

A man sitting on the park bench in the forest

A woman holding an umbrella near a sculpture.

A blurred action photo of a girl throwing a disc

an image of a girl playing with a doll

A very busy woman working on her computer.

A person is skiing on a snowy hill top.

A man riding a wave on top of a surfboard.

real   real   real   real   real   real   real

generated   generated   generated   generated   generated   generated   generated

generated   generated   generated   generated   generated   generated   generated

generated   generated   generated   generated   generated   generated   generated

Fig. 6. Examples of generated poses using a regression model. The first row shows the ground-truth pose from the validation set. The text on the top is the associated text. The three poses below each ground-truth pose are synthesized. The synthesized poses of the regression model are clearly worse than the poses synthesized by the proposed model shown in Fig. 3.

in the validation set; the second, $\bar{d}_{gt}^p$, is the distance to the ground truth pose, and the third, $\bar{d}_{all}^p$, is the average distance to all poses in the validation set. Finally, we average the distances over the generated k poses over all samples. The results are reported in Table 2. As for the validation set, we observe that the vanilla GAN struggles to generate realistic poses and WGAN-LP performs slightly better than WGAN-GP. The regression of keypoint coordinates performs also worse than the heatmap representation. Furthermore, we calculate the mean distance in the text encoding space. To this end, we obtain for each generated pose the nearest neighbor pose from the validation set. We then compute the distance between the input text and the text of the corresponding nearest neighbor pose. We average the distances over all generated poses. This measure is denoted by $\bar{d}_{p_{nn}}^t$. The differences are smaller compared to the pose distances, but it still shows that the WGANs outperform the vanilla GAN.

Table 1. Quantitative evaluation with respect to the training set. Regression indicates that the keypoint coordinates are regressed instead of being detected using a heatmap representation.

GAN model	Pose distance		
	$\bar{d}_{nn}^p$	$\bar{d}_{t_{nn}}^p$	$\bar{d}_{all}^p$
Vanilla GAN	205.2	344.9	351.1
WGAN-GP	82.9	260.5	293.8
WGAN-LP regression	98.0	268.1	291.1
WGAN-LP	77.2	253.6	287.2

Table 2. Quantitative evaluation with respect to the validation set. Regression indicates that the keypoint coordinates are regressed instead of using a heatmap representation.

GAN model	Pose distance			Text distance
	$\bar{d}_{nn}^p$	$\bar{d}_{gt}^p$	$\bar{d}_{all}^p$	$\bar{d}_{p_{nn}}^t$
Vanilla GAN	218.8	343.2	352.0	10.8
WGAN-GP	110.2	255.7	293.4	10.5
WGAN-LP regression	128.2	264.7	290.8	10.5
WGAN-LP	102.3	246.0	286.9	10.5

Interpolation Test. Another interesting qualitative measure is the interpolation between two text descriptions and observing the generated poses. If the generated poses show smooth transitions between the interpolations, we can conclude that the model learned a proper distribution instead of just having memorized the training samples [3]. Given two embedded text descriptions h_1 and h_2, we interpolate between them by $\hat{h} = w \cdot h_1 + (1-w) \cdot h_2$ with $w \in \{1, 0.75, 0.5, 0.25, 0\}$. For this experiment, we keep the noise z fixed. Figure 7 shows two interpolation examples. In the first example, we interpolate between 'The man is standing on the beach' and 'The man is holding a surfboard'. We observe that the right arm gradually moves up for the holding pose. We also observe that the full body pose is generated at the beginning, but the camera gets closer on the right hand side and only two-thirds of the person are visible. The second example interpolates between 'The boy has a tennis racket in his hands' and 'The boy is going to serve the ball'.

User Study. For the subjective evaluation, we have designed an online questionnaire in which 20 text descriptions from the validation set are taken. For each text description, a user is presented with two human poses, in which one is the real pose matching the text, and the other is synthesized by the model conditioned on this text. The 20 captions are randomly selected from the validation set and the generated poses have not been cherry-picked. The user is asked to choose which of the two poses matches the caption better or if they match the

Fig. 7. Interpolation results of text encoding. In each row, the leftmost and rightmost poses are synthesized from the captions. The three poses in the middle are synthesized from interpolations of the encodings of the two captions while z is kept fixed.

Table 3. The percentage of the users choosing the matching pose as the real pose, generated pose or "equally well".

Real pose	Generated pose	Equally well
48.81%	35.31%	15.88%

text equally well. The results are summarized in Table 3. Eighty people in total participated in the survey. The ratio between choosing generated and real poses is around 5:7. And for more than 50% of the time, the users cannot correctly distinguish the generated pose from the real one, i.e., they either choose the generated pose or rate the poses equally well.

6   Conclusion

In this work, we have addressed the task of human pose synthesis from text for highly complex poses. We have designed an effective model using a conditional Wasserstein GAN that generates plausible matching poses from text descriptions. We have demonstrated by qualitative and quantitative results on the COCO dataset that the proposed approach is effective, and additionally outperforms a vanilla GAN and a regression-based approach. We have also conducted a user study that confirmed our results. The model was also able to interpolate poses between two text descriptions. Furthermore, we have shown that the model generalizes well and can additionally generate plausible poses for unseen sentences that are not part of the COCO dataset.

Acknowledgement. The work has been funded by the Deutsche Forschungsgemeinschaft (DFG, German Research Foundation) GA 1927/5-1 and the ERC Starting Grant ARCA (677650).

References

1. Arjovsky, M., Chintala, S., Bottou, L.: Wasserstein generative adversarial networks. In: International Conference on Machine Learning (2017)
2. Bojanowski, P., Grave, E., Joulin, A., Mikolov, T.: Enriching word vectors with subword information. Trans. Assoc. Comput. Ling. **5** 23–25 (2017)
3. Borji, A.: Pros and cons of gan evaluation measures. Computer Vision and Image Understanding **179** (2019)
4. Chu, C., Zhmoginov, A., Sandler, M.: Cyclegan, a master of steganography. arXiv preprint arXiv:1712.02950 (2017)
5. Dai, B., Fidler, S., Urtasun, R., Lin, D.: Towards diverse and natural image descriptions via a conditional gan. In: IEEE International Conference on Computer Vision (2017)
6. Goodfellow, I., et al.: Generative adversarial nets. Adv. Neural Inf. Process. Syst. **27** 501–505 (2014)
7. Grabner, H., Gall, J., Van Gool, L.: What makes a chair a chair? In: IEEE Conference on Computer Vision and Pattern Recognition (2011)
8. Gulrajani, I., Ahmed, F., Arjovsky, M., Dumoulin, V., Courville, A.C.: Improved training of wasserstein gans. In: Advances in Neural Information Processing Systems (2017)
9. Gupta, A., Satkin, S., Efros, A.A., Hebert, M.: From 3d scene geometry to human workspace. In: IEEE Conference on Computer Vision and Pattern Recognition (2011)
10. Hinton, G.E.: Deep belief networks. Scholarpedia **4**(5) (2009)
11. Kingma, D.P., Welling, M.: Auto-encoding variational bayes. arXiv preprint arXiv:1312.6114 (2013)
12. Lassner, C., Pons-Moll, G., Gehler, P.V.: A generative model of people in clothing. In: IEEE International Conference on Computer Vision (2017)
13. Li, B., Qi, X., Lukasiewicz, T., Torr, P.: Controllable text-to-image generation. In: Advances in Neural Information Processing Systems (2019)
14. Li, W., Zhang, P., Zhang, L., Huang, Q., He, X., Lyu, S., Gao, J.: Object-driven text-to-image synthesis via adversarial training. In: IEEE Conference on Computer Vision and Pattern Recognition (2019)
15. Li, X., Liu, S., Kim, K., Wang, X., Yang, M.H., Kautz, J.: Putting humans in a scene: learning affordance in 3d indoor environments. In: IEEE Conference on Computer Vision and Pattern Recognition (2019)
16. Li, Y., et al.: Storygan: a sequential conditional gan for story visualization. In: IEEE Conference on Computer Vision and Pattern Recognition (2019)
17. Lin, T.Y., et al.: Microsoft coco: common objects in context. In: European Conference on Computer Vision (2014)
18. Loper, M., Mahmood, N., Romero, J., Pons-Moll, G., Black, M.J.: SMPL: a skinned multi-person linear model. ACM Trans. Graph. **34**, 432 (2015)
19. Ma, L., Jia, X., Sun, Q., Schiele, B., Tuytelaars, T., Van Gool, L.: Pose guided person image generation. In: Advances in Neural Information Processing Systems (2017)
20. Mikolov, T., Grave, E., Bojanowski, P., Puhrsch, C., Joulin, A.: Advances in pre-training distributed word representations. In: International Conference on Language Resources and Evaluation (2018)
21. Petzka, H., Fischer, A., Lukovnicov, D.: On the regularization of wasserstein gans. In: International Conference on Learning Representations (2018)

22. Qiao, T., Zhang, J., Xu, D., Tao, D.: Learn, imagine and create: Text-to-image generation from prior knowledge. Advances in Neural Information Processing Systems **32** (2019)
23. Qiao, T., Zhang, J., Xu, D., Tao, D.: Mirrorgan: learning text-to-image generation by redescription. In: IEEE Conference on Computer Vision and Pattern Recognition (2019)
24. Radford, A., Metz, L., Chintala, S.: Unsupervised representation learning with deep convolutional generative adversarial networks. In: International Conference on Learning Representations (2016)
25. Reed, S., Akata, Z., Yan, X., Logeswaran, L., Schiele, B., Lee, H.: Generative adversarial text to image synthesis. In: International Conference on Machine Learning (2016)
26. Reed, S.E., Akata, Z., Mohan, S., Tenka, S., Schiele, B., Lee, H.: Learning what and where to draw. Advances in Neural Information Processing Systems**29** (2016)
27. Salakhutdinov, R.: Learning deep generative models. Ann. Rev. Stat. Appl. **2** (2015)
28. Tan, H., Liu, X., Li, X., Zhang, Y., Yin, B.: Semantics-enhanced adversarial nets for text-to-image synthesis. In: IEEE International Conference on Computer Vision (2019)
29. Tompson, J.J., Jain, A., LeCun, Y., Bregler, C.: Joint training of a convolutional network and a graphical model for human pose estimation. In: Advances in Neural Information Processing Systems (2014)
30. Villani, C.: Optimal transport: old and new. Springer, Cham (2008)
31. Wei, S.E., Ramakrishna, V., Kanade, T., Sheikh, Y.: Convolutional pose machines. In: IEEE Conference on Computer Vision and Pattern Recognition (2016)
32. Xu, T., Zhang, P., Huang, Q., Zhang, H., Gan, Z., Huang, X., He, X.: Attngan: Fine-grained text to image generation with attentional generative adversarial networks. In: IEEE Conference on Computer Vision and Pattern Recognition (2018)
33. Zhang, H., et al.: Stackgan: text to photo-realistic image synthesis with stacked generative adversarial networks. In: IEEE International Conference on Computer Vision (2017)
34. Zhang, Y., Hassan, M., Neumann, H., Black, M.J., Tang, S.: Generating 3d people in scenes without people. In: IEEE Conference on Computer Vision and Pattern Recognition (2020)
35. Zhou, X., Huang, S., Li, B., Li, Y., Li, J., Zhang, Z.: Text guided person image synthesis. In: IEEE Conference on Computer Vision and Pattern Recognition (2019)
36. Zhu, S., Urtasun, R., Fidler, S., Lin, D., Change Loy, C.: Be your own prada: fashion synthesis with structural coherence. In: IEEE International Conference on Computer Vision (2017)

Long-Term Anticipation of Activities
with Cycle Consistency

Yazan Abu Farha[1]([✉]), Qiuhong Ke[2], Bernt Schiele[3], and Juergen Gall[1]

[1] University of Bonn, Bonn, Germany
abufarha@iai.uni-bonn.de
[2] The University of Melbourne, Melbourne, Australia
[3] MPI Informatics, Saarbrücken, Germany

Abstract. With the success of deep learning methods in analyzing activities in videos, more attention has recently been focused towards anticipating future activities. However, most of the work on anticipation either analyzes a partially observed activity or predicts the next action class. Recently, new approaches have been proposed to extend the prediction horizon up to several minutes in the future and that anticipate a sequence of future activities including their durations. While these works decouple the semantic interpretation of the observed sequence from the anticipation task, we propose a framework for anticipating future activities directly from the features of the observed frames and train it in an end-to-end fashion. Furthermore, we introduce a cycle consistency loss over time by predicting the past activities given the predicted future. Our framework achieves state-of-the-art results on two datasets: the Breakfast dataset and 50Salads.

1 Introduction

Humans spend a significant time of their life thinking about the future. Whether thinking about their future dream job, or planning for the next research project. Even unconsciously, people tend to anticipate future trajectories of moving agents in the surrounding environment and the activities that they will be doing in the near future. Such anticipation capability is considered a sign of intelligence and an important factor in determining how we interact with the environment and how we make decisions.

Since anticipation is an important intrinsic capability of human beings, researchers have recently tried to model this capability and embed it in intelligent and robotic systems. For example, several approaches have been proposed to anticipate future trajectories of pedestrians [4,18], or semantic segmentation of future frames in video [5,25]. These approaches have many applications in

Electronic supplementary material The online version of this chapter (https://doi.org/10.1007/978-3-030-71278-5_12) contains supplementary material, which is available to authorized users.

Z. Akata et al. (Eds.): DAGM GCPR 2020, LNCS 12544, pp. 159–173, 2021.
https://doi.org/10.1007/978-3-030-71278-5_12

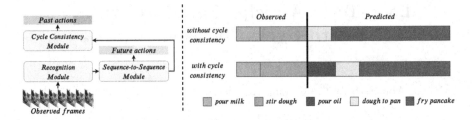

Fig. 1. (Left) Overview of the proposed approach, which is trained end-to-end and includes a cycle consistency module. (Right) Effect of the cycle consistency module. Without the cycle consistency, the network anticipates actions that are plausible based on the previous actions. However, in some cases an essential action is missing. In this example *pour oil*. By using the cycle consistency, we enforce the network to verify if all required actions have been done before. For the action *fry pancake*, *pouring oil* into the pan is required and the cycle consistency resolves this issue.

autonomous driving and navigation. Another line of research focuses on anticipating future activities [19,21,44], which has potential applications in surveillance and human-robot interaction.

While anticipating the next action a few seconds in the future has been addressed in [11,44,46], such short time horizon is insufficient for many applications. Service robots, for example, where a robot is continuously interacting with a human, require anticipating a longer time horizon, which rather includes a sequence of future activities than only predicting the next action. By anticipating longer in the future, such robots can plan ahead and accomplish their tasks more efficiently. Recent approaches, therefore, focused on increasing the prediction horizon up to several minutes and predict multiple action segments in the future [3,10,16]. While these approaches successfully predict future activities and their duration, they decouple the semantic interpretation of the observed sequence from the anticipation task using a two-step approach.

Separating the understanding of the past and the anticipation of future has several disadvantages. First, the model is not trained end-to-end, which means that the approach for temporal action segmentation is not optimized for the anticipation task. Second, if there are any mistakes in the temporal action segmentation, these mistakes will be propagated and effect the anticipated activities. Finally, the action labels do not represent all information in the observed video that is relevant for anticipating the future. In contrast to these approaches, we propose a sequence-to-sequence model that directly maps the sequence of observed frames to a sequence of future activities and their duration. We then cast the understanding of the past as an auxiliary task by proposing a recognition module, which consists of a temporal convolutional neural network and a recognition loss, that is combined with the encoder.

Furthermore, as we humans can reason about the past given the future, previous works only aim to predict the future, and it is intuitive that forcing the network to predict the past as well is helpful. To this end, we propose a cycle consistency module that predicts the past activities given the predicted future. This module verifies if, for the predicted future actions, all required actions have been done before as illustrated in Fig. 1. In this example, the actions *pour dough to pan* and *fry pancake* are plausible given the previous actions, but the action *pour oil* has been missed. The cycle consistency module then predicts from the anticipated actions, the observed actions. Since *pour dough to pan* and *fry pancake* are the inputs for the cycle consistency module, it will predict all the required preceding actions such as *pour milk, stir dough* and *pour oil*. However, as *pour oil* is not part of the observations, the cycle consistency module will have a high error, which steers the network to predict *pour oil* in the future actions.

Our contribution is thus three folded: First, we propose an end-to-end model for anticipating a sequence of future activities and their durations. Second, we show that the proposed recognition module improves the sequence-to-sequence model. Third, we propose a cycle consistency module that verifies the predictions.

We evaluate our model on two datasets with untrimmed videos containing many action segments: the Breakfast dataset [20] and 50Salads [41]. Our model is able to predict the future activities and their duration accurately achieving superior results compared to the state-of-the-art methods.

2   Related Work

Early Action Detection: The early action detection task tries to recognize an ongoing activity given only a partial observation of that activity. An initial work addressing this task by Ryoo [36] is based on a probabilistic formulation using dynamic bag-of-words of spatio-temporal features. Hoai and De la Torre [14] used a max-margin formulation. More recent approaches use recurrent neural networks with special loss functions to encourage early activity detection [26,37]. In contrast to these approaches, we anticipate future activities without even any partial observations.

Future Prediction: Predicting the future has become one of the main research topics in computer vision. Many approaches have been proposed to predict future frames [24,30], future human trajectories [4,18], future human poses [15,29,35], image semantic segmentation [5,25], or even full sentences describing future frames or steps in recipes [27,38]. However, low level representations like pixels of frames or very high level natural language sentences cannot be used directly. A complementary research direction is concentrated on anticipating activity labels. Lan *et al.* [21] predicted future actions using hierarchical representations in a max-margin framework. Koppula and Saxena [19] used a spatio-temporal graph representation to encode the observed activities and then predict object affordances, trajectories, and sub-activities. Instead of directly predicting the future action labels, several approaches were proposed to predict future visual representations and then a classifier is trained on top of the predicted representations to predict the future labels [11,12,34,39,44,46]. Predicting future representations has also been used in the literature for unsupervised representations learning [40]. Other approaches use multi-task learning to predict the future activity and its starting time [28,31], or the future activity and its location [23,42]. Miech *et al.* [32] modeled the transition probabilities between actions and combine it with a predictive model to anticipate future action labels. There is a parallel line of research addressing the anticipation task in egocentric videos [8,9]. However, these approaches are limited to very few seconds in the future. In contrast to these approaches, we address the anticipation task for a longer time horizon.

Recently, more effort has been dedicated to increase the anticipation horizon. Several methods have been proposed to anticipate activities several minutes in the future using RNNs [2,3], temporal convolutions with time-variable [16], or memory networks [10]. While these approaches manage to anticipate activities for a longer time horizon, their performance is limited. The approach of [10] relies on the ground-truth action labels of the observations, whereas the methods in [2, 3,16] follow a two-step approach. *I.e.* they first infer the activities in the observed part, then anticipate the future activities with their corresponding duration. These two steps are trained separately, which prevents the model from utilizing the visual cues in the observed frames. In contrast to these approaches, our model is trained in one step in an end-to-end fashion.

Cycle Consistency: Cycle consistency has been widely used in computer vision. It was used to learn dense correspondence [47], image-to-image translation [48], and depth estimation [13]. Recent approaches used cycle consistency in the temporal dimension [7,45]. Dwibedi *et al.* [7] introduced an approach to learn representations using video alignment as a proxy task and cycle consistency for training. In [13], the appearance consistency between consecutive video frames is used to learn representations that generalize to different tasks. Motivated by the success of cycle consistency in various applications, we apply it as an additional supervisory signal to predict the future activities.

3   The Anticipation Framework

Given a partially observed video with many activities, we want to predict all the activities that will be happening in the remainder of that video with their corresponding duration. Assuming that the observed part consists of t_o frames $X_{1:t_o} = (x_1, \ldots, x_{t_o})$ corresponding to n activities $A_{1:n} = (A_1, \ldots, A_n)$, our goal is to predict the future activities $A_{n+1:N} = (A_{n+1}, \ldots, A_N)$ and their corresponding duration $\ell_{n+1:N} = (\ell_{n+1}, \ldots, \ell_N)$, where N is the total number of activities in that video. In contrast to the previous approaches that use only the action labels of the observed frames for anticipating the future, we propose to anticipate the future activities directly from the observed frames. First, we propose a sequence-to-sequence model that maps the sequence of features from the observations to a sequence of future activities and their duration. Then, we introduce a cycle consistency module that predicts the past activities given the predicted future. The motivation of this module is to force the sequence-to-sequence module to encode all the relevant information in the observed frames

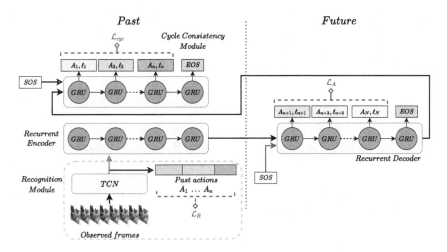

Fig. 2. Overview of the anticipation framework. The observed frames are passed through a TCN-based recognition module which produces discriminative features for the sequence-to-sequence model. The sequence-to-sequence model predicts the future activities with their duration. In addition, we enforce cycle consistency over time by predicting the past activities given the predicted future.

and verify if the predictions are plausible. Finally, we extend the sequence-to-sequence model with a recognition module that generates discriminative features that capture the relevant information for anticipating the future. An overview of the proposed model is illustrated in Fig. 2. Our framework is hence trained using an anticipation loss, a recognition loss, and a cycle consistency loss. In the following sections, we describe in detail these modules.

3.1 Sequence-to-Sequence Model

The sequence-to-sequence model maps the sequence of observed frames to a sequence of future activities and their duration. For this model, we use a recurrent encoder-decoder architecture based on gated recurrent units (GRUs).

Recurrent Encoder. The purpose of the recurrent encoder is to encode the observed frames in a single vector which will be used to decode the future activities. Given the input features X, the recurrent encoder passes these features through a single layer with a gated recurrent unit (GRU)

$$h_t^e = GRU(x_t, \ h_{t-1}^e), \tag{1}$$

where x_t is the input feature for frame t, and h_{t-1}^e is the hidden state at the previous time step. The hidden state at the last time step $h_{t_o}^e$ encodes the observed frames and will be used as an initial state for the decoder.

Recurrent Decoder. Given the output of the encoder $h_{t_o}^e$, the recurrent decoder predicts the future activities and their relative duration. The decoder also consists of a single layer with a gated recurrent unit (GRU). The hidden state at each step is updated using the GRU update rules

$$h_m^d = GRU(A_{m-1}, h_{m-1}^d), \tag{2}$$

where the input at each time step A_{m-1} is the predicted activity label for the previous step. At training time, the ground-truth label is used as input. Whereas the predicted label is used during inference time. For the first time step, a special start-of-sequence (SOS) symbol is used as input. Given the hidden state h_m^d at each time step, the future activity label A_m and its relative duration ℓ_m are predicted using a fully connected layer, *i.e.*

$$\tilde{A}_m = W_A h_m^d + b_A, \tag{3}$$

$$\hat{\ell}_m = W_\ell h_m^d + b_\ell, \tag{4}$$

where $\tilde{A}_m$ is the predicted logits for the future activity, and $\hat{\ell}_m$ is the predicted duration in log space. To get the relative duration, we apply softmax over the time steps

$$\tilde{\ell}_m = \frac{e^{\hat{\ell}_m}}{\sum_k e^{\hat{\ell}_k}}. \tag{5}$$

The decoder keeps predicting future activity labels and the corresponding duration until a special end-of-sequence (EOS) symbol is predicted. As a loss function, we use a combination of a cross entropy loss for the activity label and a mean squared error (MSE) for the predicted relative duration

$$\mathcal{L}_A = \mathcal{L}_{CE} + \mathcal{L}_{MSE}, \tag{6}$$

$$\mathcal{L}_{CE} = \frac{1}{N-n} \sum_{m=n+1}^{N} -log(\tilde{a}_{m,A}), \tag{7}$$

$$\mathcal{L}_{MSE} = \frac{1}{N-n} \sum_{m=n+1}^{N} (\tilde{\ell}_m - \ell_m)^2, \tag{8}$$

where $\mathcal{L}_A$ is the anticipation loss, $\tilde{a}_{m,A}$ is the the predicted probability for the ground truth activity label at step m, n is the number of observed action segments and N is the total number of action segments in the video.

Since the input to the encoder are frame-wise features which might be very long, the output of the encoder $h_{t_o}^e$ might not be able to capture all the relevant information. To alleviate this problem we combine the decoder with an attention mechanism using a multi-head attention module [43]. Additional details are given in the supplementary material.

3.2 Cycle Consistency

Since predicting the future from the past and the past from the future should be consistent, we propose an additional cycle consistency loss. Given the predicted future activities, we want to predict the past activities and their duration. This requires the predicted future to be good enough to predict the past. The cycle consistency loss has two benefits. First, it verifies if, for the predicted future actions, the actions that have been previously observed are plausible. Second, it encourages the recurrent decoder to keep the most important information of the observed sequence until the end, instead of storing only the information of the previous anticipated activity. In this way, a wrong prediction does not necessary propagate since the observed sequence is kept in memory.

The cycle consistency module is similar to the recurrent decoder and consists of a single layer with GRU, however, it predicts the past instead of the future. The hidden state of this GRU is initialized with the last hidden state of the recurrent decoder and at each step the hidden state is updated as follows

$$h_m^{cyc} = GRU(A_{m-1},\ h_{m-1}^{cyc}). \tag{9}$$

Given the hidden state h_m^{cyc} at step m, an activity label of the observations and its relative duration are predicted using a fully connected layer. The loss function is also similar to the recurrent decoder

$$\mathcal{L}_{cyc} = \mathcal{L}_{CE} + \mathcal{L}_{MSE}, \tag{10}$$

$$\mathcal{L}_{CE} = \frac{1}{n} \sum_{m=1}^{n} -log(\tilde{a}_{m,A}), \tag{11}$$

$$\mathcal{L}_{MSE} = \frac{1}{n} \sum_{m=1}^{n} (\tilde{\ell}_m - \ell_m)^2, \tag{12}$$

where $\mathcal{L}_{cyc}$ is the cycle consistency loss, $\mathcal{L}_{CE}$ and $\mathcal{L}_{MSE}$ are the cross entropy loss and MSE loss applied on the past activity labels and their relative duration.

While the mapping from past to future can be multi-modal, this does not limit the applicability of the cycle consistency module. Since the cycle consistency module is conditioned on the predicted future, no matter what mode is predicted, the cycle consistency makes sure it is plausible. This also applies to the inverse mapping. As there is a path from the observed frames to the cycle consistency module through the sequence-to-sequence model, there is no ambiguity in which past activities have been observed.

3.3 Recognition Module

In the sequence-to-sequence model, the input of the recurrent encoder are the frame-wise features. However, directly passing the features to the encoder is sub-optimal as the encoder might struggle to capture all the relevant information for anticipating the future activities. As past activities provide a strong signal for anticipating future activities, we use a recognition module that learns discriminative features of the observed frames. These features will then serve as an input for the sequence-to-sequence model to anticipate the future activities. Given the success of temporal convolutional networks (TCNs) in analyzing activities in videos [1,22], we use a similar network for our recognition module. Besides being a strong model for analyzing videos, TCNs are fully differentiable and can be integrated in our framework without preventing end-to-end training. For our module, we use a TCN similar to the one proposed in [1]. The TCN consists of several layers of dilated 1D convolutions where the dilation factor is doubled at each layer. The operations at each layer can be formally described as follows

$$\hat{F}_l = ReLU(W_1 * F_{l-1} + b_1), \tag{13}$$

$$F_l = F_{l-1} + W_2 * \hat{F}_l + b_2, \tag{14}$$

where F_l is the output of layer l, $*$ is the convolution operator, $W_1 \in \mathbb{R}^{3 \times K \times K}$ are the weights of the dilated convolutional filters with kernel size 3 and K is the number of the filters, $W_2 \in \mathbb{R}^{1 \times K \times K}$ are the weights of a 1×1 convolution, and $b_1, b_2 \in \mathbb{R}^K$ are bias vectors. The input of the first layer F_0 is obtained by applying a 1×1 convolution over the input features X.

The output of the last dilated convolutional layer serves as input to the subsequent modules. To make sure that these features are discriminative enough, we add a classification layer that predicts the action label at each observed frame

$$\tilde{Y}_t = Softmax(W f_{L,t} + b), \tag{15}$$

where $\tilde{Y}_t$ contains the class probabilities at time t, $f_{L,t} \in \mathbb{R}^K$ is the output of the last dilated convolutional layer at time t, $W \in \mathbb{R}^{C \times K}$ and $b \in \mathbb{R}^C$ are the weights and bias for the 1×1 convolutional layer, where C is the number of action classes. To train this module we use a cross entropy loss

$$\mathcal{L}_R = \frac{1}{t_o} \sum_{t=1}^{t_o} -log(\tilde{y}_{t,c}), \tag{16}$$

where $\tilde{y}_{t,c}$ is the predicted probability for the ground truth label c at time t, and t_o is the number of observed frames.

3.4   Loss Function

To train our framework, we sum up all the three mentioned losses

$$\mathcal{L} = \mathcal{L}_A + \mathcal{L}_R + \mathcal{L}_{cyc}, \tag{17}$$

where $\mathcal{L}_A$ is the anticipation loss, $\mathcal{L}_R$ is the recognition loss, and $\mathcal{L}_{cyc}$ is the cycle consistency loss.

Table 1. Ablation study on the Breakfast dataset. Numbers represent mean over classes (MoC) accuracy.

Observation %	20%				30%			
Prediction %	10%	20%	30%	50%	10%	20%	30%	50%
S2S	23.22	20.92	20.10	20.05	26.43	24.38	24.13	23.38
S2S + TCN	14.52	13.55	13.15	12.70	14.83	14.11	13.61	13.01
S2S + TCN + $\mathcal{L}_R$	24.72	22.43	21.70	21.77	28.35	26.29	25.01	24.47
S2S + TCN + $\mathcal{L}_R$ + $\mathcal{L}_{cyc}$	25.16	22.73	22.22	**22.01**	28.07	26.25	25.12	24.81
S2S + TCN + $\mathcal{L}_R$ + $\mathcal{L}_{cyc}$ + attn.	**25.88**	**23.42**	**22.42**	21.54	**29.66**	**27.37**	**25.58**	**25.20**

4 Experiments

We evaluate the proposed model on two datasets: the Breakfast dataset [20] and 50Salads [41]. In all experiments, we report the average of three runs.

The **Breakfast** dataset is a collection of 1,712 videos with overall 66.7 hours and roughly 3.6 million frames. Each video belongs to one out of ten breakfast related activities, such as make tea or pancakes. The video frames are annotated with fine-grained action labels like *pour milk* or *fry egg*. Overall, there are 48 different actions. On average, each video contains 6 action instances and is 2.3 minutes long. For evaluation, we use the standard 4 splits as proposed in [20] and report the average.

The **50Salads** dataset contains 50 videos showing people preparing different kinds of salad. These videos are relatively long with an average of 6.4 minutes and 20 action instances per video. The video frames are annotated with 17 fine-grained action labels like *cut tomato* or *peel cucumber*. For evaluation, we use five-fold cross-validation and report the average as in [41].

We follow the state-of-the-art evaluation protocol and report the mean over classes (MoC) accuracy for different observation/prediction percentages [3,16].

Implementation Details. For the recognition module, we used a TCN with 10 layers and 64 filters in each layer. The number of units in the GRU cells is set to 512. For each training video, we generate two training examples with 20% and 30% observation percentage. The prediction percentage is always set to 50%. All the models are trained for 80 epochs using Adam optimizer [17]. We set the learning rate to 0.001 and reduce it every 20 epochs with a factor of 0.8. For both datasets, we extract I3D [6] features for the video frames using both RGB and flow streams and sub-sample them at 5 frames per second.

4.1 Ablation Analysis

In this section, we analyze the impact of the different modules in our framework on the anticipation performance. This analysis is conducted on the Breakfast

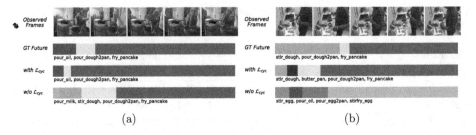

Fig. 3. Impact of the cycle consistency loss. Cycle consistency verifies if, for the predicted future actions, all required actions have been done before and no essential action is missing (a), and it further encourages the decoder to keep the important information from the observations until the end, which results in better predictions (b).

dataset and the results are shown in Table 1. Additional ablation experiments to study the impact of the input to the recurrent encoder and decoder are provided in the supplementary material.

Impact of the Recognition Module: The recognition module consists of two parts: a TCN and a recognition loss $\mathcal{L}_R$. Starting from only the sequence-to-sequence module (S2S), we can achieve a good accuracy in the range $20\% - 27\%$. By combining the sequence-to-sequence module with the recognition module (S2S + TCN + $\mathcal{L}_R$), we gain an improvement of $1\% - 2\%$ for each observation-prediction percentage. This indicates that recognition helps the anticipation task. We also evaluate the performance when the TCN is combined with the sequence-to-sequence module without the recognition loss (S2S + TCN). As shown in Table 1, the results are worse than using only the sequence-to-sequence module if we do not apply the recognition loss. This can be explained by the structure of the network. The recurrent encoder maps the features extracted by the TCN from all frames to a single vector. Without additional loss for the recognition module, the gradient vanishes and the parameters of the TCN are not well estimated. Nevertheless, by just applying the recognition loss, we get enough supervisory signal to train the TCN and improve the overall anticipation accuracy. This also highlights that the improvements from the recognition module are due to the additional recognition task and not because of having more parameters.

Impact of the Cycle Consistency Loss: The cycle consistency module predicts the past activities from the predicted future. The intuition is that to be able to predict the past activities, the predicted future activities have to be correct. As shown in Table 1, using the cycle consistency loss gives an additional improvement on the anticipation accuracy. The cycle consistency loss verifies if, for the predicted future actions, all required actions have been done before and no essential action is missing. For example in Fig. 3(a), the model observes *spoon flour, crack egg, pour milk,* and *stir dough.* Without the cycle consistency the

Table 2. Comparison between the two-step approach and ours on the Breakfast dataset. Numbers represent mean over classes (MoC) accuracy.

Observation %	20%				30%			
Prediction %	10%	20%	30%	50%	10%	20%	30%	50%
Two-Step	23.58	21.90	20.80	20.18	27.80	25.21	23.32	22.96
Ours	**25.88**	**23.42**	**22.42**	**21.54**	**29.66**	**27.37**	**25.58**	**25.20**

network did not predict the action *pour oil*, which is required to *fry pancake*. By using cycle consistency this issue is resolved. Cycle consistency also forces the decoder to remember all observed activities. As illustrated in Fig. 3 (b), the model observes *spoon flour*, *crack egg*, *pour milk*, *butter pan*, and *stir dough*. Without the cycle consistency, the network predicts the action *stirfry egg*, which would have been plausible if *spoon flour* and *pour milk* were not part of the observations. Since the cycle consistency encourages the decoder to use all observed actions for anticipation, the activity *fry pancake* is correctly anticipated.

Impact of the Attention Module: Finally, using the full model by combining the recurrent decoder with the multi-head attention module further improves the results by roughly 1%. As shown in Table 1, the gain from using the attention module is higher when the observation percentage is 30%. This is mainly because of the encoder module. Given the observed frames, the encoder tries to encode them in a single vector. This means that the encoder has to throw away more information from the long sequences compared to shorter sequences. In this case, the attention module can help in capturing some of the lost information in the encoder output by attending on the relevant information in the observations.

4.2   End-to-End vs. Two-Step Approach

To illustrate the benefits of end-to-end learning over two-step approaches, we compare our framework with its two-step counterpart. For this comparison, we first train the recognition module from our framework and then fix the weights of the TCN and train the remaining components of our model with the anticipation loss and the cycle consistency loss. Table 2 shows the results of our framework compared to the two-step approach on the Breakfast dataset with different observation and prediction percentages. As shown in the table, our framework outperforms the two-step approach with a large margin of up to 2.3%. This highlights the benefits of end-to-end approaches where the model can capture the relevant information in the observed frames to anticipate the future. On the contrary, two-step approaches can only utilize the label information of the observed frames that are not optimized for the anticipation task which is sub-optimal.

Table 3. Comparison with the state-of-the-art. Numbers represent MoC accuracy.

Observation %	20%				30%			
Prediction %	10%	20%	30%	50%	10%	20%	30%	50%
Breakfast								
RNN model [3]	18.11	17.20	15.94	15.81	21.64	20.02	19.73	19.21
CNN model [3]	17.90	16.35	15.37	14.54	22.44	20.12	19.69	18.76
RNN [3] + TCN	05.93	05.68	05.52	05.11	08.87	08.90	07.62	07.69
CNN [3] + TCN	09.85	09.17	09.06	08.87	17.59	17.13	16.13	14.42
UAAA (mode) [2]	16.71	15.40	14.47	14.20	20.73	18.27	18.42	16.86
Time-Cond. [16]	18.41	17.21	16.42	15.84	22.75	20.44	19.64	19.75
Ours	**25.88**	**23.42**	**22.42**	**21.54**	**29.66**	**27.37**	**25.58**	**25.20**
50Salads								
RNN model [3]	30.06	25.43	18.74	13.49	30.77	17.19	14.79	09.77
CNN model [3]	21.24	19.03	15.98	09.87	29.14	20.14	17.46	10.86
RNN [3] + TCN	32.31	25.51	19.10	14.15	26.14	17.69	16.33	12.97
CNN [3] + TCN	16.02	14.68	12.09	09.89	19.23	14.68	13.18	11.20
UAAA (mode) [2]	24.86	22.37	19.88	12.82	29.10	20.50	15.28	12.31
Time-Cond. [16]	32.51	27.61	21.26	**15.99**	**35.12**	**27.05**	**22.05**	15.59
Ours	**34.76**	**28.41**	**21.82**	15.25	34.39	23.70	18.95	**15.89**

4.3   Comparison with the State-of-the-Art

In this section, we compare our framework with the state-of-the-art methods on both the Breakfast dataset and 50Salads. We follow the same protocol and report results for different observation and prediction percentages. Table 3 shows the results on both datasets. All the previous approaches follow the two-step approach by inferring the action labels of the observed frames first and then use these labels to anticipate the future activities. As shown in Table 3, our framework outperforms all the state-of-the-art methods by a large margin of roughly $5\% - 8\%$ for each observation-prediction percentage pair on the Breakfast dataset. An interesting observation is that all the previous approaches achieve comparable results despite the fact that they are using different network architectures based on RNNs [2,3], CNNs [3], or even temporal convolution [16]. On the contrary, our framework clearly outperforms these approaches with the advantage that it was trained in an end-to-end fashion.

For 50Salads, our model outperforms the state-of-the-art in 50% of the cases. This is mainly because 50Salads is a small dataset. Since our model is trained end-to-end, it requires more data to show the benefits over two-step approaches.

Since the state-of-the-art methods like [3] use an RNN-HMM model [33] for recognition, we also report the results of [3] with our TCN as a recognition model. The results are shown in Table 3 (RNN [3] + TCN and CNN [3] + TCN). Our model outperforms these methods even when they are combined with TCN.

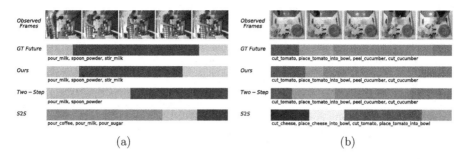

Fig. 4. Qualitative results for anticipating future activities. (a) An example from the Breakfast dataset for the case of observing 20% of the video and predicting the activities in the following 50%. (b) An example from the 50Salads dataset for the case of observing 20% of the video and predicting the activities in the following 20%.

This highlights that the improvements in our model are not only due to the TCN, but mainly because of the joint optimization of all modules for the anticipation task in an end-to-end fashion and the introduced cycle consistency loss.

Qualitative results for our model on both datasets are illustrated in Fig. 4. As shown in the figure, our model can generate accurate predictions of the future activities and their duration. We also show the results of the sequence-to-sequence (S2S) and the two-step baselines. Our model anticipates the activities better.

5   Conclusion

In this paper, we introduced a model for anticipating future activities from a partially observed video. In contrast to the state-of-the-art methods which rely on the action labels of the observations, our model directly predicts the future activities from the observed frames. We train the proposed model in an end-to-end fashion and show a superior performance compared to the previous approaches. Additionally, we introduced a cycle consistency loss for the anticipation task which further boosts the performance. Our framework achieves state-of-the-art results on two publicly available datasets.

Acknowledgments. The work has been funded by the Deutsche Forschungsgemeinschaft (DFG, German Research Foundation) – GA 1927/4-1 (FOR 2535 Anticipating Human Behavior) and the ERC Starting Grant ARCA (677650).

References

1. Abu Farha, Y., Gall, J.: MS-TCN: multi-stage temporal convolutional network for action segmentation. In: CVPR (2019)

2. Abu Farha, Y., Gall, J.: Uncertainty-aware anticipation of activities. In: ICCV Workshops (2019)
3. Abu Farha, Y., Richard, A., Gall, J.: When will you do what?-Anticipating temporal occurrences of activities. In: CVPR (2018)
4. Alahi, A., Goel, K., Ramanathan, V., Robicquet, A., Fei-Fei, L., Savarese, S.: Social LSTM: human trajectory prediction in crowded spaces. In: CVPR (2016)
5. Bhattacharyya, A., Fritz, M., Schiele, B.: Bayesian prediction of future street scenes using synthetic likelihoods. In: ICLR (2019)
6. Carreira, J., Zisserman, A.: Quo vadis, action recognition? A new model and the kinetics dataset. In: CVPR (2017)
7. Dwibedi, D., Aytar, Y., Tompson, J., Sermanet, P., Zisserman, A.: Temporal cycle-consistency learning. In: CVPR (2019)
8. Furnari, A., Battiato, S., Farinella, G.M.: Leveraging uncertainty to rethink loss functions and evaluation measures for egocentric action anticipation. In: Leal-Taixé, L., Roth, S. (eds.) ECCV 2018. LNCS, vol. 11133, pp. 389–405. Springer, Cham (2019). https://doi.org/10.1007/978-3-030-11021-5_24
9. Furnari, A., Farinella, G.M.: What would you expect? Anticipating egocentric actions with rolling-unrolling LSTMs and modality attention. In: ICCV (2019)
10. Gammulle, H., Denman, S., Sridharan, S., Fookes, C.: Forecasting future action sequences with neural memory networks. In: BMVC (2019)
11. Gammulle, H., Denman, S., Sridharan, S., Fookes, C.: Predicting the future: A jointly learnt model for action anticipation. In: ICCV (2019)
12. Gao, J., Yang, Z., Nevatia, R.: RED: reinforced encoder-decoder networks for action anticipation. In: BMVC (2017)
13. Godard, C., Mac Aodha, O., Brostow, G.J.: Unsupervised monocular depth estimation with left-right consistency. In: CVPR (2017)
14. Hoai, M., De la Torre, F.: Max-margin early event detectors. IJCV **107**(2), 191–202 (2014). https://doi.org/10.1007/s11263-013-0683-3
15. Jain, A., Zamir, A.R., Savarese, S., Saxena, A.: Structural-RNN: deep learning on spatio-temporal graphs. In: CVPR (2016)
16. Ke, Q., Fritz, M., Schiele, B.: Time-conditioned action anticipation in one shot. In: CVPR (2019)
17. Kingma, D.P., Ba, J.: Adam: a method for stochastic optimization. In: ICLR (2015)
18. Kitani, K.M., Ziebart, B.D., Bagnell, J.A., Hebert, M.: Activity forecasting. In: Fitzgibbon, A., Lazebnik, S., Perona, P., Sato, Y., Schmid, C. (eds.) ECCV 2012. LNCS, vol. 7575, pp. 201–214. Springer, Heidelberg (2012). https://doi.org/10.1007/978-3-642-33765-9_15
19. Koppula, H.S., Saxena, A.: Anticipating human activities using object affordances for reactive robotic response. TPAMI **38**(1), 14–29 (2016)
20. Kuehne, H., Arslan, A., Serre, T.: The language of actions: recovering the syntax and semantics of goal-directed human activities. In: CVPR (2014)
21. Lan, T., Chen, T.-C., Savarese, S.: A hierarchical representation for future action prediction. In: Fleet, D., Pajdla, T., Schiele, B., Tuytelaars, T. (eds.) ECCV 2014. LNCS, vol. 8691, pp. 689–704. Springer, Cham (2014). https://doi.org/10.1007/978-3-319-10578-9_45
22. Lea, C., Flynn, M.D., Vidal, R., Reiter, A., Hager, G.D.: Temporal convolutional networks for action segmentation and detection. In: CVPR (2017)
23. Liang, J., Jiang, L., Niebles, J.C., Hauptmann, A.G., Fei-Fei, L.: Peeking into the future: predicting future person activities and locations in videos. In: CVPR (2019)
24. Liang, X., Lee, L., Dai, W., Xing, E.P.: Dual motion GAN for future-flow embedded video prediction. In: ICCV (2017)

25. Luc, P., Neverova, N., Couprie, C., Verbeek, J., LeCun, Y.: Predicting deeper into the future of semantic segmentation. In: ICCV (2017)
26. Ma, S., Sigal, L., Sclaroff, S.: Learning activity progression in LSTMs for activity detection and early detection. In: CVPR (2016)
27. Mahmud, T., Billah, M., Hasan, M., Roy-Chowdhury, A.K.: Captioning near-future activity sequences. arXiv (2019)
28. Mahmud, T., Hasan, M., Roy-Chowdhury, A.K.: Joint prediction of activity labels and starting times in untrimmed videos. In: ICCV (2017)
29. Martinez, J., Black, M.J., Romero, J.: On human motion prediction using recurrent neural networks. In: CVPR (2017)
30. Mathieu, M., Couprie, C., LeCun, Y.: Deep multi-scale video prediction beyond mean square error. In: ICLR (2016)
31. Mehrasa, N., Jyothi, A.A., Durand, T., He, J., Sigal, L., Mori, G.: A variational auto-encoder model for stochastic point processes. In: CVPR (2019)
32. Miech, A., Laptev, I., Sivic, J., Wang, H., Torresani, L., Tran, D.: Leveraging the present to anticipate the future in videos. In: CVPR Workshops (2019)
33. Richard, A., Kuehne, H., Gall, J.: Weakly supervised action learning with RNN based fine-to-coarse modeling. In: CVPR (2017)
34. Rodriguez, C., Fernando, B., Li, H.: Action anticipation by predicting future dynamic images. In: ECCV Workshops. Springer (2018)
35. Ruiz, A.H., Gall, J., Moreno-Noguer, F.: Human motion prediction via spatio-temporal inpainting. In: ICCV (2019)
36. Ryoo, M.S.: Human activity prediction: early recognition of ongoing activities from streaming videos. In: ICCV (2011)
37. Sadegh Aliakbarian, M., et al.: Encouraging LSTMs to anticipate actions very early. In: ICCV (2017)
38. Sener, F., Yao, A.: Zero-shot anticipation for instructional activities. In: ICCV (2019)
39. Shi, Y., Fernando, B., Hartley, R.: Action anticipation with RBF kernelized feature mapping RNN. In: ECCV (2018)
40. Srivastava, N., Mansimov, E., Salakhudinov, R.: Unsupervised learning of video representations using LSTMs. In: ICML, pp. 843–852 (2015)
41. Stein, S., McKenna, S.J.: Combining embedded accelerometers with computer vision for recognizing food preparation activities. In: UbiComp (2013)
42. Sun, C., Shrivastava, A., Vondrick, C., Sukthankar, R., Murphy, K., Schmid, C.: Relational action forecasting. In: CVPR (2019)
43. Vaswani, A., et al.: Attention is all you need. In: NIPS (2017)
44. Vondrick, C., Pirsiavash, H., Torralba, A.: Anticipating visual representations from unlabeled video. In: CVPR (2016)
45. Wang, X., Jabri, A., Efros, A.A.: Learning correspondence from the cycle-consistency of time. In: CVPR (2019)
46. Zeng, K.H., Shen, W.B., Huang, D.A., Sun, M., Carlos Niebles, J.: Visual forecasting by imitating dynamics in natural sequences. In: ICCV (2017)
47. Zhou, T., Krahenbuhl, P., Aubry, M., Huang, Q., Efros, A.A.: Learning dense correspondence via 3d-guided cycle consistency. In: CVPR (2016)
48. Zhu, J.Y., Park, T., Isola, P., Efros, A.A.: Unpaired image-to-image translation using cycle-consistent adversarial networks. In: ICCV (2017)

Multi-stage Fusion for One-Click Segmentation

Soumajit Majumder[1]([⊠])(iD), Ansh Khurana[2](iD), Abhinav Rai[3](iD),
and Angela Yao[3](iD)

[1] University of Bonn, Bonn, Germany
majumder@cs.uni-bonn.de
[2] Indian Institute of Technology Bombay, Mumbai, India
[3] School of Computing, National University of Singapore, Singapore, Singapore

Abstract. Segmenting objects of interest in an image is an essential building block of applications such as photo-editing and image analysis. Under interactive settings, one should achieve good segmentations while minimizing user input. Current deep learning-based interactive segmentation approaches use early fusion and incorporate user cues at the image input layer. Since segmentation CNNs have many layers, early fusion may weaken the influence of user interactions on the final prediction results. As such, we propose a new multi-stage guidance framework for interactive segmentation. By incorporating user cues at different stages of the network, we allow user interactions to impact the final segmentation output in a more direct way. Our proposed framework has a negligible increase in parameter count compared to early-fusion frameworks. We perform extensive experimentation on the standard interactive instance segmentation and one-click segmentation benchmarks and report state-of-the-art performance.

1 Introduction

The widespread availability of smartphones had made taking photos easier than ever. In a typical image capturing scenario, the user taps the device touchscreen to focus on the object of interest. This tap directly locates the object in the scene and can be leveraged for segmentation. Generated segmentations are implicit, but are applicable for downstream photo applications, such as simulated 'bokeh' or other special-effects filters such as background blur (see Fig. 1). In this work, we tackle *"tap-and-shoot segmentation"* [4], a special case of interactive instance segmentation.

Interactive segmentation leverages inputs such as clicks, scribbles, or bounding boxes to help segment objects from the background down to the pixel level. Two key differences distinguish tap-and-shoot segmentation from standard interactive segmentation. *First,* tap-and-shoot uses only *"positive"* clicks marking foreground, as we assume that the user clicks (only) on the object of interest during the capture process. Standard interactive segmentation uses both positive and negative clicks [18,19,28] to respectively indicate the object of interest

© Springer Nature Switzerland AG 2021
Z. Akata et al. (Eds.): DAGM GCPR 2020, LNCS 12544, pp. 174–187, 2021.
https://doi.org/10.1007/978-3-030-71278-5_13

versus background or non-relevant foreground objects. *Secondly*, tap-and-shoot has a strong focus on maximizing the mean intersection over union (mIoU) with a single click because the target application is casual photography. In contrast, standard interactive segmentation tries to achieve some threshold mIoU (*e.g.* 85%) while minimizing the total number of clicks.

This second distinction is subtle but critical for designing and learning tap-and-shoot segmentation frameworks. Our finding is that existing approaches fare poorly with only one or two clicks – they are simply not trained to maximize performance under such settings. To make the most of the first (few) click(s), we hypothesize that user cues' guidance should be fused into the network at multiple locations rather than via early fusion. Just as gradients vanish towards the initial layers during back-propagation, input signals also diminish as it makes a forward pass through the network. The many layers of deep CNNs further exacerbate this effect [14,22]. A late fusion would allow the user interaction to have a direct and more pronounced effect on the final segmentation mask. To this end, we propose an interactive segmentation framework with multi-stage fusion and demonstrate its advantages over the common early fusion frameworks and other alternatives. Specifically, we propose a light-weight fusion block that encodes the user click transformation and allows a shorter connection from user inputs to the final segmentation layer.

Most similar in spirit to our framework is [14] and [23]. These two works also propose alternatives to early fusion but are extremely parameter heavy. For example, [14] uses two dedicated VGG [26] networks to to extract features from the image and the user interactions separately before fusing into a final instance segmentation mask (see Fig. 2(c)). [23] uses a single stream but applies a simple late fusion of element-wise multiplication on the feature maps (see Fig. 2(b)). It therefore has separate 'positive' and 'negative' feature maps and the number of weights for the following layer increases by a factor of 2. For VGG, this doubles the parameters of the ensuing '*fc6*' layer from 100 to 200 million. Compared to [23], our last-stage fusion approach is light-weight and uses less than 1.5% more trainable parameters.

Our contributions are summarized as follows:

– We propose a novel one-click interactive segmentation framework that fuses user guidance at different network stages.
– We demonstrate that multi-stage fusion is highly beneficial for propagating guidance and increasing the mIoU since it allows user interaction to have a more direct impact on the final segmentation.
– Comprehensive experiments on six benchmarks show that our approach significantly outperforms existing state-of-the-art for both tap-and-shoot and standard interactive instance segmentation.

2   Related Works

As an essential building block of image/video editing applications, interactive segmentation and dates back decades [21]. The latest methods [14,18,19,23,28]

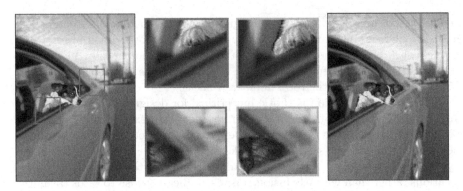

Fig. 1. Motivation. We consider the popular special-effect filter used in mobile photography - background blur. Here the user intends to blur the rest of the image barring the dog. In most existing interactive segmentation approaches [18, 19, 28], the user click (here placed on the dog) is leveraged only at the input layer and its influence diminishes through the layers. This can result in unsatisfactory image effects, e.g. portions of the dog's elbow and ear are wrongly classified as background and are mistakenly blurred (shown in enlarged red boxes). Our proposed multi-stage fusion allows user click to have a more direct effect leading to improvement in segmentation quality (shown in enlarged green boxes).

integrate deep architectures such as FCN-8s [17] or DeepLab [5,6]. Most of these approaches integrate user cues in the input stage. The clicks are transformed into 'guidance' maps and appended to the three-channel colour image input before being passed through a CNN [18, 19, 28].

Early **Interactive Instance Segmentation** methods used graph-cuts [3, 24], geodesics, or a combination [10]. These methods' performance is limited as they separate the foreground and background based on low-level colour and texture features. Consequently, for scenes where foreground and background are similar in appearance, or lighting and contrast is low, more labelling effort from the users to achieve good segmentations [28]. Recently, deep convolutional neural networks [6,17] have been incorporated into interactive segmentation frameworks. Initially, [28] used Euclidean distance-based guidance maps to represent user-provided clicks and are passed along with the input RGB image through a fully convolutional network. Subsequent works made extensions with newer CNN architectures [18], iterative training procedures [18] and structure-aware guidance maps [19]. These works share a structural similarity: the guidance maps are concatenated with the RGB image as additional channels at the first (input) layer. We refer to this form of structure as early fusion (see Fig. 2(a)). Architecture-wise, early fusion is simple and easy to train; however, user inputs' influence gets diminished through the layers.

Tap-and-Shoot Segmentation was introduced by [4], and refers to the one-click interactive setting. One assumes that during image capture, the user taps the touchscreen (once) on the foreground object of interest, from which one can directly segment the object of interest. [4] uses early fusion; it

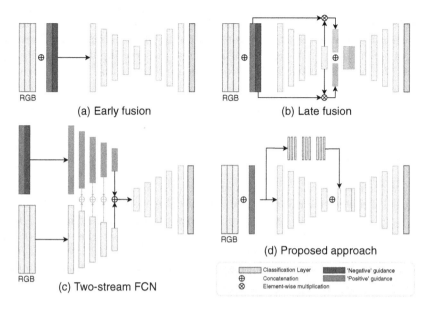

Fig. 2. (a) Existing interactive instance segmentation and "tap-and-shoot" segmentation techniques concatenate user provided cues as an extra guidance map(s) (for 'positive' and 'negative' clicks) with the RGB and pass everything through a segmentation network. (b-c) Other alternative approaches are extremely parameter heavy. (b) The work of [14] uses two dedicated VGG [26] networks for extracting features from image and user interactions *separately*. (c) The work of [23] performs late fusion via element-wise multiplication on the feature maps which requires an additional 100 million parameters. (d) We leverage user guidance at the input (early fusion) and via late fusion. Our multi-stage fusion reduces the layers of abstraction and allows user interactions to have a more direct impact on the final output.

transforms the user tap into a guidance map via two shortest-path minimizations and then concatenates the map to the input image. The authors validate only on simple datasets such as ECSSD [25] and MSRA10K [7], where the images contain a single dominant foreground object. As we show later in our benchmarks (see Table 1), these datasets are so simplistic that properly trained networks with *no* user input can also generate high-quality segmentation masks which are comparable or even surpass the results reported by [4].

Feature Fusion in Deep Architectures is an efficient way to leverage complementary information, either from different modalities [27], or different levels of abstraction [29]. Element-wise multiplication [23] and addition [14,16] are two common operations applied for fusing multiple channels. Other strategies include 'skip' connections [17], where features from earlier layers are concatenated with the features extracted from the deeper layers. Recently, a few interactive instance segmentation works have begun exploring outside of the early-fusion paradigm to integrate user guidance [14,23]. However, these approaches are heavy in their computational footprint, as they increase the number of parameters to be learned by order of hundred of millions [23]. Dilution of input information is common-place in deep CNNs as the input gets processed several blocks of convolution [22]. Feature fusion helps preserve input information by reducing the layers of abstraction between the user interaction and the segmentation output.

3   Proposed Method

3.1   Overview

We follow the conventional paradigm of [18,19,28] in which 'positive' and 'negative' user clicks are transformed into *'guidance'* maps of the same size as the input image. Unlike [18,19,28], we work within the one-click setting. The user provides a single 'positive' click on the object of interest; this click is then encoded into a single channel guidance map $\mathcal{G}$ (see Sect. 3.3). We then feed the 3-channel RGB image input and the guidance map as an additional channel into a fully convolutional network. Figure 3(a) shows an overview of our pipeline. Typically these FCNs are fine-tuned versions of semantic segmentation networks such as FCN-8s [17] or DeepLab [5].

For our base segmentation network, we use DeepLab-v2 [5]; it consists of a ResNet-101 [12] feature extraction backbone and a Pyramid Scene Parsing (PSP) module [30] acting as the prediction head. Upon receiving the input of size $h \times w \times 4$, the ResNet-101 backbone generates feature maps of dimension $h/8 \times w/8 \times 2048$ (Fig. 3(a)).

3.2   Multi-stage Fusion

The fusion module consists of 3 *Squeeze-and-Excitation* residual blocks (SE-ResNet) [13]. Proposed in [13], SE-ResNet blocks have been shown to effective for a variety of vision tasks such as image classification on ImageNet [8] and object detection on MS COCO [15]. SE-ResNet blocks incur minimal additional computational overhead as they consist of two 3×3 convolutional layers, two inexpensive fully connected layers and channel-wise scaling operation.

Each SE-ResNet block consists of a *residual* block, a *squeeze* operation which produces a channel descriptor by aggregating feature maps across their spatial operation, dimensionality reduction layer (by reduction ratio r) and an *excitation* operation which captures the channel interdependencies. The individual components of the SE-ResNet block is shown in Fig. 3(b). The residual block consists of two 3×3 convolutions, batch normalization, and a ReLU non-linearity (Fig. 3(c)). We fix the number of filter banks to be 256 for each of the 3×3 convolution. The reduction ratio r is kept as 16 [13]. The input to the fusion block is a $h/4 \times w/4 \times 256$ feature map which is obtained by processing the $h \times w \times 4$ input with 7×7 convolution operation with stride 2, batch normalization, ReLU non-linearity and a 2×2 max-pooling operation with stride 2 (*Init* block, Fig. 3(a)). The final SE-ResNet block downsamples to generate a $h/8 \times w/8 \times 256$ feature

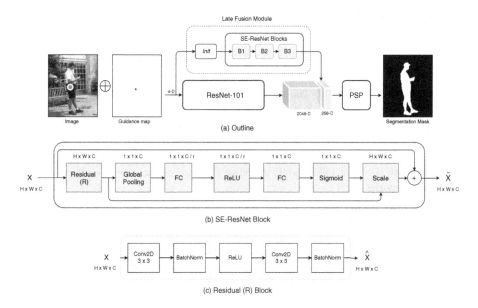

Fig. 3. (a) Overview of our pipeline. Given an image and a 'positive' user click (shown in green circle), we transform the click into a Gaussian guidance map, which is concatenated with the 3-channel image input and is fed to our segmentation network. For ease of visualization, inverted values for the Gaussian guidance map is shown in the image. The output is the segmentation mask of the selected object. (b) SE-ResNet block (c) Residual block.

map. This is concatenated with the $h/8 \times w/8 \times 2048$ obtained from the feature extraction backbone to obtain a $h/8 \times w/8 \times 2304$ feature map.

On top of these feature maps, PSP performs pooling operations at different grid scales on the feature maps to gather the global contextual prior, leading to feature maps of dimensions $h/8 \times w/8 \times 512$. The multi-scale feature pooling of PSP [30] enables the network to capture objects occurring at different image scales. Pixel-wise foreground-background classification is performed on these down-sampled feature maps. The network outputs a probability map representing whether a pixel belongs to the object of interest or not. Bi-linear interpolation is performed to up-sample the predicted probability map to have the same dimensions as the original input image.

3.3   Transforming User Click

In interactive approaches, pixel values of the guidance map are defined as a function of its distance on the image grid to the point of user interaction (Eq. 1). This includes Euclidean [14,28] and Gaussian guidance maps [18]. For each pixel position p on the image grid, the pair of distance-based guidance maps for positive (+) and negative clicks (−) can be computed as

$$\mathcal{G}_+^d(\boldsymbol{p}) = \min_{c\in\{p_+\}} d(\boldsymbol{p}, \boldsymbol{c}) \text{ and } \mathcal{G}_-^d(\boldsymbol{p}) = \min_{c\in\{p_-\}} d(\boldsymbol{p}, \boldsymbol{c}). \tag{1}$$

For Euclidean guidance maps [28], the function $d(\cdot,\cdot)$ is the Euclidean distance. For Gaussian guidance maps, the 'min' is replaced by a 'max' operator. A more recent approach advocated taking image structures such as super-pixels and region-based object proposals into consideration to generate guidance maps [19]. To generate the guidance maps, we use Gaussian transformations [18] as it offers a favourable trade-off between simplicity and performance. We initialize an image-sized all zero channel and place a Gaussian with a standard deviation of 10 pixels at the user click location. Note that we do not use 'negative' clicks in our framework.

3.4   Implementation Details

Network Optimization. We train the network to minimize the class-balanced binary cross-entropy loss,

$$\mathcal{L} = \sum_{j\in N} w_{y_j} \cdot \mathrm{BCE}(y_j, \hat{y}_j) \tag{2}$$

where N is the number of pixels in the image, $\mathrm{BCE}(\cdot)$ is the standard cross-entropy loss between the label y_j and the prediction $\hat{y}_j$ at pixel location j given by,

$$\mathrm{BCE}(y_j, \hat{y}_j) = -y_j \cdot \log\hat{y}_j - (1 - y_j) \cdot \log(1 - \hat{y}_j) \tag{3}$$

w_{y_j} is the inverse normalized frequency of labels $y_j \in \{0,1\}$ within the mini-batch. We optimize using mini-batch SGD with Nesterov momentum (with default value of 0.9) and a batch size of 5. The learning rate is fixed at 10^{-8} across all epochs and weight decay is 0.0005. For the ResNet-101 backbone, we initialize the network weights from a model pre-trained on ImageNet [8]. During training, we first update the early-fusion skeleton for 30–35 epochs. Next we freeze the weights of the early-fusion model and train the late-fusion weights for 5–10 epochs. Finally, we train the joint network for another 5 epochs.

Simulating User Clicks. Manually collecting user interactions is an expensive and arduous process [2]. In a similar vein as [4] and other interactive segmentation frameworks [18,19,28], we simulate user interactions to train and evaluate our method. During training, we use the ground truth masks of the object instances from the MSRA10K dataset. To initialize, we take the center of mass of the ground truth mask as our user click location; we then jitter the click location by $\mathcal{U}(-50, 50)$ pixels randomly. The clicked pixel location is constrained to the confines of the object ground truth mask. The random perturbation introduces variation in the training data and also allows better approximation of true user interactions.

Table 1. Ablation Study: Tap-and-Shoot Segmentation. 'res' refers to the image resolution used during training. We report average mIoU for the segmentation results after training for 16K iterations and after training convergence. The *-baseline* models receive a 3-channel RGB image as input without the guidance map $\mathcal{G}$.

Method	$\mathcal{G}$	res	GrabCut [24]	Berkeley [20]	ECSSD [25]	MSRA-10K [7]
TNS [4]	✓	256	72.3 / 79.0	55.7 / 67.0	70.3 / 76.0	81.1 / 85.0
vgg-baseline	✗	256	73.5 / 77.4	58.2 / 63.2	71.2 / 72.3	83.4 / 86.2
vgg-early	✓	256	76.2 / 80.1	62.8 / 65.3	74.8 / 76.5	87.1 / 87.5
resnet-baseline	✗	256	81.6 / 83.0	68.5 / 68.2	80.2 / 82.0	86.4 / 86.9
resnet-early	✓	256	83.3 / 84.3	75.0 / 75.3	82.0 / 83.6	88.6 / 89.6
resnet-multi	✓	256	84.1 / 85.7	75.1 / 78.4	81.9 / 85.2	91.5 / 92.1
resnet-baseline	✗	512	76.1 / 79.0	65.5 / 68.3	79.9 / 82.6	87.0 / 87.9
resnet-early	✓	512	82.9 / 84.5	76.2 / 78.1	85.6 / 85.7	91.5 / 91.4
resnet-multi	✓	512	83.1 / 86.2	80.1 / 81.3	86.8 / 87.1	92.5 / 93.1

4 Experimental Validation

4.1 Datasets

We evaluate on six publicly available datasets commonly used to benchmark interactive image segmentation [4,18,19,28]: MSRA10K [7], ECSSD [25], Grab-Cut [24], Berkeley [20], PASCAL VOC 2012 [9] and MS COCO [15]. We use mean intersection over union (mIoU) of foreground w.r.t. to the ground truth object mask across all instances to evaluate the segmentation accuracy as per existing works [4,17–19,28].

MSRA10K has 10, 000 natural images; the images are characterized by variety in the foreground objects whilst the background is relatively homogeneous. Extended complex scene saliency dataset (**ECSSD**) is a dataset of 1000 natural images with structurally complex backgrounds. **GrabCut** is a dataset consisting of 49 images with typically a distinct foreground object. It is a popular dataset for benchmarking interactive instance segmentation algorithms. **Berkeley** dataset consists of 96 natural images. **PASCAL VOC 2012** consists of 1464 training and 1449 validation images across 20 different object classes; many images contain multiple objects. **MS COCO** is a challenging large-scale image segmentation dataset with 80 different object categories, 20 of which are common with the PASCAL VOC 2012 categories.

4.2 Tap-and-Shoot Segmentation

Following [4], we use MSRA10K [7] for training and partition the dataset into three non-overlapping subsets of 8000, 1000 and 1000 images as our training, validation and test set. We report the mIoU after training for 16K iterations and again after network convergence (at 43k iterations for us, vs. 260k iterations in [4]) in Table 1. During training, we resize the images to 512×512 pixels. This choice of resolution is driven primarily by matching the resolution to that of the training images for the ResNet-101 backbone [12].

The *-baseline* models are trained using only the 3-channel RGB image and the instance ground truth mask without any user click transformations. The *-early* models use Gaussian guidance maps [18]; the network input is 3-channel RGB image and Gaussian encoding of the user's tap on the object of interest (Fig. 2(a)). The *-multi* models refer to the multi-stage fusion models with Gaussian encoding of user clicks. Note that we do not train a late-fusion model; standalone late-fusion models show inferior performance compared to their early-fusion counterparts [23].

From Table 1, we observe that our trained network converges mostly within 16K iterations. For simplistic datasets such as MSRA10K and ECSSD, the *vgg-baseline* without user click transformation compares favourably with the approach of [4] at the same training resolution of 256×256. *resnet-baseline* models trained with 512×512 images significantly outperform [4] reporting absolute mIoU gains of till 7% across the datasets. Based on this result alone, we conclude that one-click (and standard) interactive segmentation approaches should be benchmarked on more challenging datasets. Examples include PASCAL VOC 2012 and MS COCO, which feature cluttered scenes, multiple objects, occlusions and challenging lighting conditions. (see Table 3).

Furthermore, with only the Gaussian transformation and ResNet-101 backbone trained on 512×512, we are able to achieve mIoU increase in the range of 5–11% across datasets at convergence *w.r.t* [4]. Having the multi-stage fusion offers us absolute mIoU gains of 1–4% w.r.t the early fusion variant (*resnet-early vs. resnet-multi* when trained with 512×512 images). Additionally, our *resnet* models require significantly less memory; 195.8 MB (stored as 32-bit/4-byte floating point numbers) instead of the 652.45 MB required for the segmentation network of [4].

4.3   Interactive Image Segmentation

Approaches in the literature [14,18,19,28] are typically evaluated by (1) the average number of clicks needed to reach the desired level of segmentation (@85% mIoU for PASCAL VOC 2012, MS COCO, @90% mIoU for the less challenging Grabcut and Berkeley) and (2) the average mIoU *vs* the number of clicks.

The first criterion is primarily geared towards annotation tasks [18,19] where high-quality segments are desired for each instance in the scene; the fewer the number of clicks, the lower the annotation effort. In this work, we are concerned primarily with achieving high-quality segments for the object of interest given only a single click. Accordingly, given a single user click, we report the average mIoU across all instances for the GrabCut, Berkeley and the PASCAL VOC 2012 *val* dataset. For MS COCO object instances, following [28], we split the dataset into the 20 PASCAL VOC 2012 categories and the 60 additional categories, and randomly sample 10 images per category for evaluation. We also report the average mIoU across the sampled 800 MS COCO instances [14].

| Input Image with the overlaid gt mask | Euclidean guidance | Disk guidance map | Gaussian guidance |

Fig. 4. Examples of guidance maps. Given a click (shown as green circle) on the object of interest, existing approaches transform it into guidance maps and uses it as an additional input channel. For ease of visualization, inverted values for the disk guidance map and the Gaussian guidance map are shown in the image. (Color figure online)

Table 2. User Click Transformation. The best results are indicated in **bold**. COCO-20 and COCO-60 refers to the instances from 20 overlapping categories and 60 non-overlapping categories of PASCAL VOC 2012 respectively.

$\mathcal{G}$	GrabCut	Berkeley	VOC12	COCO-20	COCO-60
Euclidean [28]	82.6	82.7	75.1	63.2	46.8
Disk [2]	84.5	81.3	74.5	65.3	51.5
Gaussian [18]	84.0	82.9	78.1	64.2	49.8
Gaussian-*multi*	**86.2**(2.2 ↑)	**84.0**(1.1 ↑)	**80.8**(2.7 ↑)	64.5(0.3 ↑)	**52.3**(2.5 ↑)

For training [14,19,28], we use the ground truth masks of object instances from PASCAL VOC 2012 [9] *train* set with additional masks from Semantic Boundaries Dataset (SBD) [11] resulting in 10582 images. Note that unlike [18], we do not use the training instances from MS COCO.

Ablation Study. We perform extensive ablation studies to thoroughly analyze the effectiveness of the individual components of our one-click segmentation framework. First, to validate our choice of guidance maps, we consider the user click transformations commonly used in existing interactive segmentation algorithms - Euclidean distance maps [14,28], Gaussian distance maps [18] and disk [2]. Figure 4 shows examples of such guidance maps. For each kind of guidance map, we train separate networks to understand the impact of different user click transformations. For evaluation, we report the average mIoU over all instances in the dataset, given a single click (see Table 2). Next, we study the impact of our proposed late-fusion module (denoted by -*multi* in Table 2); we observe an average mIoU improvement of around 1.8% across different datasets.

One-Click Segmentation. We compare the segmentation performance of our method with existing interactive instance segmentation approaches (see Table 3). The approaches are grouped separately into 3 different categories - pre-deep learning approaches, deep learning-based interactive instance segmentation approaches and tap-and-shoot segmentation approaches. From Table. 3,

Table 3. Average mIoU given a single click. The approaches are grouped separately into 3 different categories - pre-deep learning approaches, deep learning-based interactive instance segmentation approaches and tap-and-shoot segmentation approaches respectively. For GC [3], GM [1], GD [10], and iFCN [28] we make use of the values provided by the authors of iFCN [28]. The mIoU improvement (in %) over existing state-of-the-art approaches is indicated using ↑.

Method	Network	GrabCut	Berkeley	VOC12	COCO-20	COCO-60
GC [3]	-	41.7	33.8	27.7	-	8.9
GM [1]	-	23.7	24.5	23.8	-	22.1
GD [10]	-	48.8	36.1	31.0	-	25.2
iFCN [28]	FCN-8s [17]	62.9	61.3	53.6		**42.9**
ITIS [18]	DeepLabv3+ [6]	82.1	-	71.0	-	-
CAG [19]	FCN-8s [17]	**83.2**	-	**74.0**	-	-
TS [14]	FCN-8s [17]	77.7	**74.5**	62.3	**42.5**	42.5
TNS [4]	FCN-8s [17]	79.0	67.0	-	-	-
Ours-best	DeepLabv2 [5]	**86.2**(3.0 ↑)	**84.0**(9.5 ↑)	**80.8**(6.8 ↑)	**64.5**(22.0 ↑)	**52.3**(9.6 ↑)

we observe that our approach outperforms the classical interactive segmentation works by a significant margin reporting 40% absolute improvement in average mIoU. We also outperform existing state-of-the-art interactive instance segmentation approaches [18,19] by a considerable margin (>3%). Additionally, we report an absolute mIoU improvement of 7.2% and 17% on Grabcut and Berkeley over the tap-and-shoot segmentation framework of [4]. We show qualitative results to demonstrate the effectiveness of our proposed algorithm (see Fig. 5). The resulting segmentations demonstrate that our approach is highly effective for the one-click segmentation paradigm.

5   User Study

Across existing state-of-the-art interactive frameworks [18,19,28], user clicks are simulated following the protocols established in [18,28]. For our user study, we consult 5 participants uninitiated to the task of interactive segmentation. We prepare a toy dataset with 50 object instances from the MSRA10K [7] dataset. We presented the image with the segmentation mask for the target object overlaid on the image and asked the users to provide their click. During training, we applied random perturbations of $\mathcal{U}(-50, 50)$ pixels to the center of mass of the object instance to obtain the final user click. Our user study found that participants placed clicks at a mean distance of 24 pixels from the center of the mask with a standard deviation of 27 pixels. This result validates our assumption that users are more likely to click in the vicinity of the object's center-of-mass. On average, we observed that users took 2.3 s with a standard deviation of 0.8 s to position their click.

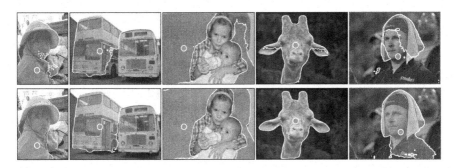

Fig. 5. Qualitative Results. Incorporating the user clicks at different stages of the network leads to an improvement in the quality of masks generated (second row) *w.r.t* the early-fusion variants (first row). Click locations are shown in green circles. The extreme right column shows a scenario where both the networks failed to generate a satisfactory mask.

6   Conclusion

In this work, we propose a one-click segmentation framework that produces high-quality segmentation masks. We validated our design choices through detailed ablation studies; we observed that having a multi-stage module improves the segmentation framework and gives the network an edge over its early-fusion variants. Via experiments, we observed that for the single click scenario, our proposed approach significantly outperforms existing state-of-the-art approaches - including the more complicated interactive instance segmentation models using state-of-the-art segmentation models [6].

However, we observe existing tap-and-shoot segmentation frameworks [4], including our proposed framework, are limited by their inability to learn from negative clicks [18,19,28]. One major drawback of such a training scenario is that the network does not have a notion of corrective clicking; if the generated segmentation mask extends beyond the object boundaries, it cannot rectify this mistake. Clicking on locations outside the object can mitigate this effect, though this then deviates from tap-and-shoot interaction.

Acknowledgment. This work was supported in part by National Research Foundation Singapore under its NRF Fellowship Programme [NRF-NRFFAI1-2019-0001] and NUS Startup Grant R-252-000-A40-133.

References

1. Bai, X., Sapiro, G.: Geodesic matting: a framework for fast interactive image and video segmentation and matting. IJCV **82**(2), 113–132 (2009)
2. Benenson, R., Popov, S., Ferrari, V.: Large-scale interactive object segmentation with human annotators. In: CVPR, pp. 11700–11709 (2019)

3. Boykov, Y.Y., Jolly, M.P.: Interactive graph cuts for optimal boundary & region segmentation of objects in nd images. In: ICCV, pp. 105–112 (2001)
4. Chen, D.J., Chien, J.T., Chen, H.T., Chang, L.W.: Tap and shoot segmentation. In: AAAI (2018)
5. Chen, L.C., Papandreou, G., Kokkinos, I., Murphy, K., Yuille, A.L.: Deeplab: semantic image segmentation with deep convolutional nets, atrous convolution, and fully connected CRFS. TPAMI **40**(4), 834–848 (2018)
6. Chen, L.C., Zhu, Y., Papandreou, G., Schroff, F., Adam, H.: Encoder-decoder with atrous separable convolution for semantic image segmentation. In: ECCV, pp. 801–818 (2018)
7. Cheng, M.M., Mitra, N.J., Huang, X., Torr, P.H., Hu, S.M.: Global contrast based salient region detection. IEEE TPAMI **37**(3), 569–582 (2014)
8. Deng, J., Dong, W., Socher, R., Li, L.J., Li, K., Fei-Fei, L.: Imagenet: a large-scale hierarchical image database. In: CVPR, pp. 248–255 (2009)
9. Everingham, M., Van Gool, L., Williams, C.K., Winn, J., Zisserman, A.: The pascal visual object classes (voc) challenge. IJCV **88**(2), 303–338 (2010)
10. Gulshan, V., Rother, C., Criminisi, A., Blake, A., Zisserman, A.: Geodesic star convexity for interactive image segmentation. In: CVPR, pp. 3129–3136 (2010)
11. Hariharan, B., Arbelaez, P., Bourdev, L., Maji, S., Malik, J.: Semantic contours from inverse detectors. In: ICCV, pp. 991–998 (2011)
12. He, K., Zhang, X., Ren, S., Sun, J.: Deep residual learning for image recognition. In: CVPR, pp. 770–778 (2016)
13. Hu, J., Shen, L., Sun, G.: Squeeze-and-excitation networks. In: CVPR, pp. 7132–7141 (2018)
14. Hu, Y., Soltoggio, A., Lock, R., Carter, S.: A fully convolutional two-stream fusion network for interactive image segmentation. Neural Netw. **109**, 31–42 (2019)
15. Lin, T.Y., Maire, M., Belongie, S., Hays, J., Perona, P., Ramanan, D., Dollár, P., Zitnick, C.L.: Microsoft COCO: common objects in context. In: ECCV, pp. 740–755 (2014)
16. Liu, D., et al.: Nuclei segmentation via a deep panoptic model with semantic feature fusion. In: AAAI, pp. 861–868 (2019)
17. Long, J., Shelhamer, E., Darrell, T.: Fully convolutional networks for semantic segmentation. In: CVPR, pp. 3431–3440 (2015)
18. Mahadevan, S., Voigtlaender, P., Leibe, B.: Iteratively trained interactive segmentation, In: BMVC (2018)
19. Majumder, S., Yao, A.: Content-aware multi-level guidance for interactive instance segmentation. In: CVPR, pp. 11602–11611 (2019)
20. McGuinness, K., O'connor, N.E.: A comparative evaluation of interactive segmentation algorithms. Pattern Recogn. **43**(2), 434–444 (2010)
21. Mortensen, E.N., Barrett, W.A.: Intelligent scissors for image composition. In: SIGGRAPH, pp. 191–198 (1995)
22. Park, T., Liu, M.Y., Wang, T.C., Zhu, J.Y.: Semantic image synthesis with spatially-adaptive normalization. In: Proceedings of the IEEE Conference on Computer Vision and Pattern Recognition, pp. 2337–2346 (2019)
23. Rakelly, K., Shelhamer, E., Darrell, T., Efros, A.A., Levine, S.: Few-shot segmentation propagation with guided networks. arXiv preprint arXiv:1806.07373 (2018)
24. Rother, C., Kolmogorov, V., Blake, A.: Grabcut: interactive foreground extraction using iterated graph cuts. ACM Trans. Graph. (TOG) **23**(3), 309–314 (2004)
25. Shi, J., Yan, Q., Xu, L., Jia, J.: Hierarchical image saliency detection on extended CSSD. IEEE TPAMI **38**(4), 717–729 (2015)

26. Simonyan, K., Zisserman, A.: Very deep convolutional networks for large-scale image recognition. arXiv preprint arXiv:1409.1556 (2014)
27. Vielzeuf, V., Pateux, S., Jurie, F.: Temporal multimodal fusion for video emotion classification in the wild. In: Proceedings of the 19th ACM International Conference on Multimodal Interaction, pp. 569–576. ACM (2017)
28. Xu, N., Price, B., Cohen, S., Yang, J., Huang, T.S.: Deep interactive object selection. In: CVPR, pp. 373–381 (2016)
29. Zhang, Y., Gong, L., Fan, L., Ren, P., Huang, Q., Bao, H., Xu, W.: A late fusion CNN for digital matting. In: CVPR, pp. 7469–7478 (2019)
30. Zhao, H., Shi, J., Qi, X., Wang, X., Jia, J.: Pyramid scene parsing network. In: CVPR, pp. 2881–2890 (2017)

Neural Architecture Performance Prediction Using Graph Neural Networks

Jovita Lukasik$^{(\boxtimes)}$ ⓘ, David Friede, Heiner Stuckenschmidt ⓘ,
and Margret Keuper ⓘ

University of Mannheim, Mannheim, Germany
{jovita,david,heiner}@informatik.uni-mannheim.de,
keuper@uni-mannheim.de

Abstract. In computer vision research, the process of automating architecture engineering, Neural Architecture Search (NAS), has gained substantial interest. Due to the high computational costs, most recent approaches to NAS as well as the few available benchmarks only provide limited search spaces. In this paper we propose a surrogate model for neural architecture performance prediction built upon Graph Neural Networks (GNN). We demonstrate the effectiveness of this surrogate model on neural architecture performance prediction for structurally unknown architectures (i.e. zero shot prediction) by evaluating the GNN on several experiments on the NAS-Bench-101 dataset.

1 Introduction

Deep learning using convolutional neural architectures has been the driving force of recent progress in computer vision and related domains. Multiple interdependent aspects such as the increasing availability of training data and compute resources are responsible for this success. Arguably, none has had as much impact as the advancement of novel neural architectures [11,19]. Thus, the focus of computer vision research has shifted from a feature engineering process to an architecture engineering process. The direct consequence is the need to automate this process using machine learning techniques.

Neural Architecture Search (NAS) [7] attends to techniques automating architecture engineering. Due to very long compute times for the recurrent search and evaluation of new candidate architectures [41], NAS research has hardly been accessible for researchers without access to large-scale compute systems. Yet, the publication of *NAS-Bench-101* [38], a dataset of over 423k fully trained neural architectures, facilitates a paradigm change in NAS research. Instead of carefully evaluating each new proposed neural architecture, NAS-Bench-101 enables to experiment with classical data-based methods such as supervised learning to evaluate neural architectures. While the impact of benchmarks such as NAS-Bench-101 on the community is high, they come at extreme computational costs. All architectures in the search space have to be extensively evaluated, which leads to practical restrictions on the search space. This calls for accurate surrogate models that enable to extrapolate expected performances to structurally

Z. Akata et al. (Eds.): DAGM GCPR 2020, LNCS 12544, pp. 188–201, 2021.
https://doi.org/10.1007/978-3-030-71278-5_14

different and larger architectures in unseen areas of the search space, i.e. *zero shot prediction*. In this paper, we first tackle the task of learning to predict the accuracy of convolutional neural architectures in a supervised way, i.e. we learn a surrogate model that enables to predict the performance of neural architectures on the CIFAR-10 image classification task. Furthermore, we evaluate our proposed model on two different zero shot prediction scenarios and show its ability to accurately predict performances in previously unseen regions of the search space.

Most current neural architectures for computer vision can be represented as directed, acyclic graphs (DAGs). Thus, we base our surrogate model on Graph Neural Networks. *Graph Neural Networks* (GNNs) [35] have proven to be very powerful comprehending local node features and graph substructures. This makes them a very useful tool to embed nodes as well as full graphs like the NAS-Bench-101 architectures into continuous spaces. Furthermore, the benefit of GNNs over Recurrent Neural Networks (RNNs) has been shown in the context of graph generating models. The model *Deep Generative Models of Graphs* (DGMG) in [22] utilizes GNNs and shows dominance over RNN methods. DGMG is able to capture the structure of graph data and its attributes in a way that probabilistic dependencies within graph nodes and edges can be expressed, yielding in learning a distribution over any graph. This makes DGMG a strong tool to map neural architectures into a feature representation which captures the complex relation within the neural architecture.

Inspired by [22], we utilize the GNN as our surrogate model for the performance prediction task.

In summary, in this paper we make the following contributions: We present a surrogate model– a graph encoder built on GNNs– for neural architecture performance prediction trained and evaluated on the NAS-Bench-101 benchmark and show that this neural performance predictor accurately predicts architecture performances in previously structurally different and unseen regions of the search space, i.e. zero shot prediction.

The remaining paper is structured as follows: Sect. 2 gives a short review of the related work. In Sect. 3 we present our proposed encoder model. Section 4 gives detailed model implementation details of the proposed surrogate model. In Sect. 5, we describe the NAS-Bench-101 dataset on which we conduct our evaluation. In Sect. 6, we present our experiments and results. Finally, we give a conclusion and outline some future directions in Sect. 7.

2   Related Work

2.1   Neural Architecture Search

Neural Architecture Search (NAS) [25,30,31,40,41], the process of designing neural network architectures in an automatic way, gained substantial attention recently. See [7] for an overview and detailed survey over recent NAS methods.

The currently most successful approaches follow different paradigms: Reinforcement learning (RL) [30,40,41] as a NAS strategy considers the neural architecture generation as the agent's action with it's reward given in terms of validation accuracy. Evolutionary Algorithm (EA) [24,31] approaches optimizing the neural architectures themselves by guiding the mutation of architectures and evaluating their fitness given in terms of validation accuracy. Bayesian optimization (BO) [16] derives kernels for architecture similarity measurements to extrapolate the search space. Gradient based methods [25,26] use continuous relaxations of neural architectures to allow for gradient-based optimization.

2.2 Neural Architecture Benchmark Datasets

NAS-Bench-101 [38] is a public dataset of 423k neural architectures and provides tabular benchmark results for a restricted cell structured architecture search space [41] with exhaustive evaluation on the CIFAR-10 image classification dataset [18]. As shown in [39], only subspaces of the architectures in NAS-Bench-101 can be used to evaluate one-shot NAS methods [25,30], motivating their proposed variant NAS-Bench-1shot1 [39].

Similarly to NAS-Bench-101, NAS-Bench-201 [6] uses a restricted, cell- structured search space, while the employed graph representation allows to evaluate discrete and one-shot NAS algorithms. The search space is even more restricted than NAS-Bench-101, providing only 6k unique evaluated architectures in total. We conduct our experiments on NAS-Bench-101, which is the largest available tabular neural architecture benchmark for computer vision problems.

2.3 Performance Predictor for Neural Architectures

The work on performance prediction models for neural architectures is very limited. [23] uses a performance predictor in an iterative manner during the search process of NAS. [1] uses features of a neural architecture, such as the validation accuracy, some architecture parameters such as the number of weights and the number of layers as well as hyperparameters, to predict learning curves during the training process by means of a SRM regressor. [26] proposes a performance prediction model learned in combination with an auto-encoder in an end-to-end manner. The neural architectures are mapped into a latent feature representation, which is then used by the predictor for performance prediction and are further decoded into new neural architectures. Recently [33] proposes a semi-supervised assessor of neural architectures. The graphs are employed by an auto-encoder to discover latent feature representations, which is then fine-tuned by means of a graph similarity measurement. Lastly, a graph convolution network is used for performance prediction.

2.4 Graph Neural Networks

Combining modern machine learning methods with graph structured data has increasingly gaining popularity. One can interpret it as an extension of deep

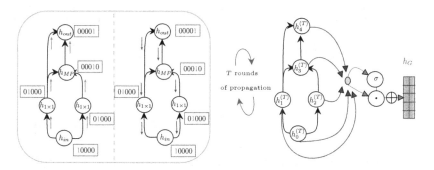

Fig. 1. Illustration of the graph encoding process: The node-level propagation using T rounds of bidirectional message passing (*left*) and the graph-level aggregation into a single graph embedding h_G (*right*). (Color figure online)

learning techniques to non-Euclidean data [4] or even as inducing relational biases within deep learning architectures to enable combinatorial generalization [3]. Because of the discrete nature of graphs, they can not trivially be optimized in differentiable learning methods that act on continuous spaces. The concept of graph neural networks is a remedy to this limitation. The idea of Graph Neural Networks as an iterative process which propagates the node states until an equilibrium is reached, was initially mentioned in 2005 [12]. Motivated by the increasing populatiry of CNNs, [5] and [15] defined graph convolutions in the Fourier domain by utilizing the graph Laplacian. The modern interpretation of GNNs was first mentioned in [17, 21, 28] where node information was inductively updated through aggregating information of each node's neighborhood. This approach was further specified and generalized by [13] and [10].

The research in GNNs enabled breakthroughs in multiple areas related to graph analysis such as computer vision [20, 36, 37], natural language processing [2], recommender systems [27], chemistry [10] and others. The capability of GNNs to accurately model dependencies between nodes makes them the foundation of our research. We utilize them to move from the discrete graph space to the continuous space.

In this paper, we want to use continuous methods, GNNs, to handle the graphs characterizing neural architectures from the NAS-Bench-101 dataset [38]. More precisely, we show the ability of GNNs to encode neural architectures such as to allow for a regression of their expected performance on an image classification problem.

3   The Graph Encoder

In this section we present our GNN-based model to encode the discrete graph space of NAS-Bench-101 into a continuous vector space. One can imagine a single GNN iteration as a two-step procedure. First, each node sends out a message to its neighbors alongside its edges. Second, each node aggregates all incoming messages to update itself. After a final amount of these iteration steps, the individual node embeddings are aggregated into a single graph embedding.

3.1   Node-Level Propagation

Let $G = (V, E)$ be a graph with nodes $v \in V$ and edges $e \in E \subseteq V \times V$. We denote $N(v) = \{u \in V \mid (u, v) \in E\}$ and $N^{out}(v) = \{u \in V \mid (v, u) \in E\}$ as the directed neighborhoods of a node $v \in V$. For each node $v \in V$, we associate an initial node embedding $h_v \in \mathbb{R}^{d_n}$. In our experiments we use a learnable look-up table based on the node types. Propagating information through the graph can be seen as an iterative *message-passing* process

$$m_{u \to v} = \Xi_{u \in N(v)}\big(M^{(t)}(h_v^{(t-1)}, h_u^{(t-1)})\big), \tag{1}$$

$$h_v^{(t)} = U^{(t)}(h_v^{(t-1)}, m_{u \to v}), \tag{2}$$

with a differentiable message module $M^{(t)}$ in (1), a differentiable update module $U^{(t)}$ in (2) and a differentiable, permutation invariant aggregation function Ξ. The message module $M^{(t)}$ is illustrated by the green arrows in Fig. 1 (left). To address the directed nature of the NAS-Bench-101 graphs, we add a reverse message module

$$m_{u \to v}^{out} = \Xi_{u \in N^{out}(v)}\big(M_{out}^{(t)}(h_v^{(t-1)}, h_u^{(t-1)}), \tag{3}$$

$$h_v^{(t)} = U^{(t)}(h_v^{(t-1)}, m_{u \to v}, m_{u \to v}^{out}). \tag{4}$$

This is outlined in Fig. 1 (left) by the red arrows and leads to so-called bidirectional message passing. The update module $U^{(t)}$ utilizes each node's incoming messages to update that node's embedding from $h_v^{(t-1)}$ to $h_v^{(t)}$.

Exploring many different choices for the message and update modules experimentally, we find that the settings similar to [22] work best for our needs. We pick a concatenation together with a single linear layer for our message modules. The update module consists of a single gated recurrent unit (GRU) where $h_v^{(t-1)}$ is treated as the hidden state. For the aggregation function, we choose the sum. To increase the capacity of our model, on the one hand, we apply multiple rounds of propagation and on the other hand, we use a different set of parameters for each round.

3.2   Graph-Level Aggregation

After the final round of message-passing, the propagated node embeddings $h = (h_v)_{v \in V}$ are aggregated into a single graph embedding $h_G \in \mathbb{R}^{d_g}$, where

$$h_G = A(h), \tag{5}$$

We obtain good results by using a linear layer combined with a gating layer that adjusts each node's fraction in the graph embedding. This aggregation layer A in (5) is further illustrated in Fig. 1 (right).

4   Model Details

In this section, we give further details on the implementation of our GNN model.

4.1   Message

The message module $M^{(t)}$ concatenates the embedding of the considered node $h_v^{(t-1)}$ as well as the incoming embedding $h_u^{(t-1)}$, each of dimension d_n. It further performs a linear transformation on the concatenated embedding. The reverse message module $M_{out}^{(t)}$ is a clone of $M^{(t)}$ initialized with its own weights,

$$M^{(t)} = \text{Lin}_{2d_n \times 2d_n}\left([h_v^{(t-1)}, h_u^{(t-1)}]\right), \tag{6}$$

$$M_{out}^{(t)} = \text{Lin}'_{2d_n \times 2d_n}\left([h_v^{(t-1)}, h_u^{(t-1)}]\right). \tag{7}$$

The message module (green) and the reverse message (red) can be seen on the left side of Fig. 1.

4.2   Update

The update module $U^{(t)}$ is a single GRU cell. First, the incoming messages $m_{u \to v}$ and $m_{u \to v}^{out}$ are added and handled as the GRU input. Second, the node embedding $h_v^{(t-1)}$ is treated as the hidden state and is updated,

$$U^{(t)} = \text{GRUCell}_{2d_n, d_n}\left(m_{u \to v} + m_{u \to v}^{out}, \; h_v^{(t-1)}\right). \tag{8}$$

4.3   Aggregation

We use two rounds of propagation before aggregating the node embeddings into a single graph embedding. This graph aggregation consists of two parts. First, a linear layer transforms the node embeddings to the required graph embedding dimension d_g. Second, another linear layer combined with a sigmoid handles each node's fraction in the graph embedding,

$$A_1 = \text{Lin}_{d_n \times d_g}(h_v^{(2)}), \tag{9}$$

$$A_2 = \sigma\left(\text{Lin}_{d_n \times 1}(h_v^{(2)})\right), \tag{10}$$

$$A = \sum_v A_1 \odot A_2. \tag{11}$$

An illustration of the aggregation module is given in Fig. 1 (right).

5   The NAS-Bench-101 Dataset

NAS-Bench-101 [38] is a public dataset of neural architectures in a restricted cell structured search space [41] evaluated on the CIFAR-10-classification set [18].

NAS-Bench-101 considers the following constraints to limit the search space: it only considers directed acyclic graphs, the number of nodes is limited to $|V| \leq 7$, the number of edges is limited to $|E| \leq 9$ and only 3 different operations are allowed $\{3 \times 3 \text{ convolution}, 1 \times 1 \text{ convolution}, 3 \times 3 \text{ max} - \text{pool}\}$. These restrictions lead to a total of 423k unique convolutional architectures, which are built from the cells in the following way: Each cell is stacked three times, followed by a max-pooling layer which reduces the feature map size by factor two. This pattern is repeated 3 times, followed by global average pooling and a dense softmax layer to produce the output. While this search space is limited it covers relevant architectures such as for example ResNet like [14] and InceptionNet like [32] models [38].

The architectures have been trained for four increasing numbers of epochs $\{4, 12, 36, 108\}$. Each of these architectures is mapped to its test, validation and training measures. In this paper we use the architectures trained for 108 epochs and aim to predict their corresponding validation and test accuracy.

6   Experiments

We conduct experiments in three complementary domains. First, we evaluate the performance prediction ability of the proposed GNN in the traditional supervised setting. Then, we conduct zero shot prediction experiments in order to show the performance of the proposed model for unseen graph structures during training. Both experiments are carried out on the validation accuracies reported in NAS-Bench101. Last, we compare our results to the recent publication by Tang et al. [33] in terms of test accuracy prediction.

Implementation Details. If not mentioned differently, we set $d_n = 250$ for the node dimensions and $d_g = 56$ for the dimension of the latent space. We split the dataset 70%/20%/10% edit-sampled into training-, test- and validation set. All our experiments are implemented using PyTorch [29] and PyTorch Geometric [9].

The hidden layers of the regressor are of size 28, 14 and 7. We used no activation function for the very last output (linear regression) and trained the joint encoder model with a learning rate of $1e^{-5}$ for 100 epochs. The hyperparameters were tuned with BOHB [8],

6.1   Performance Prediction

Supervised Performance Prediction. Here, we evaluate the latent space generated by the encoder with respect to its prediction error regarding a metric of interest of the NAS-Bench-101 graphs, i.e. the validation accuracy on CIFAR-10. For this purpose, we utilize a simple predictor, i.e. a four-layer MLP with ReLU non-linearities.

We jointly train the encoder and the predictor supervisedly end-to-end. We test for prediction as well as for zero shot prediction errors. There are a few outliers in the NAS-Bench-101 graphs that end up with a low validation accuracy

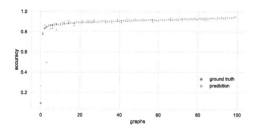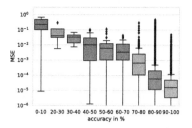

Fig. 2. (*Left*) The predicted accuracy and ground truth of 100 randomly sampled graphs from the NAS-Bench-101 dataset showing a low prediction error for graphs with high accuracy. For low accuracy architectures, our model mostly predicts low values. (*Right*) The mean and variance of the squared error of the test set performance prediction sorted by the ground truth accuracy in logarithmic scale. Predictions are very reliable for architectures in the high accuracy domain while errors are higher for very low performing architectures.

Table 1. Predictive performance of the GNN encoder in terms of RMSE on supervised validation performance prediction.

Model	Prediction
Encoder	**0.0486(±0.1%)**
RF-wide-depth feature encoding	0.061(±0.4%)
RF-one-hot encoding	0.0632(±0.01%)
MLP-one-hot encoding	0.0632(±0.02%)
RNN-one-hot encoding	0.063(±0.01%)

on the CIFAR-10 classification task. Figure 2 (left) visualizes these outliers and shows that our model is able to find them even if it cannot perfectly predict their accuracies. One can see that the model predicts the validation accuracy of well performing graphs very accurately. To further explore the loss, Fig. 2 (right) illustrates the mean and variance of the squared error of the test set partitioned in 9 bins with respect to the ground truth accuracy. The greater part of the loss arises from graphs with a low accuracy. More importantly, our model is very accurate in its prediction for graphs of interest namely graphs with high accuracy.

The rather bad prediction of graphs with low and intermediate accuracy can be explained through their low share in the dataset. Taking a look at the distribution of the individual accuracies in the overall NAS-Bench-101 dataset, as shown in Fig. 3 (left), illustrates the low share of low and intermediate accuracies in the dataset and explains therefore, the rather bad prediction behaviour of our surrogate model. Figure 3 (middle) and (right) plot the validation accuracy compared to the test accuracy of the NAS-Bench-101 dataset. This figure illustrates that predicting the best architecture on the validation set does not necessarily imply a proper prediction on the test set.

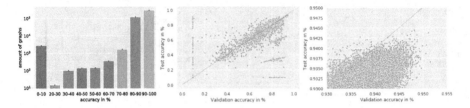

Fig. 3. Distinct properties of NAS-Bench-101; The allocation of the dataset sorted by the ground truth accuracy in logarithmic scale ~98.8% in the two last bins (*left*). NAS-Bench-101 validation and test accuracy behaviour on the CIFAR-10 image classification task. Validation accuracy in % compared to the test accuracy in % of the neural architectures in the NAS-Bench-101 dataset (*middle*). A more precise look into the areas of interest for neural architectures display that the best neural architecture by means of the test accuracy is unequal to the best accuracy by means of the validation accuracy (*right*).

We compare the results of the encoder to several baselines. Our baselines are a random forest approach and also an MLP regressor with four layers, using one-hot node feature encodings and graph depth/width feature encodings. In order to compare to an RNN baseline, we adapted the RNN-surrogate model from [23], which, in their original implementation, only handles cells of equal length. For the application to NAS-Bench-101 with cell types of different length, we do the following slight modification: we input to the LSTM a 0-padded one-hot vector of node attributes, encoding up to 7 nodes and 5 operations.

Table 1 summarises the performance prediction results on the supervised performance prediction task. All experiments are repeated 3 times and we report the mean and the relative standard deviation. The experiments show that our surrogate model is able to predict the neural architecture performances in a stable way and outperforms all baselines in terms of the RMSE by a significant margin.

Zero Shot Performance Prediction. Next, we consider the task of predicting the validation accuracy of structurally unknown graph types, i.e. zero shot prediction. The zero shot prediction task is furthermore divided into two subtasks. First, the encoder is trained on all graphs of length 2, 3, 4, 5, 7 and tested on graphs of length 6. In this scenario, the unseen architectures could be understood as interpolations of seen architectures. Second, we learn the encoder on graphs of length 2, 3, 4, 5, 6 and test it on graphs of length 7. This case is expected to be harder not only because the graphs of length 7 are the clear majority and have the highest diversity, but also because the prediction of their performance is an extrapolation out of the seen training distribution.

Table 2 summarizes the performance prediction results on the zero shot performance prediction task. All experiments are repeated 3 times and we report the mean and the relative standard deviation. As expected, the resulting RMSE is slightly higher for the extrapolation to graphs of length 7 than for the zero

Table 2. Predictive performance of the GNN encoder in terms of RMSE on the two different zero shot validation performance prediction tasks.

Model	Zero shot prediction	
	$2, 3, 4, 5, 7 - 6$	$2, 3, 4, 5, 6 - 7$
Encoder	**0.0523(±3.9%)**	**0.0573(±1.7%)**
RF-wide-depth feature encoding	0.06(±0.2%)	0.073(±0.5%)
RF-one-hot encoding	0.07(±0.04%)	0.063(±0.1%)
MLP-one-hot encoding	0.0647(±2.4%)	0.094(±12.7%)
RNN-one-hot encoding	0.062(±4.7%)	0.069(±3.3%)

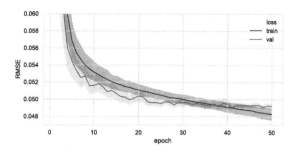

Fig. 4. Progress of loss and validation error over 50 epochs regarding performance prediction. Best validation RMSE ~0.0487.

shot prediction for graphs of length 6. Yet the overall prediction improves over all baselines by a significant margin. The higher standard deviation in comparison to the random forest baselines indicates that the performance of the GNN depends more strongly on the weight initialization than in the fully supervised case. Yet, please note that this dependence on the initialization is still significantly lower than for the MLP and RNN baselines. The experiments show that our surrogate model is able to accurately predict data that it has never seen, i.e. that it can predict the accuracies even for architectures not represented by the training distribution.

6.2 Training Behaviour

In the following, we analyse the training behaviour of our model in the different scenarios described above.

Supervised Performance Prediction. For visualisation aspects of the training behaviour of our encoder, we plot the development of the training loss against the validation loss for the supervised performance prediction from Sect. 6.2 *Supervised Performance Prediction*. Figure 4 displays this development of training loss against validation loss measured by means of the RMSE. The smallest achieved RMSE is ~0.0487 for training on 70% of the dataset, i.e. 296, 558 samples.

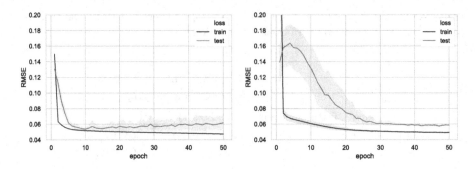

Fig. 5. Progress of loss and test error over 50 epochs regarding zero shot prediction with two distinct splits. One training set consists of all graphs of length 2, 3, 4, 5, 7 with a test set of the graphs of length 6 (left). The other consists of all graphs of length 2, 3, 4, 5, 6 with a test set of the graphs of length 7 (right).

Zero Shot Prediction. The progress of the training loss and test error for the *zero shot prediction* case of our encoder can be seen in Fig. 5. The training set containing all graphs of length 2, 3, 4, 5, 7/2, 3, 4, 5, 6 has a total amount of 361, 614/64, 542 samples. Thus the encoder is tested only on graphs of length 6/7, which corresponds to a total of 62, 010/359, 082 neural architectures. The experiments show that our model is able to accurately predict data that it has never seen before. The behaviour of the test error during the second zero shot prediction task, see Fig. 5 (right), displays interesting information. During the first epochs, the error rises before it starts decreasing and approaching the training loss asymptotically. One interpretation could be that the model first learns simple graph properties like the number of nodes before it learns more complex graph substructures that generalise to the unseen data.

6.3   Comparison to State of the Art

In this section we compare our GNN-surrogate model with the most recent state-of-the-art predictor [33]. They evaluate their predictor on the test accuracy of the NAS-Bench-101 dataset. Since predicting on the validation accuracy does not imply the same proper prediction behaviour on the test set, we evaluate our surrogate model in the same setting. In [33], an auto-encoder model is first trained on the entire NAS-Bench-101 dataset and then fine-tuned with a graph similarity metric and test accuracy labels. Because the training relies on an unsupervised pre-training, they refer to the approach as semi-supervised. To enable a direct comparison, we sample randomly 1, 000/10, 000/100, 000 graphs from the training data set and evaluate the performance prediction ability of the GNN surrogate model on all remaining 431, 624/413, 624/323, 624 graphs in the NAS-Bench-101 dataset. Please note that, at training time, the semi-supervised approach from [33] actually has access to more data than our fully supervised approach, because of the unsupervised pre-training.

Table 3. Comparison of predictive performance of surrogate models in terms of MSE on the test accuracies.

Surrogate-model	Performance prediction		
	1,000	10,000	100,000
GNN encoder	0.0044	**0.0022**	**0.0015**
Semi-supervised assessor [33]	**0.0031**	0.0026	0.0016

Table 3 shows the experimental comparison, where we report an average over three runs for our approach while the numbers of [33] are taken from their paper. The proposed GNN surrogate model surpasses the proposed semi-supervised assessor [33] when 10.000 and 100.000 training architectures are available. With only 1.000 randomly drawn training samples, the results of our approach decrease. Yet, since we do not have access to the exact training samples used in [33], the results might become less comparable the lower the number of samples drawn.

7 Conclusion

In this paper, we propose a GNN surrogate model for the prediction of the performance of neural architectures. Through multiple experiments on NAS-Bench-101, we examined various capabilities of the encoder. The GNN encoder is a powerful tool regarding supervised performance prediction and also especially in the zero-shot setup. Further research will mainly review the possibilities of neural architecture search in accordance with further performance prediction.

Acknowledgement. We thank Alexander Diete for helpful discussions and comments. This project is supported by the German Federal Ministry of Education and Research Foundation via the project DeToL.

References

1. Baker, B., Gupta, O., Raskar, R., Naik, N.: Accelerating neural architecture search using performance prediction. In: 6th International Conference on Learning Representations, ICLR 2018, Vancouver, BC, Canada, April 30 – May 3, 2018, Workshop Track Proceedings. OpenReview.net (2018). https://openreview.net/forum?id=HJqk3N1vG
2. Bastings, J., Titov, I., Aziz, W., Marcheggiani, D., Sima'an, K.: Graph convolutional encoders for syntax-aware neural machine translation. arXiv preprint arXiv:1704.04675 (2017)
3. Battaglia, P.W., et al.: Relational inductive biases, deep learning, and graph networks. arXiv preprint arXiv:1806.01261 (2018)
4. Bronstein, M.M., Bruna, J., LeCun, Y., Szlam, A., Vandergheynst, P.: Geometric deep learning: going beyond Euclidean data. IEEE Signal Process. Mag. **34**(4), 18–42 (2017)

5. Bruna, J., Zaremba, W., Szlam, A., LeCun, Y.: Spectral networks and locally connected networks on graphs. arXiv preprint arXiv:1312.6203 (2013)
6. Dong, X., Yang, Y.: NAS-Bench-102: extending the scope of reproducible neural architecture search. In: International Conference on Learning Representations (2020). https://openreview.net/forum?id=HJxyZkBKDr
7. Elsken, T., Metzen, J.H., Hutter, F.: Neural architecture search: a survey. arXiv preprint arXiv:1808.05377 (2018)
8. Falkner, S., Klein, A., Hutter, F.: BOHB: robust and efficient hyperparameter optimization at scale. In: Dy, J., Krause, A. (eds.) Proceedings of the 35th International Conference on Machine Learning. Proceedings of Machine Learning Research, vol. 80, pp. 1437–1446. PMLR, Stockholmsmässan, Stockholm Sweden, 10–15 July 2018. http://proceedings.mlr.press/v80/falkner18a.html
9. Fey, M., Lenssen, J.E.: Fast graph representation learning with PyTorch geometric. arXiv preprint arXiv:1903.02428 (2019)
10. Gilmer, J., Schoenholz, S.S., Riley, P.F., Vinyals, O., Dahl, G.E.: Neural message passing for quantum chemistry. In: Proceedings of the 34th International Conference on Machine Learning, vol. 70, pp. 1263–1272. JMLR.org (2017)
11. Goodfellow, I., et al.: Generative adversarial nets. In: Advances in Neural Information Processing Systems, pp. 2672–2680 (2014)
12. Gori, M., Monfardini, G., Scarselli, F.: A new model for learning in graph domains. In: Proceedings. 2005 IEEE International Joint Conference on Neural Networks, vol. 2, pp. 729–734. IEEE (2005)
13. Hamilton, W., Ying, Z., Leskovec, J.: Inductive representation learning on large graphs. In: Advances in Neural Information Processing Systems, pp. 1024–1034 (2017)
14. He, K., Zhang, X., Ren, S., Sun, J.: Deep residual learning for image recognition. In: CVPR (2016)
15. Henaff, M., Bruna, J., LeCun, Y.: Deep convolutional networks on graph-structured data. arXiv preprint arXiv:1506.05163 (2015)
16. Kandasamy, K., Neiswanger, W., Schneider, J., Póczos, B., Xing, E.P.: Neural architecture search with Bayesian optimisation and optimal transport. In: Advances in Neural Information Processing Systems 31: Annual Conference on Neural Information Processing Systems 2018, NeurIPS 2018, pp. 2020–2029 (2018)
17. Kipf, T.N., Welling, M.: Semi-supervised classification with graph convolutional networks. arXiv preprint arXiv:1609.02907 (2016)
18. Krizhevsky, A.: Learning multiple layers of features from tiny images. University of Toronto, May 2012
19. Krizhevsky, A., Sutskever, I., Hinton, G.E.: ImageNet classification with deep convolutional neural networks. In: Advances in Neural Information Processing Systems, pp. 1097–1105 (2012)
20. Landrieu, L., Simonovsky, M.: Large-scale point cloud semantic segmentation with superpoint graphs. In: Proceedings of the IEEE Conference on Computer Vision and Pattern Recognition, pp. 4558–4567 (2018)
21. Li, Y., Tarlow, D., Brockschmidt, M., Zemel, R.S.: Gated graph sequence neural networks. In: Bengio, Y., LeCun, Y. (eds.) 4th International Conference on Learning Representations, ICLR 2016, San Juan, Puerto Rico, 2–4 May 2016, Conference Track Proceedings (2016). http://arxiv.org/abs/1511.05493
22. Li, Y., Vinyals, O., Dyer, C., Pascanu, R., Battaglia, P.: Learning deep generative models of graphs (2018)
23. Liu, C., et al.: Progressive neural architecture search. CoRR abs/1712.00559 (2017), http://arxiv.org/abs/1712.00559

24. Liu, H., Simonyan, K., Vinyals, O., Fernando, C., Kavukcuoglu, K.: Hierarchical representations for efficient architecture search. CoRR abs/1711.00436 (2017). http://arxiv.org/abs/1711.00436
25. Liu, H., Simonyan, K., Yang, Y.: DARTS: differentiable architecture search. CoRR abs/1806.09055 (2018). http://arxiv.org/abs/1806.09055
26. Luo, R., Tian, F., Qin, T., Chen, E., Liu, T.Y.: Neural architecture optimization. In: Advances in Neural Information Processing Systems, pp. 7816–7827 (2018)
27. Monti, F., Bronstein, M., Bresson, X.: Geometric matrix completion with recurrent multi-graph neural networks. In: Advances in Neural Information Processing Systems, pp. 3697–3707 (2017)
28. Niepert, M., Ahmed, M., Kutzkov, K.: Learning convolutional neural networks for graphs. In: International Conference on Machine Learning, pp. 2014–2023 (2016)
29. Paszke, A., et al.: Automatic differentiation in PyTorch (2017)
30. Pham, H., Guan, M.Y., Zoph, B., Le, Q.V., Dean, J.: Efficient neural architecture search via parameter sharing. In: Proceedings of the 35th International Conference on Machine Learning, pp. 4092–4101 (2018)
31. Real, E., et al.: Large-scale evolution of image classifiers. In: Proceedings of the 34th International Conference on Machine Learning, pp. 2902–2911 (2017)
32. Szegedy, C., Vanhoucke, V., Ioffe, S., Shlens, J., Wojna, Z.: Rethinking the inception architecture for computer vision. In: CVPR (2016)
33. Tang, Y., et al.: A semi-supervised assessor of neural architectures (2020)
34. White, C., Neiswanger, W., Savani, Y.: BANANAS: Bayesian optimization with neural architectures for neural architecture search. arXiv preprint arXiv:1910.11858 (2019)
35. Wu, Z., Pan, S., Chen, F., Long, G., Zhang, C., Yu, P.S.: A comprehensive survey on graph neural networks. arXiv preprint arXiv:1901.00596 (2019)
36. Xu, D., Zhu, Y., Choy, C.B., Fei-Fei, L.: Scene graph generation by iterative message passing. In: Proceedings of the IEEE Conference on Computer Vision and Pattern Recognition, pp. 5410–5419 (2017)
37. Yi, L., Su, H., Guo, X., Guibas, L.J.: SyncSpecCNN: synchronized spectral CNN for 3D shape segmentation. In: Proceedings of the IEEE Conference on Computer Vision and Pattern Recognition, pp. 2282–2290 (2017)
38. Ying, C., Klein, A., Real, E., Christiansen, E., Murphy, K., Hutter, F.: NAS-bench-101: Towards reproducible neural architecture search. arXiv preprint arXiv:1902.09635 (2019)
39. Zela, A., Siems, J., Hutter, F.: NAS-Bench-1Shot1: benchmarking and dissecting one-shot neural architecture search. In: International Conference on Learning Representations (2020). https://openreview.net/forum?id=SJx9ngStPH
40. Zoph, B., Le, Q.V.: Neural architecture search with reinforcement learning. In: 5th International Conference on Learning Representations (2017)
41. Zoph, B., Vasudevan, V., Shlens, J., Le, Q.V.: Learning transferable architectures for scalable image recognition. In: Proceedings of the IEEE Conference on Computer Vision and Pattern Recognition, pp. 8697–8710 (2018)

Discovering Latent Classes
for Semi-supervised Semantic
Segmentation

Olga Zatsarynna[1]([✉]), Johann Sawatzky[1,2], and Juergen Gall[1]

[1] University of Bonn, Bonn, Germany
{s6olzats,jsawatzk,jgall}@uni-bonn.de
[2] EyewareTech, Martigny, Switzerland

Abstract. High annotation costs are a major bottleneck for the training of semantic segmentation approaches. Therefore, methods working with less annotation effort are of special interest. This paper studies the problem of semi-supervised semantic segmentation, that is only a small subset of the training images is annotated. In order to leverage the information present in the unlabeled images, we propose to learn a second task that is related to semantic segmentation but that is easier to learn and requires less annotated images. For the second task, we learn latent classes that are on one hand easy enough to be learned from the small set of labeled data and are on the other hand as consistent as possible with the semantic classes. While the latent classes are learned on the labeled data, the branch for inferring latent classes provides on the unlabeled data an additional supervision signal for the branch for semantic segmentation. In our experiments, we show that the latent classes boost the accuracy for semi-supervised semantic segmentation and that the proposed method achieves state-of-the-art results on the Pascal VOC 2012 and Cityscapes datasets.

Keywords: Semantic segmentation · Semi-supervised learning · Generative adversarial networks

1 Introduction

In recent years, deep convolutional neural networks (DCNNs) have achieved astonishing performance for the task of semantic segmentation. However, to

---

O. Zatsarynna and J. Sawatzky—Contributed equally.

Electronic supplementary material The online version of this chapter (https://doi.org/10.1007/978-3-030-71278-5_15) contains supplementary material, which is available to authorized users.

Z. Akata et al. (Eds.): DAGM GCPR 2020, LNCS 12544, pp. 202–217, 2021.
https://doi.org/10.1007/978-3-030-71278-5_15

(a) Image   (b) Latent Classes   (c) Semantic Classes

Fig. 1. Our network learns not only semantic but also latent classes that are easier to predict. The figure shows an example of latent and semantic class segmentation for an image that is not part of the training data. As it can be seen, the learned latent classes are very intuitive since the vehicles are grouped into one latent class and objects that are difficult to segment like pedestrians, bicycles, and signs are grouped into another latent class.

achieve good results, DCNN-based methods require an enormous amount of high-quality annotated training data and acquiring it takes a lot of effort and time. This problem is especially acute for the task of semantic segmentation, due to the need for per-pixel labels for every training image. To mitigate the annotation expenses, Hung et al. [14] proposed a semi-supervised algorithm that employs images without annotation during training. On labeled data, the authors train a discriminator network that distinguishes segmentation predictions and ground-truth annotations. On unlabeled data, they use the discriminator to obtain two kinds of supervision signals. First, they use an adversarial loss to enforce realism in the predictions. Second, they use the discriminator to locate regions of sufficient realism in the prediction. These regions are then annotated by the semantic class with the highest probability. Finally, the network for semantic segmentation is trained on the labeled images and the estimated regions of the unlabeled images. Recently, Mittal et al. [28] introduced an extension to [14] by improving the adversarial training and adding a semi-supervised classification module. The latter is used for refining the predictions at the inference time. Although these approaches report impressive results for semi-supervised semantic segmentation, they do not leverage the entire information which is present in the unlabeled images since they discard large parts of the images.

In this work, we propose an approach for semi-supervised semantic segmentation that does not discard any information. Our key observation is that the difficulty of the semantic segmentation task depends on the definition of the semantic classes. This means that the task can be simplified if some classes are grouped together or if the classes are defined in a different way, which is more consistent with the similarity of the instances in the feature space. If the segmentation task becomes easier, less labeled data will be required to train the network. This approach is in contrast to [14,28] that focus on regions in the unlabeled images that are easy to segment, whereas we focus to learn a simpler segmentation task with latent classes on the labeled data that is then used as additional guidance to learn the original task on the labeled and unlabeled data. Figure 1 shows an example of inferred latent classes and semantic classes.

Our network consists of two branches and is trained on labeled and unlabeled images jointly in an end-to-end fashion as illustrated in Fig. 2. While the semantic branch learns to infer the given semantic classes, the latent branch learns latent classes and infers the learned latent classes. In contrast to the semantic branch, the loss for the latent branch takes only the labeled images into account. The purpose of the latent branch is to discover latent classes that are simple enough such that they can be learned on the small set of labeled data. Without any constraints this would result in a single latent class. We therefore introduce a conditional entropy loss that minimizes the variety of semantic classes that are assigned to a particular latent class. In other words, the latent classes should be on one hand easy enough to be learned from the small set of labeled data and on the other hand they should be as consistent as possible with the semantic classes. Since the latent branch solves a simpler semantic segmentation task, we use it as additional supervision for the semantic branch on the unlabeled images. After training, the latent branch is discarded and only the semantic branch is used for inference.

We demonstrate that our model achieves state-of-the-art results on PASCAL VOC 2012 [8] and Cityscapes [6]. Additionally, we show that the learned latent classes are superior to manually defined supercategories.

2 Related Work

The expensive acquisition of pixel-wise annotated images has been recognized as a major bottleneck for the training of deep semantic segmentation models. Consequently, the community sought ways to reduce the amount of annotated images while loosing as little performance as possible.

Weakly-supervised semantic segmentation methods learn to segment images from cheaper image annotations, i.e. pixel-wise labels are exchanged for cheaper annotations for all the images in the training set. The proposed types of annotations include bounding boxes [16,23,31,41], scribbles [25,42,43] or human annotated keypoints [2]. Image level class tags have attracted special attention. A minority of works in this area first detect potential object regions and then identify the object class using the class tags [9,32,34]. The majority of approaches use class activation maps (CAMs) [49] to initially locate the classes of interest. Pinheiro et al. [33,40] pioneered in this area and several methods have improved this approach [1,3,4,10,12,13,17,30,37,43–47]. A few works leverage additional data available on the Internet. For example, [11,15,20] use videos. While the works mentioned above mainly focus on refining the localization cues obtained from the CAM, recently the task of improving the CAM itself received attention [19,20,22].

Some of the works mentioned above consider a setup where some images have pixel-wise annotations and the other images are weakly labeled. They combine fully supervised learning with weakly supervised learning. Papandreou et al. [31] proposed an expectation maximization based approach, modelling the pixel-wise labels as hidden variables and the image labels or bounding boxes as the observed ones. Lee et al. [19] introduce a sophisticated dropout method to obtain better class activation maps on unlabeled images. Earlier, Li et al. [22] improved the CAMs by automatically erasing the most discriminative parts of an object. Wei et al. [47] examine what improvement in CAMs can be achieved by dilated convolutions. Different from previous approaches, Zilong et al. [13] do not improve the CAM but focus on refining high confidence regions obtained from the CAM by deep seeded region growing. The semi-supervised setting without any additional weak supervision has been so far only addressed by [14,28].

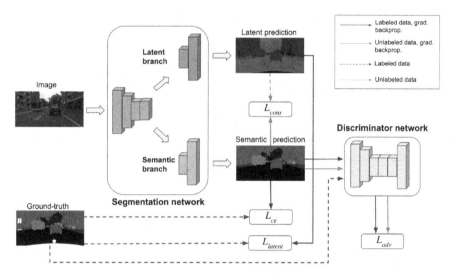

Fig. 2. Overview of the proposed method. While the semantic branch infers pixel-wise class labels, the latent branch learns latent classes and infers the learned latent classes. The latent classes are learned only on the labeled images using the latent loss L_{latent} that ensures that the latent classes are as consistent as possible with the semantic classes. The semantic branch is trained on labeled images with the cross-entropy loss L_{ce} and on unlabeled images the predictions of the latent branch are used as supervision (L_{cons}). Additionally, the semantic branch receives adversarial feedback (L_{adv}) from a discriminator network distinguishing predicted and ground truth segmentations.

While learning an easier auxiliary task as an intermediate step has been investigated in the area of domain adaptation [7,18,24,39,48], it has not been studied for semi-supervised semantic segmentation. Moreover, using latent classes to facilitate learning has been investigated for object detection [35,50], joint object detection and pose estimation [21], and weakly-supervised video segmentation [36]. However, apart from addressing a different task, these approaches focus on discovering subcategories of classes while we aim to group the classes.

3   Method

An overview of our method is given in Fig. 2. Our proposed model is a two-branch network. While the semantic branch serves to solve the final task, the purpose of the latent branch is to learn to group the semantic classes into latent classes in a data driven way as fine-grained as possible. While the fraction of annotated data is not sufficient to produce good results for the task of semantic segmentation, it is enough to learn the prediction of latent classes reasonably well, since this task is easier. Thus, the predictions of the latent branch can then serve as a supervision signal for the semantic branch on unlabeled data.

3.1   Semantic Branch

The task of the semantic branch S_c is to solve the final task of semantic segmentation, that is to predict the semantic classes for the input image. This branch is trained both on labeled and unlabeled data.

On labeled data, we optimize the semantic branch with respect to two loss terms. The first term is the cross-entropy loss:

$$L_{ce} = -\sum_{h,w,n}\sum_{c\in C} Y_n^{(h,w,c)} \log(S_c(X_n)^{(h,w,c)}) \tag{1}$$

where $X_n \in \mathbb{R}^{H\times W\times 3}$ is the image, $Y_n \in \mathbb{R}^{H\times W\times |C|}$ is the one-hot encoded ground truth for semantic classes, and S_c is the predicted probability of the semantic classes. To enforce realism in the semantic predictions, we additionally apply an adversarial loss:

$$L_{adv} = -\sum_{n,h,w} \log(D(S_c(X_n))^{(h,w)}) \tag{2}$$

Details of the discriminator network D are given in Sect. 3.4.

On unlabeled data, the loss function for the semantic branch also consists of two terms. The first one is the adversarial term (2) and the second term is the consistency loss that is described in Sect. 3.3.

3.2   Latent Branch

In order to provide additional supervision for the semantic branch on the unlabeled data, we introduce a latent branch S_l that is trained only on the labeled data. The purpose of the latent branch is to learn latent classes that are easier to distinguish than the semantic classes and that can be better learned on a small set of labeled images. Figure 1 shows an example of latent classes where for instance semantic similar classes like vehicles are grouped together. One of the latent classes often corresponds to a stuff class that includes all difficult classes. This is desirable since having several latent classes that are easy to recognize and one latent class that contains the rest results in a simple segmentation task that can be learned from a small set of labeled images. However, we have to prevent a trivial solution where a single latent class contains all semantic classes. We therefore propose a loss that ensures that the latent classes $l \in \mathcal{L}$ have to provide as much information about semantic classes $c \in \mathcal{C}$ as possible.

To this end, we use the conditional entropy as loss:

$$L_{latent} = -\sum_{l \in \mathcal{L}} \sum_{c \in \mathcal{C}} P_b(c, l) \log(P_b(c|l)). \tag{3}$$

The loss is minimized if the variety of possible semantic classes for each latent class l is as low as possible. In the best case, there is a one-to-one mapping between the latent and semantic classes. The index b denotes that the probability is calculated batchwise. We first estimate the joint probability

$$P_b(c, l) = \frac{1}{NHW} \sum_{h,w,n} S_l(X_n)^{(h,w,l)} Y_n^{(h,w,c)} \tag{4}$$

where H and W are the image height and width, N is the number of images in the batch, S_l is the predicted probability of the latent classes, and $Y_n \in \mathbb{R}^{H \times W \times |\mathcal{C}|}$ is the one-hot encoded ground truth for the semantic classes. From this, we obtain

$$P_b(c|l) = \frac{P_b(c, l)}{\sum_c P_b(c, l)}. \tag{5}$$

Obtaining the conditional entropy from multiple batches is in principle desirable, but it requires the storage of feature maps from multiple batches. Therefore we compute it per batch.

3.3 Consistency Loss

While the latent branch is trained only on the labeled data, the purpose of the latent branch is to provide additional supervision for the semantic branch on the unlabeled data. Given that the latent branch solves a simpler task than the semantic branch, we can expect that the latent classes are more accurately predicted than the semantic classes. We therefore propose a loss that measures the consistency of the prediction of the semantic branch with the prediction of the latent branch. Since the number of latent classes is less or equal than the number of semantic classes, we map the prediction of the semantic branch S_c to a probability distribution of latent classes $S_{\hat{l}_c}$:

$$S_{\hat{l}_c}(X_n)^{(h,w,l)} = \sum_{c \in \mathcal{C}} P(l|c) S_c(X_n)^{(h,w,c)}. \tag{6}$$

We estimate $P(l|c)$ from the predictions of the latent branch on the labeled data. We keep track of how often semantic and latent classes co-occur with an exponentially moving average:

$$M_{c,l}^{(i)} = (1 - \alpha) M_{c,l}^{(i-1)} + \alpha \sum_{h,w,n} Y_n^{(h,w,c)} S_l(X_n)^{(h,w,l)} \tag{7}$$

where i denotes the number of the batch. The initialization is $M_{c,l}^0 = 0$. The parameter $0 < \alpha < 1$ controls how fast we update the average. We set α to the batch size divided by the number of images in the data set. Using the acquired co-occurence matrix M, $P(l|c)$ is estimated as:

$$P(l|c) = \frac{M_{c,l}}{\sum_{k \in \mathcal{L}} M_{c,k}}. \tag{8}$$

The consistency loss is then defined by the mean cross entropy between the latent variable maps predicted by the latent branch S_l and the ones constructed based on the prediction of the semantic branch $S_{\hat{l}_c}$:

$$L_{cons} = -\frac{1}{NHW} \sum_{n,h,w} \sum_{l \in \mathcal{L}} S_l(X_n)^{(h,w,l)} \log(S_{\hat{l}_c}(X_n)^{(h,w,l)}). \tag{9}$$

The minimization of this loss forces the semantic branch to predict classes which are assigned to highly probable latent classes.

3.4   Discriminator Network

Our discriminator network D is a fully-convolutional network [27] with 5 layers and Leaky-ReLu as nonlinearity. It takes label probability maps from the segmentation network or ground-truth maps as input and predicts spatial confidence maps. Each pixel represents the confidence of the discriminator about whether the corresponding pixel in a semantic label map was sampled from the ground-truth map or the segmentation prediction. We train the discriminator network with the help of the spatial cross-entropy loss using both labeled and unlabeled data:

$$L_D = -\sum_{h,w} (1 - y_n) \log(1 - D(S_c(X_n))^{h,w}) + y_n \log(D(Y_n)^{h,w}) \tag{10}$$

where $y_n = 0$ if a sample is drawn from the segmentation network, and $y_n = 1$ if it is a ground-truth map. By minimizing such a loss, the discriminator learns to distinguish between the generated and ground-truth probability maps.

4   Experiments

4.1   Implementation Details

For a fair comparison with Hung et al. [14] and Mittal et al. [28], we choose the same backbone architecture and keep the same hyper-parameters where appropriate. For the segmentation network, we use a single scale ResNet-based DeepLab-v2 [5] architecture that is pre-trained on ImageNet [38] and MSCOCO [26]. We branch the proposed network at the last layer by applying Atrous Spatial Pyramid Pooling (ASPP) [5] two times for the semantic and latent branch.

Finally, we use bilinear upsampling to make the predictions match the initial image size.

For the discriminator network, we use a fully convolutional network, which contains 5 convolutional layers with kernels of the sizes 4×4 and 64, 128, 256, 512 and 1 channels, applied with a stride equal to 2. Each convolutional layer, except for the last one, is followed by a Leaky-ReLU with the leakage coefficient equal to 0.2.

Table 1. Comparison to the state-of-the-art on Pascal VOC 2012 using mIoU (%).

Method	Fraction of annotated images					
	1/50	1/20	1/8	1/4	1/2	Full
Hung et al. [14]	55.6	64.6	69.5	72.1	73.8	74.9
Mittal et al. [28]	**63.3**	67.2	71.4	–	–	75.6
Proposed	59.6	68.2	71.3	**72.4**	**73.9**	75.0
Proposed + Classifier	61.8	**69.3**	**72.2**	–	–	75.3

We train the segmentation network on labeled and unlabeled data jointly with $L = L_{labeled} + 0.1 \cdot L_{unlabeled}$ where the weight factor is the same as in [14]. The loss for the labeled and unlabeled data are given by

$$L_{labeled} = L_{ce} + L_{latent} + 0.01 \cdot L_{adv}, \qquad (11)$$

$$L_{unlabeled} = L_{cons} + 0.01 \cdot L_{adv}. \qquad (12)$$

The weight for the adversarial loss is also the same as in [14]. By default, we limit the number of latent classes to 20. Additional details are provided as part of the supplementary material.

We conducted our experiments on three datasets for semantic segmentation: Pascal VOC 2012 [8], Cityscapes [6] and IIT Affordances [29]. We report the results for the IIT Affordances dataset [29] in the supplementary material. The Pascal VOC 2012 dataset contains images with objects from 20 foreground classes and one background class. There are 10528 training and 1449 validation images in total. The testing of the resulting model is carried out on the validation set. The Cityscapes dataset comprises images extracted from 50 driving videos. It contains 2975, 500 and 1525 images in the training, validation and test set, respectively, with annotated objects from 19 categories. We report the results of testing the resulting model on the validation set. As an evaluation metric, we use mean-intersection-over-union (mIoU).

4.2 Comparison with the State-of-the-Art

PASCAL VOC 2012. On the PASCAL VOC 2012 dataset, we conducted our experiments on five fractions of annotated images, as shown in Table 1, where the rest of the images are used as unlabeled data. Since [14] report the results only for the latest three fractions, we evaluate the performance of their method for the unreported fractions based on the publicly available code. The improvement is especially pronounced, if we look at the sparsely labeled data fractions, such as 1/50, 1/20 and 1/8. Our method performs on par with [28] and the leading method varies from data fraction to data fraction. However, our approach of learning latent variables is complementary to [28] and we can also add a classifier for refinement as in [28]. We show some qualitative results of our method in the supplementary material.

Table 2. Comparison to the state-of-the-art on Cityscapes using mIoU (%).

Method	Pre-training	Fraction of annotated images			
		1/8	1/4	1/2	Full
Mittal et al. [28]		59.3	61.9	–	65.8
Proposed		61.0	63.1	–	64.9
Hung et al. [14]	COCO	58.8	62.3	65.7	67.7
Proposed	COCO	**63.3**	**65.4**	**66.1**	66.3

Table 3. Impact of the loss terms. The evaluation is performed on Pascal VOC 2012 where 1/8 of the data is labeled. $L_{adv}^{labeled}$ denotes that the adversarial loss is only used for the labeled images.

Loss	mIoU (%)
L_{ce}	64.1
$L_{ce} + L_{latent}$	64.6
$L_{ce} + L_{latent} + L_{cons}$	67.3
$L_{ce} + L_{adv}^{labeled}$	68.7
$L_{ce} + L_{adv}$	69.4
$L_{ce} + L_{latent} + L_{cons} + L_{adv}$	71.3

Cityscapes. For the Cityscapes dataset, we follow the semi-supervised learning protocol that was proposed in [14]. This means that 1/8, 1/4 or 1/2 of the training images are annotated and the other images are used without any annotations. We report the results in Table 2. Since [28] does not pre-train the segmentation network on COCO, we evaluated our method also without COCO pre-training. We outperform both [14] and [28] on all annotated data fractions. We show some qualitative results of our method in the supplementary material.

4.3   Ablation Experiments

In our ablation experiments, we evaluate the impact of each loss term. Then we examine the impact of the number of latent classes and show that they form meaningful supercategories of the semantic classes. Finally, we show that the learned latent classes outperform supercategories that are defined by humans.

Impact of the Loss Terms. For analyzing the impact of the loss terms L_{ce} (1), L_{adv} (2), L_{latent} (3), and L_{cons} (9), we use the Pascal VOC 2012 dataset where 1/8 of the data is labeled. The results for different combinations of loss terms are reported in Table 3.

We start using only the entropy loss L_{ce} since this loss is always required. In this setting only the semantic branch is used and trained only on the labeled data. This setting achieves 64.1% mIoU. Adding the latent loss L_{latent} improves the performance by 0.5%. In this setting, the semantic and latent branch are used, but they are both only trained on the labeled data. Adding the consistency loss L_{cons} boosts the accuracy by 2.7%. This shows that the latent branch provides additional supervision for the semantic branch on the unlabeled data.

So far, we did not use the adversarial loss L_{adv}. When we add the adversarial loss only for the labeled data $L_{adv}^{labeled}$ to the entropy loss L_{ce}, the performance grows by 4.6%. In this setting, only the labeled data is used for training. If we use the adversarial loss also for the unlabeled data, the accuracy increases by 0.7%. This shows that the adversarial loss improves semi-supervised learning, but the gain is not as high compared to additionally using the latent branch to supervise the semantic branch on the unlabeled data. In this setting, all loss terms are used and the accuracy increases further by 1.9%. Compared to the entropy loss L_{ce}, the proposed loss terms increase the accuracy by 7.2%.

Impact of Number of Latent Classes. For our approach, we need to specify the maximum number of latent classes. While we used by default 20 in our previous experiments, we now evaluate it for 2, 4, 6, 10, and 20 latent classes on Pascal VOC 2012 with 1/8 of the data being labeled. The results are reported in Table 4. The performance grows monotonically with the number of latent classes reaching its peak for 20.

In the same table, we also report the number of effective latent classes. We consider a latent class l to be effectively used at threshold t, if $P(l|c) > t$ for at least one semantic class c. We report this number for $t = 0.1$ and $t = 0.9$. The number of effective latent classes differs only slightly for these two thresholds. This shows that a latent class typically either constitutes a supercategory of at least one semantic class or it is not used at all. We observe that until 10, all latent classes are used. If we allow up to 20 latent classes, only 14 latent classes are effectively used. In practice, we recommend to set the number of maximum latent classes to the number of semantic classes. The approach will then select as many latent classes as needed. Although we assume that the number of latent classes is less or equal to the number of semantic classes, we also evaluated the approach for 50 latent classes. As expected, the accuracy drops but the approach remains stable. The number of effectively used latent classes also remains at 14. In practice, this setting should not be used since it violates the assumptions of the approach and can lead to unexpected behavior in some cases.

To see if a semantic class is typically mapped to a single latent class, we plot $P(l|c)$ for inference on Pascal VOC 2012 as well as on Cityscapes and show the results in Fig. 3(a) and Fig. 3(b), respectively. Indeed, the mapping from semantic classes to latent classes is very sparse. Typically, for each semantic class c, there is one dominant latent class l, i.e., $P(l|c) > 0.9$. If the number of latent classes increases to 20, some of the latent classes are not used. On Pascal VOC 2012, similar categories like cat and dog or cow, horse, and sheep are grouped. Some groupings are based on the common background like aeroplane and bird. The grouping bicycle, bottle, and dining table combines the most difficult classes of the dataset. However, we observed that there are small variations of the groupings for different runs when the number of latent classes is very small. On Cityscapes with 20 latent classes, the semantic classes pole, traffic light, and traffic sign; person, rider, motorcycle, and bicycle; wall and fence; truck, bus, and train are grouped together. These groupings are very intuitive.

Table 4. Impact of the number of latent classes. The evaluation is performed on Pascal VOC 2012 where 1/8 of the data is labeled. A latent class l is considered effective, if there exists a semantic class c so that $P(l|c) > t$. The third column shows this number for $t = 0.1$ and the fourth for $t = 0.9$.

Max. latent classes	mIoU (%)	Effective latent classes	
		$t = 0.1$	$t = 0.9$
2	69.7	2	2
4	70.2	4	4
6	70.3	6	6
10	70.7	10	10
20	71.3	16	14
50	70.8	18	14

Comparison of Learned Latent Classes with Manually Defined Latent Classes. Since the latent classes typically learn supercategories of the semantic classes, the question arises if the same effect can be achieved with manually defined supercategories. In this experiment, the latent classes are replaced with 10 manually defined supercategories. More details regarding these supercategories are provided in the supplementary material. In this setting, the latent branch is trained to predict these supercategories on the labeled data using the cross-entropy loss. For unlabeled data, everything remains the same as for the proposed method. We report the results in Table 5. The performance using the

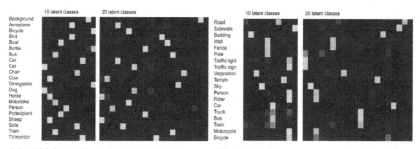

(a) $P(l|c)$ on Pascal VOC 2012 for 10 and 20 latent classes

(b) $P(l|c)$ on Cityscapes for 10 and 20 latent classes.

Fig. 3. The distribution of latent classes for both datasets is pretty sparse, essentially the latent classes form supercategories of semantic classes that are similar in appearance. The grouping bicycle, bottle, and dining table for 10 latent classes seems to be unexpected, but due to the low number of latent classes the network is forced to group additional semantic classes. In this case, the network tends to group the most difficult classes of the dataset. In case of 20 latent classes, the merged classes are very intuitive, but not all latent classes are effectively used.

Table 5. Comparison of learned latent classes with manually defined latent classes. The evaluation is performed on Pascal VOC 2012 where 1/8 of the data is labeled. In case of learned latent classes, the second column reports the maximum number of latent classes. In case of manually defined latent classes, the exact number of classes is reported.

Method	Classes	mIoU (%)
Manual	10	69.0
Learned	10	70.7
Semantic classes	21	68.5
Semantic classes (KL)	21	69.1
Learned	20	71.3

supercategories is only 69.0%, which is significantly below the proposed method for 10 latent variables.

Another approach would be to learn all semantic classes instead of the latent classes in the latent branch. In this case, both branches learn the same semantic classes. This gives 68.5%, which is also worse than the learned latent classes. If both branches predict the same semantic classes, we can also train them symmetrically. Being more specific, on labeled data they are both trained with the cross-entropy loss as well as the adversarial loss. On unlabeled data, we apply the adversarial loss to both of them and use the symmetric Kullback–Leibler divergence (KL) as a consistency loss. This approach performs better, giving 69.1%, but it is still inferior to our proposed method. Overall, this shows the necessity to learn the latent classes in a data-driven way.

5   Conclusion

In this work, we addressed the task of semi-supervised semantic segmentation, where a small fraction of the data set is labeled in a pixel-wise manner, while most images do not have any types of labeling. Our key contribution is a two-branch segmentation architecture, which uses latent classes learned in a data-driven way on labeled data to supervise the semantic segmentation branch on unlabeled data. We evaluated our approach on the Pascal VOC 2012 and the Cityscapes dataset where the proposed method achieves state-of-the-art results.

Acknowledgement. This work was funded by the Deutsche Forschungsgemeinschaft (DFG, German Research Foundation) GA 1927/5-1 and under Germany's Excellence Strategy EXC 2070 – 390732324.

References

1. Ahn, J., Kwak, S.: Learning pixel-level semantic affinity with image-level supervision for weakly supervised semantic segmentation. In: IEEE Conference on Computer Vision and Pattern Recognition (CVPR), pp. 4981–4990 (2018)
2. Bearman, A., Russakovsky, O., Ferrari, V., Fei-Fei, L.: What's the point: semantic segmentation with point supervision. In: Leibe, B., Matas, J., Sebe, N., Welling, M. (eds.) ECCV 2016. LNCS, vol. 9911, pp. 549–565. Springer, Cham (2016). https://doi.org/10.1007/978-3-319-46478-7_34
3. Briq, R., Moeller, M., Gall, J.: Convolutional simplex projection network for weakly supervised semantic segmentation (2018)
4. Chaudhry, A., Dokania, P.K., Torr, P.H.: Discovering class-specific pixels for weakly-supervised semantic segmentation. In: British Machine Vision Conference (BMVC) (2017)
5. Chen, L.C., Papandreou, G., Kokkinos, I., Murphy, K., Yuille, A.L.: DeepLab: semantic image segmentation with deep convolutional nets, atrous convolution, and fully connected CRFs. IEEE Trans. Pattern Anal. Mach. Intell. **40**(4), 834–848 (2018)
6. Cordts, M., et al.: The cityscapes dataset for semantic urban scene understanding. In: IEEE Conference on Computer Vision and Pattern Recognition (CVPR) (2016)
7. Dai, D., Sakaridis, C., Hecker, S., Van Gool, L.: Curriculum model adaptation with synthetic and real data for semantic foggy scene understanding. Int. J. Comput. Vis. **128**, 1182–1204 (2020)
8. Everingham, M., Eslami, S.M., Van Gool, L., Williams, C.K.I., Winn, J., Zisserman, A.: The pascal visual object classes challenge: a retrospective. Int. J. Comput. Vis. (IJCV) **111**(1), 98–136 (2014)
9. Fan, R., Hou, Q., Cheng, M.-M., Yu, G., Martin, R.R., Hu, S.-M.: Associating inter-image salient instances for weakly supervised semantic segmentation. In: Ferrari, V., Hebert, M., Sminchisescu, C., Weiss, Y. (eds.) ECCV 2018. LNCS, vol. 11213, pp. 371–388. Springer, Cham (2018). https://doi.org/10.1007/978-3-030-01240-3_23
10. Ge, W., Yang, S., Yu, Y.: Multi-evidence filtering and fusion for multi-label classification, object detection and semantic segmentation based on weakly supervised learning. In: IEEE Conference on Computer Vision and Pattern Recognition (CVPR), pp. 1277–1286 (2018)

11. Hong, S., Yeo, D., Kwak, S., Lee, H., Han, B.: Weakly supervised semantic segmentation using web-crawled videos. In: IEEE Conference on Computer Vision and Pattern Recognition (CVPR), pp. 2224–2232 (2017)

12. Hou, Q., Massiceti, D., Dokania, P.K., Wei, Y., Cheng, M.-M., Torr, P.H.S.: Bottom-up top-down cues for weakly-supervised semantic segmentation. In: Pelillo, M., Hancock, E. (eds.) EMMCVPR 2017. LNCS, vol. 10746, pp. 263–277. Springer, Cham (2018). https://doi.org/10.1007/978-3-319-78199-0_18

13. Huang, Z., Wang, X., Wang, J., Liu, W., Wang, J.: Weakly-supervised semantic segmentation network with deep seeded region growing. In: IEEE Conference on Computer Vision and Pattern Recognition (CVPR), pp. 7014–7023 (2018)

14. Hung, W.C., Tsai, Y.H., Liou, Y.T., Lin, Y.Y., Yang, M.H.: Adversarial learning for semi-supervised semantic segmentation. In: Proceedings of the British Machine Vision Conference (BMVC) (2018)

15. Jin, B., Segovia, M.V.O., Ssstrunk, S.: Webly supervised semantic segmentation. In: IEEE Conference on Computer Vision and Pattern Recognition (CVPR), pp. 1705–1714 (2017)

16. Khoreva, A., Benenson, R., Hosang, J., Hein, M., Schiele, B.: Simple does it: weakly supervised instance and semantic segmentation. In: IEEE Conference on Computer Vision and Pattern Recognition (CVPR), pp. 1665–1674 (2017)

17. Kolesnikov, A., Lampert, C.H.: Seed, expand and constrain: three principles for weakly-supervised image segmentation. In: Leibe, B., Matas, J., Sebe, N., Welling, M. (eds.) ECCV 2016. LNCS, vol. 9908, pp. 695–711. Springer, Cham (2016). https://doi.org/10.1007/978-3-319-46493-0_42

18. Kurmi, V.K., Bajaj, V., Venkatesh, K.S., Namboodiri, V.P.: Curriculum based dropout discriminator for domain adaptation. In: British Machine Vision Conference (BMVC) (2019)

19. Lee, J., Kim, E., Lee, S., Lee, J., Yoon, S.: Ficklenet: weakly and semi-supervised semantic image segmentation using stochastic inference. In: IEEE Conference on Computer Vision and Pattern Recognition (CVPR) (2019)

20. Lee, J., Kim, E., Lee, S., Lee, J., Yoon, S.: Frame-to-frame aggregation of active regions in web videos for weakly supervised semantic segmentation. In: IEEE International Conference on Computer Vision (ICCV) (2019)

21. Li, H., He, X., Barnes, N., Wang, M.: Learning hough transform with latent structures for joint object detection and pose estimation. In: Tian, Q., Sebe, N., Qi, G.-J., Huet, B., Hong, R., Liu, X. (eds.) MMM 2016. LNCS, vol. 9517, pp. 116–129. Springer, Cham (2016). https://doi.org/10.1007/978-3-319-27674-8_11

22. Li, K., Wu, Z., Peng, K., Ernst, J., Fu, Y.: Guided attention inference network. IEEE Trans. Pattern Anal. Mach. Intell. **42**, 2996–3010 (2019)

23. Li, Q., Arnab, A., Torr, P.H.: Weakly- and semi-supervised panoptic segmentation. In: European Conference on Computer Vision (ECCV), pp. 106–124 (2018)

24. Lian, Q., Lv, F., Duan, L., Gong, B.: Constructing self-motivated pyramid curriculums for cross-domain semantic segmentation: a non-adversarial approach. In: IEEE International Conference on Computer Vision (ICCV) (2019)

25. Lin, D., Dai, J., Jia, J., He, K., Sun, J.: Scribblesup: scribble-supervised convolutional networks for semantic segmentation. In: IEEE Conference on Computer Vision and Pattern Recognition (CVPR), pp. 3159–3167 (2016)

26. Lin, T.-Y., et al.: Microsoft COCO: common objects in context. In: Fleet, D., Pajdla, T., Schiele, B., Tuytelaars, T. (eds.) ECCV 2014. LNCS, vol. 8693, pp. 740–755. Springer, Cham (2014). https://doi.org/10.1007/978-3-319-10602-1_48

27. Long, J., Shelhamer, E., Darrell, T.: Fully convolutional networks for semantic segmentation. In: IEEE Conference on Computer Vision and Pattern Recognition (CVPR), pp. 3431–3440 (2015)
28. Mittal, S., Tatarchenko, M., Brox, T.: Semi-supervised semantic segmentation with high- and low-level consistency. IEEE Tran. Pattern Anal. Mach. Intell. (2019)
29. Nguyen, A., Kanoulas, D., Caldwell, D.G., Tsagarakis, N.: Object-based affordances detection with convolutional neural networks and dense conditional random fields. In: IEEE/RSJ International Conference on Intelligent Robots and Systems (IROS) (2017)
30. Oh, S.J., Benenson, R., Khoreva, A., Akata, Z., Fritz, M., Schiele, B.: Exploiting saliency for object segmentation from image level labels. In: IEEE International Conference on Computer Vision and Pattern Recognition (CVPR), pp. 5038–5047 (2017)
31. Papandreou, G., Chen, L.C., Murphy, K.P., Yuille, A.L.: Weakly- and semi-supervised learning of a deep convolutional network for semantic image segmentation. In: International Conference on Computer Vision (ICCV), pp. 1742–1750 (2015)
32. Pathak, D., Krähenbühl, P., Darrell, T.: Constrained convolutional neural networks for weakly supervised segmentation. In: International Conference on Computer Vision (ICCV), pp. 1796–1804 (2015)
33. Pinheiro, P.H.O., Collobert, R.: From image-level to pixel-level labeling with convolutional networks. In: IEEE Conference on Computer Vision and Pattern Recognition (CVPR), pp. 1713–1721 (2015)
34. Qi, X., Liu, Z., Shi, J., Zhao, H., Jia, J.: Augmented feedback in semantic segmentation under image level supervision. In: Leibe, B., Matas, J., Sebe, N., Welling, M. (eds.) ECCV 2016. LNCS, vol. 9912, pp. 90–105. Springer, Cham (2016). https://doi.org/10.1007/978-3-319-46484-8_6
35. Razavi, N., Gall, J., Kohli, P., van Gool, L.: Latent hough transform for object detection. In: Fitzgibbon, A., Lazebnik, S., Perona, P., Sato, Y., Schmid, C. (eds.) ECCV 2012. LNCS, vol. 7574, pp. 312–325. Springer, Heidelberg (2012). https://doi.org/10.1007/978-3-642-33712-3_23
36. Richard, A., Kuehne, H., Gall, J.: Weakly supervised action learning with RNN based fine-to-coarse modeling. In: IEEE Conference on Computer Vision and Pattern Recognition (CVPR), pp. 1273–1282 (2017)
37. Roy, A., Todorovic, S.: Combining bottom-up, top-down, and smoothness cues for weakly supervised image segmentation. In: IEEE Conference on Computer Vision and Pattern Recognition (CVPR), pp. 7282–7291 (2017)
38. Russakovsky, O., et al.: ImageNet large scale visual recognition challenge. Int. J. Comput. Vis. (IJCV) 115(3), 211–252 (2015)
39. Sakaridis, C., Dai, D., Van Gool, L.: Guided curriculum model adaptation and uncertainty-aware evaluation for semantic nighttime image segmentation. In: IEEE International Conference on Computer Vision (ICCV) (2019)
40. Shimoda, W., Yanai, K.: Distinct class-specific saliency maps for weakly supervised semantic segmentation. In: Leibe, B., Matas, J., Sebe, N., Welling, M. (eds.) ECCV 2016. LNCS, vol. 9908, pp. 218–234. Springer, Cham (2016). https://doi.org/10.1007/978-3-319-46493-0_14
41. Song, C., Huang, Y., Ouyang, W., Wang, L.: Box-driven class-wise region masking and filling rate guided loss for weakly supervised semantic segmentation. In: IEEE Conference on Computer Vision and Pattern Recognition (CVPR) (2019)

42. Tang, M., Djelouah, A., Perazzi, F., Boykov, Y., Schroers, C.: Normalized cut loss for weakly-supervised CNN segmentation. In: IEEE Conference on Computer Vision and Pattern Recognition (CVPR), pp. 1818–1827 (2018)
43. Tang, M., Perazzi, F., Djelouah, A., Ayed, I.B., Schroers, C., Boykov, Y.: On regularized losses for weakly-supervised CNN segmentation. In: Ferrari, V., Hebert, M., Sminchisescu, C., Weiss, Y. (eds.) ECCV 2018. LNCS, vol. 11220, pp. 524–540. Springer, Cham (2018). https://doi.org/10.1007/978-3-030-01270-0_31
44. Wang, X., You, S., Li, X., Ma, H.: Weakly-supervised semantic segmentation by iteratively mining common object features. In: IEEE Conference on Computer Vision and Pattern Recognition (CVPR), pp. 1354–1362 (2018)
45. Wei, Y., Feng, J., Liang, X., Cheng, M.M., Zhao, Y., Yan, S.: Object region mining with adversarial erasing: a simple classification to semantic segmentation approach. In: IEEE Conference on Computer Vision and Pattern Recognition (CVPR), pp. 6488–6496 (2017)
46. Wei, Y., et al.: STC: a simple to complex framework for weakly-supervised semantic segmentation. IEEE Trans. Pattern Anal. Mach. Intell. **39**(11), 2314–2320 (2017)
47. Wei, Y., Xiao, H., Shi, H., Jie, Z., Feng, J., Huang, T.S.: Revisiting dilated convolution: a simple approach for weakly- and semi-supervised semantic segmentation. In: IEEE Conference on Computer Vision and Pattern Recognition (CVPR), pp. 7268–7277 (2018)
48. Zhang, Y., David, P., Gong, B.: Curriculum domain adaptation for semantic segmentation of urban scenes. In: IEEE International Conference on Computer Vision (ICCV), pp. 2039–2049 (2017)
49. Zhou, B., Khosla, A., Lapedriza, A., Oliva, A., Torralba, A.: Learning deep features for discriminative localization. In: IEEE Conference on Computer Vision and Pattern Recognition (CVPR), pp. 2921–2929 (2016)
50. Zhu, X., Anguelov, D., Ramanan, D.: Capturing long-tail distributions of object subcategories, pp. 915–922 (2014)

Riemannian SOS-Polynomial Normalizing Flows

Jonathan Schwarz[1,2(✉)], Felix Draxler[1,2,3], Ullrich Köthe[3],
and Christoph Schnörr[1,2]

[1] Heidelberg Collaboratory for Image Processing, Heidelberg University,
Heidelberg, Germany
`jonathan.schwarz@iwr.uni-heidelberg.de`
[2] Image and Pattern Analysis Group, Heidelberg University, Heidelberg, Germany
[3] Visual Learning Lab, Heidelberg University, Heidelberg, Germany

Abstract. Sum-of-Squares polynomial normalizing flows have been proposed recently, without taking into account the convexity property and the geometry of the corresponding parameter space. We develop two gradient flows based on the geometry of the parameter space of the cone of SOS-polynomials. Few proof-of-concept experiments using non-Gaussian target distributions validate the computational approach and illustrate the expressiveness of SOS-polynomial normalizing flows.

Keywords: Normalizing flows · SOS polynomials · Riemannian gradient flows

1 Introduction

Optimal transport has become a central topic for mathematical modelling [19,24] and for computational approaches to data analysis and machine learning [17]. Wasserstein distances based on various transportation cost functions and their dual formulations, parametrized by deep networks, provide a framework for generative data-driven modeling [6].

A current prominent line of research initiated by [20,21] concerns the representation and estimation of so-called normalizing flows, in order to model a data distribution ν in terms of an elementary reference measure μ, typically the standard Gaussian $\mu = \mathcal{N}(0, I_n)$, as pushforward measure $\nu = T_\sharp \mu$ with respect to a transportation map (diffeomorphism) T. This framework supports a broad range of tasks like density estimation, exploring a posteriori distributions, latent variable models, variational inference, uncertainty quantification, etc. See [12,14,16] for recent surveys.

A key requirement is the ability to evaluate efficiently both T and T^{-1} along with the corresponding Jacobians. Based on classical work [11], triangular maps

This work is supported by Deutsche Forschungsgemeinschaft (DFG) under Germany's Excellence Strategy EXC-2181/1 - 390900948 (the Heidelberg STRUCTURES Excellence Cluster).

Z. Akata et al. (Eds.): DAGM GCPR 2020, LNCS 12544, pp. 218–231, 2021.
https://doi.org/10.1007/978-3-030-71278-5_16

T and their relation to optimal transport, therefore, has become a focus of research [4,8]. While the deviation from *optimal* transport, as defined by [7], can be bounded by transportation inequalities [22], merely regarding triangular maps T as diffeomorphisms (performing *non*-optimal transport) does not restrict expressiveness [3]. Accordingly, triangular maps parametrized by deep networks are nowadays widely applied.

Contribution, Organization. A basic property that ensures the invertibility of T is monotonicity, which in connection with triangular maps can be achieved by the coordinatewise integration of nonnegative functions. In a recent paper [10], Sum-of-Squares (SOS) polynomials that are nonnegative by construction, were used for this purpose, as part of the standard procedure for training deep networks. However, both the convexity properties and the geometry of the parameter space of the cone of SOS polynomials [2,13] were completely ignored. In this work, we take this geometry into account and devise computational approaches to the construction of transportation maps T. Specifically, we contribute:

- We introduce basic notions in Sect. 2 and specify the parametrization of triangular transportation maps using SOS polynomials in Sect. 3.
- Based on this parametrization, two algorithms for learning the parameters from given data are developed in Sect. 4. Algorithm 1 directly exploits the Riemannian geometry of the positive definite matrix cone. Algorithm 2 pulls back the objective function to the tangent bundle and performs ordinary gradient descent using a Krylov subspace method for approximating a related matrix-valued entire function.
- We evaluate both algorithms and the expressiveness of SOS-polynomial flows in Sect. 5 using few academical non-Gaussian distributions. To enable a clear assessment, we do not use a deep network for additional parametrization.

Our findings regarding the first algorithm are quite positive which stimulates further research on suitable extensions to large problem dimensions.

Notation. Let $n \in \mathbb{N}$, then $[n]$ denotes the set $\{1, 2, \ldots, n\}$. We denote the vector space of real multivariate polynomials in n variables of degree at most $d \in \mathbb{N}$ by $\mathbb{R}[x]_d = \mathbb{R}[x_1, \ldots, x_n]_d$. $x^\alpha = x_1^{\alpha_1} \cdots x_n^{\alpha_n} \in \mathbb{R}[x]_d$ is a monomial corresponding to $\alpha \in \mathbb{N}_d^n = \{\alpha \in \mathbb{N}^n : |\alpha| = \sum_{i \in [n]} \alpha_i \leq d\}$. The vectors

$$v_d(x) = (x^\alpha) \in \mathbb{R}^{s_n(d)}, \quad \alpha \in \mathbb{N}_d^n, \quad s_n(d) = \binom{n+d}{d}, \tag{1.1}$$

that comprise all monomials in n variables of degree not greater than d, form a basis of $\mathbb{R}[x]_d$. The number n of variables is implicitly determined by the number of arguments, and may vary. For example, if $d = 2$, then $v_2(x) = (1, x_1, \ldots, x_n, x_1^2, x_1 x_2, \ldots, x_n^2)^\top$ with $s_n(2) = \frac{1}{2}(n+1)(n+2)$. We set $t_n(d) = s_{n-1}(d) = \dim v_d(x_1, \ldots, x_{n-1}, 0)$. S^n, S_+^n and $\mathcal{P}_n$ denote the spaces of symmetric matrices, of symmetric and positive semidefinite matrices, and of symmetric and positive definite matrices, respectively, of dimension $n \times n$. $\langle a, b \rangle = a^\top b$ denotes the Euclidean inner product of $a, b \in \mathbb{R}^n$.

2 Preliminaries

2.1 Normalizing Flows

Let μ and ν denote the reference measure and the target measure supported on $\mathbb{R}^n$, respectively. Throughout this paper, we assume that $\mu = \mathcal{N}(0, I_n)$ is the standard multivariate Gaussian distribution and that ν is absolutely continuous with respect to the Lebesgue measure such that

$$d\mu(x) = p(x)dx, \qquad d\nu(y) = q(y)dy \tag{2.1}$$

with density functions p, q. Our objective is to compute a smooth diffeomorphism $T\colon \mathbb{R}^n \to \mathbb{R}^n$ such that $\nu = T_\sharp\mu$ becomes the pushforward (or image) measure of μ with respect to T, defined by

$$\nu(V) = \mu\big(T^{-1}(V)\big), \qquad V \subset \mathbb{R}^n, \tag{2.2}$$

for all measurable subsets V. In terms of the densities (2.1), Eq. (2.1) reads

$$q(y) = p\big(T^{-1}(y)\big)|\det dT^{-1}(y)|, \tag{2.3a}$$

$$p(x) = q\big(T(x)\big)|\det dT(x)|, \qquad y = T(x) \tag{2.3b}$$

with the Jacobian matrices dT, dT^{-1}. As detailed in Sects. 2.2 and 3, we consider a subclass of diffeomorphisms

$$\mathcal{T_A} = \{T_\mathcal{A} \in \mathrm{Diff}(\mathbb{R}^n)\colon \mathcal{A} \in \mathcal{P}_{n,d}\}, \tag{2.4}$$

whose elements are defined by (3.5b) and (3.6). Assuming samples

$$\{y_i\}_{i\in[N]} \sim \nu \tag{2.5}$$

from the target distribution to be given, the goal is to determine some $T_\mathcal{A} \in \mathcal{T_A}$ such that (2.2) approximately holds. To this end, following [14, Section 4], we set $S_\mathcal{A} = T_\mathcal{A}^{-1}$ and consider the KL divergence

$$\mathrm{KL}\left((S_\mathcal{A})_\sharp q\|p\right) = \mathrm{KL}\left(q\|(T_\mathcal{A})_\sharp p\right) = \mathbb{E}_q[-\log p \circ S_\mathcal{A} - \log \det dS_\mathcal{A}] + c, \tag{2.6}$$

where the constant c collects terms not depending on $S_\mathcal{A}$. Replacing the expectation by the empirical expectation defines the objective function

$$J\colon \mathcal{P}_{n,d} \to \mathbb{R}, \qquad J(\mathcal{A}) = \frac{1}{N}\sum_{i\in[N]}\Big(-\log p\big(S_\mathcal{A}(y_i)\big) - \log \det dS_\mathcal{A}(y_i)\Big). \tag{2.7}$$

After detailing the class of maps (2.4) in Sects. 2.2 and 3, the Riemannian gradient flow with respect to (2.7) will induce a *normalizing flow* of q to p (Sect. 4).

2.2   Triangular Increasing Maps

A mapping $T \colon \mathbb{R}^n \to \mathbb{R}^n$ is called *triangular* and *increasing*, respectively, if each component function T_k only depends on variables x_i with $i \leq k$ (property (2.8a)) and if each function (2.8b) is increasing in x_k.

$$T_k(x) = T_k(x_1, \ldots, x_k), \quad \forall k \in [n] \tag{2.8a}$$

$$x_k \mapsto T_k(x_1, \ldots, x_k), \quad \forall k \in [n] \tag{2.8b}$$

The existence of a triangular map $T \colon C_1 \to C_2$ for any two open solid convex subsets $C_1, C_2 \subset \mathbb{R}^n$ was shown by Knothe [11]. More generally, the existence of a unique (up to μ-equivalence) triangular increasing map T that achieves (2.2), for any given absolutely continuous probability measures μ, ν, was established by [3, Lemma 2.1]. Property (2.8a) implies that the Jacobian matrices dT and dT^{-1} are triangular, which is computationally convenient in connection with (2.3) and (2.7).

3   SOS Polynomials and Triangular Increasing Maps

In this section, we adopt the approach from [10] using SOS polynomials for the construction of increasing triangular maps. The key difference is that we will exploit the geometry and convexity of the parameter space in Sect. 4 for deriving normalizing flows.

Definition 1 (SOS polynomial) [13]. *A polynomial $p \in \mathbb{R}[x]_{2d}$ is a sum-of-squares (SOS) polynomial if there exist $q_1, \ldots, q_m \in \mathbb{R}[x]_d$ such that*

$$p(x) = \sum_{k \in [m]} q_k^2(x). \tag{3.1}$$

We denote the subset of SOS polynomials by $\Sigma[x]_{2d} \subset \mathbb{R}[x]_{2d}$.

The following basic proposition says that each SOS polynomial corresponds to a parameter matrix A on the positive definite manifold.

Theorem 1 ([2, Thm. 3.39]). *A polynomial $p(x) = \sum_{\alpha \in \mathbb{N}_{2d}^n} p_\alpha x^\alpha$ is SOS if and only if there exists a matrix A such that*

$$p(x) = \langle v_d(x), A v_d(x) \rangle, \qquad A \in \mathcal{P}_{s_n(d)}. \tag{3.2}$$

Note that $p(x) \geq 0$, $\forall x \in \mathbb{R}^n$, by construction. Next, we use (2.8) and the representation (3.2) in order to define a family (2.4) of increasing triangular maps. Based on (3.2), define the sequence of SOS polynomials

$$p_{[k]}(x) := p_{[k]}(x_1, \ldots, x_k) = \langle v_d(x_1, \ldots, x_k), A_{[k]} v_d(x_1, \ldots, x_k) \rangle \tag{3.3a}$$

$$\in \Sigma[x_1, \ldots, x_k]_{2d}, \quad A_{[k]} \in \mathcal{P}_{s_k(d)}, \quad k \in [n] \tag{3.3b}$$

and the sequence of linear forms

$$\langle c_{[k]}, v_d(x_1, \ldots, x_{k-1}, 0) \rangle, \qquad c_{[k]} \in \mathbb{R}^{t_n(d)}, \quad k \in [n] \tag{3.4}$$

that are parametrized by symmetric positive definite matrices $A_{[k]}$ and vectors $c_{[k]}$, respectively. We collectively denote these parameters by

$$\mathcal{A} := \{c_{[1]}, \ldots, c_{[n]}, A_{[1]}, \ldots, A_{[n]}\} \in \mathcal{P}_{n,d} \tag{3.5a}$$

$$\mathcal{P}_{n,d} := \mathbb{R}^{t_1(d)} \times \cdots \times \mathbb{R}^{t_n(d)} \times \mathcal{P}_{s_1(d)} \times \cdots \times \mathcal{P}_{s_n(d)}. \tag{3.5b}$$

Then the map

$$T_{\mathcal{A}} \in \mathrm{Diff}(\mathbb{R}^n), \qquad x \mapsto T_{\mathcal{A}}(x) = \left(T_{[1]}(x_1), \ldots, T_{[n]}(x_1, \ldots, x_n) \right)^{\top} \tag{3.6a}$$

$$T_{[k]}(x_1, \ldots, x_k) = \langle c_{[k]}, v_d(x_1, \ldots, x_{k-1}, 0) \rangle \tag{3.6b}$$

$$+ \int_0^{x_k} p_{[k]}(x_1, \ldots, x_{k-1}, \tau) \, d\tau \tag{3.6c}$$

is *triangular and increasing* due to the nonnegativity of the SOS polynomials $p_{[k]}$.

The inverse maps $S_{\mathcal{A}} = T_{\mathcal{A}}^{-1}$ have a similar structure and could be parametrized in the same way. The objective function (2.7) therefore is well defined.

4   Riemannian Normalizing Flows

In this section, we will develop two different gradient descent flows with respect to the objective function (2.7) that take into account the geometry of the parameter space $\mathcal{P}_{n,d}$ (3.5b). Either flow is supposed to transport the target measure ν that is only given through samples (2.5), to the reference measure μ. This will be numerically evaluated in Sect. 5.

Section 4.1 works out details of the Riemannian gradient flow leading to Algorithm 1. Section 4.2 develops a closely related flow using different numerical techniques, leading to Algorithm 2. In what follows, the tangent space to (3.5b) at $\mathcal{A}$ is given and denoted by

$$\mathcal{S}_{n,d} = T_{\mathcal{A}} \mathcal{P}_{n,d} = \mathbb{R}^{t_1(d)} \times \cdots \times \mathbb{R}^{t_n(d)} \times S^{s_1(d)} \times \cdots \times S^{s_n(d)}. \tag{4.1}$$

4.1   Riemannian Gradient

Consider the open cone of positive definite symmetric $n \times n$ matrices $\mathcal{P}_n$. This becomes a Riemannian manifold [1] with the metric

$$g_A(U, V) = \mathrm{tr}(A^{-1} U A^{-1} V), \qquad U, V \in T_A \mathcal{P}_n = S^n. \tag{4.2}$$

The Riemannian gradient of a smooth function $J \colon \mathcal{P}_n \to \mathbb{R}$ reads

$$\mathrm{grad}\, J(A) = A \left(\partial_A J(A) \right) A, \tag{4.3}$$

where $\partial J(A)$ denotes the Euclidean gradient. The exponential map is globally defined and has the form

$$\exp_A(U) = A^{\frac{1}{2}} \operatorname{expm}(A^{-\frac{1}{2}} U A^{-\frac{1}{2}}) A^{\frac{1}{2}}, \qquad A \in \mathcal{P}_n, \quad U \in S^n, \tag{4.4}$$

with the matrix exponential function $\operatorname{expm}(B) = e^B$, $B \in \mathbb{R}^{n \times n}$. Discretizing the flow using the geometric explicit Euler scheme with step size h and iteration counter $t \in \mathbb{N}$ yields

$$A_{t+1} = \exp_{A_t}\left(-h \operatorname{grad} J(A_t)\right) \tag{4.5a}$$

$$= A_t^{\frac{1}{2}} \operatorname{expm}\left(-h A_t^{\frac{1}{2}} \partial_A J(A_t) A_t^{\frac{1}{2}}\right) A_t^{\frac{1}{2}}, \qquad t \in \mathbb{N}, \quad A_0 \in \mathcal{P}_n. \tag{4.5b}$$

Applying this discretization to the respective components of (3.5) yields the following natural gradient flow for the objective function (2.7):

Algorithm 1: Riemannian SOS Flow

Initialization
Choose $\mathcal{A}_0 \in \mathcal{P}_{n,d}$ such that $T_{[1]} \approx \operatorname{id}$.
while *not converged* **do**

$\quad$ $(A_{[k]})_{t+1} =$
$\quad$ $(A_{[k]})_t^{\frac{1}{2}} \operatorname{expm}\left(-h(A_{[k]})_t^{\frac{1}{2}} \partial_{A_{[k]}} J(\mathcal{A}_t)(A_{[k]})_t^{\frac{1}{2}}\right)(A_{[k]})_t^{\frac{1}{2}}, \quad \forall k \in [n],$
$\quad$ $(c_{[k]})_{t+1} = (c_{[k]})_t - h \partial_{c_{[k]}} J(\mathcal{A}_t), \quad \forall k \in [n].$

4.2 Exponential Parameterization

Consider again first the case of a smooth objective function $J \colon \mathcal{P}_n \to \mathbb{R}$. We exploit the fact that the exponential map (4.4) is *globally* defined on the entire tangent space S^n of $(\mathcal{P}_n, g)$, which does not generally hold for Riemannian manifolds. Using

$$\exp_I(U) = \operatorname{expm}(U), \quad U \in S^n, \tag{4.6}$$

we pull back J to the vector space S^n,

$$\widetilde{J} \colon S^n \to \mathbb{R}, \qquad \widetilde{J}(U) = J \circ \operatorname{expm}(U), \tag{4.7}$$

and perform ordinary gradient descent:

$$U_{t+1} = U_t - h \partial \widetilde{J}(U_t), \qquad t \in \mathbb{N}, \quad U_0 \in S^n. \tag{4.8}$$

Denote the canonical inner product on S^n by $\langle U, V \rangle = \operatorname{tr}(UV)$. Then the gradient of $\widetilde{J}(U)$ is given by the equation

$$\frac{d}{d\tau} \widetilde{J}(U + \tau V)\big|_{\tau=0} = \langle \partial \widetilde{J}(U), V \rangle = d_A J \circ d_U \operatorname{expm}(V), \quad \forall V \in S^n, \tag{4.9}$$

where $A = \operatorname{expm}(U)$.

It remains to evaluate the differential of the matrix exponential on the right-hand side of (4.9). Using the vectorization operator vec(.), that turns matrices into vectors by stacking the column vectors, and we have the identity

$$\text{vec}(CXB^{\top}) = (B \otimes C)\,\text{vec}(X). \tag{4.10}$$

Thus, by [9, Thm. 10.13], we conclude

$$\text{vec}\left(d_U \exp\text{m}(V)\right) = K(U)\,\text{vec}(V) \tag{4.11a}$$
$$K(U) = (I \otimes e^U)\psi\left(U \oplus (-U)\right), \tag{4.11b}$$

where $\otimes$ denotes the Kronecker matrix product [23], $\oplus$ denotes the Kronecker sum

$$A \oplus B = A \otimes I_n + I_n \otimes B, \tag{4.12}$$

and ψ denotes the matrix-valued function given by the entire function

$$\psi = \frac{e^x - 1}{x} \tag{4.13}$$

with matrix argument x. Applying vec$(\cdot)$ to the left-hand side of (4.9) and substituting (4.11) in the right-hand side gives

$$\left\langle \text{vec}(\partial \tilde{J}(U)), \text{vec}(V) \right\rangle = \left\langle \text{vec}(\partial J(A)), K(U)\,\text{vec}(V) \right\rangle, \qquad \forall V \in S^n. \tag{4.14}$$

Hence, taking into account the symmetry of $K(U)$,

$$\partial \tilde{J}(U) = \text{vec}^{-1}\left(K(U)\,\text{vec}(\partial J(A))\right). \tag{4.15}$$

As a result, (4.8) becomes

$$U_{t+1} = U_t - h\,\text{vec}^{-1}\left(K(U)\,\text{vec}\left(\partial J(A_t)\right)\right), \quad A_t = \exp\text{m}(U_t), \quad U_0 \in S^n. \tag{4.16}$$

In order to evaluate iteratively this equation, the matrix $K(U)$ given by (4.11b) is never computed. Rather, based on [18], the product $K(U)\,\text{vec}\left(\partial J(A_t)\right)$ is computed by approximating the product $\psi(U \oplus (-U))\partial J(A_t)$ as follows. Using the shorthands

$$C = U \oplus (-U), \qquad b = \partial J(A_t) \tag{4.17}$$

one computes the Krylov subspace

$$\mathcal{K}_m(C, q_1) = \text{span}\{q_1, Cq_1, \ldots, C^{m-1}q_1\}, \quad q_1 = \frac{b}{\|b\|} \tag{4.18}$$

using the basic Arnoldi iteration with initial vector q_1, along with a orthonormal basis $V_m = (q_1, \ldots, q_m)$ of $\mathcal{K}_m(C, q_1)$. This yields the approximation

$$\psi\left(U \oplus (-U)\right)\partial J(A_t) \approx \psi(C)b \approx \|b\|V_m\psi(H_m)e_1, \quad H_m = V_m^{\top}CV_m, \tag{4.19}$$

where $e_1 = (1, 0, \ldots, 0)^\top$ denotes the first canonical unit vector. The right-hand side of (4.19) only involves the evaluation of ψ for the much smaller matrix H_m, which can be savely done by computing

$$\psi(H_m)e_1 = \begin{pmatrix} I_m & 0 \end{pmatrix} \operatorname{expm}\begin{pmatrix} H_m & e_1 \\ 0 & 0 \end{pmatrix} e_{m+1} \qquad (4.20)$$

and using any available routine [15] for the matrix exponential. Putting together, the iteration (4.8) is numerically carried out by computing

$$U_{t+1} = U_t - h\operatorname{vec}^{-1}\left(\|\partial J(A_t)\| (I \otimes \operatorname{expm}(U_t)) V_m \psi(H_m)e_1 \right) \qquad (4.21a)$$

$$A_t = \operatorname{expm}(U_t), \quad t \in \mathbb{N}, \quad U_0 \in S^n. \qquad (4.21b)$$

In view of (4.6), we replace the overall parametrization (3.5a) by

$$\mathcal{U} := \{c_{[1]}, \ldots, c_{[n]}, U_{[1]}, \ldots, U_{[n]}\} \in \mathcal{S}_{n,d} \qquad (4.22a)$$

$$\mathcal{S}_{n,d} := \mathbb{R}^{t_1(d)} \times \cdots \times \mathbb{R}^{t_n(d)} \times S_{s_1(d)} \times S_{s_n(d)}. \qquad (4.22b)$$

Consequently, analogous to (4.7), we denote the pulled back objective function (2.7) by $\widetilde{J}(\mathcal{U})$. Applying the procedure worked out above to each positive definite component of the overall parametrization (4.22) results in Algorithm 2.

Algorithm 2: Exponential SOS Flow

Initialization
Choose $\mathcal{A}_0 \in \mathcal{P}_{n,d}$ such that $T_{[1]} \approx \operatorname{id}$.
$U_{[k]} = \operatorname{logm}(A_{[k]}), \ \forall k \in [n]$.
while not converged do
  $(A_{[k]})_t = \operatorname{expm}\left((U_{[k]})_t\right)$
  $(U_{[k]})_{t+1} = (U_{[k]})_t - h\operatorname{vec}^{-1}\left(K\left((U_{[k]})_t\right) \operatorname{vec}\left(\partial_{A_{[k]}} J(A_t)\right) \right)$
  $(c_{[k]})_{t+1} = (c_{[k]})_t - h\,\partial_{c_{[k]}} J(A_t), \quad \forall k \in [n].$

Remark 1 (polymomial basis). The framework outlined above does not depend on the specific choice of a *monomial* basis (1.1). For example, replacing $v_d(x)$ by

$$Q v_d(x), \quad Q \in \operatorname{GL}\left(s_n(d); \mathbb{R}\right) \qquad (4.23)$$

for some linear regular transformation Q, provides a viable alternative. For instance, a polynomial basis that is orthogonal with respect to a weighted L_2 inner product makes sense, especially if prior information about the support $\operatorname{supp}\nu$ of the target measure is available.

4.3   Application: Sampling from the Target Measure

In this section, we consider the objective function (2.7) for the specific case $\mu = \mathcal{N}(0, I_n)$ and the task to generate samples $y = T_{\mathcal{A}}(x) \sim \nu$ from the estimated target measure, using samples $x \sim \mu$ that are simple to compute.

Taking into account the specific form of μ and the triangular structure of $S_{\mathcal{A}}$, the objective function (2.7) simplifies to

$$J(\mathcal{A}) = \frac{1}{N} \sum_{i \in [N]} \sum_{k \in [n]} \left(\frac{1}{2} \left(S_{[k]}(y_{i,1}, \ldots, y_{i,k}) \right)^2 - \log \partial_k S_{[k]}(y_{i,1}, \ldots, y_{i,k}) \right). \quad (4.24)$$

Both Algorithm 1 and 2 can be used to minimizer (4.24) numerically. The evaluation of the map $T_{\mathcal{A}} = S_{\mathcal{A}}^{-1}$ makes use of the triangular structure of in order to solve the equations

$$S_{\mathcal{A}}(y) = \begin{bmatrix} S_{[1]}(y_1) \\ S_{[2]}(y_1, y_2) \\ \vdots \\ S_{[n]}(y_1, \ldots, y_n) \end{bmatrix} = x. \quad (4.25)$$

for $y = T_{\mathcal{A}}(x)$ by computing recursively

$$y_k = \left(S_{[k]}(y_1, \ldots, y_{k-1}, \cdot) \right)^{-1}(x_k), \quad k \in [n]. \quad (4.26)$$

Each step involves few iterations of the one-dimensional Newton method that converges to the unique solution, thanks to the monotonicity of the triangular maps that holds by construction – cf. (3.6).

5   Numerical Experiments

In this section, we report numerical results as proof of concept and discuss the following two aspects:

- Expressiveness of polynomial SOS maps for measure transport and generative modeling (Sects. 5.2 and 5.3);
- performance and comparison of the two geometric flows approximated by Algorithms 1 and 2 (Sect. 5.4).

We point out that unlike the paper [10], no deep network was used for additional parametrization which would obscure the influence of the SOS-polynomial maps.

5.1   Implementation Details

We used the three two-dimensional densities *open-ring*, *closed-ring* and *mixture of two Gaussians* for this purpose (Fig. 1), that play the role of the data measure ν. A sample set $y_i \sim \nu$, $i \in [N]$, with $N = 2.000$, was generated as input data.

Next, either algorithm was applied in order to estimate numerically the *SOS*-parameters $\mathcal{A}$ given by (3.5a), by minimizing the objective function (4.24). We used SOS-polynomials of degrees $2d \in \{2, 4, 6\}$ for parametrizing the maps $T_{\mathcal{A}}(x)$. Taking into account the symmetry of the matrices the corresponding numbers of variables to be determined are 12, 31, 70. Finally, samples $x_i \sim \mu$

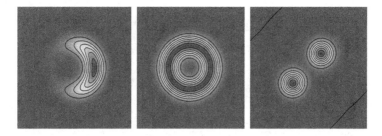

Fig. 1. Three non-Gaussian distributions used to evaluate Riemannian SOS-polynomial normalizing flows. From (*left to right*): *open ring, closed ring* and *mixture of two Gaussians* distributions.

were generated and the map $T_{\mathcal{A}} = S_{\mathcal{A}}^{-1}$ was computed (Sect. 4.3) in order to generate samples $y_i = T_{\mathcal{A}}(x_i)$. Corresponding kernel density estimates can then be compared to the plots depicted by Fig. 1.

Both Algorithms 1 and 2 were modified in a stochastic gradient descent like manner: Every update was performed using the gradient with respect to a *single random* index $i \in [N]$ of the objective (4.24), such that each index i was visited after N updates. Thus, even though the considered problem sizes are small, we modified both geometric gradient descent algorithms such that they remain efficient for larger problem sizes [5].

5.2 Riemannian SOS-Polynomial Normalizing Flows

Figure 2 displays recovered densities using the procedure described in Sect. 5.1. See also the figure caption. The low-degree SOS polynomials used to parametrize and estimate the transportation maps $T_{\mathcal{A}}$ suffice to generate samples $y_i = T(x_i)$ by pushing forward samples $x_i \sim \mathcal{N}(0, I_n)$ such that sample y_i follow the ground-truth densities ν depicted by Fig. 1 quite accurately.

We also checked the influence of changing the polynomial basis according to Remark 1 (page 8). Specifically, Hermite polynomials that are orthogonal with respect to a weighted L_2 inner product were used instead of the canonical monomial basis. Figure 4 illustrates that this did not affect the process in a noticeable way. Neither did the result for the Gaussian mixture density show any noticeable effect.

5.3 Exponential SOS-Polynomial Normalizing Flows

We repeated all experiments reported in Sect. 5.2 using Algorithm 2, instead of Algorithm 1, that is based on the parametrization detailed in Sect. 4.2. The results are shown by Fig. 3.

We generally observed fairly good density approximations even for low-degree polynomial parametrizations, that do not achieve the accuracy of the results obtained using the Riemannian flows, however (cf. Fig. 2). In particular, we

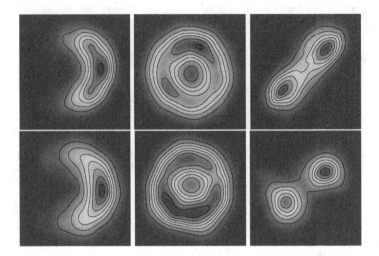

Fig. 2. *Riemannian SOS-Polynomial Normalizing Flows.* Kernel density estimate plots based on $N = 2000$ samples $y_i = T_\mathcal{A}(x_i) = S_\mathcal{A}^{-1}(x_i)$ generated by the transportation maps $T_\mathcal{A}$ corresponding to the densities shown by Fig. 1 and samples $x_i \sim \mathcal{N}(0, I_n)$. The columns correspond from (*left to right*) to the degrees $2d \in \{2, 6\}$ of the SOS-polynomials that were used to compute the increasing triangular maps $T_\mathcal{A}$. Except for the mixture of two Gaussians density, low-degree SOS-polynomals suffice to recover the densities quite accurately.

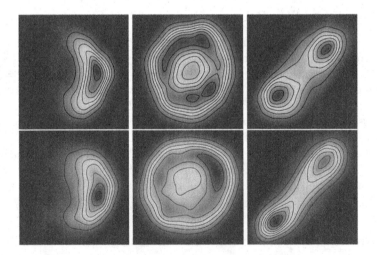

Fig. 3. *Exponential SOS-Polynomial Normalizing Flows.* Results of the experiments obtained using Algorithm 2 using the same data as for the experiments illustrated by Fig. 2. In comparison to the former results, the approximation accuracy deteriorated slightly. In addition, choosing larger polynomial degrees may not improve the result. We attribute this finding to the fact that Algorithm 2 is based on approximating the geometry of the parameter space in various ways (see text).

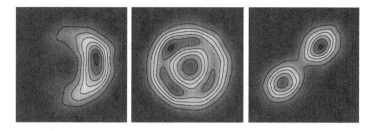

Fig. 4. *Riemannian SOS-Polynomial Normalizing Flows* using Hermite polynomials, rather than the canonical monomial basis, and using the same data as for the experiments illustrated by Fig. 2.

observed that increasing the polynomial degree did *not* systematically improve the approximation.

We attribute this negative finding to two facts: Firstly, Algorithm 2 does not exactly respect the geometry of the parameter space $P_{n,d}$ (3.5b). Secondly, the Krylov subspace approximation underlying the updates (4.21) may also affect the approximation accuracy. We leave a more detailed analysis for future work.

5.4 Comparison Between Riemannian and Exponential SOS Flow

Comparing the results discussed in Sects. 5.2 and 5.3 suggests that Riemannian SOS-polynomial normalizing flows should be preferred.

A striking difference concerns the dependency on the polynomial degree. While the Riemannian SOS flow generally yield improved density approximations when the degree is increased, this is hardly the case when using the exponential parametrization. Possible reasons were discussed in the preceding section.

In both cases, however, even small polynomial degrees enable to represent densities by transportation maps quite accurately.

6 Conclusion

We studied transportation maps for generative modeling using Sum-of-Squares polynomials for the construction of increasing triangular maps. Two parametrizations were studied along with two numerical algorithms for estimating the parameters by minimizing a sample-based objective function. Experiments show that low-degree polynomials suffice to recover basic non-Gaussian distributions quite accurately. Riemannian SOS-polynomial flows that fully respect the geometry of the parameter space perform best, whereas approximations of the geometry may cause detrimental effects. We merely regard the reported preliminary experimental results as proof of concept, conducted with low-degree parametrizations and small dimension of the underlying domain. Our future work will be devoted to geometric methods for taming the complexity of large degree parametrizations and the representation of high-dimensional generative models.

References

1. Bhatia, R.: Positive Definite Matrices. Princeton University Press, Princeton (2006)
2. Blekherman, G., Parrilo, P., Thomas, R.R. (eds.): Semidefinite Optimization and Convex Algebraic Geometry. SIAM (2013)
3. Bogachev, V.I., Kolesnikov, A.V., Medvedev, K.V.: Triangular transformations of measures. Sbornik: Math. **196**(3), 309–335 (2005)
4. Bonnotte, N.: From Knothe's rearrangement to Brenier's optimal transport map. SIAM J. Math. Anal. **45**(1), 64–87 (2013)
5. Bottou, L., Curtis, F., Nocedal, J.: Optimization methods for large-scale machine learning. SIAM Rev. **60**(2), 223–311 (2018)
6. Bousquet, O., Gelly, S., Tolstikhin, I., Simon-Gabriel, C.J., Schölkopf, B.: From optimal transport to generative modeling: the VEGAN cookbook. CoRR abs/1705.07642 (2017)
7. Brenier, Y.: Polar factorization and monotone rearrangement of vector-valued functions. Comm. Pure Appl. Math. **44**(4), 375–417 (1991)
8. Carlier, G., Galichon, A., Santambrogio, F.: From Knothe's transport to Brenier's map and a continuation method for optimal transport. SIAM J. Math. Anal. **41**(6), 2554–2576 (2010)
9. Higham, N.: Functions of Matrices: Theory and Computation. SIAM (2008)
10. Jaini, P., Selby, K.A., Yu, Y.: Sum-of-Squares Polynomial Flow. preprint arXiv:1905.02325 (2019)
11. Knothe, H.: Contributions to the theory of convex bodies. Mich. Math. J. **4**(1), 39–52 (1957)
12. Kobyzev, I., Prince, S.D., Brubaker, M.A.: Normalizing Flows: An Introduction and Review of Current Methods. preprint arXiv:1908.09257 (2019)
13. Marshall, M.: Positive Polynomials and Sum of Squares. American Mathematical Society (2008)
14. Marzouk, Y., Moselhy, T., Parno, M., Spantini, A.: An introduction to sampling via measure transport. In: Ghanem, R., Higdon, D., Owhadi, H. (eds.) Handbook of Uncertainty Quantification, pp. 1–41. Springer, Cham (2017). https://doi.org/10.1007/978-3-319-11259-6_23-1
15. Moler, C., Van Loan, C.: Nineteen dubious ways to compute the exponential of a matrix, twenty-five years later. SIAM Rev. **45**(1), 3–49 (2003)
16. Papamakarios, G., Nalisnick, E., Rezende, D., Mohamed, S., Lakshminarayanan, B.: Normalizing Flows for Probabilistic Modeling and Inference. preprint arXiv:1912.02762 (2019)
17. Peyré, G., Cuturi, M.: Computational Optimal Transport. CNRS (2018)
18. Saad, Y.: Analysis of some Krylov subspace approximations to the matrix exponential operator. SIAM J. Numer. Anal. **29**(1), 209–228 (1992)
19. Santambrogio, F.: Optimal Transport for Applied Mathematicians. Birkhäuser (2015)
20. Tabak, E.G., Turner, C.V.: A family of nonparametric density estimation algorithms. Comm. Pure Appl. Math. **66**(2), 145–164 (2013)

21. Tabak, E., Vanden-Eijnden, E.: Density estimation by dual ascent of the log-likelihood. Commun. Math. Sci. **8**(1), 217–233 (2010)
22. Talagrand, M.: Transportation cost for gaussian and other product measures. Geom. Funct. Anal. **6**, 587–600 (1996)
23. Van Loan, C.F.: The ubiquitous Kronecker product. J. Comput. Appl. Math. **123**, 85–100 (2000)
24. Villani, C.: Optimal Transport: Old and New. Springer, Heidelberg (2009). https://doi.org/10.1007/978-3-540-71050-9

Automated Water Segmentation and River Level Detection on Camera Images Using Transfer Learning

Rémy Vandaele[1,2]([✉]) [ID], Sarah L. Dance[1,3] [ID], and Varun Ojha[2] [ID]

[1] Department of Meteorology, University of Reading, Reading, UK
{r.a.vandaele,s.l.dance}@reading.ac.uk
[2] Department of Computer Science, University of Reading, Reading, UK
v.k.ojha@reading.ac.uk
[3] Department of Mathematics and Statistics, University of Reading, Reading, UK

Abstract. We investigate a deep transfer learning methodology to perform water segmentation and water level prediction on river camera images. Starting from pre-trained segmentation networks that provided state-of-the-art results on general purpose semantic image segmentation datasets ADE20k and COCO-stuff, we show that we can apply transfer learning methods for semantic water segmentation. Our transfer learning approach improves the current segmentation results of two water segmentation datasets available in the literature. We also investigate the usage of the water segmentation networks in combination with on-site ground surveys to automate the process of water level estimation on river camera images. Our methodology has the potential to impact the study and modelling of flood-related events.

1 Introduction

In recent years, the price and accessibility of surveillance cameras has greatly improved. Notably, this progress has allowed many organizations, private and public, to install surveillance cameras along rivers and other water bodies. The availability of these cameras allows the interested parties to monitor the conditions of the river as well as its surroundings for purposes such as boating, fishing, flood monitoring, etc. [14,27].

For the flood-risk management sector, the use of such cameras brings an unparalleled opportunity for the study and modelling of flood-related events. Indeed, as of now, to measure the water level of rivers, it is necessary to rely on water gauges [25]. Gauges are expensive to install and maintain, and their measurements can be unreliable when the river goes out-of-bank during a flood. Satellite data from Synthetic Aperture Radar (SAR) can provide information when the river goes out-of-bank, but the frequency of satellite overpasses is limited (currently at most once in each 12 h period) [5,17].

River cameras offer a new possibility: by using the measurements of the heights of particular landmarks or objects in the field of view of the camera, or

Z. Akata et al. (Eds.): DAGM GCPR 2020, LNCS 12544, pp. 232–245, 2021.
https://doi.org/10.1007/978-3-030-71278-5_17

by matching the camera image with light detection and ranging (LIDAR) digital surface model data [10], it becomes possible to directly estimate the water level from a camera. Such an example is given in Fig. 1. This approach is much more flexible in matter of river surveillance location choices as well as budget.

Fig. 1. Sequence of river camera images with annotated landmark heights L1 (10 m), L2 (11 m) and L3 (12 m). At T1, water (segmented in blue) water has not reached any of the landmarks: the water level is below 10 m. At T2, L1 is reached by water, but not L2 or L3: the water level is between 10 and 11 m. At T3, water has reached L2 but not L3: water level is between 11 and 12 m. At T4, water has reached all the landmarks: water level is above 12 m. (Color figure online)

When considering this approach, the water level measurements must be calculated through a complex workflow: an operator (algorithm or human) has to segment the image to find which areas/landmarks are flooded. Once the operator knows which landmarks were flooded or not, it is possible to estimate the water level: the lower bound will be the height of the highest flooded landmark, and the upper bound will be the height of the lowest not flooded landmark. However, if a human operator is considered, this process makes the water level measurement, time consuming, and possibly an unusable approach since the number of images to study (locations, extent in time, framerate) are typically large.

Our goal is to automate the process of river water segmentation by applying transfer learning (TL) on deep convolutional neural networks, and assess its potential for flood modelling. Specifically, for the datasets at our disposal, we study the relevance of using TL approaches in order to perform water segmentation and possibly use this segmentation to estimate the river levels as accurately as possible. Our paper brings three novel contributions:

1. We develop water segmentation algorithms by using TL, and demonstrate that it outperforms the current methods presented in the literature.
2. We provide an insightful comparison of several TL approaches for water segmentation.
3. We show that it is possible to use our semantic segmentation method in combination with ground survey measurements to estimate water levels for a variety of locations.

In Sect. 2, we discuss the current related methods that are used to address the problem of water segmentation on river camera images. In Sect. 3, we motivate and explain the approach we used to tackle the problem of water segmentation. In Sect. 4, we show the results of our TL approach. We compare our method with the current state-of-the-art methods and show that we are able to improve the water segmentation performance. In Sect. 5, we analyze the efficiency of water segmentation networks to estimate the river levels. We make our final observations and conclusions in Sect. 6.

2   Related Work

There have been many successful applications of deep learning to images from surveillance cameras. Some examples include deep learning for crowd counting [22,30], abnormal behavior detection [8], pedestrian detection [26] or even parking occupancy detection [1]. Until now however, most attempts that have tried to tackle the problem of water detection in the context of floods have been realized using hand-crafted features [9]. However, those algorithms remain sensitive to luminosity and water reflection problems [9].

A deep learning approach was applied to flood detection in [16]. The authors perform water detection on a home-made, accessible, dataset of 300 water images that were gathered from the web and annotated manually. The performance of three semantic segmentation networks (FCN-8 [15], Tiramisu [12] and Pix2Pix [11]) are evaluated. By training the networks from scratch, Tiramisu produces the best results, with 90.47% pixel accuracy. It is not clear however if the results are transferable to water level estimation for real cases.

In another work, water detection is performed in the context of autonomous driving for low-cost boats [23]. In this work, a deep learning architecture using a fully convolutional network based on U-Net [20] to perform the water segmentation is proposed. A pixel accuracy of 97.45% is obtained. However, the evaluation protocol used images from the same video streams (therefore very similar images) both for training and test sets, which suggests that the reported results might be overestimated.

In [28], a deep semantic segmentation network is trained from scratch for water segmentation and river level estimation. The biggest originality of this paper lies in their development of the SOFI index (the percentage of segmented water pixels in the image) to evaluate the quality of their results.

A water level estimation model based on voluntary geographic information (VGI), LIDAR data and river camera images is developed in [13]. Notably, random forests are used to develop a waterline detection algorithm [2].

3   Transfer Learning for Water Segmentation

In Sect. 2, we saw that little research has focused on water segmentation, especially in the context of flooding. Indeed, there are only a few specific water segmentation datasets.

However, semantic segmentation of natural images is an area that has been extensively studied over the past years. State-of-the-art algorithms for multipurpose semantic segmentation are based on the use of Fully Convolutional Networks (FCNs) [15].

The most well-known datasets used for the comparison of semantic segmentation algorithms are COCO-stuff [3] and ADE20k [31]. These two datasets contain large sets of images semantically annotated with 182 types of labels for COCO-stuff and 150 for ADE20k. As we show in Table 1, some label types of these two datasets correspond to water bodies. These two datasets, among others [6,7], are widely used for evaluating semantic segmentation algorithms.

Table 1. Water body related images in ADE20k and COCO-stuff datasets.

ADE20k dataset			COCO-stuff dataset		
	Training	Test		Training	Test
Water	709	75	River	2113	90
Sea	651	57	Sea	6598	292
River	320	26	Water-other	2453	79
Waterfall	80	9			

Given these observations, we decided to tackle the problem of water segmentation using transfer learning (TL).

For a supervised learning problem, the aim is to find a function $f : X \to Y$ from a dataset of N input-output pairs $B = \{(x_i, y_i)_{i=1}^N : x_i \in X, y_i \in Y\}$ such that the function f should be able to predict the output of a new (possibly unseen) input, as accurately as possible. The set X is called the input space, and Y the output space.

With TL, the aim is also to build a function f_t for a *target* problem with input space X_t, output space Y_t and possibly a dataset B_t. However, TL tries to build f_t by *transferring* knowledge from a *source* problem s with input space X_s, output space Y_s and a dataset B_s.

Inductive TL [18] is the branch of TL related to problems where we have datasets of input-output pairs in both source and target domains, and where $X_s = X_t$ and $Y_s \neq Y_t$. Typically, inductive TL is used to repurpose well known, efficient machine learning models trained on large datasets to related problems with smaller training datasets [19,21].

In our case, we want to use inductive TL where the source problem s will be the segmentation of ADE20K or COCO-stuff images, and the target problem t will be the binary water segmentation of river camera images. We think the problems of segmenting the ADE20K and COCO-stuff datasets are especially relevant in our context given the fact that they contain labels of water bodies, which makes source and target output domains fairly similar.

In the scope of this study, we chose to focus on three TL approaches. With the first TL approach, we use the pre-trained network as such, taking advantadge of the water body labels to directly create binary semantic segmentation masks. With the second approach, we consider model transfer approaches, where we fine-tune semantic segmentation networks pre-trained on either ADE20K or COCO-stuff on water segmentation datasets. We also test a third TL approach related to sample selection, where we fine-tune the pre-trained network on the subset of ADE20k and COCO-stuff images containing water bodies. While other inductive TL approaches exist and could possibly outperform our current results, we found that our methods are computationally efficient, which will be critical for potential future applications in near-real-time water level estimation.

4   Water Segmentation Experiments

In this section, we discuss the water segmentation experiments that were performed on water segmentation datasets available in the literature. We also compare our results to state-of-the-art water detection results.

4.1   Protocol

Pre-trained Networks. The purpose of our experiments is to evaluate the relevance of applying TL for water segmentation. As we explained in Sect. 3, we chose to consider two datasets for pre-training: ADE20k and COCO-stuff. We chose these datasets as they contain water-labelled images. For each of these two datasets, we study one of its best performing semantic segmentation networks.

For ADE20k, the network we considered is an FCN with a ResNet50 encoder and an UperNet decoder [31]. UperNet [29] is a model that is based on Pyramid Pooling in order to avoid the use of deconvolution layers. During training, the images are rescaled at 5 different sizes: the shorter edge of the image is rescaled to either 300, 375, 450, 525 or 600 pixels, and the bigger edge is rescaled according to the image aspect ratio. We re-used the original implementation as well as the available pre-trained network weights.

For COCO-stuff, we chose the state-of-the-art network referenced by the authors of the dataset, *DeepLab* (v2). It has a ResNet101 encoder, and an atrous spatial pyramid pooling decoder able to robustly segment objects at multiple scales [4]. We used a pytorch implementation of the model with available pre-trained COCO-stuff weights[1].

First TL Approach: Pre-trained Network Use. With this method, we use the pre-trained weights of the networks: we do not tune any layer of the network. We apply the pre-trained networks on our images, and aggregate the predictions of water body labels (lake, river, sea, water, and other similar water related labels) as the output water segmentation. Given that the networks were trained with images of water bodies, this first, simple approach should provide a baseline result for the evaluation of our other approaches. We refer to this approach as **Pre-Trained.**

---

[1] https://github.com/kazuto1011/deeplab-pytorch.

Second TL Approach: Networks Fine-Tuning. As the output of ADE20k semantic segmentation networks is not binary, the last output layers of the semantic segmentation networks could not be directly reused in our binary semantic segmentation problem. This is why we considered three fine-tuning methodologies:

- **HEAD**. With this approach, we only retrain the last output layers of the network. The rationale is that the network has already learned all the necessary filters to perform water segmentation, and it requires only to learn how to perform the binary segmentation.
- **WHOLE**. We fine-tune the entire network, with a random initialization of the last binary output layers.
- **2STEPS**. We first retrain the last layer of the network with all the other layers kept as is. Once the last layer is retrained, we fine-tune the entire network. This approach can be considered as retraining the entire network after having applied the HEAD approach.

Third TL Approach: Sample Selection. As we show in Table 1, the two datasets on which our networks were pre-trained contain images with water related labels. We thus consider a *sample selection* approach algorithm in order to perform TL: we extract all the images containing water labels from the ADE20k and COCO-stuff dataset, and fine-tune the two pre-trained networks on this new dataset with binary masks. In our experiments, we will refer to this approach as **Sample Selection**. We then fine-tuned the network using the WHOLE approach. HEAD and 2STEPS were also tested during our experiments, but for clarity purposes, we chose to only present the results using the approach providing the best results.

Relevance of Using TL. In order to understand the relative performance of these TL approaches, we also considered what results could be obtained with the same networks trained from scratch (only using the training images of the dataset). We will refer to this approach as **Scratch**. For the same purpose, we also compared our TL approach with the water semantic segmentation results obtained in the literature [16,23].

Training. We trained the networks using the parameters recommended by the authors [4,31]. For the *fine-tuning* and *scratch* approaches, we increased the number of epochs to 300 in order to ensure full convergence for all the networks. For the approaches WHOLE and 2STEPS, we used an initial learning rate value 10 times smaller than its recommended value (0.001) in order to start with less aggressive updates.

4.2 Datasets

Our experiments are performed on two datasets used for water segmentation in the literature, and which we use for evaluating the performance of our TL methodology.

- **INTCATCH**, an available dataset of RGB images annotated with binary semantic segmentation water/not-water masks [23]. The images come from a camera positioned on a boat. It was designed for waterline detection for driving autonomous boats. The dataset consists of 144 training images and 39 test images. We noticed that the images in training and test come from two video streams with relatively high frame-rates, which makes training and test datasets look similar.
- **LAGO** (named after the main author [16]), an accessible dataset of RGB images with binary semantic segmentation of water/not-water labelled pixels. The dataset was created through manual collection of camera images having a field-of-view capturing riverbanks. This dataset was used for river segmentation [16]. The dataset is made of 300 images, with 225 used in training, and 75 in test.

Sample images of the datasets are shown in Fig. 2.

Fig. 2. Sample images from the datasets used for the water segmentation experiments.

4.3   Performance Criteria

Let $I \in [0, 255]^{H \times W \times 3}$ be a typical 8-bit RGB, image of height H and width W. Let $S \in [0, 1]^{H \times W}$ be its corresponding, ground-truth, pixel-wise, semantic segmentation mask, and $\hat{S} \in [0, 1]^{H \times W}$ be the estimation (prediction) of this segmentation made by our semantic segmentation algorithm. The two performance criteria used for the evaluation of semantic segmentation methods are defined as follows:

Pixel Accuracy (Acc). In (1), we define the pixel accuracy as the percentage of pixels correctly estimated by our algorithm.

$$Acc = \frac{\sum_{y=1}^{H} \sum_{x=1}^{W} 1 - |(S(y, x) - \hat{S}(y, x))|}{H \times W} \tag{1}$$

Mean Intersection over Union (MIoU). The Intersection over Union (IoU) represents the percentage of overlap between the ground truth and its estimation. The $MIoU$ criteria defined in (2) is the average of the IoU over all the pixel labels. In our case, the pixel label types are water and background (not-water). Thus, we need to consider the binary case:

$$MIoU = \frac{1}{2} \sum_{y=1}^{H} \sum_{x=1}^{W} \frac{S(y,x)\hat{S}(y,x)}{S(y,x) + \hat{S}(y,x) - S(y,x)\hat{S}(y,x)}$$
$$+ \frac{1}{2} \sum_{y=1}^{H} \sum_{x=1}^{W} \frac{(1 - S(y,x))(1 - \hat{S}(y,x))}{(1 - S(y,x)) + (1 - \hat{S}(y,x)) - (1 - S(y,x))(1 - \hat{S}(y,x))}$$
$$(2)$$

The advantage of using $MIoU$ over Acc is that it is less sensitive to class imbalance within the image. However, one of the works we are comparing with provide their results for Acc only. For the sake of transparency, we provide our results using both criteria.

4.4   Results and Analysis

The results of the water segmentation approaches are presented in Table 2.

Table 2. Results of the water segmentation approaches on LAGO and INTCATCH test datasets.

			LAGO		INTCATCH	
			MIoU	*Acc*	*MIoU*	*Acc*
Gonzalez et al. [16]			81.91	90.2	–	–
Steccanella et al. [23]			–	–	–	97.5
ResNet50-UperNet	*Pre-trained*		90.2	95.4	97.4	98.7
Pre-trained on ADE20k	*Fine-tuning*	HEAD	89.06	94.37	98.06	99.03
		WHOLE	93.32	96.50	98.94	99.47
		2STEPS	93.09	96.44	99	99.5
	Sample Selection		92.2	96.95	98.95	99.48
	Scratch		83.41	91.74	96.09	98.02
DeepLab	*Pre-trained*		90.34	95.52	97.70	98.85
Pre-trained on COCO-stuff	*Fine-tuning*	HEAD	92.19	96.04	99.07	99.54
		WHOLE	93.74	96.76	99.19	99.59
		2STEPS	93.72	96.75	99.16	99.56
	Sample Selection		91.69	96.31	98.59	99.3
	Scratch		80.70	89.95	98.73	99.36

As explained in Sect. 4.2, we noticed that the images contained in the INTCATCH training and test sets are largely similar to each other as they are frames randomly sampled from the same two videos. This is how we explain the excellent performance of the different networks tested on these images.

On both LAGO and INTCATCH datasets, we can observe that, for both networks, the pre-trained networks, and TL approaches provide better results than the methods presented in the literature [16,23]. We can also see that the networks retrained from *scratch* obtain results similar to the ones of the state-of-the-art on the respective datasets. This shows that the use of semantic segmentation networks that are first trained on large multi-purpose datasets can improve the performance. Indeed, even without any kind of fine-tuning, the pre-trained networks already outperform the state-of-the-art.

For the three datasets and both networks, fine-tuning the entire networks (WHOLE and 2STEPS) or using *sample selection* always provides better results than the *pre-trained* networks. This shows that it is possible to further improve the performance of the segmentation by fine-tuning the networks weights. Between *sample selection* and *fine-tuning*, the *fine-tuning* approaches seems to provide the best results.

We also observe that HEAD provides results relatively close to or inferior to the *pre-trained* approach. Furthermore, 2STEPS approach always obtains better results than HEAD. This implies that it is necessary to fine-tune the entire networks rather than retraining only its output layer.

5   River Level Estimation Experiments

In this section, our goal is to describe the experiments that we performed to evaluate whether the semantic segmentation networks assessed in the previous section can be used in the context of river level estimation.

5.1   Datasets

Our river level estimation datasets consist of RGB images coming from video streams of Farson Digital river cameras located at 4 different locations in the U.K. [27]. For each location, the camera position and orientation is fixed, which means that the field of view stays the same for all of the images for a location.

Each location is annotated with landmarks for which heights were manually measured during a ground survey. The images composing the datasets were all sampled from the camera video streams with the purpose of observing a specific flood event. On each sampled image, the landmarks were annotated with binary information flooded/unflooded, that could be used to estimate the water levels in the images (see Fig. 1).

From our first location, Diglis Lock, we extracted 141 images and used 7 landmarks. For the second location, Evesham, we extracted 134 images and used 13 landmarks. For the third location, Strensham Lock, we extracted 144 images and used 24 landmarks. For the fourth location, Tewkesbury Marina, we extracted 144 images and used 4 landmarks. In our nomenclature, **Farson** corresponds to the union of the images collected from the 4 mentioned locations. Sample images for each of the locations are given in Fig. 3.

Fig. 3. Images from Farson river camera datasets [27], with their landmarks in red dots. (Color figure online)

5.2   Protocol

Performance Criteria. As explained in Sect. 5.1, only specific landmark (pixel) locations were annotated on those images. This is why we chose to use the balanced accuracy classification score $BAcc$ defined as:

$$BAcc = 100 \times (\frac{1}{2}\frac{TF}{F} + \frac{1}{2}\frac{TU}{U}), \tag{3}$$

where F is the number of actual flooded landmarks, TF the number of correctly classified flooded landmarks, U the number of actual unflooded landmarks and TU the number of correctly classified unflooded landmarks. Given that the extracted time periods of the river camera datasets might create an imbalance between flooded or unflooded landmarks, therefore, we think (3) is a relevant performance criteria to consider.

Experimental Design. We reused the networks that were trained in Sect. 4 to produce binary segmentation masks of the river camera images using the fully trained/fine-tuned networks. A landmark is predicted as flooded if its pixels location is predicted as water, and unflooded otherwise.

A TL approach trying to directly output the water level from the camera images could have been considered. However, this approach requires water-level annotated images for each location as the water levels will vary. Thus, we assess that our landmark classification approach is more relevant.

Table 3. Balanced accuracies (see Sect. 5.2) of landmark classification using TL on the Farson dataset. Note that Pre-Trained and Sample Selection do not need to be fine-tuned over LAGO or INTCATCH datasets.

			Network trained/fine-tuned on		
			–	LAGO	INTCATCH
			BAcc	*BAcc*	*BAcc*
ResNet50-UperNet	*Pre-trained*		83.4		
Pre-trained on ADE20k	*Fine-tuning*	HEAD		87.96	78.89
		WHOLE		93.06	88.03
		2STEPS		93.29	88.25
	Sample selection		90.97		
	Scratch			91.56	80.47
DeepLab	*Pre-trained*		87.41		
Pre-trained on COCO-stuff	*Fine-tuning*	HEAD		93.41	91.65
		WHOLE		95.04	94.31
		2STEPS		95.06	93.78
	Sample Selection		91.32		
	Scratch			87.1	85.55

5.3 Results and Analysis

The results are presented in Table 3. The *scratch* approach, which does not use TL, tends to perform worse than the *pre-trained* networks without any kind of fine-tuning. Training the network from scratch on LAGO dataset seems to be the most favorable case. We explain this by the fact that the scratch approach overfits its training dataset, and the LAGO dataset is focusing on river images similar to the Farson dataset.

We can observe that fine-tuning the networks on either LAGO or INTCATCH allows improvement in the landmark classification performance. The WHOLE and 2STEPS approaches that fine-tune the entire networks, obtain the best overall performance. Only retraining the last layer (HEAD) has varying impacts on the performance: while it is always better than retraining the network from scratch, it does not always reach the performance of using the pre-trained network.

The *sample selection* approach provide good performance on both networks. However, when comparing the TL methods, it is always outranked by *fine-tuning* the entire networks (WHOLE and 2STEPS) over LAGO, which is a dataset containing river images. Note that in the context of reusing the semantic segmentation networks for landmark detection over the Farson dataset, the *sample selection* approach is similar to the WHOLE fine-tuning approaches, the difference being the dataset on which they are fine-tuned.

We can also observe that DeepLab network seems to obtain better results than ResNet50-UperNet overall. From what we have seen on the segmentation results, we believe that the choice of landmark locations played a significant role, and that these results should not be directly correlated to the quality of the segmentation: for example, we observed that while DeepLab seemed to be able to make better distinctions between the edges of the river (where the landmarks are typically located), it was also making more mistakes than ResNet50-UperNet elsewhere in the image (the sky was sometimes considered as water, reflections in the water were not always considered as water). In our case of water segmentation in time series images, several post-segmentation filtering approaches could be considered to improve the landmark detection results: if the N images before and after the current image have segmented landmark X as water/not-water, it is likely that landmark X is also water/not-water in the current image. The information regarding the landmark height could also be used to perform filtering: if N landmarks located at higher locations are segmented as water, it is likely that the lower landmarks should also be segmented as water.

6   Conclusion

In this paper, we have explored the possibilities of using TL in the context of water segmentation, especially for river level detection.

We have shown that TL approaches were able to outperform the current literature in water segmentation on two different datasets. We have also proven that using fine-tuning and/or sample selection could further improve the water segmentation performance. These networks obtained significantly worse performance once retrained from scratch.

We have supplied quantified and encouraging results to demonstrate the utility of our proposed TL approaches in the context of flood modelling, able to predict flood situations with high accuracy.

Future research will focus on an in-depth analysis of our results for practical flood modelling studies, with the aim to provide more advanced statistics helpful to hydrologists, but that are going beyond the scope of this current study [24].

More practically, we would like to consider merging river camera images with LIDAR digital surface model data [10], which can allow to obtain surface elevation of the terrain on a 1 m grid. In theory, this could allow our approach to rely on more landmarks for the estimation of water levels, while avoiding the tedious work of performing ground surveys to measure the heights of those landmarks.

Acknowledgement. This work was funded by the UK EPSRC EP/P002331/1. The datasets used in this study are all available as described in references [16,23,27].

References

1. Amato, G., Carrara, F., Falchi, F., Gennaro, C., Meghini, C., Vairo, C.: Deep learning for decentralized parking lot occupancy detection. Expert Syst. Appl. **72**, 327–334 (2017)
2. Breiman, L.: Random forests. Mach. Learn. **45**(1), 5–32 (2001)

3. Caesar, H., Uijlings, J., Ferrari, V.: Coco-stuff: thing and stuff classes in context. In: Proceedings of the IEEE Conference on Computer Vision and Pattern Recognition, pp. 1209–1218 (2018)
4. Chen, L.C., Papandreou, G., Kokkinos, I., Murphy, K., Yuille, A.L.: DeepLab: Semantic image segmentation with deep convolutional nets, atrous convolution, and fully connected CRFs. IEEE Trans. Pattern Anal. Mach. Intell. **40**(4), 834–848 (2017)
5. Cooper, E.S., Dance, S.L., García-Pintado, J., Nichols, N.K., Smith, P.: Observation operators for assimilation of satellite observations in fluvial inundation forecasting. Hydrol. Earth Syst. Sci. **23**, 2541–2559 (2019)
6. Cordts, M., et al.: The cityscapes dataset for semantic urban scene understanding. In: Proceedings of the IEEE Conference on Computer Vision and Pattern Recognition, pp. 3213–3223 (2016)
7. Everingham, M., Van Gool, L., Williams, C.K., Winn, J., Zisserman, A.: The pascal visual object classes (VOC) challenge. Int. J. Comput. Vis. **88**(2), 303–338 (2010)
8. Fang, Z., et al.: Abnormal event detection in crowded scenes based on deep learning. Multimed. Tools Appl. **75**(22), 14617–14639 (2016)
9. Filonenko, A., Hernández, D.C., Seo, D., Jo, K.H., et al.: Real-time flood detection for video surveillance. In: IECON 2015–41st Annual Conference of the IEEE Industrial Electronics Society, pp. 004082–004085. IEEE (2015)
10. Hirt, C.: Digital Terrain Models, pp. 1–6. Springer, Cham (2014). https://doi.org/10.1007/978-3-319-02370-0_31-1
11. Isola, P., Zhu, J.Y., Zhou, T., Efros, A.A.: Image-to-image translation with conditional adversarial networks. In: Proceedings of the IEEE Conference on Computer Vision and Pattern Recognition, pp. 1125–1134 (2017)
12. Jégou, S., Drozdzal, M., Vazquez, D., Romero, A., Bengio, Y.: The one hundred layers tiramisu: fully convolutional densenets for semantic segmentation. In: Proceedings of the IEEE Conference on Computer Vision and Pattern Recognition Workshops, pp. 11–19 (2017)
13. Lin, Y.T., Yang, M.D., Han, J.Y., Su, Y.F., Jang, J.H.: Quantifying flood water levels using image-based volunteered geographic information. Remote Sens. **12**(4), 706 (2020)
14. Lo, S.W., Wu, J.H., Lin, F.P., Hsu, C.H.: Visual sensing for urban flood monitoring. Sensors **15**(8), 20006–20029 (2015)
15. Long, J., Shelhamer, E., Darrell, T.: Fully convolutional networks for semantic segmentation. In: Proceedings of the IEEE Conference on Computer Vision and Pattern Recognition, pp. 3431–3440 (2015)
16. Lopez-Fuentes, L., Rossi, C., Skinnemoen, H.: River segmentation for flood monitoring. In: 2017 IEEE International Conference on Big Data (Big Data), pp. 3746–3749. IEEE (2017)
17. Mason, D.C., Dance, S.L., Vetra-Carvalho, S., Cloke, H.L.: Robust algorithm for detecting floodwater in urban areas using synthetic aperture radar images. J. Appl. Remote Sens. **12**(4), 045011 (2018)
18. Pan, S.J., Yang, Q.: A survey on transfer learning. IEEE Trans. Knowl. Data Eng. **22**(10), 1345–1359 (2009)
19. Reyes, A.K., Caicedo, J.C., Camargo, J.E.: Fine-tuning deep convolutional networks for plant recognition. CLEF (Work. Notes) **1391**, 467–475 (2015)
20. Ronneberger, O., Fischer, P., Brox, T.: U-Net: convolutional networks for biomedical image segmentation. In: Navab, N., Hornegger, J., Wells, W.M., Frangi, A.F. (eds.) MICCAI 2015. LNCS, vol. 9351, pp. 234–241. Springer, Cham (2015). https://doi.org/10.1007/978-3-319-24574-4_28

21. Sabatelli, M., Kestemont, M., Daelemans, W., Geurts, P.: Deep transfer learning for art classification problems. In: Leal-Taixé, L., Roth, S. (eds.) ECCV 2018. LNCS, vol. 11130, pp. 631–646. Springer, Cham (2019). https://doi.org/10.1007/978-3-030-11012-3_48
22. Sam, D.B., Surya, S., Babu, R.V.: Switching convolutional neural network for crowd counting. In: 2017 IEEE Conference on Computer Vision and Pattern Recognition (CVPR), pp. 4031–4039. IEEE (2017)
23. Steccanella, L., Bloisi, D., Blum, J., Farinelli, A.: Deep learning waterline detection for low-cost autonomous boats. In: Strand, M., Dillmann, R., Menegatti, E., Ghidoni, S. (eds.) IAS 2018. AISC, vol. 867, pp. 613–625. Springer, Cham (2019). https://doi.org/10.1007/978-3-030-01370-7_48
24. Stephens, E., Schumann, G., Bates, P.: Problems with binary pattern measures for flood model evaluation. Hydrol. Process. **28**(18), 4928–4937 (2014)
25. Tauro, F., et al.: Measurements and observations in the XXI century (MOXXI): innovation and multi-disciplinarity to sense the hydrological cycle. Hydrol. Sci. J. **63**(2), 169–196 (2018)
26. Tian, Y., Luo, P., Wang, X., Tang, X.: Pedestrian detection aided by deep learning semantic tasks. In: Proceedings of the IEEE Conference on Computer Vision and Pattern Recognition, pp. 5079–5087 (2015)
27. Vetra-Carvalho, S., et al.: Collection and extraction of water level information from a digital river camera image dataset. Data Brief **33**, 106338 (2020)
28. Moy de Vitry, M., Kramer, S., Wegner, J.D., Leitão, J.P.: Scalable flood level trend monitoring with surveillance cameras using a deep convolutional neural network. Hydrol. Earth Syst. Sci. **23**(11), 4621–4634 (2019)
29. Xiao, T., Liu, Y., Zhou, B., Jiang, Y., Sun, J.: Unified perceptual parsing for scene understanding. In: Ferrari, V., Hebert, M., Sminchisescu, C., Weiss, Y. (eds.) ECCV 2018. LNCS, vol. 11209, pp. 432–448. Springer, Cham (2018). https://doi.org/10.1007/978-3-030-01228-1_26
30. Zhang, C., Li, H., Wang, X., Yang, X.: Cross-scene crowd counting via deep convolutional neural networks. In: Proceedings of the IEEE Conference on Computer Vision and Pattern Recognition, pp. 833–841 (2015)
31. Zhou, B., Zhao, H., Puig, X., Fidler, S., Barriuso, A., Torralba, A.: Scene parsing through ade20k dataset. In: Proceedings of the IEEE Conference on Computer Vision and Pattern Recognition, pp. 633–641 (2017)

Does SGD Implicitly Optimize
for Smoothness?

Václav Volhejn[(✉)] and Christoph Lampert

IST Austria, Klosterneuburg, Austria
vvolhejn@student.ethz.ch

Abstract. Modern neural networks can easily fit their training set perfectly. Surprisingly, despite being "overfit" in this way, they tend to generalize well to future data, thereby defying the classic bias–variance trade-off of machine learning theory. Of the many possible explanations, a prevalent one is that training by stochastic gradient descent (SGD) imposes an implicit bias that leads it to learn simple functions, and these simple functions generalize well. However, the specifics of this implicit bias are not well understood.

In this work, we explore the *smoothness conjecture* which states that SGD is implicitly biased towards learning functions that are smooth. We propose several measures to formalize the intuitive notion of smoothness, and we conduct experiments to determine whether SGD indeed implicitly optimizes for these measures. Our findings rule out the possibility that smoothness measures based on first-order derivatives are being implicitly enforced. They are supportive, though, of the smoothness conjecture for measures based on second-order derivatives.

1 Introduction

Classical machine learning wisdom suggests that the expressive power of a model class (its *capacity*) should be carefully balanced with the amount of available training data: if the capacity is too low, learned models will *underfit* and not manage to fit the training set, let alone the test set. If the capacity is too high, learned models do fit the training set, but they *overfit* to spurious patterns and fail to represent the underlying trend, again failing to generalize well to the test set. Thus, the learned models generalize best when the capacity is in a sweet-spot somewhere between underfitting and overfitting. This observation is also known as *bias–variance trade-off*.

Several researchers have observed that neural networks seem to defy the bias–variance trade-off: increasing model capacity often improves generalization performance, even if the network is already apparently "overfit". This phenomenon had first been reported more than 20 years ago, e.g. [5,15], but it has only begun

---

Electronic supplementary material The online version of this chapter (https://doi.org/10.1007/978-3-030-71278-5_18) contains supplementary material, which is available to authorized users.

Z. Akata et al. (Eds.): DAGM GCPR 2020, LNCS 12544, pp. 246–259, 2021.
https://doi.org/10.1007/978-3-030-71278-5_18

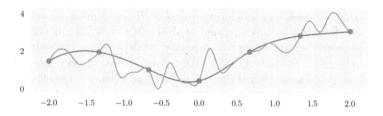

Fig. 1. A non-smooth function (orange) and smooth function (blue) interpolating a one-dimensional dataset. (Color figure online)

receiving wider attention in recent years. This started with the work of [22], who showed that plotting the test loss as a function of model capacity (represented by the hidden layer size) does not yield the U-shaped curve predicted by the bias–variance trade-off, but starts to decrease again for high model class capacities. The authors then conjecture that the surprising generalization performance of "overfit" neural networks might be due to implicit regularization in the training process: while the training objective only penalizes the prediction quality, the optimization process nevertheless prefers solutions that have "small complexity" and therefore generalize well.

It is still an open question, though, what exactly the implicitly regularized complexity measure is. In this work, we explore the conjecture (put forward, e.g., in [19]) that it is "smoothness" which is implicitly regularized: stochastic gradient descent tends to produce functions that are not needlessly "rough" or "bumpy". For an illustration, see Fig. 1. While this *smoothness conjecture* is intuitively appealing, it is not clear so far how the intuitive concept of "smoothness" would be correctly formalized mathematically. This is especially a problem because in high dimensions, as common in machine learning, there are many possible notions of smoothness for a function.

Our goal in this work is to make progress towards a formal analysis of the smoothness conjecture. Specifically, our main steps are the following:

– We define four measures that express "smoothness" of a trained neural network; two rely on first-order information, two on second-order information.
– We introduce two experimental settings that allow us to assess compatible the smoothness conjecture for each of these measures is with empirical observations.
– Based on our experimental results, we argue that first-order smoothness measures can be excluded as candidates for SGD's implicit regularization, whereas second-order methods are promising candidates.

2   Related Work

To the best of our knowledge, the first modern paper that observed the unexpected generalization behavior of neural networks is [22], where the authors focus on the fact that the test loss keeps decreasing with the number of hidden units. This was later followed by more refined analyses (e.g. [2,4,27]), which observed a *"double descent"* behavior: for low-capacity model classes, the standard reasoning of over- and underfitting holds. For model classes of very high capacity, though, where the training error can be reduced to zero, i.e. the data is interpolated, higher model class capacity corresponds to further reductions of the test loss. Later work [21] confirmed the findings in extensive experiments and observed that a double descent occurs not only as a function of model size but also the number of training epochs.

In [28] it was shown that modern deep convolutional neural networks are able to fit datasets even with random labels, and it is thereby easy to find models of small training error that do not generalize well. Consequently, the explanation for the unexpectedly good generalization behavior cannot be that all interpolating models generalize equally well. Instead, it must be the specific solutions found by standard network training (using stochastic gradient descent) that have these favorable properties.

A popular form of explanation introduced already by [22] is that the training procedure is implicitly biased towards solutions with low complexity. Subsequently, most works concentrated on the question which property of trained neural networks it could be that would make them generalize well. Suggestions include the *sharpness* of the reached loss function minimum [13], *distance from initialization* [20], *Fisher–Rao norm* [18], as well as various measures based on parameter norms [3,23]). However, an extensive empirical comparison in [10] showed that many of the proposed measures are not positively correlated with generalization quality. Therefore, the question of how to *enforce* generalization for high-complexity model classes remains so far unsolved.

Some of the smoothness measures that we discuss later have been studied previously in other contexts. We postpone the discussion of this related work to Sect. 3.4, after we have presented the measures in technical form.

In this work, we do not try to solve the question of which complexity measure should best be minimized for neural networks to generalize well, but the question which such measure SGD actually implicitly regularizes, if any. Our approach is inspired by [19], who observe that under certain conditions, training shallow ReLU networks in the one-dimensional setting using gradient descent yields "simple" functions that are close to a piecewise-linear interpolation of the training data. The author do not explore analogs for real networks with high-dimensional inputs, though. Another work that is related to our analysis is the recent preprint [14], where also the smoothness of trained neural networks is studied. The authors find that overparametrized networks interpolate almost linearly between the samples, which is consistent with our findings. The work does not answer the question if smoothness is actively minimized by SGD, though.

3   Does SGD Implicitly Optimize for Smoothness?

We study the implicit regularization properties of stochastic gradient descent training for neural networks in a standard setup of supervised learning. We adopt a regression setting with input set $\mathbb{R}^d$. And output set $\mathbb{R}$. Assuming a fixed but unknown data distribution $\mathbb{P}$, the goal is to learn a function $f \colon \mathbb{R}^d \to \mathbb{R}$ with low *expected loss*,

$$\mathcal{L}(f) = \mathbb{E}_{(x,y) \sim \mathbb{P}} (f(x) - y)^2. \tag{1}$$

While the data distribution is unknown, we do have access to a training set $D = \{(x_1, y_1), \ldots, (x_n, y_n)\}$ consisting of independent and identically distributed (i.i.d.) samples from $\mathbb{P}$. This allows us to define the *training loss*

$$L(f) = \frac{1}{n} \sum_{i=1}^{n} (f(x_i) - y_i)^2. \tag{2}$$

For a fixed model class, $\mathcal{F}$, e.g. the set of ReLU networks of a fixed architecture, we want to select (*learn*) a function $f \in \mathcal{F}$ which minimizes the training loss. We are primarily interested in models that perfectly fit or *interpolate* the training data, meaning the learned function satisfies $L(f) = 0$. For numerical reasons, we only require $L(f) < \varepsilon$ with a small ε (e.g. 10^{-5}) in practice. Typically, if the model class is rich enough to contain any model that fulfills this condition, then it contains many of them. The informal *smoothness conjecture* is:

> *When the model class is a set of neural networks that is rich enough to interpolate the training data and we use stochastic gradient descent for training, then the resulting trained model is not an arbitrary minimizer of the training loss, but among the smoothest possible ones.*

In this work, we aim towards a better understanding of the validity of this conjecture. First, we formalize several smoothness measures, which makes it possible to treat the above conjecture as a mathematical rather than just an informal statement. Then, we provide experimental evidence that support the smoothness conjecture for some smoothness measures while refuting it for others.

3.1   Measuring the Smoothness of a Function

In machine learning, the notion of smoothness of a function is often used intuitively (e.g. [4,11,14]) and it is rarely defined formally. In this section we formulate four measures that assign scalar smoothness values to functions $f \colon \mathbb{R}^d \to \mathbb{R}$. To be precise, the measures we define quantify *roughness* or the *absence of smoothness*, as we will use the convention that small values (close to 0) indicate smooth functions, whereas large values indicate functions with little smoothness. This convention is, unfortunately, necessary to be compatible with most of the prior literature. The non-negativity reflects the qualitative use of the term smoothness as a single-sided bounded measure: there is a limit on how smooth a functions can be (e.g. attained by constant functions), but there is no obvious limit to how non-smooth it could be.

The measures we discuss can be classified into two categories: *first-order* and *second-order* smoothness measures. First-order measures are based on properties of the first-order derivatives (gradients) or differences between function values of f. Second-order measures are based on second-order derivatives, or differences between gradients of f.

3.2   First-Order Smoothness Measures

Inspired by the common procedure for linear models, a simple way to formalize smoothness is to identify it with *steepness*: if a function is very steep (has a large gradient magnitude), then it is not very smooth. For non-linear functions, which we are interested in, the gradient varies for different input arguments. To obtain a scalar measure, we take the expected value of the Euclidean norm of the function's gradient with respect to the underlying data distribution.

Definition 1 (Gradient norm). *Let $\mathbb{P}_X$ be a probability distribution over $\mathbb{R}^d$ and let $f \colon \mathbb{R}^d \to \mathbb{R}$ be a function that is differentiable almost everywhere with respect to $\mathbb{P}_X$. We define the gradient norm of f with respect to $\mathbb{P}_X$ as*

$$\mathrm{GN}(f) = \mathbb{E}_{x \sim \mathbb{P}_X} \|\nabla_x f(x)\|. \tag{3}$$

GN is non-negative and it is 0 for those functions whose gradient is zero almost everywhere. This, in particular, includes constant functions, but also piece-wise constant functions as long as the set where f changes values has measure zero according to $\mathbb{P}_X$.

Because it is defined as an expected value over the data distribution, we can approximate $\mathrm{GN}(f)$ by random sampling: let $x_1, \ldots, x_N$ be N data samples that were not used for training f, then we set

$$\widehat{\mathrm{GN}}(f) = \frac{1}{N} \sum_{i=1}^{N} \|\nabla_x f(x_i)\|. \tag{4}$$

For our experiments we use $N = 1000$ and we use automatic differentiation to compute the gradient. The value of N was chosen heuristically as a compromise between accuracy and computational efficiency.

An alternative approach for characterizing smoothness that avoids the condition of differentiability is to study changes of the function values along one-dimensional line segments. For this, we define

Definition 2. *Let $f \colon \mathbb{R}^d \to \mathbb{R}^k$ be a (potentially vector-valued) function and let $a, b \in \mathbb{R}^d$. We define a* line segment *of f from a to b to be a function $f_{[a,b]} \colon [0, 1] \to \mathbb{R}^k$ defined as*

$$f_{[a,b]}(t) = f((1 - t)a + tb). \tag{5}$$

Studying the curve induced by the function values on any such line segment, we obtain an intuitive measure of smoothness. If the curve is straight and short, the underlying function is smoother than if the curve is wrinkled and long. Mathematically, we define the *function path length* as the expected value of the *total variation* over all line segments of f with end points distributed according to the data distribution:

Definition 3 (Function path length). *Let* $\mathbb{P}_X$ *be a probability distribution over* $\mathbb{R}^d$ *and let* $f \colon \mathbb{R}^d \to \mathbb{R}$ *be a function. We define the* function path length *of* f *with respect to* $\mathbb{P}_X$ *as*

$$\mathrm{FPL}(f) = \mathbb{E}_{a,b \sim \mathbb{P}_X} \mathrm{TV}(f_{[a,b]}) \tag{6}$$

where the total variation of a function $g \colon [0,1] \to \mathbb{R}^k$ *is defined as*

$$\mathrm{TV}(g) = \sup_{P \in \mathcal{P}} \sum_{i=1}^{|P|} \|g(t_i) - g(t_{i-1})\| \tag{7}$$

with $\mathcal{P}$ *denoting the set of all partitions of the interval* $[0,1]$:

$$\mathcal{P} = \{P = (t_0, t_1, \dots, t_{|P|}) \,\big|\, 0 = t_0 < t_1 \cdots < t_{|P|} = 1\}. \tag{8}$$

FPL is non-negative by construction and minimized (with value 0) by all constant functions.

As GN before, the fact that FPL is defined in terms of an expectation operation over the data distribution makes it possible to derive a sampling-based approximation. Let $(a_i, b_i)_{i=1,\dots,N}$ be N pairs of data points that were not used during the training of f. Then we set

$$\widehat{\mathrm{FPL}}(f) = \frac{1}{N} \sum_{i=1}^{N} \widehat{\mathrm{TV}}(f_{[a_i,b_i]}) \tag{9}$$

where $\widehat{\mathrm{TV}}$ approximates TV using a regular subdivision of the input interval:

$$\widehat{\mathrm{TV}}(f_{[a,b]}) = \sum_{i=1}^{n-1} |f(t_i) - f(t_{i-1})| \tag{10}$$

with $t_i = \frac{i}{n-1}a + \left(1 - \frac{i}{n-1}\right)b$ for $i \in \{0, \dots, n-1\}$. For our experiments, we use $N = 1000$ and $n = 100$.

While first-order smoothness measures are intuitive and efficient, they also have some shortcomings. In particular, neither the gradient norm nor the function path length can distinguish between some functions which we would not consider equally smooth. For example, take $f(x) = x$ on $[0,1]$ and $g(x) = 0$ on $[0, \frac{1}{2}]$ and $g(x) = 2x - 1$ on $[\frac{1}{2}, 1]$ under a uniform data distribution. Both functions have identical function path length and gradient norm, even though intuitively one would consider f smoother than g. This problem can be overcome by looking at measures that take second-order information (i.e. curvature) into account.

3.3 Second-Order Smoothness Measures

A canonical choice for a second-order smoothness measure would be to compute properties (e.g. the Frobenius norm or operator norm) of the Hessian matrix.

Unfortunately, this is not tractable in practice, because of the high computational effort of computing the Hessian matrix many times, as well as the memory requirements, which are quadratic in the number of input dimensions.

Instead, the first measure we propose relies on an analog of the construction used for the function path length, now applied to the function's gradient instead of its values.

Definition 4 (Gradient path length). *Let $\mathbb{P}_X$ be a probability distribution over $\mathbb{R}^d$ and let $f\colon \mathbb{R}^d \to \mathbb{R}$ be a differentiable function. We define the gradient path length as*

$$\text{GPL}(f) = \mathbb{E}_{a,b \sim \mathbb{P}_X} \text{TV}((\nabla_x f)_{[a,b]}) \tag{11}$$

GPL is non-negative by construction. It vanishes on constant functions, but also on linear (more precisely: affine) ones. To approximate GPL in practice, we use the same construction as for FPL, where the occurring gradients are computed using automatic differentiation.

A special situation emerges for two-layer ReLU networks, i.e. functions of the form

$$f(x) = \langle w^{(2)}, a \rangle + b^{(2)} \text{ with } a_i = \text{ReLU}\big[\langle w_i^{(1)}, x \rangle + b_i^{(1)}\big] \quad \text{for } i = 1, \ldots, h, \tag{12}$$

where h is the number of hidden units in the network and $w_1^{(1)}, \ldots, w_h^{(1)}$ and $w^{(2)}$ are weight vectors of suitable dimensions and $b_1^{(1)}, \ldots, b_h^{(1)}$ and $b^{(2)}$ are scalar bias terms. For these, we can compute a measure of second-order smoothness explicitly from the parameter values.

Definition 5 (Weights product). *Let $f_\theta\colon \mathbb{R}^d \to \mathbb{R}$ be a two-layer ReLU network with parameters $\theta = (W^{(1)}, b^{(1)}, w^{(2)}, b^{(2)})$, where $W^{(1)} = (w_1^{(1)}, \ldots, w_h^{(1)})$ with $w_i^{(1)} \in \mathbb{R}^d$ and $b^{(1)} = (b_1^{(1)}, \ldots, b_h^{(1)})$ with $b_i^{(1)} \in \mathbb{R}$ for $i = 1, \ldots, h$, as well as $w^{(2)} \in \mathbb{R}^h$ and $b^{(2)} \in \mathbb{R}$. We define the* weights product *measure as*

$$\text{WP}(f_\theta) = \sum_{i=1}^{h} |w_i^{(2)}| \cdot \|w_i^{(1)}\| \tag{13}$$

where $w_i^{(2)}$ indicates the i-th entry of the vector $w^{(2)}$ for any $i = 1, \ldots, h$.

WP is non-negative by construction, and takes the value 0 on networks where for each neuron in the hidden layer either all incoming weights or the outgoing weight are zero, with arbitrary values of the bias terms. From Eq. (12) one sees that all constant functions can be expressed this way.

A small computation establishes that for one-dimensional inputs, WP is equal to the total variation of the derivative, under the assumption that the positions at which the hidden units switch between deactivation and activation $(-b_i^{(1)}/w_i^{(1)})$ are unique. In higher dimensions, each summand in (13) is still the norm of the difference of the gradients on the two sides of the ReLU activation function. Thus WP is a second-order measure, based on the changes of the gradient.

In contrast to the previous smoothness measures, WP is only defined for two-layer ReLU networks and does not take the underlying data distribution into account. Its advantage, though, is that it can easily be computed exactly, without having to rely on sampling-based approximations as for the other measures.

3.4 Smoothness Measures in Related Work

The measure we call *gradient norm*, as well as minor variants, were explored in multiple prior works, e.g. [6, 24–26]. Generally, the findings are that a small average norm of the Jacobian, i.e. the gradient in the scalar setting, can lead to improved generalization. In [17], a measure of "rugosity" (roughness) is proposed based on the learned function's Hessian matrix. The authors also discuss a Monte Carlo approximation of this quantity, which resembles our notion of *gradient path length*, with the main difference being that it uses local perturbations instead of a line segment. Even closer is the smoothness measure in [14] which also measures how the gradient of a function changes when interpolating between two samples.

The *weight product* measure is an analog of the *path-regularizer* of [23] for two-layer ReLU networks. In that work, the measure is proposed for training-time regularization, not as a post-hoc smoothness measure. To our knowledge, the *function path length* measure has not been used in the context of neural networks, but a similar constructions was suggested, e.g., for audio signals [9].

4 Experiments

We report on our experiments that shed light on the validity of the smoothness conjecture in general, and with respect to the four proposed smoothness measures in particular. Note that the naive approach of simply checking the numeric values of the smoothness measures is not possible, because we do not know what reference value to compare them to. Ideally, this would be the smallest achievable smoothness value for any network of the studied class on the provided data. Unfortunately, we cannot easily compute these on high-dimensional data, only derive some lower bound (see Table 2).

Instead, we use two proxy setups that we consider contributions of potentially independent interest, as they would also be applicable to other measures besides smoothness. First, we study how monotonically the measures behave when networks are trained with increasing amounts of data. If the smoothness conjecture is fulfilled, one would expect perfect monotonicity, see the discussion in Sect. 4.2. Second, we analyze whether substantially smoother models exist than the one produced by SGD that nevertheless interpolate the data. Under the smoothness conjecture, this should not be the case, see Sect. 4.3.

Before reporting on the results of the experiments, though, we introduce the experimental setup.

4.1   Experimental Setup

The surprising generalization abilities have been observed for networks of all sizes. We restrict our own analysis to small networks, because the more efficient experiments allows us to try more different settings and perform multiple reruns to gain statistical power. Specifically, we use fully connected ReLU networks with one hidden layer of size $h = 256$. We train the networks using mini-batch stochastic gradient descent with a batch size of 64, and, unless specified otherwise, a learning rate of 0.01. For network initialization, the network's bias terms are initially set to 0. The weights in each of the two layers are initialized by drawing uniformly from the interval $[-\ell, \ell]$ with $\ell = \sqrt{\frac{6\alpha}{n_{in}+n_{out}}}$. n_{in} is the number of input units of the layer, n_{out} is the number of output units, and α is an *initialization scale*. For $\alpha = 1$, this initializer would reduce to the widely used Glorot uniform initializer [7]. We use $\alpha = 0.01$ instead, as it has been observed in [19] that a smaller initialization scale generally leads to smoother learned functions, and this is also consistent with our own findings.

Many existing works use the number of training epochs as stopping criterion. This is not ideal for our setting, as we observed that training models of different complexity, e.g. with different regularization terms, for the same number of epochs leads to large differences in the achieved training losses and how close to convergence the models actually are. Instead, we use a threshold of 10^{-5} on the training loss as the stopping criterion. This choice ensures that the models fit the training set almost perfectly and have converged to a comparable level.

All experiments were performed using the TensorFlow framework [1], assisted by the *Sacred* package [8] for enhanced reproducibility. Our code is available at https://github.com/vvolhejn/neural_network_smoothness.

As a data source, we use the MNIST dataset of handwritten digits [16] in the following way. For any pair of digits (a, b) with $0 \le a < b \le 9$, we construct a training set by filtering MNIST to only the images of digit a or digit b. We then select the first 10,000 images in the filtered dataset so that dataset sizes are equal among choices of (a, b). This results in $\binom{10}{2} = 45$ regression problems, which we call the *MNIST-binary* problem set. By solving multiple small regression problems instead of a single large one, we hope to reduce variance and gain more confidence that the observed trends are not just due to randomness. As data for computing the smoothness measures, we use subsets of the MNIST test set with the corresponding digits.

4.2   Monotonicity

As a first test of the hypothesis that smoothness is implicitly enforced during neural network training, we use the following observation. Imagine two training sets, D and D', where D' is identical to D, except that some more data points have been added to it. Denote by f and f' the smoothest possible functions in a hypothesis set that interpolate the data in D and D', respectively. Then f' cannot be smoother than f, because adding training samples means adding constraints to the interpolation problem, and the minimizer over a more constrained set can only achieve a higher or equal objective value.

Our experiments verify empirically if this phenomenon indeed occurs for different model classes and the smoothness measures of Sect. 3.1. We first select a dataset $D = \{(x_1, y_1), \ldots, (x_N, y_N)\}$ that all studied model classes are able to interpolate. Then we construct an increasing sequence of datasets $D_1, \ldots, D_n$ with $D_i \subsetneq D_{i+1}$ for $i = 1, \ldots, n-1$, and train models on each of these datasets, obtaining functions $f_1, \ldots, f_n$. For any smoothness measure S (with the convention that a lower value of S means a smoother function), we expect to obtain

$$S(f_1) \leq S(f_2) \leq \cdots \leq S(f_n) \tag{14}$$

if the smoothness conjecture holds for S. As a quantitative measure of how close we are to all inequalities holding, we use the Kendall rank correlation coefficient, $\tau \in [-1, 1]$, which reflects the number of inversions in the sequence, see [12]. A value of $\tau = 1$ means perfect accordance with (14).

Table 1. Monotonicity score (Kendall's τ) for the 45 MNIST-binary datasets. Standard deviation is not listed for WP and GPL because these measures reach the maximum value of τ for every dataset.

Smoothness measure	GN	FPL	GPL	WP
Kendall's τ	0.16 ± 0.32	0.07 ± 0.46	1.0	1.0

For each of the 45 MNIST-binary tasks, we use training set sizes $N \in \{64, 128, 256, 512, 1024, 2048, 4096, 8192\}$. We fix an ordering of the dataset and simply select its first N elements. To lower the variance, we repeat the experiment three times for each dataset. Therefore, we obtain a total of $3 * 45$ values of τ.

Table 1 summarizes the results as mean and standard deviation over the obtained τ values. We see that for the first-order smoothness measures, GN and FPL, the change in function smoothness is highly fluctuating and only weakly correlated with growing dataset size. In contrast, the rank correlation is consistently at its maximum value for the second-order measures, GPL and WP.

4.3   Optimality

In this section, we take a second look at the question of whether smoothness is implicitly optimized by SGD training and if yes, which notion of smoothness exactly that is. For this, we take an exclusion approach: we can be sure that a complexity measure S is *not* being regularized implicitly during training, if we are able to find another model that performs equally well on the training set but is substantially smoother according to this measure than that found by SGD.

To search for such smoother models, we rely on *explicit* regularization. During network training, we replace the original loss function L with a regularized version L_{reg}, in which we penalize high values of S:

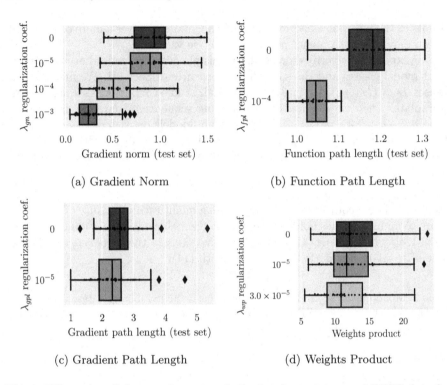

(a) Gradient Norm

(b) Function Path Length

(c) Gradient Path Length

(d) Weights Product

Fig. 2. Effect of explicit smoothness regularization for training on MNIST data for different smoothness measures (subfigures) and regularization strengths (y-axis). Lower values indicate smoother models.

$$L_{reg}(f) = L(f) + \lambda S(f) \tag{15}$$

where $\lambda > 0$ is a regularization coefficient.

When S is expensive to compute, network training is slowed down considerably, as the measure and its gradient have to be evaluated in every training step. We therefore use stochastic variants of the smoothness measures that always use the current training batch as data points. Evaluation of the trained models is done on the test data, without using these stochastic variants. Furthermore, we use a learning rate of 0.1 for the two path length measures instead of the default value of 0.01. This is merely for practical reasons as it ensures training converges in a reasonable time; we did not observe any negative side effects of this change.

Figure 2 shows the results of all experiments as box-whisker plots for the 45 MNIST-binary datasets. The first row in each plot ($\lambda = 0$) corresponds to training the unregularized objective, i.e. plain SGD training. The other rows reflect training with different amounts of regularization strength. Setups in which the explicit regularization was too strong to reach the interpolation regime (i.e. training error remained above 10^{-5}) are not reported and not included in the analysis.

Table 2. Numeric summary of explicit regularization experiment results. For each measure, we report the results for the highest regularization coefficient for which we were still able to train the models to achieve 10^{-5} training loss. For an explanation of the rows, see the main text.

Smoothness measure	GN	FPL	GPL	WP
Unregularized mean μ_{unreg}	0.93	1.17	2.62	12.84
Regularized mean μ_{reg}	0.29	1.05	2.35	11.57
Lower bound l_S	0	1	0	0
Normalized ratio r_{norm}	0.31 ± 0.13	0.33 ± 0.27	0.89 ± 0.06	0.90 ± 0.03

Table 2 contains a numeric summary of these results. The columns *Unregularized mean* and *Regularized mean* show the mean value of the respective smoothness measure across the 45 MNIST-binary tasks (the full score distribution was already provided in Fig. 2). The regularized mean is computed only from the models with the largest reported regularization coefficient. *Lower bound*, l_S, is a bound on the smallest value that the corresponding measure, S, can take on an interpolating model from the studied model class. For GN, GPL and WP, only the trivial bound 0 is readily available. For FPL, we know that data pairs of identical output value contribute 0 to Eq. (9), while data pairs of opposite output values contribute at least 2, so a lower bound on FPL for balanced data is 1. The normalized ratio is computed as

$$r_{norm} = \frac{1}{45} \sum_{i=1}^{45} \frac{S(f_i^{reg}) - l_S}{S(f_i^{unreg}) - l_S}, \tag{16}$$

where f_i^{reg} and f_i^{unreg} are the results of training models on the i-th task with and without regularization, respectively.

The plots and table show a clear trend: for the first-order smoothness measures, adding explicit regularization to the training objective results in models that have equally small training loss yet much higher smoothness (GN and FPL are reduced by approximately 70%). Consequently, we can reject the conjecture that SGD implicitly optimizes for these measures. Note that this finding is not incompatible with results in the literature that enforcing a small norm of the gradient can positively impact generalization [24–26], as that is just a sufficient criterion, not a necessary one.

For the second-order smoothness measures the results show the opposite effect. By including explicit regularization, we were not able to substantially increase the models' smoothness (GPL and WP are reduced by approximately 10%). Formally, our result cannot be taken as proof that no substantially smoother models exist. After all, we might just not have been able to find them using the explicit regularization procedure. Nevertheless, the results do support the conjecture that SGD does have a regularizing effect on neural network training, and they concretize the formulation of the smoothness conjecture: the enforced smoothness is likely of second-order type.

5   Conclusion

In this work, we empirically studied the conjecture that training neural networks by stochastic gradient descent results in models that do not only have a small training loss, but that at the same time are very smooth, even when the training objective does not explicitly enforce the latter property. If correct, the conjecture would be a major milestone towards a better understanding of the generalization properties of neural networks.

After introducing four different smoothness measures, two of first-order and two of second-order type, we reported on experiments showing that there is no support for the smoothness conjecture with respect to the first-order smoothness measures. However, our findings are quite well aligned with SGD enforcing second-order smoothness, thereby adding credibility to this instantiation of the conjecture.

For future work, it would be interesting to see if our results also transfer to deeper networks and larger datasets, as well as other network architectures, e.g. convolutional or recurrent networks. One could also now use theoretical tools to determine which second-order smoothness measure exactly is being minimized and by what mechanism.

References

1. Abadi, M., et al.: Tensorflow: a system for large-scale machine learning. In: USENIX Symposium on Operating Systems Design and Implementation (OSDI) (2016)
2. Advani, M.S., Saxe, A.M.: High-dimensional dynamics of generalization error in neural networks. CoRR abs/1710.03667 (2017). http://arxiv.org/abs/1710.03667
3. Bartlett, P.L., Foster, D.J., Telgarsky, M.: Spectrally-normalized margin bounds for neural networks. In: Conference on Neural Information Processing Systems (NeurIPS) (2017)
4. Belkin, M., Hsu, D., Ma, S., Mandal, S.: Reconciling modern machine-learning practice and the classical bias-variance trade-off. Proc. Natl. Acad. Sci. **116**(32), 15849–15854 (2019)
5. Caruana, R., Lawrence, S., Giles, C.L.: Overfitting in neural nets: backpropagation, conjugate gradient, and early stopping. In: Conference on Neural Information Processing Systems (NeurIPS) (2000)
6. Drucker, H., Le Cun, Y.: Improving generalization performance using double backpropagation. IEEE Trans. Neural Netw. (T-NN) **3**(6), 991–997 (1992)
7. Glorot, X., Bengio, Y.: Understanding the difficulty of training deep feedforward neural networks. In: Conference on Uncertainty in Artificial Intelligence (AISTATS) (2010)
8. Greff, K., Klein, A., Chovanec, M., Hutter, F., Schmidhuber, J.: The sacred infrastructure for computational research. In: Proceedings of the Python in Science Conferences-SciPy Conferences (2017)
9. Holopainen, R.: Smoothness under parameter changes: derivatives and total variation. In: Sound and Music Computing Conference (SMC) (2013)

10. Jiang, Y., Neyshabur, B., Mobahi, H., Krishnan, D., Bengio, S.: Fantastic generalization measures and where to find them. In: International Conference on Learning Representations (ICLR) (2020)
11. Kawaguchi, K., Kaelbling, L.P., Bengio, Y.: Generalization in deep learning. CoRR abs/1710.05468 (2017). http://arxiv.org/abs/1710.05468
12. Kendall, M.G.: A new measure of rank correlation. Biometrika **30**(1/2), 81–93 (1938)
13. Keskar, N.S., Mudigere, D., Nocedal, J., Smelyanskiy, M., Tang, P.T.P.: On large-batch training for deep learning: Generalization gap and sharp minima. In: International Conference on Learning Representations (ICLR) (2017)
14. Kubo, M., Banno, R., Manabe, H., Minoji, M.: Implicit regularization in over-parameterized neural networks. CoRR abs/1903.01997 (2019). http://arxiv.org/abs/1903.01997
15. Lawrence, S., Gilesyz, C.L., Tsoi, A.C.: What size neural network gives optimal generalization? Convergence properties of backpropagation. Tech. rep., Institute for Advanced Computer Studies, University of Maryland (1996)
16. LeCun, Y., Bottou, L., Bengio, Y., Haffner, P.: Gradient-based learning applied to document recognition. Proc. IEEE **86**(11), 2278–2324 (1998)
17. LeJeune, D., Balestriero, R., Javadi, H., Baraniuk, R.G.: Implicit rugosity regularization via data augmentation. CoRR abs/1905.11639 (2019). http://arxiv.org/abs/1905.11639
18. Liang, T., Poggio, T.A., Rakhlin, A., Stokes, J.: Fisher-Rao metric, geometry, and complexity of neural networks. In: Conference on Uncertainty in Artificial Intelligence (AISTATS) (2019)
19. Maennel, H., Bousquet, O., Gelly, S.: Gradient descent quantizes ReLU network features. CoRR abs/1803.08367 (2018). http://arxiv.org/abs/1803.08367
20. Nagarajan, V., Kolter, J.Z.: Generalization in deep networks: the role of distance from initialization. In: Conference on Neural Information Processing Systems (NeurIPS) (2019)
21. Nakkiran, P., Kaplun, G., Bansal, Y., Yang, T., Barak, B., Sutskever, I.: Deep double descent: where bigger models and more data hurt. In: International Conference on Learning Representations (ICLR) (2020)
22. Neyshabur, B., Tomioka, R., Srebro, N.: In search of the real inductive bias: on the role of implicit regularization in deep learning. In: International Conference on Learning Representations (ICLR) (2014)
23. Neyshabur, B., Tomioka, R., Srebro, N.: Norm-based capacity control in neural networks. In: Workshop on Computational Learning Theory (COLT) (2015)
24. Novak, R., Bahri, Y., Abolafia, D.A., Pennington, J., Sohl-Dickstein, J.: Sensitivity and generalization in neural networks: an empirical study. In: International Conference on Learning Representations (ICLR) (2018)
25. Rifai, S., Vincent, P., Muller, X., Glorot, X., Bengio, Y.: Contractive auto-encoders: explicit invariance during feature extraction. In: International Conference on Machine Learning (ICML) (2011)
26. Sokolic, J., Giryes, R., Sapiro, G., Rodrigues, M.R.D.: Robust large margin deep neural networks. IEEE Trans. Signal Process. (T-SP) **65**(16), 4265–4280 (2017)
27. Spigler, S., Geiger, M., d'Ascoli, S., Sagun, L., Biroli, G., Wyart, M.: A jamming transition from under- to over-parametrization affects generalization in deep learning. J. Phys. A: Math. Theor. **52**(47), 474001(2019)
28. Zhang, C., Bengio, S., Hardt, M., Recht, B., Vinyals, O.: Understanding deep learning requires rethinking generalization. In: International Conference on Learning Representations (ICLR) (2017)

Looking Outside the Box: The Role of Context in Random Forest Based Semantic Segmentation of PolSAR Images

Ronny Hänsch[✉]

SAR Technology, German Aerospace Center (DLR), Cologne, Germany
rww.haensch@gmail.com

Abstract. Context - i.e. information not contained in a particular measurement but in its spatial proximity - plays a vital role in the analysis of images in general and in the semantic segmentation of Polarimetric Synthetic Aperture Radar (PolSAR) images in particular. Nevertheless, a detailed study on whether context should be incorporated implicitly (e.g. by spatial features) or explicitly (by exploiting classifiers tailored towards image analysis) and to which degree contextual information has a positive influence on the final classification result is missing in the literature. In this paper we close this gap by using projection-based Random Forests that allow to use various degrees of local context without changing the overall properties of the classifier (i.e. its capacity). Results on two PolSAR data sets - one airborne over a rural area, one space-borne over a dense urban area - show that local context indeed has substantial influence on the achieved accuracy by reducing label noise and resolving ambiguities. However, increasing access to local context beyond a certain amount has a negative effect on the obtained semantic maps.

1 Introduction

Context refers to information not contained in an individual measurement but in its local proximity or at a larger (even global) range. For image analysis, this can refer to a spatial (i.e. pixels close to each other), temporal (measurements with a small time difference), or spectral (measurements taken at similar wavelengths) neighborhood. In this paper, context refers to the *spatial* neighborhood of a pixel.

In contrast to the (semantic) analysis of close-range photography, for a long time context had played only a minor role in remote sensing, in particular for data sources such as HyperSpectral Imagery (HSI) or Synthetic Aperture Radar (SAR). One reason is the historical approach and the scientific communities that pioneered in the analysis of images from both domains. The similarity of color photographs to the early stages of the human visual cortex (e.g. being based on angular measurements of the light intensity of primary colors), inspired to model also subsequent stages according to this biological role model for which it is well known that context (spatial as well as temporal) plays a vital role for the understanding of the image input [21]. HSI and SAR images, on the other hand, are too dissimilar to human perception to have inspired a similar

© Springer Nature Switzerland AG 2021
Z. Akata et al. (Eds.): DAGM GCPR 2020, LNCS 12544, pp. 260–274, 2021.
https://doi.org/10.1007/978-3-030-71278-5_19

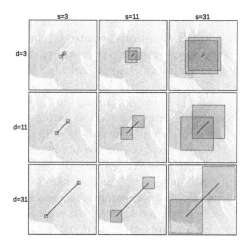

Fig. 1. We investigate the role of (visual spatial) context by varying the size of the spatial projections within the framework of projection-based Random Forests (pRFs), i.e. the size r_s and distance r_d of regions sampled relative to the patch center and used by the internal node tests of the decision trees to determine the semantic class.

approach during the early years of automated image analysis. On the contrary, early attempts to remote sensing image interpretation were often carried out by the same groups that built the corresponding sensors. Consequently, they took a rather physics-based approach and developed statistical models that aim to capture the complex relations between geo-physical and biochemical properties of the imaged object and the measured signal. Even today, approaches that aim to model the interaction of electro-magnetic waves with a scatterer with certain geometric and electro-physical properties are still in use for SAR image processing (see e.g. [7,10]). Another reason is that the information contained in a single RGB pixel of a close-range photograph is rarely sufficient to make any reliable prediction of the semantic class this pixel might belong to. On the other hand, the information contained in a single HSI or PolSAR pixel does allow to make such predictions with a surprisingly high accuracy if processed and analysed correctly.

As a consequence, although there were early attempts to incorporate context (see e.g. [24,28]) into the semantic analysis of remote sensing images, many classification methods ignored relations between spatially adjacent pixels and process each pixel independently (e.g. as in [6] for HSI and [16] for SAR data, respectively). This means in particular, that a random permutation of all pixels within the image would not effect classification performance during automatic image interpretation (quite in contrast to a visual interpretation by humans). However, neighboring pixels do contain a significant amount of information which should be exploited. On the one hand, adjacent pixels are usually correlated due to the image formation process. On the other hand, the depicted objects are usually large (with respect to the pixel size) and often rather homogeneous.

There are two distinct yet related concepts of context in images, i.e. visual context and semantic context. **Semantic context** refers to relationships on object level such as co-occurrence relations (e.g. a ship usually occurs together with water) for example modelled via Latent Dirichlet Allocations [23] or concept occurence vectors [27] and topological relations (e.g. trees are more likely to be next to a road than on a road) capturing distances and directions (see e.g. [2]). This type of context is usually exploited during the formulation of the final decision rule, e.g. by applying a context-independent pixel-wise classification followed by a spatial regularization of the obtained semantic maps [5] or by applying Markov Random Fields (MRFs, see e.g. [11,25] for usage of MRFs for the classification of SAR images). **Visual context** refers to relationships on the measurement level allowing for example to reduce the noise of an individual measurement (e.g. by local averaging) or to estimate textural properties. For example, visual context is implicitly considered during SAR speckle filtering. Another common example are approaches that combine spectral and spatial information in a pixel-wise feature vector and then apply pixel-based classification methods (e.g. [9,22]). More recent approaches move away from the use of predefined hand-crafted features and use either variants of shallow learners that have been tailored towards the analysis of image data (such as projection-based Random Forests [12]) or deep neural networks. In particular the latter have gained on importance and are often the method of choice for the (semantic) analysis of remote sensing images in general (see e.g. [15] for an overview) and SAR data in particular [31].

In this paper we address the latter type, i.e. visual context, for the special case of semantic segmentation on polarimetric SAR images. In particular, we are interested whether different data representations that implicitly integrate context are helpful and in analysing how much local context is required or sufficient to achieve accurate and robust classification results. To the best of the authors knowledge, such an investigation is missing in the current literature of PolSAR processing. Corresponding works either stop at low-level pre-processing steps such as speckle reduction [4,8] or simply assume that any amount of available contextual information leads to an improved performance.

Mostly to be able to efficiently vary available context information while keeping model capacity fixed, we use projection-based Random Forests (pRFs, [12]) which are applied to image patches and apply spatial projections (illustrated in Fig. 1) that sample regions of a certain size and distance to each other. Increasing the region size allows to integrate information over larger areas and thus adaptively reduce noise, while a larger region distance enables the RF to access information that is further away from the patch center without increasing the computational load (very similar to dilated convolutions in convolution networks [30]). Thus, the contribution of this paper is three-fold: First, we extend the general framework of [12] to incorporate node tests that can be directly applied to polarimetric scattering vectors; Second, we compare the benefits and limitations of using either scattering vectors or polarimetric sample covariance matrices for the semantic segmentation of PolSAR images; and third, we analyse how much context information is helpful to increase classification performance.

2 Projection-Based Random Forests

Traditional machine-learning approaches for semantic segmentation of PolSAR images either rely on probabilistic models aiming to capture the statistical characteristics of the scattering processes (e.g. [3,29]) or apply a processing chain that consists of pre-processing, extracting hand-crafted features, and estimating a mapping from the feature space to the desired target space by a suitable classifier (e.g. [1,26]). Modern Deep Learning approaches offer the possibility to avoid the computation of hand-crafted features by including feature extraction into the optimization of the classifier itself (see e.g. [17–20]). These networks are designed to take context into account by using units that integrate information over a local neighborhood (their receptive field). In principle, this would allow to study the role of context for the semantic segmentation of remotely sensed images with such networks. However, an increased receptive field usually corresponds to an increase of internal parameters (either due to larger kernels or deeper networks) and thus an increased capacity of the classifier.

This is why we apply projection-based Random Forests (pRFs [12]) which offer several advantages for the following experiments: Similar to deep learning approaches, pRFs learn features directly from the data and do not rely on hand-crafted features. Furthermore, they can be applied to various input data without any changes to the overall framework. This allows us to perform experiments on PolSAR data which are either represented through polarimetric scattering vectors $\mathbf{s} \in \mathbb{C}^k$ or polarimetric sample covariance matrices $\mathbf{C} \in \mathbb{C}^{k \times k}$

$$\mathbf{C} = \langle \mathbf{s}\mathbf{s}^\dagger \rangle_{w_C} \tag{1}$$

where $(\cdot)^\dagger$ denotes conjugate transpose and $\langle \cdot \rangle_{w_C}$ a spatial average over a $w_C \times w_C$ neighborhood.

Every internal node of a tree (an example of such a tree is shown in Fig. 2(a)) in a RF performs a binary test $t : D \to \{0,1\}$ on a sample $\mathbf{x} \in D$ that has reached this particular node and propagates it either to the left ($t(\mathbf{x}) = 0$) or right child node ($t(\mathbf{x}) = 1$). The RF in [12] defines the test t as

$$t(x) = \begin{cases} 0 \text{ if } d(\phi(\psi_1(\mathbf{x})), \phi(\psi_2(\mathbf{x}))) < \theta, \\ 1 \qquad \text{otherwise.} \end{cases} \tag{2}$$

where $\psi(\cdot)$ samples a region from within a patch that has a certain size r_s and distance r_d to the patch center, $\phi(\cdot)$ selects a pixel within this region, $d(\cdot)$ is a distance function, and θ is the split threshold (see Fig. 2(b) for an illustration). Region size r_s and distance r_d to the patch center are randomly sampled from a user defined range. They define the maximal possible patch size $w = 2r_d + r_s$ and thus the amount of local context that can be exploited by the test. To test whether a multi-scale approach is beneficial for classification performance, we allow the region distance to be scaled by a factor α which is randomly drawn by a user defined set of possible scales.

The pixel selection function ϕ as well as the distance function are data type dependent. The RF in [12] proposes test functions that apply to $w \times w$ patches

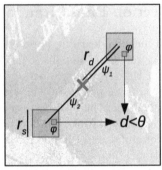

(a) A typical decision tree within a pRF (visualized via [14]) that was trained on the *OPH* data set (see Section 3.1). Leaf colors represent the dominant class in a leaf (see Figure 3(c)); leaf size represents the number of samples in this leaf.

(b) A node test t uses spatial projections of a patch centered at $\mathbf{x}$ that sample regions of size r_s and distance r_d to the patch center via $\psi()$ from which $\phi()$ selects a single pixel. The distance d between these two pixel values is then compared to the split threshold θ.

Fig. 2. Visualisation of a single decision tree of a trained pRF (left) as well as the applied spatial node projections (right).

of polarimetric covariance matrices, (i.e. $D = \mathbb{C}^{w \times w \times k \times k}$). In this case, ϕ either computes the average over the region or selects the covariance matrix within a given region with minimal, maximal, or medium span r_s, polarimetric entropy H, or anisotropy A, i.e.

$$S = \sum_{i=1}^{k} \lambda_i \ , \quad H = \sum_{i=1}^{k} \frac{\lambda_i}{S} \log\left(\frac{\lambda_i}{S}\right), \quad A = \frac{\lambda_2 - \lambda_3}{\lambda_2 + \lambda_3} \tag{3}$$

where $\lambda_1 > \lambda_2 > \lambda_3$ are the Eigenvalues of the covariance matrix. Note, that for $k = 2$, i.e. dual-polarimetric data, the covariance matrix has only two Eigenvalues which means that the polarimetric anisotropy cannot be computed.

Any measure of similarity between two Hermitian matrices P, Q (see [13] for an overview) can serve as distance function d, e.g. the Bartlett distance

$$d(P, Q) = ln\left(\frac{|P + Q|^2}{|P||Q|}\right). \tag{4}$$

We extend this concept to polarimetric scattering vectors $\mathbf{s} \in \mathbb{C}^k$ by adjusting ϕ to select pixels with minimal, maximal, or medium total target power ($\sum_i |s_i|$). Note that polarimetric scattering vectors are usually assumed to follow a complex Gaussian distribution with zero mean which means that the local sample average tends to approach zero and thus does not provide a reasonable projection. While it would be possible to use polarimetric amplitudes only, we want to work as closely to the data as possible. Extracting predefined features and using corresponding projections is possible within the pRF framework but beyond the scope of the paper. As distance $d(p, q)$ we use one of the following distance measures between polarimetric scattering vectors $p, q \in \mathbb{C}^k$:

$$\text{Span distance:} \quad d(p, q) = \sum_{i=1}^{k} |p_i| - \sum_{i=1}^{k} |q_i| \tag{5}$$

$$\text{Channel intensity distance:} \quad d(p, q) = |p_i| - |q_i| \tag{6}$$

$$\text{Phase difference:} \quad d(p, q) = \arg(p_i) - \arg(q_i) \tag{7}$$

$$\text{Ratio distance:} \quad d(p, q) = \left| \log\left(\frac{|p_i|}{|p_j|}\right) \right| - \left| \log\left(\frac{|q_i|}{|q_j|}\right) \right| \tag{8}$$

$$\text{Euclidean distance:} \quad d(p, q) = \sqrt{\sum_{i=1}^{k} |p_i - q_i|^2}, \tag{9}$$

where $\arg(z)$ denotes the phase of z.

An internal node creates multiple such test functions by randomly sampling their parameters (i.e. which ψ defined by region size and position, which ϕ, and which distance function d including which channel for channel-wise distances) and selects the test that maximises the information gain (i.e. maximal drop of class impurity in the child nodes).

3   Experiments

3.1   Data

We use two very different data sets to evaluate the role of context on the semantic segmentation of PolSAR images. The first data set (shown in Fig. 3(a), 3(c)) is a fully polarimetric SAR image acquired over Oberpfaffenhofen, Germany, by the E-SAR sensor (DLR, L-band). It has 1390×6640 pixels with a resolution of approximately $1.5\,\text{m}$. The scene contains rather large homogeneous object regions. Five different classes have been manually marked, namely City (red), Road (blue), Forest (dark green), Shrubland (light green), and Field (yellow).

The second data set (shown in Fig. 3(b)) is a dual-polarimetric image of size 6240×3953 acquired over central Berlin, Germany, by TerraSAR-X (DLR, X-band, spotlight mode). It has a resolution of approximately $1\,\text{m}$. The scene contains a dense urban area and was manually labelled into six different categories, namely Building (red), Road (cyan), Railway (yellow), Forest (dark green), Lawn (light green), and Water (blue) (see Fig. 3(d)).

The results shown in the following sections are obtained by dividing the individual image into five vertical stripes. Training data (i.e. 50,000 pixels) are drawn by stratified random sampling from four stripes, while the remaining stripe is used for testing only. We use Cohen's κ coefficient estimated from the test data and averaged over all five folds as performance measure.

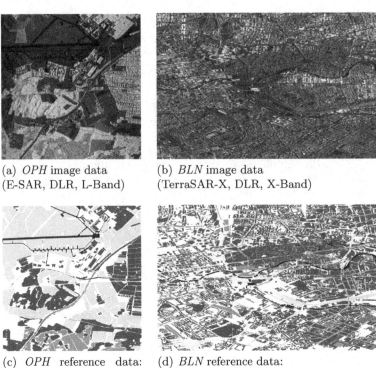

(a) *OPH* image data
(E-SAR, DLR, L-Band)

(b) *BLN* image data
(TerraSAR-X, DLR, X-Band)

(c) *OPH* reference data:
City (red), Road (blue),
Forest (dark green), Shrubland (light green), Field
(yellow), unlabelled pixels
in white

(d) *BLN* reference data:
Building (red), Road (cyan), Railway (yellow), Forest (dark green), Lawn (light
green), Water (blue), unlabelled pixels in
white.

Fig. 3. False color composite of the used PolSAR data (top) as well as color-coded reference maps (bottom) of the Oberpfaffenhofen (*OPH*, left) and Berlin (*BLN*, right) data sets. Note: Images have been scaled for better visibility. (Color figure online)

3.2   Polarimetric Scattering Vectors

As a first step we work directly on the polarimetric scattering vectors by using the projections described in Sect. 2 with $r_d, r_s \in \{3, 11, 31, 101\}$. Figure 4 shows the results when using the polarimetric scattering vectors directly without any pre-processing (i.e. no presumming, no speckle reduction, etc.). The absolute accuracy (in terms of the kappa coefficient) differs between the air- ($\kappa \in [0.64, 0.80]$) and space-borne ($\kappa \in [0.29, 0.44]$) PolSAR data. There are several reasons for this difference. One the one hand, the *OPH* data was acquired by an fully-polarimetric airborne sensor while the *BLN* data was acquired by a dual-polarimetric space-borne sensor. As a consequence, the *OPH* data contains more information (one more polarimetric channel) and has in general a better signal to noise ratio. On the other hand, the scene is simpler in terms of semantic classes, i.e. the

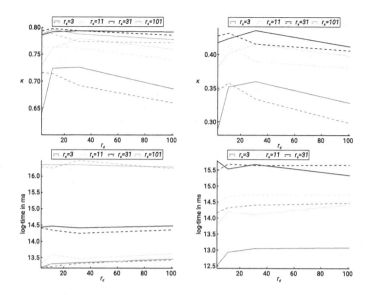

Fig. 4. Achieved κ (top) and prediction time (bottom) for *OPH* (left) and *BLN* (right) using polarimetric scattering vectors. The solid lines denote the single scale ($\alpha = 1$), the dashed lines the multi-scale ($\alpha \in \{1, 2, 5, 10\}$) case.

reference data contains less classes and object instances are rather large, homogeneous segments. In contrast, the *BLN* data contains fine grained object classes such as buildings and roads in a dense urban area.

Despite the difference in the absolute values for both data sets, the relative performance between the different parameter settings is very similar. In general, larger region sizes lead to a better performance. While the difference between 3×3 and 11×11 regions are considerable, differences between 11×11 and 31×31 regions are significantly smaller. Large regions of 101×101 pixels lead to worse results than moderate regions of 31×31. Larger regions allow to locally suppress speckle and noise and are better able to integrate local context. However, beyond a certain region size, the patches start to span over multiple object instances which makes it impossible to distinguish between the different classes.

A similar although less pronounced effect can be seen for increasing region distances. At first, performance does increase with larger distance. However, the improvement soon saturates and for very large distances even deteriorates. This effect is strongest in combination with small region sizes as the distance relative to the region size is much smaller for tests with large regions, i.e. for a test with a region distance of $r_d = 11$, regions of $r_s = 31$ still overlap.

The optimal parameter combination in terms of accuracy is $r_s = r_d = 31$, i.e. patches with $w = 93$ (note, that this only determines the maximal patch size while the actually used size depends on the specific tests selected during node optimisation). Interestingly, this seems to be independent of the data set.

A large region size has the disadvantage of an increased run time during training and prediction (the latter is shown in Fig. 4). The run time per node test increases quadratically with the region size r_s but is independent of r_d. The overall run time also depends on the average path length within the trees which might in- or decrease depending on the test quality (i.e. whether a test is able to produce a balanced split of the data with a high information gain). In general, an increased region size leads to a much longer prediction time, while an increased region distance has only a minor effect. As a consequence, if computation speed is of importance in a particular application, it is recommendable to increase sensitivity to context by setting a larger region distance than increasing the region size (at the cost of a usually minor loss in accuracy).

The dashed lines in Fig. 4 show the results when access to context is increased beyond the current local region by scaling the region distance by a factor α which is randomly selected from the set $R = \{1, 2, 5, 10\}$ (e.g. if r_d is originally selected as $r_d = 5$ and α is selected as $\alpha = 10$, the actually used region distance is 50). If the original region distance is set to a small value (i.e. $r_d = 3$) using the multi-scale approach leads to an increased performance for all region sizes. For a large region size of $r_s = 101$ this increase is marginal, but for $r_s = 3$ the increase is substantial (e.g. from $\kappa = 0.64$ to 0.72 for OPH). However, even for medium region distances ($r_d = 11$) the effect is already marginal and for large distances the performance actually decreases drastically. The prediction time is barely affected by re-scaling the region distance. In general, this reconfirms the results of the earlier experiments (a too large region distance leads inferior results) and shows that (at least for the used data sets) local context is useful to solve ambiguities in the classification decision, but global context does rarely bring further benefits. On the one hand, this is because local homogeneity is a very dominant factor within remote sensing images, i.e. if the majority of pixels in a local neighborhood around a pixel belong to a certain class, the probability is high that this pixel belongs to the same class. On the other hand, typical objects in remote sensing images (i.e. such as the here investigated land cover/use classes) are less constrained in their spatial co-occurrence than close range objects (e.g. a road can go through an urban area, through agricultural fields as well as through forest or shrubland and can even run next to a river).

3.3   Estimation of Polarimetric Sample Covariance Matrices

In a second experiment, we use the projections described in Sect. 2, i.e. the RF is applied to polarimetric sample covariance matrices instead of scattering vectors. While in contrast to scattering vectors, covariance matrices can be locally averaged, we exclude node tests that perform local averaging in order to be better comparable to the experiments on scattering vectors.

As covariance matrices are computed by locally averaging the outer product of scattering vectors, they implicitly exploit context. In particular distributed targets can be statistically described only by their second moments. Another effect is that large local windows increases the quality of the estimate considerably. However, too large local windows will soon go beyond object borders and include pixels that belong to a different physical process, i.e. in the worst case to a different semantic class, reducing the inter-class variance of the samples.

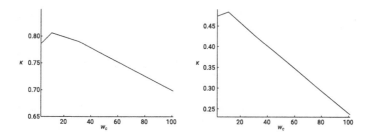

Fig. 5. Achieved κ (top) and prediction time (bottom) for *OPH* (left) and *BLN* (right) using covariance matrices computed over local windows of size w_C.

Figure 5 shows that performance barely changes for medium window sizes but degrades drastically for larger windows. A reasonable choice is $w_C = 11$, which is used in the following experiments. Note that covariance matrices are precomputed and thus do not influence computation times of the classifier.

3.4 Polarimetric Sample Covariance Matrices

In the last set of experiments, we fix the local window for computing the local polarimetric covariance matrix to $w_C = 11$ and vary region distance r_d and size r_s in the same range as for the experiments based on the scattering vector, i.e. $r_d, r_s \in \{3, 11, 31, 101\}$. The results are shown in Fig. 6. Compared to using scattering vectors directly, the achieved performance increased from $\kappa \in [0.64, 0.798]$ to $\kappa \in [0.786, 0.85]$ for *OPH* and from $\kappa \in [0.288, 0.436]$ to $\kappa \in [0.448, 0.508]$ for *BLN* which demonstrates the benefits of speckle reduction and the importance to use second-order moments. The relative performance among different settings for region size and distance, however, stays similar. Large regions perform in general better than small regions. An interesting exception can be observed for $r_s = 3$ and $r_s = 11$: While for small distances ($d \le 11$) the larger $r_s = 11$ leads to better results, the accuracy for $r_s = 3$ surpasses the one for $r_s = 11$ if $r_d = 31$. In general the results follow the trend of the experiments based on scattering vectors: First, the performance increases with increasing distance, but then declines if the region distance is too large. This is confirmed as well by the experiments with upscaled distances: While for $r_d = 3$ the results of the scaled distance is often superior to the results achieved using the original distance, the performance quickly decreases for $d > 11$.

3.5   Summary

Figure 7 shows qualitative results by using projections that allow 1) a minimal amount of context (being based on scattering vectors with $r_d = r_s = 3$ and no scaling), 2) the optimal (i.e. best κ in the experiments) amount of context (being based on covariance matrices with $r_d = r_s = 31$ and no scaling); and 3) a large amount of context (being based on covariance matrices with $r_d = 101$, $r_s = 31$ and scaling with $\alpha \in \{1, 2, 5, 10\}$). There is a significant amount of label noise if only a small amount of local context is included but even larger structures tend to be misclassified if they are locally similar to other classes. By increasing the amount of context, the obtained semantic maps become considerably smoother. Note, that these results are obtained without any post-processing. Too much context, however, degrades the results as the inter-class differences decrease leading to misclassifications in particular for smaller structures.

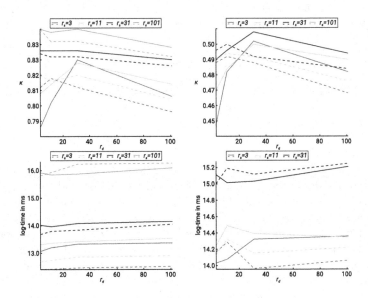

Fig. 6. Achieved κ (top) and prediction time (bottom) for *OPH* (left) and *BLN* (right) using polarimetric sample covariance matrices. The solid lines denote the single scale ($\alpha = 1$), the dashed lines the multi-scale ($\alpha \in \{1, 2, 5, 10\}$) case.

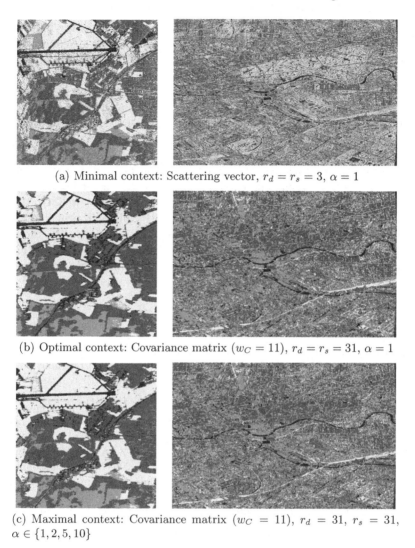

(a) Minimal context: Scattering vector, $r_d = r_s = 3$, $\alpha = 1$

(b) Optimal context: Covariance matrix ($w_C = 11$), $r_d = r_s = 31$, $\alpha = 1$

(c) Maximal context: Covariance matrix ($w_C = 11$), $r_d = 31$, $r_s = 31$, $\alpha \in \{1, 2, 5, 10\}$

Fig. 7. Obtained semantic maps (stitching of corresponding test sets) by exploiting different amounts of spatial context. Note: Images have been scaled for better visibility.

4 Conclusion and Future Work

This paper extended the set of possible spatial projections of pRFs by exploiting distance functions defined over polarimetric scattering vectors. This allows a time- and memory efficient application of pRFs directly to PolSAR images without any kind of preprocessing. However, the experimental results have shown that usually a better performance (in terms of accuracy) can be obtained by using polarimetric sample covariance matrices. We investigated the influence of the size of the spatial neighborhood over which these matrices are computed and showed that medium sized neighborhoods lead to best results where the relative performance changes were surprisingly consistent between two very different data sets. Last but not least we investigated the role context plays by varying the region size and distance of the internal node projections of pRFs. Results show that the usage of context is indeed essential to improve classification results but only to a certain extent after which performance actually drastically decreases.

Future work will confirm these findings for different sensors, i.e. HSI and optical images, as well as for different classification tasks. Furthermore, while this paper focused on visual context (i.e. on the measurement level), semantic context (i.e. on the level of the target variable) is of interest as well. On the one hand, the test selection of the internal nodes of pRFs allows in principle to take semantic context into account during the optimisation process. On the other hand, post processing steps such as MRFs, label relaxation, or stacked Random Forests should have a positive influence on the quality of the final semantic maps.

References

1. Aghababaee, H., Sahebi, M.R.: Game theoretic classification of polarimetric SAR images. Eur. J. Remote Sens. **48**(1), 33–48 (2015)
2. Aksoy, S., Koperski, K., Tusk, C., Marchisio, G., Tilton, J.C.: Learning Bayesian classifiers for scene classification with a visual grammar. IEEE Trans. Geosci. Remote Sens. **43**(3), 581–589 (2005)
3. Anfinsen, S.N., Eltoft, T.: Application of the matrix-variate Mellin transform to analysis of polarimetric radar images. IEEE Trans. Geosci. Remote Sens. **49**(6), 2281–2295 (2011)
4. Bouchemakh, L., Smara, Y., Boutarfa, S., Hamadache, Z.: A comparative study of speckle filtering in polarimetric radar SAR images. In: 2008 3rd International Conference on Information and Communication Technologies: From Theory to Applications, pp. 1–6 (2008)
5. Bovolo, F., Bruzzone, L.: A context-sensitive technique based on support vector machines for image classification. In: Pal, S.K., Bandyopadhyay, S., Biswas, S. (eds.) Pattern Recogn. Mach. Intell., pp. 260–265. Springer, Heidelberg (2005). https://doi.org/10.1007/11590316_36
6. Camps-Valls, G., Bruzzone, L.: Kernel-based methods for hyperspectral image classification. IEEE Trans. Geosci. Remote Sens. **43**(6), 1351–1362 (2005)
7. Deng, X., López-Martínez, C., Chen, J., Han, P.: Statistical modeling of polarimetric SAR data: a survey and challenges. Remote Sens. **9**(4) (2017). https://doi.org/10.3390/rs9040348
8. Farhadiani, R., Homayouni, S., Safari, A.: Impact of polarimetric SAR speckle reduction on classification of agriculture lands. ISPRS - Int. Arch. Photogrammetry Remote Sens. Spatial Inf. Sci. XLII-4/W18, 379–385 (2019). https://doi.org/10.5194/isprs-archives-XLII-4-W18-379-2019
9. Fauvel, M., Chanussot, J., Benediktsson, J.A., Sveinsson, J.R.: Spectral and spatial classification of hyperspectral data using SVMs and morphological profiles. In: 2007 IEEE International Geoscience and Remote Sensing Symposium, pp. 4834–4837 (2007)

10. Fischer, G., Papathanassiou, K.P., Hajnsek, I.: Modeling and compensation of the penetration bias in InSAR DEMs of Ice sheets at different frequencies. IEEE J. Selected Topics Appl. Earth Observ. Remote Sens. **13**, 2698–2707 (2020)

11. Fjortoft, R., Delignon, Y., Pieczynski, W., Sigelle, M., Tupin, F.: Unsupervised classification of radar images using hidden Markov chains and hidden Markov random fields. IEEE Trans. Geosci. Remote Sens. **41**(3), 675–686 (2003)

12. Hänsch, R., Hellwich, O.: Skipping the real world: classification of PolSAR images without explicit feature extraction. ISPRS J. Photogrammetry Remote Sens. **140**, 122–132 (2017). https://doi.org/10.1016/j.isprsjprs.2017.11.022

13. Hänsch, R., Hellwich, O.: A comparative evaluation of polarimetric distance measures within the random forest framework for the classification of PolSAR images. In: IGARSS 2018–2018 IEEE International Geoscience and Remote Sensing Symposium, pp. 8440–8443. IEEE, July 2018. https://doi.org/10.1109/IGARSS.2018.8518834

14. Hänsch, R., Wiesner, P., Wendler, S., Hellwich, O.: Colorful trees: visualizing random forests for analysis and interpretation. In: 2019 IEEE Winter Conference on Applications of Computer Vision (WACV), pp. 294–302. IEEE, January 2019. https://doi.org/10.1109/WACV.2019.00037

15. Hoeser, T., Kuenzer, C.: Object detection and image segmentation with deep learning on earth observation data: a review-part I: evolution and recent trends. Remote Sens. **12**, 1667 (2020)

16. Jong-Sen Lee, Grunes, M.R., Pottier, E., Ferro-Famil, L.: Unsupervised terrain classification preserving polarimetric scattering characteristics. IEEE Trans. Geosci. Remote Sens. **42**(4), 722–731 (2004)

17. Ley, A., D'Hondt, O., Valade, S., Hänsch, R., Hellwich, O.: Exploiting GAN-based SAR to optical image transcoding for improved classification via deep learning. In: EUSAR 2018; 12th European Conference on Synthetic Aperture Radar, pp. 396–401. VDE, June 2018

18. Liu, X., Jiao, L., Tang, X., Sun, Q., Zhang, D.: Polarimetric convolutional network for polSAR image classification. IEEE Trans. Geosci. Remote Sens. **57**(5), 3040–3054 (2019)

19. Mohammadimanesh, F., Salehi, B., Mahdianpari, M., Gill, E., Molinier, M.: A new fully convolutional neural network for semantic segmentation of polarimetric SAR imagery in complex land cover ecosystem. ISPRS J. Photogramm. Remote. Sens. **151**, 223–236 (2019)

20. Mullissa, A., Persello, C., Stein, A.: Polsarnet: a deep fully convolutional network for polarimetric sar image classification. IEEE J. Selected Topics Appl. Earth Observ. Remote Sens. **12**(12), 5300–5309 (2019)

21. Paradiso, M.A., Blau, S., Huang, X., MacEvoy, S.P., Rossi, A.F., Shalev, G.: Lightness, filling-in, and the fundamental role of context in visual perception. In: Visual Perception, Progress in Brain Research, vol. 155, pp. 109–123. Elsevier (2006)

22. Pesaresi, M., Benediktsson, J.A.: A new approach for the morphological segmentation of high-resolution satellite imagery. IEEE Trans. Geosci. Remote Sens. **39**(2), 309–320 (2001)

23. Tang, H.H., et al.: A multiscale latent Dirichlet allocation model for object-oriented clustering of VHR panchromatic satellite images. IEEE Trans. Geosci. Remote Sens. **51**(3), 1680–1692 (2013)

24. Tilton, J.C., Swain, P.H.: Incorporating spatial context into statistical classification of multidimensional image data. LARS (Purdue University. Laboratory for Applications of Remote Sensing), vol. 072981 (1981)

25. Tison, C., Nicolas, J., Tupin, F., Maitre, H.: A new statistical model for Markovian classification of urban areas in high-resolution SAR images. IEEE Trans. Geosci. Remote Sens. **42**(10), 2046–2057 (2004)

26. Uhlmann, S., Kiranyaz, S.: Integrating color features in polarimetric SAR image classification. IEEE Trans. Geosci. Remote Sens. **52**(4), 2197–2216 (2014)

27. Vogel, J., Schiele, B.: Semantic modeling of natural scenes for content-based image retrieval. Int. J. Comput. Vis. **72**, 133–157 (2007). https://doi.org/10.1007/s11263-006-8614-1

28. Watanabe, T., Suzuki, H.: An experimental evaluation of classifiers using spatial context for multispectral images. Syst. Comput. Japan **19**(4), 33–47 (1988)

29. Wu, Y., Ji, K., Yu, W., Su, Y.: Region-based classification of polarimetric SAR images using wishart MRF. IEEE Geosci. Remote Sens. Lett. **5**(4), 668–672 (2008)

30. Yu, F., Koltun, V.: Multi-scale context aggregation by dilated convolutions (2015)

31. Zhu, X.X., et al.: Deep learning meets SAR (2020)

Haar Wavelet Based Block Autoregressive Flows for Trajectories

Apratim Bhattacharyya[1]([✉]), Christoph-Nikolas Straehle[2], Mario Fritz[3], and Bernt Schiele[1]

[1] Max Planck Institute for Informatics, Saarbrücken, Germany
abhattac@mpi-inf.mpg.de
[2] Bosch Center for Artificial Intelligence, Renningen, Germany
[3] CISPA Helmholtz Center for Information Security, Saarbrücken, Germany

Abstract. Prediction of trajectories such as that of pedestrians is crucial to the performance of autonomous agents. While previous works have leveraged conditional generative models like GANs and VAEs for learning the likely future trajectories, accurately modeling the dependency structure of these multimodal distributions, particularly over long time horizons remains challenging. Normalizing flow based generative models can model complex distributions admitting exact inference. These include variants with split coupling invertible transformations that are easier to parallelize compared to their autoregressive counterparts. To this end, we introduce a novel Haar wavelet based block autoregressive model leveraging split couplings, conditioned on coarse trajectories obtained from Haar wavelet based transformations at different levels of granularity. This yields an exact inference method that models trajectories at different spatio-temporal resolutions in a hierarchical manner. We illustrate the advantages of our approach for generating diverse and accurate trajectories on two real-world datasets – Stanford Drone and Intersection Drone.

1 Introduction

Anticipation is a key competence for autonomous agents such as self-driving vehicles to operate in the real world. Many such tasks involving anticipation can be cast as trajectory prediction problems, e.g. anticipation of pedestrian behaviour

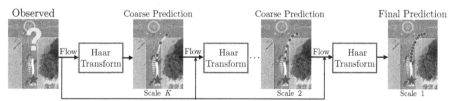

Fig. 1. Our normalizing flow based model uses a Haar wavelet based decomposition to block autoregressively model trajectories at K coarse-to-fine scales.

Electronic supplementary material The online version of this chapter (https://doi.org/10.1007/978-3-030-71278-5_20) contains supplementary material, which is available to authorized users.

Z. Akata et al. (Eds.): DAGM GCPR 2020, LNCS 12544, pp. 275–288, 2021.
https://doi.org/10.1007/978-3-030-71278-5_20

in urban driving scenarios. To capture the uncertainty of the real world, it is crucial to model the distribution of likely future trajectories. Therefore recent works [3,5,27,36] have focused on modeling the distribution of likely future trajectories using either generative adversarial networks (GANs, [15]) or variational autoencoders (VAEs, [22]). However, GANs are prone to mode collapse and the performance of VAEs depends on the tightness of the variational lower bound on the data log-likelihood which is hard to control in practice [9,20]. This makes it difficult to accurately model the distribution of likely future trajectories.

Normalizing flow based exact likelihood models [12,13,23] have been considered to overcome these limitations of GANs and VAEs in the context of image synthesis. Building on the success of these methods, recent approaches have extended the flow models for density estimation of sequential data e.g. video [25] and audio [21]. Yet, VideoFlow [25] is autoregressive in the temporal dimension which results in the prediction errors accumulating over time [26] and reduced efficiency in sampling. Furthermore, FloWaveNet [21] extends flows to audio sequences with odd-even splits along the temporal dimension, encoding only *local* dependencies [4,20,24]. We address these challenges of flow based models for trajectory generation and develop an exact inference framework to accurately model future trajectory sequences by harnessing long-term spatio temporal structure in the underlying trajectory distribution.

In this work, we propose *HBA-Flow*, an exact inference model with coarse-to-fine block autoregressive structure to encode long term spatio-temporal correlations for multimodal trajectory prediction. The advantage of the proposed framework is that multimodality can be captured over long time horizons by sampling trajectories at coarse-to-fine spatial and temporal scales (Fig. 1). Our contributions are: 1. we introduce a block autoregressive exact inference model using Haar wavelets where flows applied at a certain scale are conditioned on coarse trajectories from previous scale. The trajectories at each level are obtained after the application of Haar wavelet based transformations, thereby modeling long term spatio-temporal correlations. 2. Our HBA-Flow model, by virtue of block autoregressive structure, integrates a multi-scale block autoregressive prior which further improves modeling flexibility by encoding dependencies in the latent space. 3. Furthermore, we show that compared to fully autoregressive approaches [25], our HBA-Flow model is computationally more efficient as the number of sampling steps grows logarithmically in trajectory length. 4. We demonstrate the effectiveness of our approach for trajectory prediction on Stanford Drone and Intersection Drone, with improved accuracy over long time horizons.

2   Related Work

Pedestrian Trajectory Prediction. Work on traffic participant prediction dates back to the Social Forces model [18]. More recent works [1,18,35,38] consider the problem of traffic participant prediction in a social context, by taking into account interactions among traffic participants. Notably, Social LSTM [1] introduces a social pooling layer to aggregate interaction information of nearby traffic participants. An efficient extension of the social pooling operation is developed in [10] and alternate instance and category layers to model interactions in [28]. Weighted interactions are proposed in [7]. In contrast, a multi-agent tensor fusion scheme is proposed in [40] to capture interactions. An attention based model to effectively integrate visual cues in path prediction tasks is proposed in [37]. However, these methods mostly assume a deterministic future and do not directly deal with the challenges of uncertainty and multimodality.

Generative Modeling of Trajectories. To deal with the challenges of uncertainty and multimodality in anticipating future trajectories, recent works employ either conditional VAEs or GANs to capture the distribution of future trajectories. This includes, a conditional VAE based model with a RNN based refinement module [27], a VAE based model [14] that "personalizes" prediction to individual agent behavior, a diversity enhancing "Best of Many" loss [5] to better capture multimodality with VAEs, an expressive normalizing flow based prior for conditional VAEs [3] among others. However, VAE based models only maximize a lower bound on the data likelihood, limiting their ability to effectively model trajectory data. Other works, use GANs [16,36,40] to generate socially compliant trajectories. GANs lead to missed modes of the data distribution. Additionally, [11,34] introduce push-forward policies and motion planning for generative modeling of trajectories. Determinantal point processes are used in [39] to better capture diversity of trajectory distributions. The work of [29] shows that additionally modeling the distribution of trajectory end points can improve accuracy. However, it is unclear if the model of [29] can be used for predictions across variable time horizons. In contrast to these approaches, in this work we directly maximize the exact likelihood of the trajectories, thus better capturing the underlying true trajectory distribution.

Autoregressive Models. Autoregressive exact inference models like Pixel-CNN [31] have shown promise in generative modeling. Autoregressive models for sequential data includes a convolutional autoregressive model [30] for raw audio and an autoregressive method for video frame prediction [25]. In particular, for sequential data involving trajectories, recent works [32] propose an autoregressive method based on visual sources. The main limitation of autoregressive approaches is that the models are difficult to parallelize. Moreover, in case of sequential data, errors tend to accumulate over time [26].

Normalizing Flows. Split coupling normalizing flow models with affine transformations [12] offer computationally efficient tractable Jacobians. Recent methods [13,23] have therefore focused on split coupling flows which are easier to parallelize. Flow models are extended in [13] to multiscale architecture and the modeling capacity of flow models is further improved in [23] by introducing 1×1 convolution. Recently, flow models with more complex invertible components [8,19] have been leveraged for generative modeling of images. Recent works like FloWaveNet [21] and VideoFlow [21] adapt the multi-scale architecture of Glow [23] with sequential latent spaces to model sequential data, for raw audio and video frames respectively. However, these models still suffer from the limited modeling flexibility of the split coupling flows. The "squeeze" spatial pooling operation in [23] is replaced with a Haar wavelet based downsampling scheme in [2] along the spatial dimensions. Although this leads to improved results on image data, this operation is not particularly effective in case of sequential data as it does not influence temporal receptive fields for trajectories – crucial for modeling long-term temporal dependencies. Therefore, Haar wavelet downsampling of [2] does not lead to significant improvement in performance on sequential data (also observed empirically). In this work, instead of employing Haar wavelets as a downsampling operation for reducing spatial resolution [2] in split coupling flows, we formulate a coarse-to-fine block autoregressive model where Haar wavelets produce trajectories at different spatio-temporal resolutions.

3   Block Autoregressive Modeling of Trajectories

In this work, we propose a coarse-to-fine block autoregressive exact inference model, *HBA-Flow*, for trajectory sequences. We first provide an overview of conditional normalizing flows which form the backbone of our HBA-Flow model. To extend normalizing flows for trajectory prediction, we introduce an invertible transformation based on Haar wavelets which decomposes trajectories into K coarse-to-fine scales (Fig. 1). This is beneficial for expressing long-range spatio-temporal correlations as coarse trajectories provide global context for the subsequent finer scales. Our proposed HBA-Flow framework integrates the coarse-to-fine transformations with invertible split coupling flows where it block autoregressively models the transformed trajectories at K scales.

3.1   Conditional Normalizing Flows for Sequential Data

We base our HBA-Flow model on normalizing flows [12] which are a type of exact inference model. In particular, we consider the transformation of the conditional distribution $p(\mathbf{y}|\mathbf{x})$ of trajectories $\mathbf{y}$ to a distribution $p(\mathbf{z}|\mathbf{x})$ over $\mathbf{z}$ with conditional normalizing flows [2,3] using a sequence of n transformations $g_i : \mathbf{h}_{i-1} \mapsto \mathbf{h}_i$, with $\mathbf{h}_0 = \mathbf{y}$ and parameters θ_i,

$$\mathbf{y} \xleftrightarrow{g_1} \mathbf{h}_1 \xleftrightarrow{g_2} \mathbf{h}_2 \cdots \xleftrightarrow{g_n} \mathbf{z}. \tag{1}$$

Given the Jacobians $\mathbf{J}_{\theta_i} = \partial \mathbf{h}_i / \partial \mathbf{h}_{i-1}$ of the transformations g_i, the exact likelihoods can be computed with the change of variables formula,

$$\log p_\theta(\mathbf{y}|\mathbf{x}) = \log p(\mathbf{z}|\mathbf{x}) + \sum_{i=1}^{n} \log |\det \mathbf{J}_{\theta_i}|, \tag{2}$$

Given that the density $p(\mathbf{z}|\mathbf{x})$ is known, the likelihood over $\mathbf{y}$ can be computed exactly. Recent works [12,13,23] consider invertible split coupling transformations g_i as they provide a good balance between efficiency and modeling flexibility. In (conditional) split coupling transformations, the input $\mathbf{h}_i$ is split into two halves $\mathbf{l}_i$, $\mathbf{r}_i$, and g_i applies an invertible transformation only on $\mathbf{l}_i$ leaving $\mathbf{r}_i$ unchanged. The transformation parameters of $\mathbf{l}_i$ are dependent on $\mathbf{r}_i$

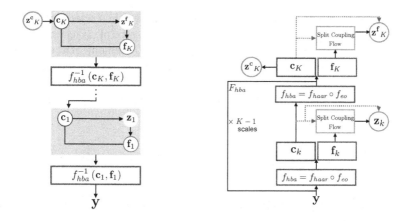

Fig. 2. Left: *HBA-Flow* generative model with the Haar wavelet [17] based representation F_{hba}. Right: Our multi-scale *HBA-Flow* model with K scales of Haar based transformation.

and $\mathbf{x}$, thus $\mathbf{h}_{i+1} = [g_{i+1}(\mathbf{l}_i|\mathbf{r}_i, \mathbf{x}), \mathbf{r}_i]$. The main advantage of (conditional) split coupling flows is that both inference and sampling are parallelizable when the transformations g_{i+1} have an efficient closed form expression of the inverse g_{i+1}^{-1}, e.g. affine [23] or non-linear squared [41] and unlike residual flows [8].

As most of the prior work, e.g. [2,12,13,23], considers split coupling flows g_i that are designed to deal with fixed length data, these models are not directly applicable to data of variable length such as trajectories. Moreover, recall that for variable length sequences, while VideoFlow [25] utilizes split coupling based flows to model the distribution at each time-step, it is still fully autoregressive in the temporal dimension, thus offering limited computational efficiency. FloWaveNets [21] split $\mathbf{l}_i$ and $\mathbf{r}_i$ along even-odd time-steps for audio synthesis. This even-odd formulation of the split operation along with the inductive bias [4,20,24] of split coupling based flow models is limited when expressing local and global dependencies which are crucial for capturing multimodality of the trajectories over long time horizons. Next, we introduce our invertible transformation based on Haar wavelets to model trajectories at various coarse-to-fine levels to address the shortcomings of prior flow based methods [21,25] for sequential data.

3.2 Haar Wavelet Based Invertible Transform

Haar wavelet transform allows for a simple and easy to compute coarse-to-fine frequency decomposed representation with a finite number of components unlike alternatives e.g. Fourier transformations [33]. In our HBA-Flow framework, we construct a transformation F_{hba} comprising of mappings f_{hba} recursively applied across K scales. With this transformation, trajectories can be encoded at different levels of granularity along the temporal dimension. We now formalize invertible function f_{hba} and its multi-scale Haar wavelet based composition F_{hba}.

Single Scale Invertible Transformation. Consider the trajectory at scale k as $\mathbf{y}_k = [\mathbf{y}_k^1, \cdots, \mathbf{y}_k^{T_k}]$, where T_k is the number of timesteps of trajectory $\mathbf{y}_k$. Here, at scale $k = 1$, $\mathbf{y}_1 = \mathbf{y}$ is the input trajectory. Each element of the trajectory is a vector, $\mathbf{y}_k^j \in \mathbb{R}^d$ encoding spatial information of the traffic participant. Our proposed invertible transformation f_{hba} at any scale k is a composition, $f_{hba} = f_{haar} \circ f_{eo}$. First, f_{eo} transforms the trajectory into even ($\mathbf{e}_k$) and odd ($\mathbf{o}_k$) downsampled trajectories,

$$f_{eo}(\mathbf{y}_k) = \mathbf{e}_k, \mathbf{o}_k \text{ where, } \mathbf{e}_k = [\mathbf{y}_k^2, \cdots, \mathbf{y}_k^{T_k}] \text{ and } \mathbf{o}_k = [\mathbf{y}_k^1, \cdots, \mathbf{y}_k^{T_k-1}]. \quad (3)$$

Next, f_{haar} takes as input the even ($\mathbf{e}_k$) and odd ($\mathbf{o}_k$) downsampled trajectories and transforms them into coarse ($\mathbf{c}_k$) and fine ($\mathbf{f}_k$) downsampled trajectories using a scalar "mixing" parameter α. In detail,

$$f_{haar}(\mathbf{e}_k, \mathbf{o}_k) = \mathbf{f}_k, \mathbf{c}_k \text{ where, } \mathbf{c}_k = (1-\alpha)\mathbf{e}_k + \alpha \mathbf{o}_k \text{ and}$$
$$\mathbf{f}_k = \mathbf{o}_k - \mathbf{c}_k = (1-\alpha)\mathbf{o}_k + (\alpha-1)\mathbf{e}_k \quad (4)$$

where, the coarse ($\mathbf{c}_k$) trajectory is the element-wise weighted average of the even ($\mathbf{e}_k$) and odd ($\mathbf{o}_k$) downsampled trajectories and the fine ($\mathbf{f}_k$) trajectory is the element-wise difference to the coarse downsampled trajectory. The coarse trajectories ($\mathbf{c}_k$) provide global context for finer scales in our block autoregressive approach, while the fine trajectories ($\mathbf{f}_k$) encode details at multiple scales. We now discuss the invertibilty of this transformation f_{hba} and compute the Jacobian.

Lemma 1. *The generalized Haar transformation $f_{hba} = f_{haar} \circ f_{eo}$ is invertible for $\alpha \in [0, 1)$ and the determinant of the Jacobian of the transformation $f_{hba} = f_{haar} \circ f_{eo}$ for sequence of length T_k with $\mathbf{y}_k^j \in \mathbb{R}^d$ is $\det \mathbf{J}_{hba} = (1-\alpha)^{(d \cdot T_k)/2}$.*

We provide the proof in the supplementary material. This property allows our HBA-Flow model to exploit f_{hba} for spatio-temporal decomposition of the trajectories $\mathbf{y}$ while remaining invertible with a tractable Jacobian for exact inference. Next, we use this transformation f_{hba} to build the coarse-to-fine multi-scale Haar wavelet based transformation F_{hba} and discuss its properties.

Multi-scale Haar Wavelet Based Transformation. To construct our generalized Haar wavelet based transformation F_{hba}, the mapping f_{hba} is applied recursively at K scales (Fig. 2, left). The transformation f_{hba} at a scale k applies a low and a high pass filter pair on the input trajectory $\mathbf{y}_k$ resulting in the coarse trajectory $\mathbf{c}_k$ and the fine trajectory $\mathbf{f}_k$ with high frequency details. The coarse (spatially and temporally sub-sampled) trajectory ($\mathbf{c}_k$) at scale k is then further decomposed by using it as the input trajectory $\mathbf{y}_{k+1} = \mathbf{c}_k$ to f_{hba} at scale $k + 1$. This is repeated at K scales, resulting in the complete Haar wavelet transformation $F_{hba}(\mathbf{y}) = [\mathbf{f}_1, \cdots, \mathbf{f}_K, \mathbf{c}_K]$ which captures details at multiple (K) spatio-temporal scales. The finest scale $\mathbf{f}_1$ models high-frequency spatio-temporal information of the trajectory $\mathbf{y}$. The subsequent scales $\mathbf{f}_k$ represent details at coarser levels, with c_K being the coarsest transformation which expresses the "high-level" spatio-temporal structure of the trajectory (Fig. 1).

Next, we show that the number of scales K in F_{hba} is upper bounded by the logarithm of the length of the sequence. This implies that F_{hba}, when integrated in the multi-scale block auto-regressive model provides a computationally efficient setup for generating trajectories.

Lemma 2. *The number of scales K of the Haar wavelet based representation F_{hba} is $K \leq \log(T_1)$, for an initial input sequence $\mathbf{y}_1$ of length T_1.*

Proof. The Haar wavelet based transformation f_{hba} halves the length of trajectory $\mathbf{y}_k$ at each level k. Thus, for an initial input sequence $\mathbf{y}_1$ of length T_1, the length of the coarsest level K in $F_{hba}(\mathbf{y})$ is $|c_K| = {}^{T_1}/_{2^K} \geq 1$. Thus, $K \leq \log(T_1)$.

3.3   Haar Block Autoregressive Framework

HBA-Flow Model. We illustrate our HBA-Flow model in Fig. 2. Our HBA-Flow model first transforms the trajectories $\mathbf{y}$ using F_{hba}, where the invertible transform f_{hba} is recursively applied on the input trajectory $\mathbf{y}$ to obtain $\mathbf{f}_k$ and $\mathbf{c}_k$ at scales $k \in \{1, \cdots, K\}$. Therefore, the log-likelihood of a trajectory $\mathbf{y}$ under our HBA-Flow model can be expressed using the change of variables formula as,

$$
\begin{aligned}
\log(p_\theta(\mathbf{y}|\mathbf{x})) &= \log(p_\theta(\mathbf{f}_1, \mathbf{c}_1|\mathbf{x})) + \log|\det(\mathbf{J}_{hba})_1| \\
&= \log(p_\theta(\mathbf{f}_1, \cdots, \mathbf{f}_K, \mathbf{c}_K|\mathbf{x})) + \sum_{i=1}^{K} \log|\det(\mathbf{J}_{hba})_i|.
\end{aligned}
\tag{5}
$$

Next, our HBA-Flow model factorizes the distribution of fine trajectories w.l.o.g. such that $\mathbf{f}_k$ at level k is conditionally dependent on the representations at scales $k+1$ to K,

$$
\begin{aligned}
\log(p_\theta(\mathbf{f}_1, \cdots, \mathbf{f}_K, \mathbf{c}_K|\mathbf{x})) &= \log(p_\theta(\mathbf{f}_1|\mathbf{f}_2, \cdots, \mathbf{f}_K, \mathbf{c}_K, \mathbf{x})) + \cdots \\
&\quad + \log(p_\theta(\mathbf{f}_K|\mathbf{c}_K, \mathbf{x})) + \log(p_\theta(\mathbf{c}_K|\mathbf{x})).
\end{aligned}
\tag{6}
$$

Finally, note that $[\mathbf{f}_{k+1}, \cdots, \mathbf{f}_K, \mathbf{c}_K]$ is the output of the (bijective) transformation $F_{hba}(\mathbf{c}_k)$ where f_{hba} is recursively applied to $\mathbf{c}_k = \mathbf{y}_{k+1}$ at scales $\{k+1, \cdots, K\}$. Thus HBA-Flow equivalently models $p_\theta(\mathbf{f}_k|\mathbf{f}_{k+1}, \cdots, \mathbf{c}_K, \mathbf{x})$ as $p_\theta(\mathbf{f}_k|\mathbf{c}_k, \mathbf{x})$,

$$
\begin{aligned}
\log(p_\theta(\mathbf{y}|\mathbf{x})) &= \log(p_\theta(\mathbf{f}_1|\mathbf{c}_1, \mathbf{x})) + \cdots + \log(p_\theta(\mathbf{f}_K|\mathbf{c}_K, \mathbf{x})) \\
&\quad + \log(p_\theta(\mathbf{c}_K|\mathbf{x})) + \sum_{i=1}^{K} \log|\det(\mathbf{J}_{hba})_i|.
\end{aligned}
\tag{7}
$$

Therefore, as illustrated in Fig. 2 (right), our HBA-Flow models the distribution of each of the fine components $\mathbf{f}_k$ block autoregressively conditioned on the coarse representation $\mathbf{c}_k$ at that level. The distribution $p_\theta(\mathbf{f}_k|\mathbf{c}_k, \mathbf{x})$ at each scale k is modeled using invertible conditional split coupling flows (Fig. 2, right)

[21], which transform the input distribution to the distribution over latent "priors" z_k. This enables our framework to model variable length trajectories. The log-likelihood with our HBA-Flow approach can be expressed using the change of variables formula as,

$$\log(p_\theta(\mathbf{f}_k|\mathbf{c}_k, \mathbf{x})) = \log(p_\phi(\mathbf{z}_k|\mathbf{c}_k, \mathbf{x})) + \log|\det(\mathbf{J}_{sc})_k| \tag{8}$$

where, $\log|\det(\mathbf{J}_{sc})_k|$ is the log determinant of Jacobian $(\mathbf{J}_{sc})_k$ of the split coupling flow at level k. Thus, the likelihood of a trajectory $\mathbf{y}$ under our HBA-Flow model can be expressed exactly using Eqs. (7) and (8).

The key advantage of our approach is that after spatial and temporal downsampling of coarse scales, it is easier to model long-term spatio-temporal dependencies. Moreover, conditioning the flows at each scale on the coarse trajectory provides global context as the downsampled coarse trajectory effectively increases the spatio-temporal receptive field. This enables our HBA-Flows better capture multimodality in the distribution of likely future trajectories.

HBA-Prior. Complex multimodel priors can considerably increase the modeling flexibility of generative models [3,21,25]. The block autoregressive structure of our HBA-Flow model allows us introduce a Haar block autoregressive prior (HBA-Prior) over $\mathbf{z} = [\mathbf{z}_1, \cdots, \mathbf{z}^{\mathbf{f}}_K, \mathbf{z}^{\mathbf{c}}_K]$ in Eq. (8), where $\mathbf{z}_k$ is the latent representation for scales $k \in \{1, \cdots, K-1\}$ and $\mathbf{z}^{\mathbf{f}}_K, \mathbf{z}^{\mathbf{c}}_K$ are the latents for the coarse and fine representations scales K. The log-likelihood of the prior factorizes as,

$$\begin{aligned}\log(p_\phi(\mathbf{z}|\mathbf{x})) = &\log(p_\phi(\mathbf{z}_1|\mathbf{z}_2, \cdots, \mathbf{z}^{\mathbf{f}}_K, \mathbf{z}^{\mathbf{c}}_K, \mathbf{x})) + \cdots \\ &+ \log(p_\phi(\mathbf{z}^{\mathbf{f}}_K|\mathbf{z}^{\mathbf{c}}_K, \mathbf{x})) + \log(p_\phi(\mathbf{z}^{\mathbf{c}}_K|\mathbf{x})).\end{aligned} \tag{9}$$

Each coarse level representation $\mathbf{c}_k$ is the output of a bijective transformation of the latent variables $[\mathbf{z}_{k+1}, \cdots, \mathbf{z}^{\mathbf{f}}_K \mathbf{z}^{\mathbf{c}}_K]$ through the invertible split coupling flows and the transformations f_{hba} at scales $\{k+1, \cdots, K\}$. Thus, HBA-Prior models $p_\phi(\mathbf{z}_k|\mathbf{z}_{k+1}, \cdots, \mathbf{z}^{\mathbf{f}}_K, \mathbf{z}^{\mathbf{c}}_K, \mathbf{x})$ as $p_\phi(\mathbf{z}_k|\mathbf{c}_k, \mathbf{x})$ at every scale (Fig. 2, left). The log-likelihood of the prior can also be expressed as,

$$\begin{aligned}\log(p_\phi(\mathbf{z}|\mathbf{x})) = &\log(p_\phi(\mathbf{z}_1|\mathbf{c}_1, \mathbf{x})) + \cdots + \log(p_\phi(\mathbf{z}_{K-1}|\mathbf{c}_{K-1}, \mathbf{x})) \\ &+ \log(p_\phi(\mathbf{z}^{\mathbf{f}}_K|\mathbf{c}_K, \mathbf{x})) + \log(p_\phi(\mathbf{z}^{\mathbf{c}}_K|\mathbf{x})).\end{aligned} \tag{10}$$

We model $p_\phi(\mathbf{z}_k|\mathbf{c}_k, \mathbf{x})$ as conditional normal distributions which are multimodal as a result of the block autoregressive structure. In comparison to the fully autoregressive prior in [25], our HBA-Prior is efficient as it requires only $\mathcal{O}(\log(T_1))$ sampling steps.

Analysis of Sampling Time. From Eq. (6) and Fig. 2 (left), our HBA-Flow model autoregressively factorizes across the fine components $\mathbf{f}_k$ at K scales. From Lemma 2, $K \leq \log(T_1)$. At each scale our HBA-Flow samples the fine components $\mathbf{f}_k$ using split coupling flows, which are easy to parallelize. Thus, given enough parallel resources, our HBA-Flow model requires maximum $K \leq \log(T_1)$ i.e. $\mathcal{O}(\log(T_1))$ sampling steps and is significantly more efficient compared to fully autoregressive approaches e.g. VideoFlow [25], which require $\mathcal{O}(T_1)$ steps.

Table 1. Five fold cross validation on the Stanford Drone dataset. Lower is better for all metrics. Visual refers to additional conditioning on the last observed frame. Top: state of the art, Middle: Baselines and ablations, Bottom: Our HBA-Flow.

Method	Visual	Er @ 1sec	Er @ 2sec	Er @ 3sec	Er @ 4sec	-CLL	Speed
"Shotgun" [32]	–	0.7	1.7	3.0	4.5	91.6	–
DESIRE-SI-IT4 [27]	✓	1.2	2.3	3.4	5.3	–	–
STCNN [32]	✓	1.2	2.1	3.3	4.6	–	–
BMS-CVAE [5]	✓	0.8	1.7	3.1	4.6	126.6	58
CF-VAE [3]	–	0.7	1.5	2.5	3.6	84.6	47
CF-VAE [3]	✓	0.7	1.5	2.4	3.5	84.1	88
Auto-regressive [25]	–	0.7	1.5	2.6	3.7	86.8	134
FloWaveNet [21]	–	0.7	1.5	2.5	3.6	84.5	**38**
FloWaveNet [21] + HWD [2]	–	0.7	1.5	2.5	3.6	84.4	**38**
FloWaveNet [21]	✓	0.7	1.5	2.4	3.5	84.1	77
HBA-Flow (Ours)	–	**0.7**	1.5	2.4	3.4	84.1	41
HBA-Flow + Prior (Ours)	–	**0.7**	**1.4**	**2.3**	3.3	83.4	43
HBA-Flow + Prior (Ours)	✓	**0.7**	**1.4**	**2.3**	**3.2**	**83.1**	81

4 Experiments

We evaluate our approach for trajectory prediction on two challenging real world datasets – Stanford Drone [35] and Intersection Drone [6]. These datasets contain trajectories of traffic participants including pedestrians, bicycles, cars recorded from an aerial platform. The distribution of likely future trajectories is highly multimodal due to the complexity of the traffic scenarios e.g. at intersections.

Evaluation Metrics. We are primarily interested in measuring the match of the learned distribution to the true distribution. Therefore, we follow [3,5,27,32] and use Euclidean error of the top 10% of samples (predictions) and the (negative) conditional log-likelihood (-CLL) metrics. The Euclidean error of the top 10% of samples measures the coverage of all modes of the target distribution and is relatively robust to random guessing as shown in [3].

Architecture Details. We provide additional architecture details in the supplemental material.

4.1 Stanford Drone

We use the standard five-fold cross validation evaluation protocol [3,5,27,32] and predict the trajectory up to 4 s into the future. We use the Euclidean error of the top 10% of predicted trajectories at the standard (1/5) resolution using 50 samples and the CLL metric in Table 1. We additionally report sampling time for a batch of 128 samples in milliseconds.

We compare our HBA-Flow model to the following state-of-the-art models: The handcrafted "Shotgun" model [32], the conditional VAE based models of [3,5,27] and the autoregressive STCNN model [32]. We additionally include the

	Mean Top 10%	FloWaveNet [21]	*HBA-Flows* (Ours)
Observed	B - GT, Y -[21], R - Ours	Predictions	Predictions

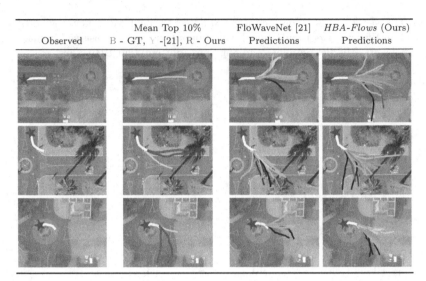

Fig. 3. Mean top 10% predictions (Blue - Groudtruth, Yellow - FloWaveNet [21], Red - Our *HBA-Flow* model) and predictive distributions on Intersection Drone dataset. The predictions of our HBA-Flow model are more diverse and better capture the multimodality the future trajectory distribution. (Color figure online)

various exact inference baselines for modeling trajectory sequences: the autoregressive flow model of VideoFlow [25], FloWaveNet [21] (without our Haar wavelet based block autoregressive structure), FloWaveNet [21] with the Haar wavelet downsampling of [2] (FloWaveNet + HWD), our HBA-Flow model with a Gaussian prior (without our HBA-Prior). The FloWaveNet [21] baselines serves as ideal ablations to measure the effectiveness of our block autoregressive HBA-Flow model. For fair comparison, we use two scales (levels) $K = 2$ with eight non-linear squared split coupling flows [41] each, for both our HBA-Flow and FloWaveNet [21] models. Following [3,32] we additionally experiment with conditioning on the last observed frame using a attention based CNN (indicated by "Visual" in Table 1).

We observe from Table 1 that our HBA-Flow model outperforms both state-of-the-art models and baselines. In particular, our HBA-Flow model outperforms the conditional VAE based models of [3,5,27] in terms of Euclidean distance and -CLL. Further, our HBA-Flow exhibits competitive sampling speeds. This shows the advantage of exact inference in the context of gen-

Method	mADE ↓	mFDE ↓
SocialGAN [16]	27.2	41.4
MATF GAN [40]	22.5	33.5
SoPhie [36]	16.2	29.3
Goal Prediction [11]	15.7	28.1
CF-VAE [3]	12.6	22.3
HBA-Flow + Prior (Ours)	**10.8**	**19.8**

Table 2. Evaluation on the Stanford Drone using the split of [11,36,40].

erative modeling of trajectories – leading to better match to the groundtruth distribution. Our HBA-Flow model generates accurate trajectories compared to the VideoFlow [25] baseline. This is because unlike VideoFlow, errors do not accumulate in the temporal dimension of HBA-Flow. Our HBA-Flow model outperforms the FloWaveNet model of [21] with comparable sampling speeds demonstrating the effectiveness of the coarse-to-fine block autoregressive structure of our HBA-Flow model in capturing long-range spatio-temporal dependencies. This is reflected in the predictive distributions and the top 10% of predictions of our HBA-Flow model in comparison with FloWaveNet [21] in Fig. 3. The predictions of our HBA-Flow model are more diverse and can more effectively capture the multimodality of the trajectory distributions especially at complex traffic situations e.g. intersections and crossings. We provide additional examples in the supplemental material. We also observe in Table 1 that the addition of Haar wavelet downsampling [2] to FloWaveNets [21] (FloWaveNet + HWD) does not significantly improve performance. This illustrates that Haar wavelet downsampling as used in [2] is not effective in case of sequential trajectory data as it is primarily a spatial pooling operation for image data. Finally, our ablations with Gaussian priors (HBA-Flow) additionally demonstrate the effectiveness of our HBA-Prior (HBA-Flow + Prior) with improvements with respect to accuracy. We further include a comparison using the evaluation protocol of [11,35–37] in Table 2. Here, only a single train/test split is used. We follow [3,11] and use the minimum average displacement error (mADE) and minimum final displacement error (mFDE) as evaluation metrics. Similar to [3,11] the minimum is calculated over 20 samples. Our HBA-Flow model outperforms the state-of-the-art demonstrating the effectiveness of our approach.

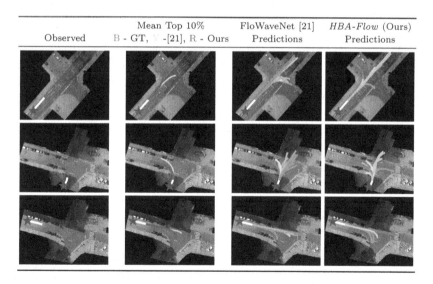

Fig. 4. Mean top 10% predictions (Blue - Groudtruth, Yellow - FloWaveNet [21], Red - Our *HBA-Flow* model) and predictive distributions on Intersection Drone dataset. The predictions of our HBA-Flow model are more diverse and better capture the modes of the future trajectory distribution. (Color figure online)

Table 3. Five fold cross validation on the Intersection Drone dataset.

Method	Er @ 1sec	Er @ 2sec	Er @ 3sec	Er @ 4sec	Er @ 5sec	-CLL
BMS-CVAE [5]	0.25	0.67	1.14	1.78	2.63	26.7
CF-VAE [3]	0.24	0.55	0.93	1.45	2.21	21.2
FloWaveNet [21]	0.23	0.50	0.85	1.31	1.99	19.8
FloWaveNet [21] + HWD [2]	0.23	0.50	0.84	1.29	1.96	19.5
HBA-Flow + Prior (Ours)	**0.19**	**0.44**	**0.82**	**1.21**	**1.74**	**17.3**

4.2   Intersection Drone

We further include experiments on the Intersection Drone dataset [6]. The dataset consists of trajectories of traffic participants recorded at German intersections. In comparison to the Stanford Drone dataset, the trajectories in this dataset are typically longer. Moreover, unlike the Stanford Drone dataset which is recorded at a University Campus, this dataset covers more "typical" traffic situations. Here, we follow the same evaluation protocol as in Stanford Drone dataset and perform a five-fold cross validation and evaluate up to 5 s into the future.

We report the results in Table 3. We use the strongest baselines from Table 1 for comparison to our HBA-Flow + Prior model (with our HBA-Prior), with three scales, each having eight non-linear squared split coupling flows [41]. For fair comparison, we compare with a FloWaveNet [21] model with three levels and eight non-linear squared split coupling flows per level. We again observe that our HBA-Flow leads to much better improvement with respect to accuracy over the FloWaveNet [21] model. Furthermore, the performance gap between HBA-Flow and FloWaveNet increases with longer time horizons. This shows that our approach can better encode spatio-temporal correlations. The qualitative examples in Fig. 4 from both models show that our HBA-Flow model generates diverse trajectories and can better capture the modes of the future trajectory distribution, thus demonstrating the advantage of the block autoregressive structure of our HBA-Flow model. We also see that our HBA-Flow model outperforms the CF-VAE model [3], again illustrating the advantage of exact inference.

5   Conclusion

In this work, we presented a novel block autoregressive *HBA-Flow* framework taking advantage of the representational power of autoregressive models and the efficiency of invertible split coupling flow models. Our approach can better represent the multimodal trajectory distributions capturing the long range spatio-temporal correlations. Moreover, the block autoregressive structure of our approach provides for efficient $\mathcal{O}(\log(T))$ inference and sampling. We believe that accurate and computationally efficient invertible models that allow exact likelihood computations and efficient sampling present a promising direction of research of anticipation problems in autonomous systems.

References

1. Alahi, A., Goel, K., Ramanathan, V., Robicquet, A., Fei-Fei, L., Savarese, S.: Social LSTM: human trajectory prediction in crowded spaces. In: CVPR (2016)
2. Ardizzone, L., Lüth, C., Kruse, J., Rother, C., Köthe, U.: Guided image generation with conditional invertible neural networks. arXiv preprint arXiv:1907.02392 (2019)
3. Bhattacharyya, A., Hanselmann, M., Fritz, M., Schiele, B., Straehle, C.N.: Conditional flow variational autoencoders for structured sequence prediction. In: BDL@NeurIPS (2019)
4. Bhattacharyya, A., Mahajan, S., Fritz, M., Schiele, B., Roth, S.: Normalizing flows with multi-scale autoregressive priors. In: CVPR (2020)
5. Bhattacharyya, A., Schiele, B., Fritz, M.: Accurate and diverse sampling of sequences based on a "best of many" sample objective. In: CVPR (2018)
6. Bock, J., Krajewski, R., Moers, T., Vater, L., Runde, S., Eckstein, L.: The ind dataset: a drone dataset of naturalistic vehicle trajectories at german intersections. arXiv preprint arXiv:1911.07602 (2019)
7. Chandra, R., Bhattacharya, U., Bera, A., Manocha, D.: Traphic: trajectory prediction in dense and heterogeneous traffic using weighted interactions. In: CVPR (2019)
8. Chen, T.Q., Behrmann, J., Duvenaud, D.K., Jacobsen, J.H.: Residual flows for invertible generative modeling. In: NeurIPS (2019)
9. Cremer, C., Li, X., Duvenaud, D.: Inference suboptimality in variational autoencoders. In: ICML (2018)
10. Deo, N., Trivedi, M.M.: Convolutional social pooling for vehicle trajectory prediction. In: CVPR Workshop (2018)
11. Deo, N., Trivedi, M.M.: Scene induced multi-modal trajectory forecasting via planning. In: ICRA Workshop (2019)
12. Dinh, L., Krueger, D., Bengio, Y.: Nice: non-linear independent components estimation. In: ICLR (2015)
13. Dinh, L., Sohl-Dickstein, J., Bengio, S.: Density estimation using real NVP. In: ICLR (2017)
14. Felsen, P., Lucey, P., Ganguly, S.: Where will they go? Predicting fine-grained adversarial multi-agent motion using conditional variational autoencoders. In: ECCV (2018)
15. Goodfellow, I.J., et al.: Generative adversarial nets. In: NIPS (2014)
16. Gupta, A., Johnson, J., Fei-Fei, L., Savarese, S., Alahi, A.: Social gan: socially acceptable trajectories with generative adversarial networks. In: CVPR (2018)
17. Haar, A.: Zur theorie der orthogonalen funktionensysteme. Math. Ann. **69**(3), 331–371 (1910)
18. Helbing, D., Molnar, P.: Social force model for pedestrian dynamics. Phys. Rev. E **51**, 4282 (1995)
19. Ho, J., Chen, X., Srinivas, A., Duan, Y., Abbeel, P.: Flow++: improving flow-based generative models with variational dequantization and architecture design. In: ICML (2019)
20. Huang, C.W., Dinh, L., Courville, A.: Augmented normalizing flows: bridging the gap between generative flows and latent variable models. arXiv preprint arXiv:2002.07101 (2020)
21. Kim, S., Lee, S.G., Song, J., Kim, J., Yoon, S.: Flowavenet: a generative flow for raw audio. In: ICML (2019)

22. Kingma, D.P., Welling, M.: Auto-encoding variational bayes. In: ICLR (2014)
23. Kingma, D.P., Dhariwal, P.: Glow: generative flow with invertible 1x1 convolutions. In: NeurIPS (2018)
24. Kirichenko, P., Izmailov, P., Wilson, A.G.: Why normalizing flows fail to detect out-of-distribution data. arXiv preprint arXiv:2006.08545 (2020)
25. Kumar, M., et al.: Videoflow: a flow-based generative model for video. In: ICLR (2020)
26. Lee, A.X., Zhang, R., Ebert, F., Abbeel, P., Finn, C., Levine, S.: Stochastic adversarial video prediction. arXiv preprint arXiv:1804.01523 (2018)
27. Lee, N., Choi, W., Vernaza, P., Choy, C.B., Torr, P.H., Chandraker, M.: Desire: distant future prediction in dynamic scenes with interacting agents. In: CVPR (2017)
28. Ma, Y., Zhu, X., Zhang, S., Yang, R., Wang, W., Manocha, D.: Trafficpredict: trajectory prediction for heterogeneous traffic-agents. In: AAAI (2019)
29. Mangalam, K., Girase, H., Agarwal, S., Lee, K.-H., Adeli, E., Malik, J., Gaidon, A.: It Is not the journey but the destination: endpoint conditioned trajectory prediction. In: Vedaldi, A., Bischof, H., Brox, T., Frahm, J.-M. (eds.) ECCV 2020. LNCS, vol. 12347, pp. 759–776. Springer, Cham (2020). https://doi.org/10.1007/978-3-030-58536-5_45
30. van den Oord, A., et al.: Wavenet: a generative model for raw audio. In: ISCA Speech Synthesis Workshop (2016)
31. van den Oord, A., Kalchbrenner, N., Espeholt, L., Kavukcuoglu, K., Vinyals, O., Graves, A.: Conditional image generation with PixelCNN decoders. In: NIPS (2016)
32. Pajouheshgar, E., Lampert, C.H.: Back to square one: probabilistic trajectory forecasting without bells and whistles. In: NeurIPs Workshop (2018)
33. Porwik, P., Lisowska, A.: The haar-wavelet transform in digital image processing: its status and achievements. Machine graphics and vision **13**(1/2), 79–98 (2004)
34. Rhinehart, N., Kitani, K.M., Vernaza, P.: R2p2: a reparameterized pushforward policy for diverse, precise generative path forecasting. In: ECCV (2018)
35. Robicquet, A., Sadeghian, A., Alahi, A., Savarese, S.: Learning social etiquette: human trajectory understanding in crowded scenes. In: Leibe, B., Matas, J., Sebe, N., Welling, M. (eds.) ECCV 2016. LNCS, vol. 9912, pp. 549–565. Springer, Cham (2016). https://doi.org/10.1007/978-3-319-46484-8_33
36. Sadeghian, A., Kosaraju, V., Sadeghian, A., Hirose, N., Rezatofighi, S.H., Savarese, S.: Sophie: an attentive gan for predicting paths compliant to social and physical constraints. In: CVPR (2019)
37. Sadeghian, A., Legros, F., Voisin, M., Vesel, R., Alahi, A., Savarese, S.: Car-net: clairvoyant attentive recurrent network. In: ECCV (2018)
38. Yamaguchi, K., Berg, A.C., Ortiz, L.E., Berg, T.L.: Who are you with and where are you going? In: CVPR (2011)
39. Yuan, Y., Kitani, K.: Diverse trajectory forecasting with determinantal point processes. In: ICLR (2020)
40. Zhao, T., et al.: Multi-agent tensor fusion for contextual trajectory prediction. In: CVPR (2019)
41. Ziegler, Z.M., Rush, A.M.: Latent normalizing flows for discrete sequences. In: ICML (2019)

Center3D: Center-Based Monocular 3D Object Detection with Joint Depth Understanding

Yunlei Tang[1(✉)], Sebastian Dorn[2], and Chiragkumar Savani[2]

[1] Technical University of Darmstadt, Darmstadt, Germany
harryyunlei@gmail.com
[2] Ingolstadt, Germany

Abstract. Localizing objects in 3D space and understanding their associated 3D properties is challenging given only monocular RGB images. The situation is compounded by the loss of depth information during perspective projection. We present Center3D, a one-stage anchor-free approach and an extension of CenterNet, to efficiently estimate 3D location and depth using only monocular RGB images. By exploiting the difference between 2D and 3D centers, we are able to estimate depth consistently. Center3D uses a combination of classification and regression to understand the hidden depth information more robustly than each method alone. Our method employs two joint approaches: (1) **LID**: a classification-dominated approach with sequential **L**inear **I**ncreasing **D**iscretization. (2) **DepJoint**: a regression-dominated approach with multiple Eigen's transformations [6] for depth estimation. Evaluating on KITTI dataset [8] for moderate objects, Center3D improved the AP in BEV from 29.7% to **43.5%**, and the AP in 3D from 18.6% to **40.5%**. Compared with state-of-the-art detectors, Center3D has achieved a better speed-accuracy trade-off in realtime monocular object detection.

1 Introduction and Related Work

3D object detection is currently one of the most challenging topics for both industry and academia. Applications of related developments can easily be found in the areas of robotics, autonomous driving [4,18,21] etc. The goal is to have agents with the ability to identify, localize, react, and interact with objects in their surroundings. 2D object detection approaches [11,17,26] achieved impressive results in the last decade. In contrast, inferring associated 3D properties from a 2D image turned out to be a challenging problem in computer vision, due to the intrinsic scale ambiguity of 2D objects and the lack of depth information. Hence many approaches involve additional sensors like LiDAR [20,23] or radar [22] to measure depth. However, there are reasons to prefer monocular-based approaches too. LiDAR has reduced range in adverse weather conditions, while visual information of a simple RGB camera is more dense and also more

Electronic supplementary material The online version of this chapter (https://doi.org/10.1007/978-3-030-71278-5_21) contains supplementary material, which is available to authorized users.

robust under rain, snow, etc. Another reason is that cameras are currently significantly more economical than high precision LiDARs and are already available in *e.g.* robots, vehicles, etc. Additionally, the processing of single RGB images is much more efficient and faster than processing 3D point clouds in terms of CPU and memory utilization.

These compelling reasons have led to research exploring the possibility of 3D detection solely from monocular images [1,2,9,12,14,16,19]. The network structure of most 3D detectors starts with a 2D region proposal based (also called anchor based) approach, which enumerates an exhaustive set of predefined proposals over the image plane and classifies/regresses only within the region of interest (ROI). MonoGRNet [16] consists of parameter-specific subnetworks. All further regressions are guided by the detected 2D bounding box. M3D-RPN [1] demonstrates a single-shot model with a standalone 3D RPN, which generates 2D and 3D proposals simultaneously. Additionally, the specific design of depth-aware convolutional layers improved the network's 3D understanding. With the help of an external network, Multi-Fusion [24] estimates a disparity map and subsequently a LiDAR point cloud to improve 3D detection. Due to multi-stage or anchor-based pipelines, most of them perform slowly.

Most recently, to overcome the disadvantages above, 2D anchor-free approaches have been used by researchers [2,5,11,26]. They model objects with keypoints like centers, corners or points of interest of 2D bounding boxes. Anchor-free approaches are usually one-stage, thus eliminating the complexity of designing a set of anchor boxes and fine tuning hyperparameters. Our paper is also an extension of one of these works CenterNet: Objects as Points [26], which proposed a possibility to associate a 2D anchor free approach with a 3D detection.

Nevertheless, the performance of CenterNet is still restricted by the fact that a 2D bounding box and a 3D cuboid are sharing the same center point. In this paper we show the difference between the center points of 2D bounding boxes and the projected 3D center points of objects, which are almost never at the same image position. Comparing CenterNet, our main contributions are as follows: 1. We additionally regress the 3D centers from 2D centers to locate the objects in the image plane and in 3D space more properly. 2. By examining depth estimation in monocular images, we show that a combination of classification and regression explores visual clues better than using only a single approach. An overview of our approach is shown in Fig. 1.

We introduce two approaches to validate the second conclusion: (1) Motivated by DORN [7] we consider depth estimation as a sequential classification with residual regression. According to the statistics of the instances in the KITTI dataset, a novel discretization strategy is used. (2) We divide the complete depth range of objects into two bins, foreground and background, either with overlap or associated. Classifiers indicate which depth bin or bins the object belongs to. With the help of Eigen's transformation [6], two regressors are trained to gather specific features for closer and farther away objects, respectively. For illustration see the depth part in Fig. 1.

Compared to CenterNet, our approach improved the AP of easy, moderate, hard objects in BEV from 31.5, 29.7, 28.1 to **56.7**, **43.5**, **41.2**, in 3D space from 19.5, 18.6, 16.6 to **52.5**, **40.5**, **34.9**, which is comparable with state-of-the-art

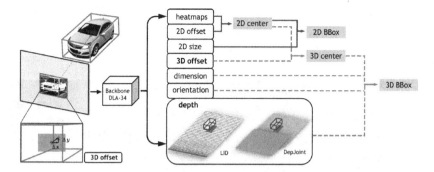

Fig. 1. Overview of Center3D. A monocular input image is fed to the backbone *DLA-34*, which generates feature maps. *Heatmaps* and *2D offset* are subsequently used to detect the *2D center* [26]. The latter is relocated by *3D offset* to propose the *3D center*, which is illustrated in the bottom-left of the figure. By applying a combination of regression and classification, *DepJoint* or *LID*, Center3D is inferring the *depth* of the associated *3D center*. *Depth*, together with regressed *dimensions*, *orientation*, and *3D center* are finally used to propose the *3D BBox*. Our contributions comparing CenterNet are indicated in bold or with dashed lines.

approaches. Center3D achieves a better speed-accuracy trade-off on the KITTI dataset in the field of monocular 3D object detection. Details are given in Table 1 and discussed in Sect. 3.

2   Center3D

2.1   CenterNet Baseline

The 3D detection approach of CenterNet described in [26] is the basis of our work. It models an object as a single point: the center of its 2D bounding box. For each input monocular RGB image, the original network produces a heatmap for each category, which is trained with focal loss [13]. The heatmap describes a confidence score for each location, the peaks in this heatmap thus represent the possible keypoints of objects. All other properties are then regressed and captured directly at the center locations on the feature maps respectively. For generating a complete 2D bounding box, in addition to width and height, a local offset will be regressed to capture the quantization error of the center point caused by the output stride. For 3D detection and localization, the additional abstract parameters, i.e. depth, 3D dimensions and orientation, will be estimated separately by adding a head for each of them. Following the output transformation of Eigen et al. [6] for depth estimation, CenterNet converts the feature output into an exponential area to suppress the depth space.

2.2   Regressing 3D Center Points

The 2D performance of CenterNet is very good, while the APs in 3D perform poorly, as the first row shown in Table 1. This is caused by the difference between

Fig. 2. 3D bounding box estimation on KITTI validation set. *first row:* the output of CenterNet. Projected 3D bounding boxes located around estimated 2D centers. The position of centers is generated by the peak of the Gaussian kernel on the heatmap. *second row:* the ground truth of input images. Here the 2D (red) and 3D (green) projected bounding boxes with their center points are shown. *third row:* the output of Center3D. The 3D cuboid is based on 3D center points shifted from 2D space with offset. More qualitative results can be found in the supplementary material. (Color figure online)

the center point of the visible 2D bounding box in the image and the projected center point of the complete object from physical 3D space. This is illustrated in the first two rows of Fig. 2. A center point of the 2D bounding box for training and inference is enough for detecting and decoding 2D properties, *e.g.* width and height, while all additionally regressed 3D properties, *e.g.* depth, dimension and orientation, should be consistently decoded from the projected 3D center of the object. The gap between 2D and 3D center points decreases for faraway objects and for objects which appear in the center area of the image plane. However the gap becomes significant for objects that are close to the camera or on the image boundary. Due to perspective projection, this offset will increase as vehicles get closer. Close objects are especially important for technical functions based on perception (e.g. in autonomous driving or robotics).

Hence we split the 2D and 3D tasks into separate parts, as shown in Fig. 1. Assuming that the centers of 2D bounding boxes is $\mathbf{c}_{2D}^{i} = (x_{2D}^{i}, y_{2D}^{i})$, and the 3D projected center points of cuboids from physical space is $\mathbf{c}_{3D}^{i} = (x_{3D}^{i}, y_{3D}^{i})$. We still locate an object with $\mathbf{c}_{2D}^{i}$, which is definitively included in the image, and determine the 2D bounding box of the visible part with w^{i} and h^{i}. For the 3D task we relocate $\mathbf{c}_{3D}^{i}$ by adding two head layers on top of the backbone and regress the offset $\mathbf{\Delta c}^{i} = (x_{3D}^{i} - x_{2D}^{i}, y_{3D}^{i} - y_{2D}^{i})$ from 2D to 3D centers. Given the projection matrix $\mathbf{P}$ in KITTI, we now determine the 3D location $\mathbf{C} = (X, Y, Z)$ by converting the transformation in homogeneous coordinates.

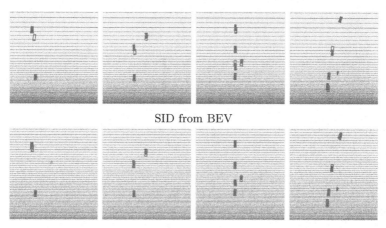

SID from BEV

LID from BEV

Fig. 3. The comparison of the discretization strategies LID (*first row*) and SID (*second row*) from BEV between 0 m and 54 m, with a setting of $d_{min} = 1$ m, $d_{max} = 91$ m and $N = 80$. The solid red lines indicate the threshold of each bin, the solid rectangles represent the ground truth vehicles in BEV, while blue rectangles represent the estimations.

2.3 Enriching Depth Information

This section introduces two novel approaches to infer depth cues over monocular images: First, we adapt the advanced DORN [7] approach from pixel-wise to instance-wise depth estimation. We introduce a novel linear-increasing discretization (**LID**) strategy to divide continuous depth values into discrete ones, which distributes the bin sizes more evenly than spacing-increasing discretization (SID) in DORN. Additionally, we employ a residual regression for refinement of both discretization strategies. Second, with the help of a reference area (**RA**) we describe the depth estimation as a joint task of classification and regression (**DepJoint**) in exponential range.

LID Usually a faraway object with higher depth value and less visible features will induce a higher loss, which could dominate the training and increases uncertainty. On the other hand these targets are usually less important for functions based on object detection. To this end, DORN solves the ordinal regression problem by quantizing depth into discrete bins with SID strategy. It discretizes the given continuous depth interval $[d_{min}, d_{max}]$ in *log* space and hence down-weight the training loss in faraway regions with higher depth values, see Eq. 1. However, such a discretization often yields too dense bins within unnecessarily close range, where objects barely appear (as shown in Fig. 3 first row). According to the histogram in Fig. 4 most instances of the KITTI dataset are between 5 m and 80 m. Assuming that we discretize the range between $d_{min} = 1$ m and $d_{max} = 91$ m into $N = 80$ sub-intervals, 29 bins will be involved within just 5 m.

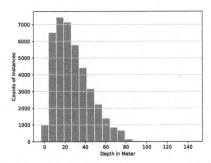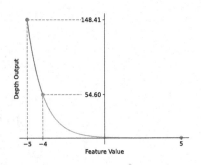

Fig. 4. *left*: Histogram of the depth. The analysis is based on instances in the KITTI dataset. *right*: Transformation of Eigen et al. [6] according to depth estimation. The x-axis indicates the feature output, and the y-axis is the depth output after transformation (given in meter).

Thus, we use the LID strategy to ensure the lengths of neighboring bins increase linearly instead of *log*-wise. For this purpose, assume the length of the first bin is δ. Then the length of the next bin is always δ longer than the previous bin. Now we can encode an instance depth d in $l_{\text{int}} = \lfloor l \rfloor$ ordinal bins according to LID and SID respectively. Additionally, we reserve and regress the residual decimal part $l_{\text{res}} = l - l_{\text{int}}$ for both discretization strategies:

$$
\begin{aligned}
\text{SID:} \quad & l = N\frac{\log d - \log d_{\max}}{\log d_{\max} - \log d_{\min}}, \\
\text{LID:} \quad & l = -0.5 + 0.5\sqrt{1 + \frac{8(d - d_{\min})}{\delta}}, \quad \delta = \frac{2(d_{\max} - d_{\min})}{N(1 + N)}.
\end{aligned}
\tag{1}
$$

During the inference phase, DORN counts the number of activated bins with probability higher than 0.5, as estimated by the ordinal label $\hat{l}_{\text{int}}$, and uses the median value of the $\hat{l}_{\text{int}}$-th bin as the estimated depth in meters. The notation of symbols with $\hat{\ }$ denotes the output of estimation. However, relying on discrete median values of bins only is not precise enough for instance localization. Hence we modify the training to be a combination of classification and regression. For classification we follow the ordinal loss with binary classification and add a shared layer to regress the residuals l_{res} additionally. Given an input RGB image $I \in \mathbb{R}^{W \times H \times 3}$, where W represents the width and H the height of I, we generate a depth feature map $\hat{D} \in \mathbb{R}^{\frac{W}{R} \times \frac{H}{R} \times (2N+1)}$, where R is the output stride. Backpropagation is only applied on the centers of 2D bounding boxes located at $\hat{c}_{2D}^i$, where $i \in \{0, 1, ..., K-1\}$ indicates the instance number of total K instances over the image. The final loss $\mathcal{L}_{\text{dep}}$ is defined as the sum of the residual loss $\mathcal{L}_{\text{res}}^i$ and ordinal loss $\mathcal{L}_{\text{ord}}^i$:

$$\mathcal{L}_{\text{dep}} = \sum_{i=0}^{K-1} (\mathcal{L}_{\text{res}}^i + \mathcal{L}_{\text{ord}}^i), \quad \mathcal{L}_{\text{res}}^i = \text{SmL1}(\hat{l}_{\text{res}}^i, l_{\text{res}}^i),$$

$$\mathcal{L}_{\text{ord}}^i = -\left(\sum_{n=0}^{l^i-1} \log \mathcal{P}_n^i + \sum_{n=l^i}^{N-1} \log(1 - \mathcal{P}_n^i) \right), \quad \mathcal{P}_n^i = \mathcal{P}\left(\hat{l}^i > n\right), \tag{2}$$

where $\mathcal{P}_n^i$ is the probability that the i-th instance is farther away than the n-th bin, and $SmL1$ represents the smooth L1 loss function [15]. During inference, the amount of activated bins will be counted up as $\hat{l}_{\text{int}}^i$. We refine the result by taking into account the residual part, $\hat{l} = \hat{l}_{\text{int}}^i + \hat{l}_{\text{res}}^i$, and decode the result by inverse-transformation of Eq. 1.

DepJoint The transformation described by Eigen et al. [6] converts the output depth to an exponential scale. It generates a depth feature map $\hat{D} \in \mathbb{R}^{\frac{W}{R} \times \frac{H}{R} \times 1}$. The output $\hat{d}$ at the estimated center point of a 2D bounding box $\hat{\mathbf{c}}_{\text{2D}}^i$ is converted to $\Phi(\hat{d}) = e^{-\hat{d}}$. This enriches the depth information for closer objects by putting more feature values into smaller ranges. As shown on the right panel of Fig. 4, the feature map values between -4 and 5 correspond to a depth up to 54.60 m, while feature values corresponding to more distant objects up to 148.41 m account for only 10% of the feature output range $[-5, 5]$. The transformation is reasonable, since closer objects are of higher importance. Eigen's transformation shows an impressive precision on closer objects but disappoints on objects which are farther away. To improve on the latter, we introduce the DepJoint approach, which treats the depth estimation as a joint classification and regression. Compared to using Eigen's transformation solely, it also emphasizes the distant field. DepJoint divides the depth range $[d_{\min}, d_{\max}]$ in two bins with scale parameter α and β:

$$\begin{aligned} \text{Bin } 1 &= [d_{\min}, (1-\alpha)d_{\min} + \alpha d_{\max}], \\ \text{Bin } 2 &= [(1-\beta)d_{\min} + \beta d_{\max}, d_{\max}]. \end{aligned} \tag{3}$$

Each bin will only be activated during training when the object lies within the appropriate interval. The first bin is used to regress the absolute value of depth d^i, while the second bin is used to regress the residual value of depth $\tilde{d}^i = d_{\max} - d^i$. With this transformation, a larger depth value will be supported with more features. We use the binary Cross-Entropy loss $CE_b(\cdot)$ for classification of each bin b and regress d^i and $\tilde{d}^i = d_{\max} - d^i$ with L1 loss $L1(\cdot)$ subsequent to an output transformation $\Phi(\cdot)$. Hence the output of the depth head is $\hat{D} \in \mathbb{R}^{\frac{W}{R} \times \frac{H}{R} \times 6}$ and the loss for training is defined as:

$$\mathcal{L}_{\text{dep}} = \sum_{i=0}^{K-1} \left(\mathcal{L}_{\text{cls}}^i + \mathcal{L}_{\text{reg}}^i \right), \quad \mathcal{L}_{\text{cls}}^i = \sum_b CE_b(d^i),$$

$$\mathcal{L}_{\text{reg}}^i = \mathbb{1}_1\left(d^i\right) \cdot L1\left(d^i, \Phi\left(\hat{d}_1^i\right)\right) + \mathbb{1}_2\left(d^i\right) \cdot L1\left(\tilde{d}^i, \Phi\left(\hat{d}_2^i\right)\right), \tag{4}$$

where $\hat{d}_b^i$ represents the regression output for the b-th bin and i-th instance. The indicator function $\mathbb{1}_b\left(d^i\right)$ will only be activated, when d^i stays in b-th Bin. Training is only applied on 2D centers of bounding boxes. During inference the weighted average will be decoded as the final result:

$$\hat{d}^i = \mathcal{P}_{\text{Bin 1}}^i\left(\hat{d}^i\right) \cdot \Phi\left(\hat{d}_1^i\right) + \mathcal{P}_{\text{Bin 2}}^i\left(\hat{d}^i\right) \cdot \left(d_{\max} - \Phi\left(\hat{d}_2^i\right)\right), \qquad (5)$$

where $P_{\text{Bin b}}^i$ denotes the normalized probability of $\hat{d}^i$.

2.4 Reference Area

Conventionally the regressed values of a single instance will be trained and accessed only on a single center point, which reduces the calculation. However, it also restricts the perception field of regression and affects reliability. To overcome these disadvantages, we apply the concept used by Eskil et al. [9] and Krishna et al. [10]. Instead of relying on a single point, a **Reference Area (RA)** based on the 2D center point is defined within the 2D bounding box, whose width and height are set accordingly with a proportional value γ. All values within this area contribute to regression and classification. If RAs overlap, the area closest to the camera dominates, since only the closest instance is completely visible on the monocular image. During inference all predictions in the related RA will be weighted equally. Supplementary material contain additional details.

3 Experiments

3.1 Implementation Details

We performed experiments on the KITTI object detection benchmark [8], which contains 7481 training images and 7518 testing images. All instances are divided into easy, moderate and hard targets according to visibility in the image [8]. To numerically compare our results with other approaches we use intersection over union (IoU) based on 2D bounding boxes (AP), bounding boxes in Bird's-eye view (BEV AP) and in 3D space (3D AP). Most recently, the KITTI evaluation benchmark has been using 40 instead of 11 recalls. However, many methods only evaluated the average precision on 11 recalls (AP_{11}) in percentage. For fair comparison, we show here firstly AP_{11} on the validation set and then AP_{40} on the official test set.

Like most previous works, and in particular CenterNet, we firstly only consider the "Car" category and follow the standard training/validation split strategy in [3], which leads to 3712 images for training and 3769 images for validation. In particular, we keep the modified Deep Layer Aggregation (DLA)-34 [25] as the backbone. Regarding different approaches, we add specific head layers, which consist of one 3×3 convolutional layer with 256 channels, ReLu activation and a 1×1 convolution with desired output channels at the end. We trained the network from scratch in PyTorch [15] on 2 GPUs (1080Ti) with batch sizes 7 and

Table 1. $AP_{11}(\%)$ on KITTI validation set at 0.5 IoU threshold. We focus on the car detection result here. RT indicates runtime in *ms*. *ct3d* denotes CenterNet with 3D center points instead of 2D center points. *eigen* represents the original Eigen's transformation in CenterNet, while *lid* refers to the LID and *dj* represents the DepJoint approach. *ra* indicates a reference area supporting regression tasks. The best result is marked in bold, the second best is underlined. E, M and H indicate Easy, Moderate and Hard instances.

Approach	RT (ms)	2D AP			BEV / 3D AP		
		E	M	H	E	M	H
CenterNet [26]	**43**	**97.1**	87.9	79.3	31.5 / 19.5	29.7 / 18.6	28.1 / 16.6
CenterNet(ct3d)	-	87.1	85.6	69.8	46.8 / 39.9	37.9 / 31.4	32.7 / 30.1
Mono3D [2]	-	92.3	**88.7**	79.0	30.5 / 25.2	22.4 / 18.2	19.2 / 15.5
MonoGRNet [16]	60	-	-	-	- / 50.5	- / 37.0	- / 30.8
Multi-Fusion [24]	120	-	-	-	55.0 / 47.9	36.7 / 29.5	31.3 / 26.4
M3D-RPN [1]	161	90.2	83.7	67.7	55.4 / 49.0	42.5 / 39.6	35.3 / 33.0
Center3D(+eigen)	47	96.7	88.0	79.4	47.6 / 38.0	37.6 / 30.8	32.4 / 29.4
Center3D(+lid)	53	96.9	87.5	79.0	51.3 / 44.0	39.3 / 35.0	33.9 / 30.6
Center3D(+dj)	54	96.1	86.8	78.2	55.4 / 49.7	41.7 / 38.1	35.6 / 32.9
Center3D(+dj+ra)	56	96.8	88.2	79.6	56.7/ 52.5	43.5/ 40.5	41.2/ 34.9

Table 2. $AP_{40}(\%)$ on KITTI test set at 0.7 IoU threshold. We show the car, pedestrain and cyclist detection results here. E, M and H indicate Easy, Moderate and Hard instances.

Approach	Car								
	2D AP			AOS			BEV/3D AP		
	E	M	H	E	M	H	E	M	H
M3D-RPN [1]	89.0	**85.1**	69.3	88.4	**82.8**	67.1	**21.0/14.8**	13.7/9.7	10.2/7.4
Center3D	**95.1**	**85.1**	**73.1**	**93.1**	82.5	**70.8**	18.9/12.0	**14.0/9.3**	**12.4/8.1**

BEV/3D AP	Pedestrain			Cyclist		
	E	M	H	E	M	H
M3D-RPN [1]	**5.7/4.9**	**4.1/3.5**	3.3/**2.9**	1.3/0.9	0.8/0.7	0.8/0.5
Center3D	**5.7/4.9**	3.7/3.4	**3.5**/2.8	**5.3/4.3**	**2.8/2.4**	**2.7/2.1**

9. We trained the network for 70 epochs with an initial learning rate of $1.25e^{-4}$ or $2.4e^{-4}$, which drops by a factor of 10 at 45 and 60 epochs if not specified otherwise.

3.2   Center3D

We can bridge the gap between 2D and 3D center points by adding 2 specific layers to regress the offset $\Delta \mathbf{c}^i$. For demonstration we perform an experiment, which is indicated as *CenterNet(ct3d)* in Table 1. It models the object as a projected 3D center point with 4 distances to boundaries. The visible object, whose 3D center point is out of the image, is ignored during training. As Table 1 shows, for easy targets *ct3d* increases the BEV AP by 48.6% and the 3D AP by 104.6% compared to the baseline of CenterNet. This is achieved by the proper

Table 3. Experimental results of LID. We show the comparison between SID and LID, the influence of different bins. *-res* indicates no regression of residuals as an ablation study. APs are given in percentage.

	Bin	BEV AP			3D AP		
		Easy	Mode	Hard	Easy	Mode	Hard
Eigen	–	47.6	37.6	32.4	38.0	30.8	29.4
SID	40	33.4	27.6	26.9	26.7	24.4	21.1
LID	40	31.5	25.6	24.9	24.7	22.7	19.4
SID	100	47.6	37.5	32.3	39.6	33.8	29.4
LID	100	<u>50.2</u>	<u>39.2</u>	**33.9**	<u>41.5</u>	**35.6**	**31.3**
SID	80	48.7	37.9	<u>32.9</u>	40.4	34.3	30.0
LID	80	**51.3**	**39.3**	**33.9**	**44.0**	<u>35.0</u>	<u>30.6</u>
LID/-res	80	37.1	33.0	29.2	31.9	26.7	25.8

decoding of 3D parameters based on an appropriate 3D center point. However, simply modeling an object with a 3D center will hurt 2D performance, since some 3D centers are not attainable, although the object is still partly visible in the image.

In contrast, the Center3D approach is able to balance the trade-off between a tightly fitting 2D bounding box and a proper 3D location. The regression of offsets is regularizing the spatial center, while also preserving the stable precision in 2D (the 7th row in Table 1). BEV AP for moderate targets improves from 29.7% to 37.6%, and 3D AP increases from 18.6% to 30.8%, which performs comparably to the state of the art. Since Center3D is also the basis for all further experiments, we treat the performance as our new baseline for comparison.

3.3   LID

We first implement and adjust the DORN [7] approach for depth estimation instance-wise rather than pixel-wise. Following DORN we add a shift ξ to both distance extremum $d^*_{\min}$ and $d^*_{\max}$ to ensure $d_{\min} = d^*_{\min} + \xi = 1.0$. In addition, we perform experiments for our LID approach to demonstrate its effectiveness. We set the number of bins to 80, and add a single head layer to regress the residuals of the discretization. Hence, for depth estimation, we add head layers to generate the output features $\hat{D} \in \mathbb{R}^{\frac{W}{R} \times \frac{H}{R} \times 161}$, while CenterNet generates an output feature $\hat{D} \in \mathbb{R}^{\frac{W}{R} \times \frac{H}{R} \times 1}$. Here the depth loss weight $\lambda_{\text{dep}} = 0.1$ yields the best performance. We compare the results with our new baseline Center3D with the same learning rate of $1.25e^{-4}$. The best result is shown as *Center3D(+lid)* in Table 1.

More detailed, Table 3 shows both LID and SID with different number of bins improved even instance-wise with additional layers for ordinal classification, when a proper number of bins is used. Our discretization strategy LID shows a considerably higher precision in 3D evaluation, comprehensively when

Table 4. Experimental results of DepJoint. The regression part of depth estimation is supported by RA. The dependence on α/β is shown, which represents the threshold scale of first/second bins regarding to $d_{\max}$. The left column shows the results of associated strategy, while the right column shows the results of overlapping strategy. APs are given in percentage.

	BEV AP			3D AP				BEV AP			3D AP		
α/β	E	M	H	E	M	H	α/β	E	M	H	E	M	H
0.7/0.3	52.9	40.6	35.0	44.5	**37.5**	**32.7**	0.2/0.2	48.6	37.5	33.3	42.1	32.9	30.9
0.6/0.4	49.9	39.6	34.1	44.7	35.5	30.8	0.3/0.3	**53.2**	**41.2**	**35.1**	**47.4**	36.4	32.3
0.8/0.2	47.9	38.7	33.6	40.3	31.9	30.9	0.4/0.4	51.5	40.3	34.1	45.5	36.0	31.2
0.9/0.1	48.4	37.8	32.7	41.4	31.2	29.4	0.5/0.5	42.9	35.9	31.9	37.7	29.3	28.3

80 and 100 bins are used. A visualization of inferences of both approaches from BEV is shown in Fig. 3. LID only preforms worse than SID in the 40 bin case, where the number of intervals is not enough for instance-wise depth estimation. Furthermore we verify the necessity of the regression of residuals by comparing the last two rows in Table 3. The performance of LID in 3D will deteriorate drastically if this refinement module is removed.

3.4 DepJoint and Reference Area

In this section, we evaluate the performance of DepJoint approach. Firstly, only the regression part of depth estimation is supported by RA, which is sensitive according to its size. We set $\gamma = 0.4$ as default for RA, which yields mostly the best result. The supporting experimental results can be found in the supplementary material. Additionally, we apply $d_{\min} = 0\,\mathrm{m}$, $d_{\max} = 60\,\mathrm{m}$ and $\lambda_{\mathrm{dep}} = 0.1$ for all experiments. Table 4 shows the experimental results. As introduced in Sect. 2.3, we can divide the whole depth range into two overlapping or associated bins. For overlapping strategy, the overlapping area should be defined properly. When the overlapping area is too small, the overlapping strategy actually converts to the associated strategy. On the other hand, if the overlapping area is too wide, the two independent bins tend to capture the general feature instead of specific feature of objects in panoramic depth. In that case, the two bins would not focus on objects in foreground and background respectively anymore, since the input objects during training for both bins are almost the same. This can also explain why the threshold choices of 0.7/0.3 and 0.6/0.4 result in a better accuracy in 3D space comparing with 0.9/0.1 and 0.8/0.2. For the associated strategy, the thresholds α/β of 0.3/0.3 and 0.4/0.4 show the best performance for the following reason: usually more distant objects show less visible features in the image. Hence, we want to set both thresholds a little lower, to assign more instances to the distant bins and thereby suppress the imbalance of visual clues between the two bins.

Furthermore, the ablation study (without RA) verifies the effectiveness of DepJoint in comparison with Eigen's transformation. For DepJoint approach, α/β are set to 0.7/0.3 and $\lambda_{\mathrm{dep}} = 0.5$. As *Center3D(+eigen)* and *(+dj)* shown in Table 1, DepJoint approach has a considerably higher AP in both BEV and 3D space.

3.5 Comparison to the State of the Art

Table 1 shows the comparison with state-of-the-art monocular 3D detectors on the validation dataset. *Center3D(+lid)* and *(+dj)* follow the settings described above. However, *Center3D(+dj+ra)* is supported here by RA in the regression of 3D offset, dimension, rotation and 2D width/height comprehensively ($\gamma = 0.4$). The corresponding loss weightings are all set to 0.1, except $\lambda_{\mathrm{dep}} = 0.5$ and $\lambda_{\mathrm{rot}} = 1$. The learning rate is $2.4e^{-4}$.

As Table 1 shown, all Center3D models perform at comparable 3D performance with respect to the best approaches currently available. Both LID and DepJoint approach for depth estimation have a higher AP than simply applying the Eigen's transformation in 3D task. Especially, *Center3D(+dj+ra)* achieved state-of-the-art performance with BEV APs of 56.7% and 43.5% for easy and moderate targets, respectively. For hard objects in particular, it outperforms all other approaches with BEV AP of 41.2% and with 3D AP of 34.9%.

Table 2 shows the AP_{40} of *Center3D(+dj+ra)* on the KITTI test set. In comparison with M3D-RPN, Center3D outperforms in 2D AP and average orientation similarity (AOS) obviously with comparable BEV and 3D APs. Besides, Center3D performs particularly better for hard object detection and Cyclist detection. More qualitative results can be found in the supplementary material.

Center3D preserves the advantages of an anchor-free approach. It performs better than most other approaches in 2D AP, especially on easy objects. Most importantly, it infers on the monocular input image with the highest speed (around three times faster than M3D-RPN, which performs similarly to Center3D in 3D). Therefore, Center3D is able to fulfill the requirement of a real-time detection.

4   Conclusion

In this paper we introduced Center3D, a one-stage anchor-free monocular 3D object detector, which models and detects objects with center points of 2D bounding boxes. We recognize and highlight the importance of the difference between centers in 2D and 3D by regressing the offset directly, which transforms 2D centers to 3D centers. In order to improve depth estimation, we further explored the effectiveness of joint classification and regression when only monocular images are given. Both classification-dominated (LID) and regression-dominated (DepJoint) approaches enhance the AP in BEV and 3D space. Finally, we employed the concept of RAs by regressing in predefined areas, to overcome the sparsity of the feature map in anchor-free approaches. Center3D performs comparably to state-of-the-art monocular 3D approaches with significantly improved runtime during inference. Center3D achieved a better trade-off between 3D precision and inference speed.

References

1. Brazil, G., Liu, X.: M3d-rpn: Monocular 3d region proposal network for object detection. In: Proceedings of the IEEE International Conference on Computer Vision. pp. 9287–9296 (2019)
2. Chen, X., Kundu, K., Zhang, Z., Ma, H., Fidler, S., Urtasun, R.: Monocular 3d object detection for autonomous driving. In: Proceedings of the IEEE Conference on Computer Vision and Pattern Recognition. pp. 2147–2156 (2016)
3. Chen, X., et al.: 3d object proposals for accurate object class detection. In: Advances in Neural Information Processing Systems. pp. 424–432 (2015)
4. Chen, X., Ma, H., Wan, J., Li, B., Xia, T.: Multi-view 3d object detection network for autonomous driving. In: Proceedings of the IEEE Conference on Computer Vision and Pattern Recognition. pp. 1907–1915 (2017)
5. Duan, K., Bai, S., Xie, L., Qi, H., Huang, Q., Tian, Q.: Centernet: Keypoint triplets for object detection. In: Proceedings of the IEEE International Conference on Computer Vision. pp. 6569–6578 (2019)
6. Eigen, D., Puhrsch, C., Fergus, R.: Depth map prediction from a single image using a multi-scale deep network. In: Advances in neural information processing systems. pp. 2366–2374 (2014)
7. Fu, H., Gong, M., Wang, C., Batmanghelich, K., Tao, D.: Deep ordinal regression network for monocular depth estimation. In: Proceedings of the IEEE Conference on Computer Vision and Pattern Recognition. pp. 2002–2011 (2018)
8. Geiger, A., Lenz, P., Urtasun, R.: Are we ready for autonomous driving? the kitti vision benchmark suite. In: 2012 IEEE Conference on Computer Vision and Pattern Recognition. pp. 3354–3361. IEEE (2012)
9. Jörgensen, E., Zach, C., Kahl, F.: Monocular 3d object detection and box fitting trained end-to-end using intersection-over-union loss. arXiv preprint arXiv:1906.08070 (2019)
10. Krishnan, A., Larsson, J.: Vehicle detection and road scene segmentation using deep learning. Chalmers University of Technology (2016)
11. Law, H., Deng, J.: Cornernet: Detecting objects as paired keypoints. In: Proceedings of the European Conference on Computer Vision (ECCV). pp. 734–750 (2018)
12. Li, B., Ouyang, W., Sheng, L., Zeng, X., Wang, X.: Gs3d: An efficient 3d object detection framework for autonomous driving. In: Proceedings of the IEEE Conference on Computer Vision and Pattern Recognition. pp. 1019–1028 (2019)
13. Lin, T.Y., Goyal, P., Girshick, R., He, K., Dollár, P.: Focal loss for dense object detection. In: Proceedings of the IEEE international conference on computer vision. pp. 2980–2988 (2017)
14. Mousavian, A., Anguelov, D., Flynn, J., Kosecka, J.: 3d bounding box estimation using deep learning and geometry. In: Proceedings of the IEEE Conference on Computer Vision and Pattern Recognition. pp. 7074–7082 (2017)
15. Paszke, A., et al.: Automatic differentiation in pytorch (2017)

16. Qin, Z., Wang, J., Lu, Y.: Monogrnet: A geometric reasoning network for monocular 3d object localization. Proceedings of the AAAI Conference on Artificial Intelligence. **33**, 8851–8858 (2019)
17. Ren, S., He, K., Girshick, R., Sun, J.: Faster r-cnn: Towards real-time object detection with region proposal networks. In: Advances in neural information processing systems. pp. 91–99 (2015)
18. Rey, D., Subsol, G., Delingette, H., Ayache, N.: Automatic detection and segmentation of evolving processes in 3d medical images: Application to multiple sclerosis. Medical image analysis **6**(2), 163–179 (2002)
19. Roddick, T., Kendall, A., Cipolla, R.: Orthographic feature transform for monocular 3d object detection. arXiv preprint arXiv:1811.08188 (2018)
20. Shin, K., Kwon, Y.P., Tomizuka, M.: Roarnet: A robust 3d object detection based on region approximation refinement. In: 2019 IEEE Intelligent Vehicles Symposium (IV). pp. 2510–2515. IEEE (2019)
21. Surmann, H., Nüchter, A., Hertzberg, J.: An autonomous mobile robot with a 3d laser range finder for 3d exploration and digitalization of indoor environments. Robotics and Autonomous Systems **45**(3–4), 181–198 (2003)
22. Vasile, A.N., Marino, R.M.: Pose-independent automatic target detection and recognition using 3d laser radar imagery. Lincoln laboratory journal **15**(1), 61–78 (2005)
23. Wang, Z., Jia, K.: Frustum convnet: Sliding frustums to aggregate local point-wise features for amodal 3d object detection. arXiv preprint arXiv:1903.01864 (2019)
24. Xu, B., Chen, Z.: Multi-level fusion based 3d object detection from monocular images. In: Proceedings of the IEEE conference on computer vision and pattern recognition. pp. 2345–2353 (2018)
25. Yu, F., Wang, D., Shelhamer, E., Darrell, T.: Deep layer aggregation. In: Proceedings of the IEEE conference on computer vision and pattern recognition. pp. 2403–2412 (2018)
26. Zhou, X., Wang, D., Krähenbühl, P.: Objects as points. In: arXiv preprint arXiv:1904.07850 (2019)

Constellation Codebooks for Reliable Vehicle Localization

Isabell Hofstetter[(⊠)], Malte Springer, Florian Ries, and Martin Haueis

Mercedes-Benz AG, Sindelfingen, Germany
{isabell.hofstetter,malte.springer,florian.ries,
martin.haueis}@daimler.com

Abstract. Safe feature-based vehicle localization requires correct and reliable association between detected and mapped localization landmarks. Incorrect feature associations result in faulty position estimates and risk integrity of vehicle localization. Depending on the number and kind of available localization landmarks, there is only a limited guarantee for correct data association due to various ambiguities.

In this work, a new data association approach is introduced for feature-based vehicle localization which relies on the extraction and use of unique geometric patterns of localization features. In a preprocessing step, the map is searched for unique patterns that are formed by localization landmarks. These are stored in a so called *codebook*, which is then used online for data association. By predetermining constellations that are unique in a given map section, an online guarantee for reliable data association can be given under certain assumptions on sensor faults.

The approach is demonstrated on a map containing cylindrical objects which were extracted from LiDAR data. The evaluation of a localization drive of about 10 min using various codebooks both demonstrates the feasibility as well as limitations of the approach.

Keywords: Vehicle localization · Geometric hashing · Data association

1 Introduction

High definition (HD) maps are essential to enable safe highly automated driving. Consequently, reliable and accurate vehicle localization in such maps is inevitable. In the past years, various feature-based localization approaches have been introduced. These methods utilize a separate map layer containing localization features, also called *landmarks*, which they aim to associate with detected features in the vehicle's surrounding. The resulting association constraints pose an optimization problem, which is solved for an accurate pose estimate.

State-of-the-art methods differ mainly in the kind of localization features that are used. Whereas some approaches rely on dense point clouds such as visual point features [17] or RADAR point clouds [16], other methods make use of landmarks such as road markings [11,15], or geometric primitives [4,7].

© Springer Nature Switzerland AG 2021
Z. Akata et al. (Eds.): DAGM GCPR 2020, LNCS 12544, pp. 303–315, 2021.
https://doi.org/10.1007/978-3-030-71278-5_22

However, all these methods imply some kind of feature association. Especially in those cases where localization features are sparse and do not have an expressive descriptor, data association represents one of the main challenges for reliable vehicle localization. To satisfy integrity requirements, a guarantee for correct data association is desirable.

This contribution focuses on the extraction and use of unique landmark patterns, hereafter referred to as *constellation codebooks*, for reliable data association. Landmarks used for localization often appear in periodic orders, so that many ambiguities arise in the data association step. This contribution suggests to identify that specific geometric information which breaks the symmetry and to cluster landmarks such that the resulting feature patterns reflect the uniqueness of the environment. This enables reliable feature association even in sparse feature maps.

The proposed approach consists of a two-step process: In a first step, a given map of localization features is searched for such unique geometric patterns of landmarks. In this work, these constellations are referred to as *codewords*. All extracted codewords are stored in a *codebook*, which can then be used online for vehicle localization. For easy and fast search and access, this work suggests to store the codebook in a hash table format.

During the localization drive, the geometric patterns of detected localization features are compared to the codebook content. As soon as one of the codewords is detected online, incorrect associations can be excluded and a certain guarantee can be given that the resulting associations are correct. Hereby, the suggested data association approach, which is based on Geometric Hashing [19], does not make any assumptions on prior vehicle poses. It allows for feature association based only on the geometric information provided by the extracted features, while ambiguities are eliminated due to the preprocessing step.

The method is derived and tested on a dataset representing an urban loop of 3.8 km in Sindelfingen, Germany, where pole-like features are used as localization landmarks. The method, however, is not limited to this specific kind of features and can easily be adapted to arbitrary localization features. Experimental results verify the potential of the proposed approach.

2   Related Work

In order to provide reliable localization that meets high integrity requirements, most approaches rely on outlier detection and exclusion [1, 21] or pattern matching techniques [2, 7] to avoid incorrect associations between map and sensor data. By accumulating sensor features and matching a group of landmarks with the map, these approaches try to reduce or even eliminate the number of ambiguities in data association and exclude associations that are found to be outliers. A major drawback of most approaches is the assumption that a certain number of correct associations is found and only a small percentage of outliers needs to be eliminated. Also, a guarantee that all outliers are excluded or that the correct alignment with the map is found cannot be given.

Brenner et al. suggest a pattern matching algorithm in [4,14] for vehicle localization. By deriving a descriptor for cylindrical localization features based on other features in the near surrounding, they notice that such patterns can be ambiguous as well and not every accumulated pattern of localization landmarks provides enough unique information for a reliable alignment with the map. The existence of ambiguities in data association is also discussed in [6], where a method based on Geometric Hashing [8] for the extraction of such patterns is suggested. Applying the same Geometric Hashing algorithm for the actual feature association step has been introduced in [5] and a similar technique is used in this work. The main contribution of this work, however, focuses on the generation and use of constellation codebooks, i.e. unique geometric patterns of localization features.

Extracting unique information from the environment for localization has been considered in works like [9]. Here, a method is proposed for creating unique identifiers, called fingerprint sequences, for visually distinct locations using panoramic color images. Also, object recognition methods like Bag of Words approaches as utilized in [10,20] are well known and provide a similar kind of vocabulary, however, without satisfying any uniqueness requirements. Furthermore, these methods have only been applied in the field of image data processing. To the authors' knowledge, the concept of identifying unique patterns in feature maps and the use of codebooks for localization has not been studied before.

3   Codebook Generation

In the following, a *codebook* is defined as a set of feature patterns, hereafter referred to as *codewords*. The codebook is computed for a given map or map section and used for feature-based vehicle localization. The proposed feature constellations, which are contained in a codebook must satisfy certain constraints in order to enable reliable vehicle localization. The main requirement is the uniqueness of the extracted constellations in a given map section. If a feature pattern is known to be unique in a map even under various sensor or map faults, this information can be used for reliable data association.

In the following subsections, the codebook generation is described in detail. This involves the derivation of various requirements for the constellations of interest, as well as a search algorithm for the identification of such feature patterns.

3.1   Definition of Uniqueness

Feature patterns of interest are desired to be unique in a sense that they are uniquely associable in a given map section. This must still be true, even if the measurements of the landmarks are affected by noise. In other words: For a unique feature constellation, there exists no other approximative matching in the map section of interest than the correct one.

To describe these assumptions in a mathematical way, the distance d between single landmarks $p = (x, y) \in \mathbb{R}^2$ and sets of landmarks $G = \{(x_1, y_1), \ldots, (x_N, y_N)\}$, $(x_i, y_i) \in \mathbb{R}^2$, $i = 1, \ldots, N$, is defined as

$$d(p, G) := \min\{\|p - g\|_2 \mid g \in G\}, \tag{1}$$

which is the distance of p to its nearest neighbor in G. Thus, the mean point-to-point distance between two sets of landmarks G_1 and G_2 with cardinality $|G_1| = |G_2|$ can be defined as

$$D(G_1, G_2) := \frac{1}{|G_1|} \sum_{g \in G_1} d(g, G_2). \tag{2}$$

Let $\mathcal{T}$ be the set of all euclidean transformations, i.e. all combinations of rotation and translation. Then, a pattern of landmarks C is unique in a map section M if

$$\forall\, T_{eucl} \in \mathcal{T} \,\nexists\, F \subset M \text{ with } |F| = |C| : D(T_{eucl}(C), F) < \epsilon \tag{3}$$

for $F \neq C$ and a minimum distance ϵ.

This means that considering any arbitrary transformation $T_{eucl} \in \mathcal{T}$, no subset $F \subset M$ with $|F| = |C|$ exists, whose mean distance to $T_{eucl}(C)$ is smaller than ϵ. Therefore, there exists only one transformation, which allows an approximative matching [3] of the constellation with the map, which is the identity transformation.

In the following, ϵ is also called *uniqueness threshold*, because the choice of ϵ is decisive for the classification of feature patterns in unique or ambiguous patterns. For large ϵ, less patterns will be identified as unique. On the other hand, for small ϵ, patterns that differ only slightly in their geometrical characteristics will be classified as unique, which results in less ambiguous patterns.

3.2 Codeword Extraction

In practice, it is not feasible to find unique feature patterns in large maps of entire cities or even countries. However, a reasonable assumption is that an initial, very rough position estimate within a map is given. Therefore, the patterns of interest do not need to be unique within an entire map, but rather in a map section of limited size. In the following, the uniqueness requirement (3) will be limited to certain map tiles T of size R. Thus, $R = \infty$ refers to the entire map.

Also, to facilitate the detection of codewords C during the localization drive, the spatial expansion of codewords, i.e. the diameter of the smallest enclosing circle containing a codeword, should be limited. This can be described as

$$d(p_i, p_j) \leq L \quad \forall\, p_i, p_j \in C, \ i \neq j, \tag{4}$$

with a given maximum diameter L.

Algorithm 1. CODEBOOK GENERATION

Input: Map M, tile radius R, max. expansion L, uniqueness threshold ϵ.
Output: Codebook CB containing all unique constellations with cardinality s_{cw}.

1. Generate overlapping map tiles $\{T_1, \ldots, T_n\}$ of size R covering the whole map M.
2. Gather map features for each tile $T_i = \{p_1, \ldots, p_{|T_i|}\}$, $i = 1, \ldots, n$.
3. For each tile T_i, construct all possible feature combinations C of size s_{cw} fulfilling expansion requirement (4) for the given maximum expansion L.
4. Check uniqueness requirement (3) with given uniqueness threshold ϵ for every constellation C regarding the current map tile.
5. If C fulfills (3), add it to the codebook CB. Discard C otherwise.

Now, in order to generate a constellation codebook, landmark patterns with characteristics (3) and (4) can be identified in a map M. The method is summarized in Algorithm 1.

In steps 1) and 2), the map is divided into map tiles of a given radius R. This parameter should be chosen in a way that the current map tile can always be found reliably, for example using a low-cost GPS module. Then, each tile is regarded separately.

First, all possible feature combinations of the desired cardinality s_{cw} and expansion L are constructed. These are the candidate patterns which need to be classified into unique or ambiguous patterns.

Regarding a tile T_i containing N_f features, there are a total of $\binom{N_f}{s_{cw}}$ constellations of s_{cw} features. As an example, for $N_f = 50$ and $s_{cw} = 3$, there exist $\binom{50}{3} = 19.600$ constellations in total. For larger patterns of cardinality $s_{cw} = 6$, there are almost 16 million different patterns. However, by limiting the constellation expansion to L as suggested by Eq. (4), this number of candidate patterns can be reduced noticeably.

All candidate patterns found in step 3) are then compared to each other and uniqueness requirement (3) is checked for a given uniqueness threshold ϵ. Here, scan matching methods like the well-known Iterative Closest Point (ICP) algorithm [13] can be used. If the constellation proves to fulfill the uniqueness requirement, it is added to the codebook. If it is not unique according to (3), the pattern is discarded. Finally, steps 3) – 5) are repeated for each previously constructed map tile T_i.

Figure 1 shows example constellations for the first map tile T_0 with tile radius $R = 200$ m. Map features are plotted in gray. Codewords are visualized by colored lines connecting the features belonging to the codeword.

4   Localization Using Constellation Codebooks

The offline generated constellation codebook can now be used for the purpose of data association for vehicle localization. In order to allow an easy and fast recognition of codewords and a quick search for associations, in this work, a Geometric

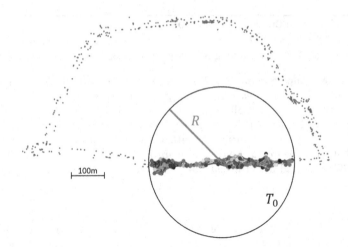

Fig. 1. Map representing a loop of 3.8 km of roads. Localization features are drawn in gray. Colored lines connect those features that belong to one codeword in tile T_0 of size $R = 200$ m.

Hashing based method is proposed for the representation of codebooks as well as the actual data association step. Geometric Hashing was initially introduced in [8] for object recognition purposes. It is a popular technique for the association of geometric features that have undergone transformations. The method consists of a training phase as well as the recognition phase. During the training phase, features of interest are represented in a variety of coordinate systems that are defined by the features themselves and stored in a quickly searchable, tabular format. Online, a given input sample S can be searched for the objects of interest with the help of the previously generated hash table.

The following subsections will discuss a modified hashing algorithm which is suggested in combination with constellation codebooks as they were derived in Sect. 3. For further details on Geometric Hashing, the authors refer to [8,19] or [12].

4.1   Hash Table Generation

For the generation of the hash table, which is used for codeword recognition during localization, each codeword is hashed separately. Details on the hashing procedure as well as the noise model will be discussed briefly.

Hashing of Codewords. Let CB be a codebook that contains a number of codewords $CB = \{c_1, c_2, \ldots, c_N\}$. Each codeword c_i consists of s_{cw} features $c_i = \{p_1^i, \ldots, p_{s_{cw}}^i\}$, $i = 1, \ldots, N$. Each ordered pair of features (p_n^i, p_m^i) with $p_n^i, p_m^i \in c_i$ and $n \neq m$ represents a *basis pair*. Let K_i be the set of basis pairs of codeword c_i. The cardinality of this basis set is $|K_i| = s_{cw}(s_{cw} - 1)$. Each basis $k = (p_n^i, p_m^i) \in K_i$ defines a local coordinate frame $O_{(p_n^i, p_m^i)}$, called the *geometric* *basis.* For each basis $k \in K_i$, the remaining features $c_i \setminus k$ are transformed into the corresponding basis frame and quantized according to a grid. Finally, a hash function is applied to each feature location and the hashes are stored in a hash table H. The main parameters used in the hashing procedure are visualized in Fig. 2.

Noise Model. If the measured features are affected by noise, the positional error in the hash locations has to be described by a valid noise model. Obviously, the positional noise on the feature locations in the hash frame depends heavily on the chosen basis pair as well as the position of the features within the geometric basis. A small distance between basis pair features will enhance the noise in angular direction, while the noise in radial direction stays constant. Therefore, the authors suggest to estimate the resulting noise distribution in polar coordinates. For the underlying sensor noise, a Gaussian distribution with standard deviation σ is assumed, which is centered at the "true" location of the feature. A Monte Carlo Simulation is performed to generate noisy measurements, which are then transformed into the basis frame. For each geometric basis and each feature within that basis, the standard deviation in angular σ_θ and radial σ_r component are estimated using these simulated measurements and stored in the hash table for the online use.

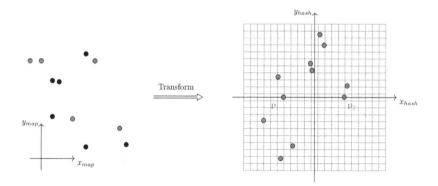

Fig. 2. Visualization of the Geometric Hashing procedure. An example codeword of size $s_{cw} = 5$ (green) is hashed according to the basis frame $O_{(x_{hash}, y_{hash})}$ defined by the features p_1 and p_2. Other map features not belonging to the codeword (gray) are discarded in the hashing procedure. (Color figure online)

4.2   Feature Association

Given a sample of detected features D and the previously generated hash table H, feature associations can now be determined by searching the hash table. For more detail on data association for vehicle localization based on Geometric

Hashing the authors refer to [5]. In the following, the method will be briefly recapitulated.

First, two arbitrary detections $p_1, p_2 \in D$ are chosen, which define a geometric basis $O_{(p_1,p_2)}$ just as described before. All feature detections D are then transformed into this basis by a rigid transformation, quantized, and hashed accordingly. For each computed hash value, appropriate bins in the hash table H are accessed and a vote is given to each basis found there. In this step, not only the exact bin is accessed but also its direct neighborhood to ensure that noisy measurements that fell into neighboring bins can be associated as well.

Then, the set of bases B is determined that received as many votes as the number of features belonging to the codeword, namely s_{cw}. These bases represent those codewords that could potentially be matched. Finally, the resulting candidate associations are validated based on the Mahalanobis distance between feature measurement in the hash frame and noise distribution, which is stored in the hash table.

If valid associations were found, the matching codeword is returned. Otherwise, the previously described steps can be repeated for each available basis pair $k = (p_i, p_j)$ with $p_i, p_j \in D, i \neq j$. In practice, not every generated basis will necessarily have a matching basis in the hash table. If a basis pair was chosen, which is not part of a codeword, the resulting basis cannot be matched with the hash table. Therefore, multiple bases have to be generated in order to find one that is part of a codeword.

5   Experimental Results

5.1   Dataset

The vehicle *BerthaONE* [18] was used for the acquisition of the dataset utilized to demonstrate and evaluate the proposed approach. A loop of about 3.8 km in an urban area in Sindelfingen, Germany, which results in about 10 min of driving time, was used for the evaluation. The corresponding localization map layer contains 593 cylinders such as traffic signs, tree trunks, and traffic lights as localization features, which were extracted from LiDAR data as described in [7]. Each cylinder is described by its center point $(x, y) \in \mathbb{R}^2$ on the ground plane. The feature locations in map coordinates were shown in Fig. 1. A reference solution for validation is provided by [17].

5.2   Localization Framework

The localization framework used to demonstrate the approach is visualized in Fig. 3. A State-of-the-Art localization framework consisting of map, feature detectors, feature association, and optimizer is adapted and extended by a codebook generator and hashing functionalities. The main contribution of this work, namely the codebook generation and feature association, is highlighted in green. The hashing functionalities are highlighted in gray.

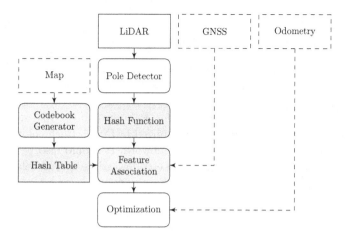

Fig. 3. Overview of the suggested localization framework using Constellation Codebooks and a Geometric Hashing algorithm for data association.

Additional sensors, such as a GNSS receiver and an odometry unit, are added to the framework in dashed lines. These are optional components that are not needed for the data association step itself. However, the GNSS input can be used to determine the current map tile and odometry measurements could be utilized to improve the continuity of the localization solution whenever the suggested method is unavailable.

However, one of the biggest advantages of the proposed approach lies in the fact that no prior pose propagation is needed to find correct associations. Therefore, such information is completely neglected in this work.

5.3   Constellation Codebooks

To evaluate the described approach, a variety of codebooks is generated for the before mentioned map and used for localization. Parameters for the codebook generation that were examined in this work are the tile size R, the uniqueness threshold ϵ as well as the number of resulting codewords. All generated codebooks contain feature constellations of size $s_{cw} \in \{3, 4\}$ with a maximum constellation expansion of $L = 35$ m, which corresponds to the sensor range. Constellations of size $s_{cw} > 4$ could also be extracted using the same method. However, this is something that was not evaluated, since no enhancement of the system availability is expected. Also, it is desirable to extract the minimal number of landmarks required to provide a reliable localization solution.

Depending on the chosen tile radius R and the uniqueness threshold ϵ, a different number of feature patterns was identified as unique patterns in the map ranging in between 0 up to 550.000 codewords. The numbers can be taken from Fig. 4. Here, codebooks were generated for tiles of radius $R \in \{\infty,\ 300\text{m},\ 200\text{m}\}$ and uniqueness threshold $\epsilon \in \{0.25\text{m},\ 0.5\text{m},\ 0.75\text{m},\ 1.0\text{m},\ 1.25\text{m}\}$. For large uniqueness thresholds, the number of unique feature patterns decreases as expected. While there are more than 400.000 codewords using a tile size of $R = 300\,\text{m}$ and $\epsilon = 0.25\,\text{m}$, only 129 codewords are available for $\epsilon = 1.25\,\text{m}$ and the same tile size. In the case of $R = \infty$, no codewords were found at all for $\epsilon > 0.75\,\text{m}$. Also, for decreasing tile sizes, the number of codewords increases. Looking at $\epsilon = 0.5\,\text{m}$, there are 17.802 codewords if the uniqueness is required in the whole map. For $R = 300\,\text{m}$, this number increases to 62.939, and for $R = 200\,\text{m}$ to 104.491. As an example, a codebook for $R = \infty$ and $\epsilon = 0.5\,\text{m}$ containing 17.802 codewords of 3 or 4 features is visualized in Fig. 5. Each color represents the features that are part of a different codeword. It should also be noted that there are some parts in the map where the codeword density is relatively low. This is due to the low feature density in those regions.

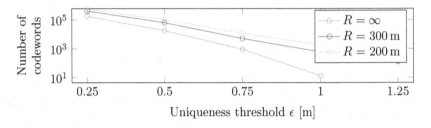

Fig. 4. Number of extracted codewords of size $s_{cw} \in \{3, 4\}$ for various tile sizes R and uniqueness thresholds ϵ.

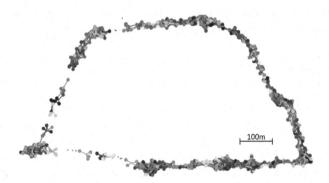

Fig. 5. Example codebook with feature patterns of size $s_{cw} \in \{3, 4\}$ for tile radius $R = \infty$ and uniqueness threshold $\epsilon = 0.5\,\text{m}$. Each color represents one codeword. Features that belong to one codeword are connected by lines.

5.4   Localization Results

The potential of the proposed approach is mainly assessed by the availability of the system and the percentage of correct associations. The association

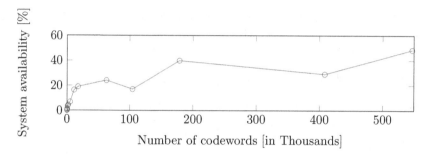

Fig. 6. Overall availability of feature associations vs. total number of codewords contained in the codebooks used for localization.

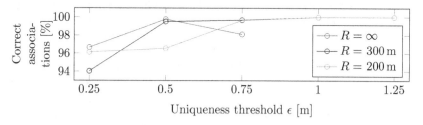

Fig. 7. Percentage of correct associations during localization with the use of various codebooks that were generated for various tile sizes R and uniqueness thresholds ϵ.

availability obviously depends very strongly on the codebook used and especially on the number of codewords contained in the codebook. This correlation can be observed in Fig. 6. The use of codebooks that contain a few hundred codewords results in a low availability of less than 5 %. For larger codebooks with more than 500.000 codewords, an association availability of about 48 % is reached. This is due to the fact that map features often are occluded or not yet in sensor range. Only if all features of one codeword are detected, the association is possible and the system is available. Therefore, the system availability increases with the number of codewords that are contained in the codebook used. The percentage of correct associations is visualized in Fig. 7. Here, the uniqueness threshold ϵ, which was chosen to extract the codewords, plays an important role. For a relatively small choice of $\epsilon = 0.25$ m, codewords were correctly associated in up to 97 % of the time, depending on the tile size R. By increasing the uniqueness threshold up to 1 m, 100 % correct associations are found independent of the tile size $R \in \{200\text{m}, 300\text{m}\}$. For $R = \infty$ and $\epsilon > 0.75$ m no codewords are available, therefore the system is not available.

Overall, there exists a trade-off between system availability and reliability. In order to reach a high availability, it is desirable to extract many codewords. This can be achieved using small tiles and small uniqueness thresholds. On the other hand, for reliable associations, a high uniqueness threshold ϵ should be chosen which results in fewer codewords, but 100% correct associations. In practice, it is conceivable to use multiple codebooks with different uniqueness thresholds to boost availability, where no high integrity solution is available.

6 Conclusions

In this work, a new approach for reliable data association for feature-based vehicle localization was proposed. The method suggests to preprocess a given map in order to identify and extract unique geometric landmark patterns, called *codewords*, which are suitable for reliable feature association. These unique feature constellations are hashed and stored in a hash table format, the *constellation codebook*, for quick and easy search and access during localization.

The exclusive use of unique landmark patterns for data association enables, under certain assumptions on sensor faults, a guarantee for correct feature association and, therefore, provides reliable vehicle localization.

Experimental results were discussed for a feature map of about 3.8 km of roads containing about 600 cylindrical localization landmarks. A variety of codebooks with different parameter choices was constructed for this map. The evaluation of a localization drive of about 10 min in the before mentioned map using these previously generated codebooks demonstrates a trade-off between system availability and reliability.

Future work will focus on the use of multiple constellation codebooks that satisfy different uniqueness requirements for a continuous and accurate localization solution.

References

1. Alsayed, Z., Bresson, G., Verroust-Blondet, A., Nashashibi, F.: Failure detection for laser-based SLAM in urban and peri-urban environments. In: 2017 IEEE 20th International Conference on Intelligent Transportation Systems (ITSC), pp. 1–7 (2017)
2. Ascani, A., Frontoni, E., Mancini, A., Zingaretti, P.: Feature group matching for appearance-based localization. In: 2008 IEEE/RSJ International Conference on Intelligent Robots and Systems (IROS), pp. 3933–3938 (2008)
3. Bishnu, A., Das, S., Nandy, S.C., Bhattacharya, B.B.: A simple algorithm for approximate partial point set pattern matching under rigid motion. In: Rahman, M.S., Fujita, S. (eds.) WALCOM 2010. LNCS, vol. 5942, pp. 102–112. Springer, Heidelberg (2010). https://doi.org/10.1007/978-3-642-11440-3_10
4. Brenner, C.: Global localization of vehicles using local pole patterns. In: Denzler, J., Notni, G., Süße, H. (eds.) DAGM 2009. LNCS, vol. 5748, pp. 61–70. Springer, Heidelberg (2009). https://doi.org/10.1007/978-3-642-03798-6_7
5. Hofstetter, I., Sprunk, M., Ries, F., Haueis, M.: Reliable data association for feature-based vehicle localization using geometric hashing methods. In: 2020 IEEE International Conference of Robotics and Automation (ICRA). IEEE (2020)
6. Hofstetter, I., Sprunk, M., Schuster, F., Ries, F., Haueis, M.: On ambiguities in feature-based vehicle localization and their a priori detection in maps. In: 2019 IEEE Intelligent Vehicles Symposium (IV), pp. 1192–1198. IEEE (2019)

7. Kümmerle, J., Sons, M., Poggenhans, F., Lauer, M., Stiller, C.: Accurate and efficient self-localization on roads using basic geometric primitives. In: 2019 International Conference on Robotics and Automation (ICRA), pp. 5965–5971. IEEE (2019)

8. Lamdan, Y., Wolfson, H.J.: Geometric hashing: A general and efficient model-based recognition scheme (1988)

9. Lamon, P., Nourbakhsh, I., Jensen, B., Siegwart, R.: Deriving and matching image fingerprint sequences for mobile robot localization. In: 2020 IEEE International Conference of Robotics and Automation (ICRA), vol. 2, pp. 1609–1614. IEEE (2001)

10. Lavoué, G.: Combination of bag-of-words descriptors for robust partial shape retrieval. Vis. Comput. **28**(9), 931–942 (2012)

11. Poggenhans, F., Salscheider, N.O., Stiller, C.: Precise localization in high-definition road maps for urban regions. In: 2018 IEEE/RSJ International Conference on Intelligent Robots and Systems (IROS), pp. 2167–2174 (2018)

12. Rigoutsos, I.: Massively Parallel Bayesian Object Recognition. New York University, Department of Computer Science, Technical Report (1992)

13. Rusinkiewicz, S., Levoy, M.: Efficient variants of the icp algorithm. In: Proceedings third international conference on 3-D digital imaging and modeling, pp. 145–152. IEEE (2001)

14. Schlichting, A., Brenner, C.: Localization using automotive laser scanners and local pattern matching. In: 2014 IEEE Intelligent Vehicles Symposium Proceedings, pp. 414–419. IEEE (2014)

15. Schreiber, M., Knöppel, C., Franke, U.: Laneloc: Lane marking based localization using highly accurate maps. In: IEEE Intelligent Vehicles Symposium (IV), pp. 449–454 (2013)

16. Schuster, F., Keller, C.G., Rapp, M., Haueis, M., Curio, C.: Landmark based radar SLAM using graph optimization. In: 2016 IEEE 19th International Conference on Intelligent Transportation Systems (ITSC), pp. 2559–2564 (2016)

17. Sons, M., Lauer, M., Keller, C.G., Stiller, C.: Mapping and localization using surround view. In: IEEE Intelligent Vehicles Symposium (IV), pp. 1158–1163 (2017)

18. Taş, Ö.Ş., Salscheider, N.O., et al.: Making bertha cooperate-team annieways entry to the 2016 grand cooperative driving challenge. IEEE Trans. Intell. Trans. Syst. **19**(4), 1262–1276 (2017)

19. Wolfson, H.J., Rigoutsos, I.: Geometric hashing: An overview. IEEE Comput. Sci. Eng. **4**(4), 10–21 (1997)

20. Yuan, J., Wu, Y., Yang, M.: Discovery of collocation patterns: from visual words to visual phrases. In: 2007 IEEE Conference on Computer Vision and Pattern Recognition, pp. 1–8. IEEE (2007)

21. Zinoune, C., Bonnifait, P., Ibañez-Guzmán, J.: Sequential FDIA for autonomous integrity monitoring of navigation maps on board vehicles. IEEE Trans. Intell. Transp. Syst. **17**(1), 143–155 (2016)

Towards Bounding-Box Free Panoptic Segmentation

Ujwal Bonde[1(✉)], Pablo F. Alcantarilla[1], and Stefan Leutenegger[1,2]

[1] SLAMcore Ltd., London, UK
{Ujwal,Pablo,Stefan}@slamcore.com
[2] Imperial College London, London, UK

Abstract. In this work we introduce a new Bounding-Box Free Network (BBFNet) for panoptic segmentation. Panoptic segmentation is an ideal problem for proposal-free methods as it already requires per-pixel semantic class labels. We use this observation to exploit class boundaries from off-the-shelf semantic segmentation networks and refine them to predict instance labels. Towards this goal BBFNet predicts coarse watershed levels and uses them to detect large instance candidates where boundaries are well defined. For smaller instances, whose boundaries are less reliable, BBFNet also predicts instance centers by means of Hough voting followed by mean-shift to reliably detect small objects. A novel triplet loss network helps merging fragmented instances while refining boundary pixels. Our approach is distinct from previous works in panoptic segmentation that rely on a combination of a semantic segmentation network with a computationally costly instance segmentation network based on bounding box proposals, such as Mask R-CNN, to guide the prediction of instance labels using a Mixture-of-Expert (MoE) approach. We benchmark our proposal-free method on Cityscapes and Microsoft COCO datasets and show competitive performance with other MoE based approaches while outperforming existing non-proposal based methods on the COCO dataset. We show the flexibility of our method using different semantic segmentation backbones and provide video results on challenging scenes in the wild in the supplementary material.

1 Introduction

Panoptic segmentation is the joint task of predicting semantic scene segmentation together with individual instances of objects present in the scene. Historically this has been explored under different umbrella terms of scene understanding [37] and scene parsing [32]. In [17], Kirillov *et al.* coined the term and gave a more concrete definition by including the suggestion from Forsyth *et al.* [10] of splitting the objects categories into *things* (countable objects like persons, cars, *etc.*.) and *stuff* (uncountable like sky, road, *etc.*.) classes. While *stuff*

Electronic supplementary material The online version of this chapter (https://doi.org/10.1007/978-3-030-71278-5_23) contains supplementary material, which is available to authorized users.

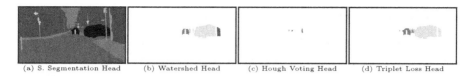

(a) S. Segmentation Head   (b) Watershed Head   (c) Hough Voting Head   (d) Triplet Loss Head

Fig. 1. BBFNet gradually refines the class boundaries of the semantic segmentation network to predict panoptic segmentation. The Watershed head detects candidates for large instances whereas the Hough voting head detects small object instances. The Triplet Loss network refines and merges the detection to obtain the final instance labels.

classes require only semantic label prediction, *things* need both the semantic and instance labels. Along with this definition, *Panoptic Quality* (PQ) measure was proposed to benchmark different methods. Since then, there has been a more focused effort towards panoptic segmentation with multiple datasets [7,23,24] supporting it.

Existing methods for panoptic segmentation can be broadly classified into two groups. The first group uses a proposal based approach for predicting *things*. Traditionally these methods use completely separate instance and scene segmentation networks. Using a MoE approach, the outputs are combined either heuristically or through another sub-network. Although, more recent works propose sharing a common feature backbone for both networks [16,27], this split of tasks restricts the backbone network to the most complex branch. Usually this restriction is imposed by the instance segmentation branch ([13]).

The second group of work uses a proposal free approach for instance segmentation allowing for a more efficient design. An additional benefit of these methods is that they do not need bounding-box predictions. While bounding-box detection based approaches have been popular and successful, they require predicting auxiliary quantities like scale, width and height which do not directly contribute to instance segmentation. Furthermore, the choice of bounding-boxes for object-detection had been questioned in the past [28]. We believe panoptic segmentation to be an ideal problem for a bounding-box free approach since it already contains structured information from semantic segmentation.

In this work, we exploit this using a flexible panoptic segmentation head that can be added to any off-the-shelf semantic segmentation network. We coin this as *Bounding-Box Free Network* (BBFNet) which is a proposal free network and predicts *things* by gradually refiningclass boundaries predicted by the base network. To achieve this we exploit previous works in non-proposal based methods for instance segmentation [2,4,26]. Based on the output of a semantic segmentation network, BBFNet first detects noisy and fragmented large instance candidates using a watershed-level prediction head (see Fig. 1). These candidate regions are clustered and their boundaries improved with a triplet loss based head. The remaining smaller instances, with unreliable boundaries, are detected using a Hough voting [3] head that predicts the offsets to the center of the instance. Without using MoE our method produces comparable results to proposal based approaches while outperforming proposal-free methods on the COCO dataset.

2   Related Work

Most current works in panoptic segmentation fall under the proposal based approach for detecting *things*. In [17], Kirillov *et al.* use separate networks for semantic segmentation (*stuff*) and instance segmentation (*things*) with a heuristic MoE fusion of the two results for the final prediction. Realising the duplication of feature extractors in the two related tasks, [16,18,21,27,35] propose using a single backbone feature extractor network. This is followed by separate branches for the two sub-tasks with a heuristic or learnable MoE head to combine the results. While panoptic Feature Pyramid Networks (FPN) [16] uses Mask R-CNN [13] for the *things* classes and fills in the *stuff* classes using a separate FPN branch, UPSNet [35] combines the resized logits of the two branches to predict the final output. In AUNet [21], attention masks predicted from the Region Proposal Network (RPN) parallizable and the instance segmentation head help fusing the results of the two tasks. Instead of relying only on the instance segmentation branch, TASCNet [18] predicts a coherent mask for the *things* and *stuff* classes using both branches. All these methods rely on Mask R-CNN [13] for predicting *things*. Mask R-CNN is a two-stage instance segmentation network which uses a RPN to predict initial candidates for instance. The two-stage serial approach makes Mask R-CNN accurate albeit computationally expensive and inflexible thus slowing progress towards real-time panoptic segmentation.

In FPSNet [12], the authors replace Mask R-CNN with a computationally less expensive detection network and use its output as a soft attention mask to guide the prediction of *things* classes. This trade off is at a cost of considerable reduction in accuracy while continuing to use a computationally expensive backbone (ResNet50 [14]). In [20] the authors make up for the reduced accuracy by using an affinity network but this is at the cost of computational complexity. Both these methods still use bounding-boxes for predicting *things*. In [31], the detection network is replaced with an object proposal network which predicts instance candidates. In contrast, we propose a flexible panoptic segmentation head that relies only on a semantic segmentation network which, when replaced with faster networks [29,30] allows for a more efficient solution.

A parallel direction gaining increased popularity is the use of proposal-free approach for predicting *things*. In [33], the authors predict the direction to the center and replace bounding box detection with template matching using these predicted directions as a feature. Instead of template matching, [1,19] use a dynamically initiated conditional random field graph from the output of an object detector to segment instances. In the more recent work of Gao *et al.* [11], cascaded graph partitioning is performed on the predictions of a semantic segmentation network and an affinity pyramid computed within a fixed window for each pixel. Cheng *et al.* [5] simplify this process by adopting a parallelizable grouping algorithm for *thing* pixels. In comparison, our flexible panoptic segmentation head predicts *things* by refining the segmentation boundaries obtained from any backbone semantic segmentation network. Furthermore, our post-processing steps are computationally more efficient compared to other proposal-free approaches while outperforming them on multiple datasets.

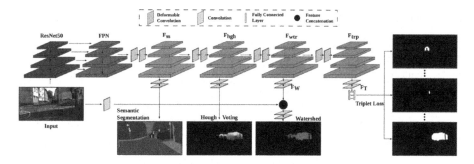

Fig. 2. BBFNet gradually refines the class boundaries of the backbone semantic segmentation network to predict panoptic segmentation. The watershed head predicts quantized watershed levels (shown in different colours) which is used to detect large instance candidates. For smaller instances we use Hough voting with fixed bandwidth. The output shows offsets (X_{off}, Y_{off}) colour-coded to represent the direction of the predicted vector. Triplet head refines and merges the detection to obtain the final instance labels. We show the class probability (colour-map *hot*) for different instances with their center pixels used as f_a. Table 1 lists the components of individual heads while Sect. 3 explains them in detail.

3 Panoptic Segmentation

In this section we introduce our non-bounding box approach to panoptic segmentation. Figure 2 shows the various blocks of our network and Table 1 details the main components of BBFNet. The backbone semantic segmentation network consists of a ResNet50 followed by an FPN [22]. In FPN, we only use the P2, P3, P4 and P5 feature maps which contain 256 channels each and are 1/4, 1/8, 1/16 and 1/32 of the original scale respectively. Each feature map then passes through the same series of eight Deformable Convolutions (DC) [8]. Intermediate features after every couple of DC are used to predict semantic segmentation (Sect. 3.1), Hough votes (Sect. 3.2), watershed energies (Sect. 3.3) and features for the triplet loss [34] network. We first explain each of these components and their corresponding training loss.

3.1 Semantic Segmentation

The first head in BBFNet is used to predict semantic segmentation. This allows BBFNet to quickly predict *things* (C_{things}) and *stuff* (C_{stuff}) labels while the remainder of BBFNet improves *things* boundaries using semantic segmentation features F_{seg}. We use per-pixel cross-entropy loss to train this head given by:

$$L_{ss} = \sum_{c \in \{C_{stuff}, C_{thing}\}} y_c \log(p_c^{ss}), \tag{1}$$

where y_c and p_c^{ss} are respectively the one-hot ground truth label and predicted softmax probability for class c.

3.2 Hough Voting

The Hough voting head is similar to the semantic segmentation head and is used to refine F_{ss} to give Hough features F_{hgh}. These are then used to predict offsets for the center of each *things* pixel. We use a *tanh* non-linearity to squash the predictions and obtain normalised offsets ($\hat{X}_{\text{off}}$ and $\hat{Y}_{\text{off}}$). Along with the centers we also predict the uncertainty in the two directions (σ_x and σ_y) making the number of predictions from the Hough voting head equal to $4 \times C_{things}$. The predicted center for each pixel (x, y), is then given by:

$$
\begin{aligned}
\hat{X}_{\text{center}}(x, y) &= \hat{x} + \hat{X}_{\text{off}}^{C(x,y)}(x, y), \\
\hat{Y}_{\text{center}}(x, y) &= \hat{y} + \hat{Y}_{\text{off}}^{C(x,y)}(x, y),
\end{aligned}
\tag{2}
$$

where C is the predicted class and $(\hat{x}, \hat{y})$ are image normalised pixel location.

Hough voting is inherently noisy [3] and requires clustering or mode seeking methods like mean-shift [6] to predict the final object centers. As instances could have different scales, tuning clustering hyper-parameters is difficult. For this reason we use Hough voting primarily to detect small objects and to filter predictions from other heads. We also observe that the dense loss from the Hough voting head helps convergence of deeper heads in our network.

The loss for this head is only for the *thing* pixels and is given by:

$$
L_{hgh} = w \left(\frac{(X_{\text{off}} - \hat{X}_{\text{off}})^2}{\sigma_x} + \frac{(Y_{\text{off}} - \hat{Y}_{\text{off}})^2}{\sigma_y} \right) - \frac{1}{2} \left(\log(\sigma_x) + \log(\sigma_y) \right), \tag{3}
$$

where X_{off} and Y_{off} are ground truth offsets and w is the per pixel weight. To avoid bias towards large objects, we inversely weigh the instances based on the number of pixels. This allows it to accurately predict the centers for objects of all sizes. Note that we only predict the centers for the visible regions of an instance and do not consider its occluded regions.

3.3 Watershed Energies

Our watershed head is inspired from DWT [2]. Similar to that work, we quantise the watershed levels into fixed number of bins ($K = 4$). The lowest bin ($k = 0$) corresponds to background and regions that are within 2 pixels inside the instance boundary. Similarly, $k = 1$, $k = 2$ are for regions that are within 5 and 15 pixels away from the instance boundary, respectively, while $k = 3$ is for the remaining region inside the instance.

In DWT, the bin corresponding to $k = 1$ is used to detect large instance boundaries. While this does reasonably well for large objects, it fails for smaller objects producing erroneous boundaries. Furthermore, occluded instances that are fragmented cannot be detected as a single object. For this reason we use this head only for predicting large object candidates which are filtered and refined using predictions from other heads.

Due to the fine quantisation of watershed levels, rather than directly predicting the upsampled resolution, we gradually refine the lower resolution feature

Table 1. Architecture of BBFNet. dc, $conv$, ups and cat stand for deformable convolution [8], 1×1 convolution, upsampling and concatenation respectively. The two numbers that follow dc and $conv$ are the input and output channels to the blocks.* indicates that more processing is done on these blocks as detailed in Sect. 3.3 and Sect. 3.4.

Input	Blocks	Output
FPN	dc-256-256, dc-256-128	F_{ss}
F_{ss}	ups, cat, conv-512-$(C_{stuff}+C_{thing})$, ups	Segmentation
F_{ss}	$2\times$dc-128-128	F_{hgh}
F_{hgh}	ups, cat, conv-512-128, conv-128-$(4\times C_{thing})$, ups	Hough
F_{hgh}	$2\times$dc-128-128	F_{wtr}
F_{wtr}	ups, cat, conv-512-128, conv-128-16, ups	F_W^*
F_{wtr}	$2\times$dc-128-128	F_{trp}
F_{trp}	ups, cat, conv-512-128, conv-128-128, ups	F_T^*

maps while also merging higher resolution features from the backbone semantic segmentation network. F_{hgh} is first transformed into F_{wtr} followed by further refining into F_W as detailed in Table 1. Features from the shallowest convolution block of ResNet are then concatenated with F_W and further refined with two 1×1 convolution to predict the four watershed levels.

We use a weighted cross-entropy loss to train this given by:

$$L_{wtr} = \sum_{k \in (0,3)} w_k W_k \, \log(p_k^{wtr}), \qquad (4)$$

where W_k is the one-hot ground truth for k^{th} watershed level, p_k^{wtr} its predicted probability and w_k its weights.

3.4 Triplet Loss Network

The triplet loss network is used to refine and merge the detected candidate instances in addition to detecting new instances. Towards this goal, a popular choice is to formulate it as an embedding problem using triplet loss [4]. This loss forces features of pixels belonging to the same instance to group together while pushing apart features of pixels from different instances. Margin-separation loss is usually employed for better instance separation and is given by:

$$L(f_a, f_p, f_n) = \max\left((f_a - f_p)^2 - (f_a - f_n)^2 + \alpha, 0\right), \qquad (5)$$

where f_a, f_p, f_n are the anchor, positive and negative pixel features respectively and α is the margin. Choosing α is not easy and depends on the complexity of the feature space [25]. Instead, we opt for a fully-connected network to classify the pixel features and formulate it as a binary classification problem:

$$T(f_a, f_*) = \begin{cases} 1 & \text{if, } f_* = f_p, \\ 0 & \text{if, } f_* = f_n, \end{cases} \qquad (6)$$

We use the cross-entropy loss to train this head:

$$L_{trp} = \sum_{c \in (0,1)} T_c \, \log(p_c^{trp}), \tag{7}$$

T_c is the ground truth one-hot label for the indicator function and p^{trp} the predicted probability.

The pixel feature used for this network is a concatenation of F_T (see Table 1), its normalised position in the image (x, y) and the outputs of the different heads $(p^{seg}, p^{wtr}, \hat{X}_{\text{off}}, \hat{Y}_{\text{off}}, \sigma_x$ and $\sigma_y)$.

3.5   Training and Inference

We train the whole network along with its heads in using the weighted loss function:

$$L_{\text{total}} = \alpha_1 \, L_{ss} + \alpha_2 \, L_{hgh} + \alpha_3 \, L_{wtr} + \alpha_4 \, L_{trp}. \tag{8}$$

For the triplet loss network, training with all pixels is prohibitively expensive. Instead we randomly choose a fixed number of anchor pixels N_a for each instance. Hard positive examples are obtained by sampling from the farthest pixels to the object center and correspond to watershed level $k = 0$. For hard negative examples, neighbouring instances' pixels closest to the anchor and belonging to the same class are given higher weight. Only half of the anchors use hard example mining while the rest use random sampling.

We observe that large objects are easily detected by the watershed head while Hough voting based center prediction does well when objects are of the same scale. To exploit this observation, we detect large object candidates $(I_{L'})$ using connected components on the watershed predictions correspond to $k \geq 1$ bins. We then filter out candidates whose predicted Hough center $(I_{L'}^{\text{center}})$ does not fall within their bounding boxes $(BB_{L'})$. These filtered out candidates are fragmented regions of occluded objects or false detections. Using the center pixel of the remaining candidates $(I_{L''})$ as anchors points, the triplet loss network refines them over the remaining pixels allowing us to detect fragmented regions while also improving their boundary predictions.

After the initial watershed step, the unassigned *thing* pixels corresponding to $k = 0$ and primarily belong to small instances. We use mean-shift clustering with fixed bandwidth (B) to predict candidate object centers, I_S^{center}. We then back-trace pixels voting for their centers to obtain the Hough predictions I_S.

Finally, from the remaining unassigned pixels we randomly pick an anchor point and test it with the other remaining pixels. We use this as candidates regions that are filtered (I_R) based on their Hough center predictions, similar to the watershed candidates. The final detections are the union of these predictions. We summarise these steps in algorithm provided in the supplementary material.

4   Experiments

In this section we evaluate the performance of BBFNet and present the results we obtain. We first describe the datasets and the evaluation metrics used. In

Table 2. (a) Performance of different heads (W- Watershed, H- Hough Voting and T- Triplet Loss Network) on Cityscapes validation set. BBFNet exploits the complimentary performance of watershed (large objects $> 10k$ pixels) and Hough voting head (small objects $< 1k$ pixels) resulting in higher accuracy. PQ_s, PQ_m and PQ_l are the PQ scores for small, medium and large objects respectively. Bold is for best results. (b) Performance of Hough voting head (H) with varying B for different sized objects, s-small $< 1k$ pixels, l-large $> 10k$ pixels and m-medium sized instances. For reference we also plot the performance of Watershed+Triplet loss (W+T) head (see Table 2).

W	H	T	PQ	SQ	PQ$_s$	PQ$_m$	PQ$_l$
✓	✗	✗	44.4	75.7	1.3	24.1	57.9
✗	✓	✗	49.7	78.8	11.6	37.4	44.5
✗	✗	✓	55.9	79.4	10.4	45.5	72.0
✓	✓	✓	**56.6**	**80.0**	**12.6**	**47.7**	**72.5**

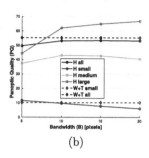

(a)   (b)

Sect. 4.1 we describe the implementation details of our network. Section 4.2 then discusses the performance of individual heads and how its combination helps improve the overall accuracies. We presents both the qualitative and quantitative results in Sect. 4.3 and show the flexibility of BBFNet in Sect. 4.4. We end this section by presenting some of the failure cases in Sect. 4.5 and comparing them with other MoE+BB based approaches.

The Cityscapes dataset [7] contains 2975 densely annotated images of driving scenes for training and a further 500 validation images. For the panoptic challenge, a total of 19 classes are split into 8 *things* and 11 *stuff* classes. Microsoft COCO [23] is a large scale object detection and segmentation dataset with over $118k$ training (2017 edition) and $5k$ validation images. The labels consists of 133 classes split into 80 *things* and 53 *stuff*.

We benchmark using the Panoptic Quality (PQ) measure which was proposed in [16]. Were avilabel we also provide the IoU score.

4.1 Implementation Details

We use the pretrained ImageNet [9] models for ResNet50 and FPN and train the BBFNet head from scratch. We keep the backbone fixed for initial epochs before training the whole network jointly. In the training loss (*Eq.* 8), we set $\alpha_1, \alpha_2, \alpha_3$ and α_4 parameters to 1.0, 0.1, 1.0 and 0.5 respectively, since we found this to be a good balance between the different losses. The mean-shift bandwidth is set to reduced pixels of $B = 10$ to help the Hough voting head detect smaller instances. In the watershed head, the number of training pixels decreases with K and needs to be offset by higher w_k. We found the weights $0.2, 0.1, 0.05, 0.01$ to work best for our experiments. Moreover, these weights help the network

Table 3. Panoptic segmentation results on the Cityscapes and COCO dataset. All methods use the same pretraining (ImageNet) and backbone (ResNet50+FPN), except those with * (ResNet101) and ± (Xception-71). Bold is for overall best results and underscore is the best result in non-BB based methods.

Method	BB	Cityscapes				COCO				
		PQ	PQ^{Th}	PQ^{St}	IoU	PQ	PQ^{Th}	PQ^{St}	IoU	PQ^{test}
FPSNet [12]	✓	55.1	48.3	60.1	-	-	-	-	-	-
Li *et al.* [20]	✓	**61.4**	54.7	**66.6**	77.8	**43.4**	48.6	**35.5**	53.7	**47.2**[*]
Porzi *et al.* [27]	✓	60.3	**56.1**	63.6	77.5	-	-	-	-	-
TASCNet [18]	✓	55.9	50.5	59.8	-	-	-	-	-	40.7
AUNet [21]	✓	56.4	52.7	59.0	73.6	39.6	49.1	25.2	45.1	45.2[*]
P. FPN [16]	✓	57.7	51.6	62.2	75.0	39.0	45.9	28.7	41.0	40.9[*]
AdaptIS [31]	✓	59.0	55.8	61.3	75.3	35.9	29.3	40.3	-	42.8[*]
UPSNet [35]	✓	59.3	54.6	62.7	75.2	42.5	48.5	33.4	54.3	46.6[*]
DIN [19]	✗	53.8	42.5	<u>62.1</u>	71.6	-	-	-	-	-
DeeperLab [36]	✗	56.5±	-	-	-	33.8±	-	-	-	34.4±
SSAP [11]	✗	56.6	49.2	-	75.1	36.5[*]	-	-	-	36.9[*]
P. DeepLab [5]	✗	<u>59.7</u>	-	-	**80.5**	35.1	-	-	-	41.4±
BBFNet	✗	56.6	<u>49.9</u>	61.1	76.5	<u>37.1</u>	<u>42.9</u>	<u>28.5</u>	**54.9**	<u>42.9</u>[*]

focus on detecting pixels corresponding to lower bins on whom the connected-component is performed. To train the triplet-loss network head we set the number of pixels per object $N_a = 1000$. To improve robustness we augment the training data by randomly cropping the images and adding alpha noise, flipping and affine transformations. No additional augmentation was used during testing. All experiments were performed on NVIDIA Titan 1080Ti.

A common practice during inference is to remove prediction with low detection probability to avoid penalising twice (FP and FN) [35]. In BBFNet, these correspond to regions with poor segmentation. We remove regions with low mean segmentation probability (<0.65). Furthermore, we also observe boundaries shared between multiple objects to be frequently predicted as different instances. We filter these by having a threshold (0.1) on the IoU between the segmented prediction and its corresponding bounding box.

4.2 Ablation Studies

We conduct ablation studies here to show the advantage of each individual head and how BBFNet exploits them. Table 2(a) shows the results of our experiments on Cityscapes. We use the validation sets for all our experiments. We observe that watershed or Hough voting heads alone do not perform well. In the case of watershed head this is because performing connected component analysis on $k = 1$ level (as proposed in [2]) leads to poor SQ. Note that performing the watershed cut at $k = 0$ is also not optimal as this leads to multiple instances that share boundaries being grouped into a single detection. By combining the

Fig. 3. Sample qualitative results of BBFNet on Cityscapes (first row) and COCO dataset. BBFNet can handle different object classes with multiple instances.

Watershed head with a refining step from the triplet loss network we observe over 10 point improvement in accuracy.

On the other hand, the performance of the Hough voting head depends on the bandwidth B that is used. Table 2(b) plots its performance with varying B. As B increases from 5 to 20 pixels we observe an initial increase in overall PQ before it saturates. This is because while the performance increases on large objects ($>10k$ pixels), it reduces on small ($<1k$ pixels) and medium sized objects. However, we observe that at lower B it outperforms the Watershed+triplet loss head on smaller objects. We exploit this in BBFNet (see Sect. 3.5) by using the watershed+triplet loss head for larger objects while using Hough voting head primarily for smaller objects.

4.3   Experimental Results

Table 3 benchmarks the performance of BBFNet with existing methods on the Cityscapes and COCO datasets. As all state-of-the-art methods report results with ResNet50+FPN networks while using the same pre-training dataset (ImageNet) we also follow this convention and report our results with this setup except where highlighted. Multi-scale testing along with horizontal-flipping were used in some works but we omit those results here as this can be applied to any existing work including BBFNet to improve performance. From the results we observe that BBFNet, without using an MoE or BB, has comparable performance to other MoE+BB based methods while outperforming non-BB based methods on the more complicated COCO dataset. Figure 3 shows some qualitative results on the Cityscapes and COCO dataset.

Table 4. Panoptic segmentation results showing the trade-off between performance and efficiency with different semantic segmentation backbones on the Cityscapes dataset. For efficiency we use flops, parameters, inference time (T_I) and the post-processing time (T_{PP}). We compare this with a baseline proposal based network (UPSNet) and a proposal free network (SSAP).

Network	Backbone	PQ	SQ	RQ	IoU	Flops (T)	Params (M)	T_I (sec)	T_{PP} (sec)
BBFNet	ERFNet [29] (w/o DC)	46.8	76.2	58.7	69.0	**0.07**	**2.58**	**0.12**	**0.15**
	MobileNetV2 [30]	48.2	77.0	60.3	70.1	0.1	4.26	0.22	0.15
	ResNet50 [14] (w/o DC)	50.4	76.3	62.4	68.9	0.49	28.2	0.16	0.15
	ResNet50 [14]	56.6	80.0	69.3	76.5	0.38	29.5	0.32	0.15
	ResNet101 [14]	57.8	**80.7**	70.2	**78.6**	0.53	48.49	0.34	0.15
SSAP [11]	ResNet50 [14]	56.6	-	-	75.1	-	-	-	≥0.26
UPSNet [35]	ResNet50 [14]	**59.3**	79.7	**73.0**	75.2	0.425	44.5	0.2	0.33

4.4 Flexibility and Efficiency

To highlight BBFNets ability to work with different segmentation backbones we compare its generalisation with different segmentation networks. As it is expected we observe an increase in performance with more complex backbones and with DC's but at a cost of reduced efficiency (see Table 4). For reference we also show the performance of a baseline proposal-based approach (UPSNet) and a proposal-free approach (SSAP). We used the author provided code of UPSNet[1] for computing efficiency figures. Note, that since UPSNet uses Mask R-CNN its backbone cannot be replaced and it is not as flexible as BBFNet.

As BBFNet does not use a separate instance segmentation head, it is computationally more efficient using only $\approx 29.5M$ parameters compared to $44.5M$ UPSNet. We find a similar pattern when we compare the number of FLOPs on a 1024×2048 image with BBFNet taking 0.38 TFLOPs compared to 0.425 TFLOPs of UPSNet. The authors of SSAP [11] do not provide details about their number of parameters, FLOPs and inference time. However, they provide timing information for their post-processing step which is a cascaded graph partitioning approach that uses the predictions of a semantic segmentation network and an affinity pyramid network. This cascaded graph partition module solves a multicut optimisation problem [15] and takes between $0.26 - 1.26$ seconds depending on the initial resolution for the cascaded graph partition. We believe that BBFNet post-processing step is simpler and presumably faster than in SSAP.

4.5 Error Analysis

We discuss the reasons for performance difference between our bounding-box free method and ones that use bounding-box proposals. UPSNet [35] is used as

---

[1] Source code available from https://github.com/uber-research/UPSNet.

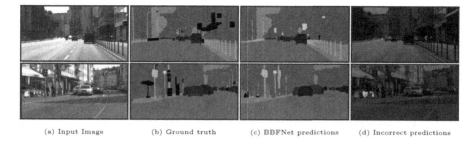

(a) Input Image   (b) Ground truth   (c) BBFNet predictions   (d) Incorrect predictions

Fig. 4. Sample results where BBFNet fails. First row shows an example where low confidence of semantic segmentation network leads to missed detection while the second row shows examples of false positives due to wrong class label prediction. Without MoE these errors from the semantic segmentation network cannot be corrected by BBFNet.

a benchmark as it shares common features with other methods. Table 5 depicts the number of predictions made for different sized objects in the Cityscapes validation dataset. We report the True Positive (TP), False Positive (FP) and the False Negative (FN) values.

One of the areas where BBFNet performs poorly is the number of small object detections. BBFNet detects 2/3 of the smaller objects compared to UPSNet. Poor segmentation (wrong class label or inaccurate boundary prediction) also leads to a relatively higher FP for medium and large sized objects. Figure 4 shows some sample examples. The multi-head MoE approach helps addressing these issues but at the cost of additional complexity and computation time (Sect. 4.3). For applications where time or memory are more critical compared to detecting smaller objects, BBFNet would be a more suited solution.

Table 5. Performance comparison of BBFNet with an MoE+BB method (UPSNet). Due to a non-MoE approach, errors from the backbone semantic segmentation network (low TP-small and high FP-medium, large) cannot be corrected by BBFNet.

Network	Small			Medium			Large		
	TP	FP	FN	TP	FP	FN	TP	FP	FN
UPSNet	1569	722	2479	3496	401	954	1539	49	82
BBFNet	1067	666	2981	3446	680	1004	1527	82	94

5   Conclusions and Future Work

We presented an efficient bounding-box free panoptic segmentation method called BBFNet. Unlike previous methods, BBFNet does not use any instance segmentation network to predict *things*. It instead refines the boundaries from the semantic segmentation output obtained from any off-the-shelf segmentation network. This allows us to be flexible while out-performing proposal-free methods on the more complicated COCO benchmark.

In the next future we would work on making the network end-to-end trainable and improving the efficiency by removing the use of DCN while maintaining similar accuracy.

Acknowledgment. We would like to thank Prof. Andrew Davison and Dr. Alexandre Morgand for their critical feedback during the course of this work.

References

1. Arnab, A., Torr, P.H.: Pixelwise instance segmentation with a dynamically instantiated network. In: IEEE Conference on Computer Vision and Pattern Recognition (CVPR) (2017)
2. Bai, M., Urtasun, R.: Deep watershed transform for instance segmentation. In: IEEE Conference on Computer Vision and Pattern Recognition (CVPR), pp. 2858–2866 (2017)
3. Ballard, D.H.: Generalizing the hough transform to detect arbitrary shapes. Pattern Recogn. **13**(2), 111–122 (1981)
4. Brabandere, B.D., Neven, D., Gool, L.V.: Semantic instance segmentation with a discriminative loss function. arXiv preprint arXiv:1708.02551 (2017)
5. Cheng, B., Collins, M., Zhu, Y., Liu, T., Huang, T., Adam, H., Chen, L.: Panoptic-deeplab: a simple, strong, and fast baseline for bottom-up panoptic segmentation. In: IEEE Conference on Computer Vision and Pattern Recognition (CVPR) (2020)
6. Cheng, Y.: Mean shift, mode seeking, and clustering. IEEE Trans. Pattern Anal. Machine Intell. **17**(8), 790–799 (1995)
7. Cordts, M., Omran, M., Ramos, S., Rehfeld, T., Enzweiler, M., Benenson, R., Franke, U., Roth, S., Schiele, B.: The cityscapes dataset for semantic urban scene understanding. In: IEEE Conf. on Computer Vision and Pattern Recognition (CVPR) (2016)
8. Dai, J., Qi, H., Xiong, Y., Li, Y., Zhang, G., Hu, H., Wei, Y.: Deformable convolutional networks. In: International Conference on Computer Vision (ICCV) (2017)
9. Deng, J., Dong, W., Socher, R., Li, L.J., Li, K., Fei-Fei, L.: Imagenet: a large-scale hierarchical image database. In: IEEE Conference on Computer Vision and Pattern Recognition (CVPR) (2009)
10. Forsyth, D., et al.: Finding pictures of objects in large collections of images. In: International Workshop on Object Representation in Computer Vision (1996)
11. Gao, N., et al.: SSAP: single-shot instance segmentation with affinity pyramid. In: International Conference on Computer Vision (ICCV) (2019)
12. de Geus, D., Meletis, P., Dubbelman, G.: Fast panoptic segmentation network. arXiv preprint arXiv:1910.03892 (2019)
13. He, K., Gkioxari, G., Dollár, P., Girshick, R.: Mask R-CNN. In: International Conference on Computer Vision (ICCV) (2017)
14. He, K., Zhang, X., Ren, S., Sun, J.: Deep residual learning for image recognition. In: IEEE Conference on Computer Vision and Pattern Recognition (CVPR) (2015)
15. Keuper, M., Levinkov, E., Bonneel, N., Lavoue, G., Brox, T., Andres, B.: Efficient decomposition of image and mesh graphs by lifted multicuts. In: International Conference on Computer Vision (ICCV) (2015)

16. Kirillov, A., Girshick, R., He, K., Dollár, P.: Panoptic feature pyramid networks. In: IEEE Conference on Computer Vision and Pattern Recognition (CVPR) (2019)

17. Kirillov, A., He, K., Girshick, R., Rother, C., Dollár, P.: Panoptic segmentation. In: IEEE Conference on Computer Vision and Pattern Recognition (CVPR) (2019)

18. Li, J., Raventos, A., Bhargava, A., Tagawa, T., Gaidon, A.: Learning to fuse things and stuff. arXiv preprint arXiv:1812.01192 (2019)

19. Li, Q., Arnab, A., Torr, P.H.S.: Weakly- and semi-supervised panoptic segmentation. In: Ferrari, V., Hebert, M., Sminchisescu, C., Weiss, Y. (eds.) ECCV 2018. LNCS, vol. 11219, pp. 106–124. Springer, Cham (2018). https://doi.org/10.1007/978-3-030-01267-0_7

20. Li, Q., Qi, X., Torr, P.: Unifying training and inference for panoptic segmentation. In: IEEE Conference on Computer Vision and Pattern Recognition (CVPR) (2020)

21. Li, Y., et al.: Attention-guided unified network for panoptic segmentation. In: IEEE Conference on Computer Vision and Pattern Recognition (CVPR) (2019)

22. Lin, T., Dollár, P., Girshick, R., He, K., Hariharan, B., Belongie, S.: Feature pyramid networks for object detection. In: IEEE Conference on Computer Vision and Pattern Recognition (CVPR) (2017)

23. Lin, T.-Y., et al.: Microsoft COCO: common objects in context. In: Fleet, D., Pajdla, T., Schiele, B., Tuytelaars, T. (eds.) ECCV 2014. LNCS, vol. 8693, pp. 740–755. Springer, Cham (2014). https://doi.org/10.1007/978-3-319-10602-1_48

24. Neuhold, G., Ollmann, T., Bulò, S.R., Kontschieder, P.: The Mapillary Vistas dataset for semantic understanding of street scenes. In: International Conference on Computer Vision (ICCV) (2017). https://www.mapillary.com/dataset/vistas

25. Neven, D., Brabandere, B.D., Proesmans, M., Gool, L.V.: Instance segmentation by jointly optimizing spatial embeddings and clustering bandwidth. In: IEEE Conference on Computer Vision and Pattern Recognition (CVPR) (2019)

26. Neven, D., Brabandere, B.D., Georgoulis, S., Proesmans, M., Gool, L.V.: Fast scene understanding for autonomous driving. arXiv preprint arXiv:1708.02550 (2017)

27. Porzi, L., Bulò, S.R., Colovic, A., Kontschieder, P.: Seamless scene segmentation. In: IEEE Conference on Computer Vision and Pattern Recognition (CVPR) (2019)

28. Redmon, J., Farhadi, A.: Yolov3: An incremental improvement. arXiv preprint arXiv:1804.02767 (2018)

29. Romera, E., Álvarez, J.M., Bergasa, L.M., Arroyo, R.: ErfNet: efficient residual factorized convnet for real-time semantic segmentation. IEEE Trans. Intell. Transp. Syst. **19**, 263–272 (2018)

30. Sandler, M., Howard, A., Zhu, M., Zhmoginov, A., Chen, L.: Mobilenetv 2: inverted residuals and linear bottlenecks. In: IEEE Conference on Computer Vision and Pattern Recognition (CVPR) (2018)

31. Sofiiuk, K., Barinova, O., Konushin, A.: Adaptis: adaptive instance selection network. In: International Conference on Computer Vision (ICCV) (2019)

32. Tighe, J., Niethammer, M., Lazebnik, S.: Scene parsing with object instances and occlusion ordering. In: IEEE Conference on Computer Vision and Pattern Recognition (CVPR) (2014)

33. Uhrig, J., Cordts, M., Franke, U., Brox, T.: Pixel-level encoding and depth layering for instance-level semantic labeling. In: German Conference on Pattern Recognition (GCPR) (2016)

34. Weinberger, K.Q., Saul, L.K.: Distance metric learning for large margin nearest neighbor classification. J. Mach. Learn. Res. (2009)

35. Xiong, Y., et al.: UPSNet: a unified panoptic segmentation network. In: IEEE Conference on Computer Vision and Pattern Recognition (CVPR) (2019)

36. Yang, T., et al.: Deeperlab: single-shot image parser. arXiv preprint arXiv:1902.05093 (2019)
37. Yao, J., Fidler, S., Urtasun, R.: Describing the scene as a whole: joint object detection. In: IEEE Conference on Computer Vision and Pattern Recognition (CVPR) (2012)

Proposal-Free Volumetric Instance Segmentation from Latent Single-Instance Masks

Alberto Bailoni[1], Constantin Pape[2], Steffen Wolf[1], Anna Kreshuk[2], and Fred A. Hamprecht[1(✉)]

[1] HCI/IWR, Heidelberg University, 69120 Heidelberg, Germany
{alberto.bailoni,steffen.wolf,fred.hamprecht}@iwr.uni-heidelberg.de
[2] EMBL, 69117 Heidelberg, Germany
{constantin.pape,anna.kreshuk}@embl.de

Abstract. This work introduces a new proposal-free instance segmentation method that builds on single-instance segmentation masks predicted across the entire image in a sliding window style. In contrast to related approaches, our method concurrently predicts all masks, one for each pixel, and thus resolves any conflict jointly across the entire image. Specifically, predictions from overlapping masks are combined into edge weights of a signed graph that is subsequently partitioned to obtain all final instances concurrently. The result is a parameter-free method that is strongly robust to noise and prioritizes predictions with the highest consensus across overlapping masks. All masks are decoded from a low dimensional latent representation, which results in great memory savings strictly required for applications to large volumetric images. We test our method on the challenging CREMI 2016 neuron segmentation benchmark where it achieves competitive scores.

1 Introduction

Instance segmentation is the computer vision task of assigning each pixel of an image to an instance, such as individual car, person or biological cell. There are two main types of successful deep learning approaches to instance segmentation: *proposal-based* and *proposal-free* methods. Recently, there has been a growing interest in the latter. Proposal-free methods do not require object detection and are preferred in imagery as studied here, in which object instances cannot be approximated by bounding boxes and are much larger than the field of view of the model.

Some recent successful proposal-free approaches [11,17,19] tackle instance segmentation by predicting, for a given patch of the input image, whether or not each pixel in the patch is part of the instance that covers the central pixel of the patch. This results a probability mask, which from now on we call *central instance mask*. These masks are then repeatedly predicted across the entire

---

Electronic supplementary material The online version of this chapter (https:// doi.org/10.1007/978-3-030-71278-5_24) contains supplementary material, which is available to authorized users.

Z. Akata et al. (Eds.): DAGM GCPR 2020, LNCS 12544, pp. 331–344, 2021.
https://doi.org/10.1007/978-3-030-71278-5_24

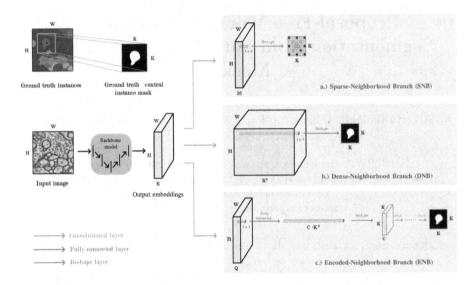

Fig. 1. Comparison between the proposed method and the strong baseline representing the current state-of-the-art. **Left**: At the top-left corner, an example of binary central instance mask for a given ground truth label image; below, the backbone model predicts feature maps with the spatial dimensions of the input image. **Right**: a) *Sparse-neighborhood branch* used in the baseline model to predict affinities for a given sparse neighborhood structure; b) Simple generalization of the *sparse-neighborhood branch* to predict dense central instance masks; c) Proposed *encoded-neighborhood branch* predicting central instance masks in a low-dimensional latent space.

image, either in a sliding window style or by starting from a seed and then shifting the field of view depending on the previously predicted masks. Final object-instances are then obtained by aggregating predictions from overlapping masks which is in itself a nontrivial and interesting problem.

In this work, we introduce a new proposal-free segmentation method that is also based on predicting central instance masks[1]. However, our approach comes with four main advantages compared to previous methods. Firstly, our model concurrently predicts all central instance masks, one for each pixel, by using a fully-convolutional approach with much smaller computational footprint than previous methods, which iteratively predict one instance at the time, one mask after the other [11,19]. Secondly, our approach predicts central instance masks in a low dimensional latent representation (see Fig. 1c), which results in great memory savings that are strictly required to apply the method to large volumetric images. Thirdly, the proposed approach aggregates predictions from overlapping central instance masks without the need for any extra parameter or threshold and outputs predictions with associated uncertainty; and, finally, all final object-instances are obtained concurrently, as opposed to previous methods predicting them one-by-one with subsequent conflict resolution.

---

[1] For interesting, closely related but independent work, see [10].

Additionally, we systematically compare the proposed model with the current state-of-the-art proposal-free method both on natural and biological images [3,8,15,18,31]. This strong baseline consists of a fully-convolutional network predicting, for each pixel, an arbitrary predefined set of short- and long-range affinities, i.e. neighborhood relations representing how likely it is for a pair of pixels to belong to the same object instance (see Fig. 1a).

Our method achieves competitive scores on the challenging CREMI 2016 neuron segmentation benchmark. In our set of validation experiments, we show how predicting encoded central instance masks always improves accuracy. Moreover, when predictions from overlapping masks are combined into edge weights of a graph that is subsequently partitioned, the result is a method that is strongly robust to noise and gives priority to predictions sharing the highest consensus across predicted masks. This parameter-free algorithm, for the first time, outperforms super-pixel based methods, which have so far been the default choice on the challenging data from the CREMI competition challenge.

2   Related Work

Many of the recent successful instance segmentation methods on natural images are *proposal-based*: they first perform object detection, for example by predicting anchor boxes [24], and then assign a class and a binary segmentation mask to each detected bounding box [9,23]. Proposal-Free methods on the other hand directly group pixels into instances. Recent approaches use metric learning to predict high-dimensional associative pixel embeddings that map pixels of the same instance close to each other, while mapping pixels belonging to different instances further apart, e.g. [13,14]. Final instances are then retrieved by applying a clustering algorithm. A post-processing step is needed to merge instances that are larger then the field of view of the network.

Aggregating Central Instance Masks – The line of research closest to ours predicts overlapping central instance masks in a sliding window style across the entire image. The work of [17] aggregates overlapping masks and computes intersection over union scores between them. In neuron segmentation, floodfilling networks [11] and MaskExtend [19] use a CNN to iteratively grow one instance/neuron at a time, merging one mask after the other. Recently, the work of [20] made the process more efficient by employing a combinatorial encoding of the segmentation, but the method remains orders of magnitude slower as compared to the convolutional one proposed here, since in our case all masks are predicted at the same time and for all instances at once. The most closely related work to ours is the independent preprint [10], where a very similar model is applied to the BBBC010 benchmark microscopy dataset of *C. elegans* worms. However, here we propose a more efficient model that scales to 3D data, and we provide an extensive comparison to related models predicting long-range pixel-pair affinities.

Predicting Pixel-Pair Affinities – Instance-aware edge detection has experienced recent progress due to deep learning, both on natural images and biological data [3,8,15,18,22,27,31,33]. Among these methods, the most recent ones also predict long-range affinities between pixels and not only direct-neighbor relationships [8,15,18]. Other related work [7,28] approach boundary detection via a structured learning approach. In neuron segmentation, boundaries predicted by a CNN are converted to final instances with subsequent postprocessing and superpixel-merging. Some methods define a graph with both positive and negative weights and formulate the problem in a combinatorial framework, known as multicut or correlation clustering problem [4]. In neuron segmentation and connectomics, exact solvers can tackle problems of considerable size [1], but accurate approximations [21,32] and greedy agglomerative algorithms [3,16,30] are required on larger problems.

3   Model and Training Strategy

In this section, we first define central instance masks in Sect. 3.1. Then, in Sect. 3.2, we present our first main contribution, a model trained end-to-end to predict encoded central instance masks, one for each pixel of the input image.

3.1   Local Central Instance Masks

This work proposes to distinguish between different object instances based on instance-aware pixel-pair affinities in the interval $[0, 1]$, which specify whether or not two pixels belong to the same instance or not. Given a pixel of the input image with coordinates $\boldsymbol{u} = (u_x, u_y)$, a set of affinities to neighboring pixels within a $K \times K$ window is learned, where K is an odd number. We define the $K \times K$-neighborhood of a pixel as: $\mathcal{N}_{K \times K} \equiv \mathcal{N}_K \times \mathcal{N}_K$, where $\mathcal{N}_K \equiv \left\{ -\frac{K-1}{2}, \ldots, \frac{K-1}{2} \right\}$ and represent the affinities relative to pixel $\boldsymbol{u}$ as a central instance mask $\mathcal{M}_{\boldsymbol{u}} : \mathcal{N}_{K \times K} \to [0, 1]$.

We represent the associated training targets as binary ground-truth masks $\hat{\mathcal{M}}_{\boldsymbol{u}} : \mathcal{N}_{K \times K} \to \{0, 1\}$, which can be derived from a ground-truth instance label image $\hat{L} : H \times W \to \mathbb{N}$ with dimension $H \times W$:

$$\forall \boldsymbol{u} \in H \times W, \quad \forall \boldsymbol{n} \in \mathcal{N}_{K \times K} \quad \hat{\mathcal{M}}_{\boldsymbol{u}}(\boldsymbol{n}) = \begin{cases} 1, & \text{if } \hat{L}(\boldsymbol{u}) = \hat{L}(\boldsymbol{u} + \boldsymbol{n}) \\ 0, & \text{otherwise.} \end{cases} \quad (1)$$

We actually use similar definitions in 3D, but use 2D notation here for simplicity.

3.2   Training Encoded Central Instance Masks End-To-End

In several related work approaches [3,8,15,18,31], affinities between pairs of pixels are predicted for a predefined sparse stencil representing a set of N short- and long-range neighborhood relations for each pixel ($N = 8$ *sparse-neighborhood branch* of Fig. 1a). The N output feature maps are then trained with a binary classification loss.

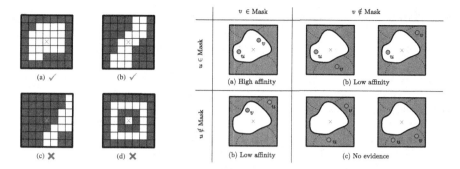

Fig. 2. Examples of expected (**a-b**) and not expected (**c-d**) binary 2D central instance masks.

Fig. 3. Computing instance-aware affinity between pixels u and v from instance masks associated to the central pixel in the patch (orange cross).

On paper, this training method can be easily generalized to output a feature map of size $K^2 \times H \times W$ and thus predict a full $K \times K$ central instance mask for each pixel of the input image (see *dense-neighborhood branch* in Fig. 1b). Nevertheless, in practice, this model has prohibitively large memory requirements for meaningful values of K, precluding application to 3D data of interest here.

However, among the $2^{K \cdot K}$ conceivable binary masks $\hat{\mathcal{M}}_u : \mathcal{N}_{K^2} \to \{0, 1\}$, in practice only a tiny fraction corresponds to meaningful instance masks (see some examples in Fig. 2). This suggests that it is possible to find a compact representation that spans the manifold of expected instance shapes.

As our first main contribution, we test this assumption by training a model end-to-end to predict, for each pixel $u \in H \times W$ of the input image, a latent vector $z_u \in \mathbb{R}^Q$ encoding the $K \times K$ central instance mask $\mathcal{M}_u$ centered at pixel u (see *encoded-neighborhood branch* in Fig. 1c). The backbone model is first trained to output a more compact $Q \times H \times W$ feature map and then a tiny convolutional decoder network is applied to each pixel of the feature map to decode masks. During training, decoding one mask for each pixel in the image would be too memory consuming. Thus, we randomly sample R pixels with coordinates $u_1, \ldots, u_R$ and only decode the associated masks $\mathcal{M}_{u_1}, \ldots, \mathcal{M}_{u_R}$. Given the ground-truth central instance masks $\hat{\mathcal{M}}_{u_i}$ defined in Eq. 1, the training loss is then defined according to the Sørensen-Dice coefficient formulated for fuzzy set membership values, similarly to what was done in [31]. Ground-truth labels are not always pixel-precise and it is often impossible to estimate the correct label for pixels that are close to a ground-truth label transition. Thus, in order to avoid noise during training, we predict completely empty masks for pixels that are less than two pixels away from a label transition, so that the model is trained to predict single-pixel clusters along the ground-truth boundaries. In our experiments, this approach performed better than masking the training loss along the boundaries.

Algorithm 1: Affinities from Aggregated Central Instance Masks

---

Input: Graph $\mathcal{G}(V, E)$; central instance masks $\mathcal{M}_u : \mathcal{N}_{K \times K} \to [0, 1]$
Output: Affinities $\bar{a}_e \in [0, 1]$ with variance σ_e^2 for all edges $e \in E$

1: **for** each edge $e = (u, v) \in E$ in graph $\mathcal{G}$ **do**
2:    Get coordinates $u = (u_x, u_y)$ and $v = (v_x, v_y)$ of pixels linked by edge e
3:    Collect all T masks $\mathcal{M}_{c_1}, \ldots, \mathcal{M}_{c_T}$ including both pixel u and pixel v
4:    Init. vectors $[a_1, \ldots, a_T] = [w_1, \ldots, w_T] = 0$ for affinities and evidence weights
5:    **for** $i \in 1, \ldots, T$ **do**
6:       Get relative coords. of u and v with respect to the central pixel c_i
7:       $a_i \leftarrow \min\left(\mathcal{M}_{c_i}(u - c_i), \mathcal{M}_{c_i}(v - c_i)\right)$ ▷ Fuzzy-AND: both values active
8:       $w_i \leftarrow \max\left(\mathcal{M}_{c_i}(u - c_i), \mathcal{M}_{c_i}(v - c_i)\right)$ ▷ Fuzzy-OR: at least one value active
9:    Get weighted affinity average $\bar{a}_e = \sum_i a_i w_i \,/\, \sum_i w_i$
10:    Get weighted affinity variance $\sigma_e^2 = \sum_i w_i (a_i - \bar{a}_e)^2 \,/\, \sum_i w_i$
11: **return** $\bar{a}_e, \sigma_e^2$ for each $e \in E$

3.3   Predicting Multi-scale Central Instance Masks

Previous related work [8,15,18] shows that predicting long-range affinities between distant pixels improves accuracy as compared to predicting only short-range ones. However, predicting large central instance masks would translate to a bigger model that, on 3D data, would have to be trained on a small 3D input field of view. This, in practice, usually decreases accuracy because of the reduced 3D context available to the network. Thus, we instead predict multiple central instance masks of the same window size $7 \times 7 \times 5$ but at different resolutions, so that the lower the resolution the larger the size of the associated patch in the input image. These multiple masks at different resolutions are predicted by adding several *encoded-neighborhood branches* along the hierarchy of the decoder in the backbone model, which in our case is a 3D U-Net [5,26] (see Fig. S7). In this way, the encoded central instance masks at higher and lower resolutions can be effectively learned at different levels in the feature pyramid of the U-Net.

4   Affinities with Uncertainty from Aggregated Masks

In order to obtain an instance segmentation from the predictions of the model presented in Sect. 3, we now compute instance-aware pixel-pair affinities for a given sparse N-neighborhood structure (see Table S3 in Supplementary Material for details about the structure) and use them as edge weights of a pixel grid-graph $\mathcal{G}(V, E)$, such that each node represents a pixel/voxel of the image. The graph is then partitioned to obtain object instances.

In this section, we propose an algorithm that, without the need of any threshold parameter, aggregates predictions from overlapping central instance masks and outputs edge weights with associated uncertainty. Related work either

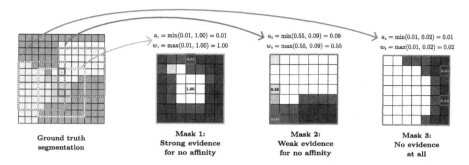

Fig. 4. Proposed method to average overlapping masks and compute the affinity between pixel u and pixel v (highlighted in red in the ground-truth segmentation on the left). For simplicity, we only consider three masks among all the ones including both pixels u and v. In *Mask 1*, only v is part of the mask, so there is a strong evidence for no affinity between u and v; in *Mask 2*, u is predicted to be part of the mask only with a low confidence, so the contribution of this mask in the final average will be weak; in *Mask 3*, both pixels are not part of the central instance mask, so there is no evidence about their affinity. The final affinity value of edge (u, v) is given by the weighted average of the collected affinities a_i weighted with the evidence weights w_i: $\bar{a}_e = \sum_{i=1}^{3} a_i w_i / \sum_i w_i$ (Color figure online)

thresholds the predicted central instance masks [10,11,19] or computes Intersection over Union (IoU) scores for overlapping patches [17]. However, an advantage of predicting pixel-pair affinities/pseudo-probabilities as compared to IoU scores is that affinities can easily be translated into attractive and repulsive interactions in the grid-graph and a parameter-free partitioning algorithm can be employed to yield instances.

Here, we propose a simple algorithm to aggregate predictions from multiple patches: Fig. 4 shows a simplified example of how Algorithm 1 computes the affinity for an edge e linking a pair of pixels u and v. As a first step, the algorithm loops over all predicted central instance masks including both u and v. However, not all these masks are informative, as we visually explain in Fig. 3: a mask $\mathcal{M}_{c_i}$ centered at pixel c_i provides any evidence about the affinity between pixels u and v only if at least one of the two pixels belongs to the mask (fuzzy OR operator at line 8 in Algorithm 1). If both pixels do not belong to it, we cannot say anything about whether they belong to the same instance (see Fig. 3c). We model this with an evidence weight $w_i \in [0, 1]$, which is low when both pixels do not belong to the mask. On the other hand, when at least one of the two pixels belongs to the mask, we distinguish two cases (fuzzy AND operator at line 7 in Algorithm 1): i) both pixels belong to the mask (case in Fig. 3a), so by transitivity we conclude they should be in the same instance and their affinity a_i should tend to one; ii) only one pixel belongs to the mask (case in Fig. 3b), so that according to this mask they are in different instances and their affinity should tend to zero.

At the end, we compute a weighted average $\bar{a}_e$ and variance σ_e^2 of the collected affinities from all overlapping masks, such that masks with more evidence will contribute more on average, and the obtained variance is a measure of how consistent were the predictions across masks. The algorithm was implemented on GPU using PyTorch and the variance was computed via Welford's online stable algorithm [29].

5   Experiments on Neuron Segmentation

We evaluate and compare our method on the task of neuron segmentation in electron microscopy (EM) image volumes. This application is of key interest in connectomics, a field of neuro-science with the goal of reconstructing neural wiring diagrams spanning complete central nervous systems. We test our method on the competitive CREMI 2016 EM Segmentation Challenge [6]. We use the second half of CREMI sample C as validation set for our comparison experiments in Table 1 and then we train a final model on all the three samples with available ground truth labels to submit results to the leader-board in Table 2. Results are evaluated using the CREMI score, which is given by the geometric mean of Variation of Information Score (VOI split + VOI merge) and Adapted Rand-Score (Rand-Score) [2]. See Sect. S7.4 in Supplementary Material for more details on data augmentation, strongly inspired by related work.

5.1   Architecture Details of the Tested Models

As a backbone model we use a 3D U-Net consisting of a hierarchy of four feature maps with anisotropic downscaling factors $(\frac{1}{2}, \frac{1}{2}, 1)$, similarly to [14,15,31]. Models are trained with the Adam optimizer and a batch size equal to one. Before applying the loss, we slightly crop the predictions to prevent training on borders where not enough surrounding context is provided. See Sect. S7.2 and Fig. S7 in Supplementary Material for all details about the used architecture.

Baseline Model (SNB) – As a strong baseline, we re-implement the current state-of-the-art and train a model to predict affinities for a sparse neighborhood structure (Fig. 1a). We perform deep supervision by attaching three *sparse-neighborhood branches* (SNB) at different levels in the hierarchy of the UNet decoder and train the coarser feature maps to predict longer range affinities. Details about the used neighborhood structures and the architecture can be found in Table S3 and Fig. S7 in Supplementary Material.

Proposed Model (ENB) – We then train a model to predict encoded central instance masks (Fig. 1c). Similarly to the baseline model, we provide deep supervision by attaching four *encoded-neighborhood branches* (ENB) to the backbone U-Net. As explained in Sect. 3.3, all branches predict 3D masks of shape $7 \times 7 \times 5$, but at different resolutions $(1, 1, 1)$, $(\frac{1}{4}, \frac{1}{4}, 1)$ and $(\frac{1}{8}, \frac{1}{8}, 1)$, as we show in the architecture in Fig. S7. A visualization of the learned latent spaces is given in Fig. S8.

Combined Model (SNB+ENB) – Finally, we also train a combined model to predict both central instance masks and a sparse neighborhood of affinities, by providing deep supervision both via *encoded-neighborhood* and *sparse-neighborhood branches*. The backbone of this model is then trained with a total of seven branches: three branches equivalent to the ones used in the baseline model SNB, plus four additional ones like those in the ENB model (see Fig. S7).

5.2   Graph Partitioning Methods

Given the predicted encoded central instance masks, we compute affinities a_e either with the average aggregation method introduced in Sect. 4 (**MaskAggr**) or the efficient approach described in Sect. S7.3. The result of either is a signed pixel grid-graph, i.e. a graph with positive and negative edge weights that needs to be partitioned into instances. The used neighborhood connectivity of the graph is given in Table S3. Positive and negative edge weights w_e are computed by applying the additive transformation $w_e = a_e - 0.5$ to the predicted affinities.

To obtain final instances, we test different partitioning algorithms. The Mutex Watershed (**MWS**) [31] is an efficient algorithm to partition graphs with both attractive and repulsive weights without the need for extra parameters. It can easily handle the large graphs considered here with up to 10^8 nodes/voxels and 10^9 edges[2].

Then, we also test another graph partitioning pipeline that has often been applied to neuron segmentation because of its robustness. This method first generates a 2D super-pixel over-segmentation from the model predictions and then partitions the associated region-adjacency graph to obtain final instances. Super-pixels are computed with the following procedure: First, the predicted direct-neighbor affinities are averaged over the two isotropic directions to obtain a 2D neuron-membrane probability map; then, for each single 2D image in the stack, super-pixels are generated by running a watershed algorithm seeded at the maxima of the boundary-map distance transform (**WSDT**). Given this initial over-segmentation, a 3D region-adjacency graph is built, so that each super-pixel is represented by a node in the graph. Edge weights of this graph are computed by averaging short- and long-range affinities over the boundaries of neighboring super-pixels. Finally, the graph is partitioned by applying the average agglomeration algorithm proposed in [3] (**GaspAvg**).

5.3   Results and Discussion

Pre-Training of the Encoded Space – The proposed model based on an *encoded-neighborhood branch* can be properly trained only if the dimension Q of the latent space is large enough to accommodate all possible occurring neighborhood patterns. To find a small but sufficiently large Q, we trained a convolutional

---

[2] Among all edges given by the chosen neighborhood structure, we add only 10% of the long-range ones, since the Mutex Watershed was shown to perform optimally in this setup [3,31].

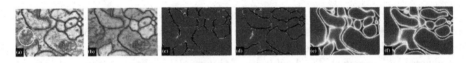

Fig. 5. Comparison between different affinities and their robustness to noise. **(a-b)** Raw data and ground-truth labels. **(c-d)** Affinities predicted by the *sparse-neighborhood branch*, which is trained with a dense binary classification loss (high affinities are red). **(e-f)** Affinities computed by averaging overlapping masks as explained in Sect. 4 (MaskAggr). Affinities from averaged masks are smoother and present a more consistent boundary evidence in the noisy region highlighted by the red circle in **(a)**. Here we show affinities along the horizontal $(-4, 0, 0)$ and vertical $(0, -4, 0)$ directions.

Variational Auto-encoder (VAE) [12,25] to compress binary ground-truth central instance masks $\hat{\mathcal{M}}_u$ into latent variables $z_u \in \mathbb{R}^Q$ and evaluated the quality of the reconstructed binary masks via the reconstruction loss. We concluded that $Q = 32$ is large enough to compress the masks considered here consisting of $7 \times 7 \times 5 = 245$ pixels. As a first experiment, we tried to make use of this VAE-pretrained latent space to train the proposed *encoded-neighborhood branch* and predict encoded masks directly in this space by using an L2 loss on the encoded vectors. However, similarly to the findings of [10], this approach performed worse than directly training the full model end-to-end as described in Sect. 3.2.

Training Encoded Masks – As we show in our validation experiments in Table 1, models trained to predict encoded central instance masks (ENB) achieved better scores than the current state-of-the-art method predicting affinities for a sparse neighborhood structure (SNB). Our interpretation of this result is that using the encoding process to predict central instance masks encourages the model to predict segment shapes that are consistent in a larger neighborhood, which can be helpful to correctly segment the most difficult regions of the data.

Aggregating Overlapping Masks – In our validation experiments of Table 1, we also test the affinities computed by averaging over overlapping masks (Mask-Aggr), as described in Sect. 4. We then partition the resulting signed graph by using the Mutex Watershed, which has empirical linearithmic complexity in the number of edges. Our experiments show that, for the first time on this type of more challenging neuron segmentation data, the Mutex Watershed (MWS) achieves better scores than the super-pixel-based methods (WSDT+GaspAvg), which have so far been known to be more robust to noise but also require the user to tune more hyper-parameters.

We also note that the MWS achieves competitive scores only with affinities computed from aggregating overlapping masks (MaskAggr). This shows that the MWS algorithm can take full advantage of the central instance aggregation process by assigning the highest priority to the edges with largest attractive and repulsive weights that were consistently predicted across overlapping masks.

Table 1. Comparison experiments on our CREMI validation set. Training encoded central instance masks (ENB) achieved better scores than the current state-of-the-art approach training only affinities for a sparse neighborhood (SNB). The model that performed best was the one using the method proposed in Sect. 4 to average overlapping masks (MaskAggr).

Train Sparse Neighbor. (SNB)	Train Encoded Masks (ENB)	Aggregate Overlapping Masks (MaskAggr)	Partitioning algorithm	No superpixels required	CREMI-Score (lower is better)	VI-merge (lower is better)
✗	✗	✗	MWS	✗	**0.153**	0.272
□	✗	✗	MWS	✗	0.184	0.273
□	✗	□	MWS	✗	0.419	0.302
✗	✗	□	MWS	✗	0.532	0.447
✗	□	□	MWS	✗	1.155	0.874
□	✗	□	WSDT+GaspAvg	□	0.173	**0.234**
✗	✗	□	WSDT+GaspAvg	□	0.237	0.331
✗	□	□	WSDT+GaspAvg	□	0.254	0.355
✗	✗	✗	WSDT+GaspAvg	□	0.334	0.388
□	✗	✗	WSDT+GaspAvg	□	0.357	0.391

On the other hand, most of the affinities trained with the *sparse-neighborhood branch* and a dense binary classification loss are almost binary, i.e. they present values either really close to zero or really close one (see comparison between different types of affinities in Fig. 5). This is not an ideal setup for the MWS, which is a greedy algorithm merging and constraining clusters according to the most attractive and repulsive weights in the graph. In fact, in this setting the MWS can often lead to over-segmentation and under-segmentation artifacts like those observed in the output segmentations of the (SNB+ENB+MWS) and (SNB+MWS) models. Common causes of these mistakes can be for example inconsistent predictions from the model and partially missing boundary evidence, which are very common in this type of challenging application (see Fig. 5 for an example).

Finally, we also note that superpixel-based methods did not perform equally well on affinities computed from aggregated masks and the reason is that these methods were particularly tailored to perform well with the more *binary-like* classification output of the *sparse-neighborhood branch*.

Training Both Masks and a Sparse Neighborhood – In our validation experiments, the combined model, which was trained to predict both a sparse neighborhood (SNB) and encoded central instance masks (ENB), achieved the best scores and yielded sharper and more accurate mask predictions. In general, providing losses for multiple tasks simultaneously has often been proven beneficial in a supervised learning setting. Moreover, the dense gradient of the *encoded-neighborhood branch*, which focuses on locally correct predictions, nicely complements the sparse gradient[3] of the *encoded-neighborhood branch*, which focuses on predictions that are consistent in a larger neighborhood. We expect this to be

---

[3] The gradient of the *encoded-neighborhood branch* is sparse, due to GPU-memory constraints as explained in Sect. 3.2.

Table 2. Representative excerpt of the published methods currently part of the CREMI leaderboard [6] (July 2020). The best method proposed in this work achieves competitive scores and is based on an efficient parameter-free algorithm that does not rely on superpixels. For more details about the partitioning algorithms used by related work, see references in the first column.

Model	Train Sparse Neighbor. (SNB)	Train Encoded Masks (ENB)	Aggregate Overlapping Masks (MaskAggr)	Partitioning algorithm	No superpixels required	CREMI-Score (lower is better)
GaspUNet[3]	✗	☐	☐	WSDT+LMulticut	☐	**0.221**
PNIUNet[15]	✗	☐	☐	Z-Watershed+Agglo	☐	0.228
GaspUNet[3]	✗	☐	☐	GaspAvg	✗	0.241
OurUNet	✗	✗	✗	MWS	✗	0.246
OurUNet	☐	✗	☐	WSDT+GaspAvg	☐	0.268
MALAUNet[7]	✗	☐	☐	WSDT+Multicut	☐	0.276
OurUNet	✗	✗	☐	WSDT+GaspAvg	☐	0.280
CRUNet[33]	☐	☐	☐	3D-Watershed	☐	0.566
LFC[22]	✗	☐	☐	Z-Watershed+Agglo	☐	0.616

another reason why the combination of affinities and central instance masks performed best in our experiments.

Results on Test Samples – The evaluation on the three test samples presented in Table 2 confirms our findings from the validation experiments: among the methods tested in this work, the best scores are achieved by the combined model (ENB+SNB) and by using the Mutex Watershed algorithm (MWS) on affinities averaged over overlapping masks (MaskAggr). Our method achieves comparable scores to the only other method in the leader-board that does not rely on super-pixels (line 3 in Table 2). This method uses the average agglomeration algorithm GaspAvg proposed in [3] instead of the MWS. GaspAvg has been shown to be more robust to noise than Mutex Watershed, however it is also considerably more computationally expensive to run on large graphs like the ones considered here.

6   Conclusions

We have presented a new proposal-free method predicting encoded central instance masks in a sliding window style, one for each pixel of the input image, and introduced a parameter-free approach to aggregate predictions from overlapping masks and obtain all instances concurrently. When applied to large volumetric biological images, the resulting method proved to be strongly robust to noise and compared favorably to competing methods that need super-pixels and hence more hyper parameters. The proposed method also endows its predictions with an uncertainty measure, depending on the consensus of the overlapping central instance masks. In future work, we plan to use these uncertainty measures to estimate the confidence of individual instances, which could help facilitate the subsequent proof-reading step still needed in neuron segmentation.

Acknowledgements. Funded by the Deutsche Forschungsgemeinschft (DFG, German Research Foundation) - Projektnummer 240245660 - SFB 1129.

References

1. Andres, B., et al.: Globally optimal closed-surface segmentation for connectomics. In: Fitzgibbon, A., Lazebnik, S., Perona, P., Sato, Y., Schmid, C. (eds.) ECCV 2012. LNCS, vol. 7574, pp. 778–791. Springer, Heidelberg (2012). https://doi.org/10.1007/978-3-642-33712-3_56
2. Arganda-Carreras, I., et al.: Crowdsourcing the creation of image segmentation algorithms for connectomics. Front. Neuroanat. **9**, 142 (2015)
3. Bailoni, A., Pape, C., Wolf, S., Beier, T., Kreshuk, A., Hamprecht, F.A.: A generalized framework for agglomerative clustering of signed graphs applied to instance segmentation. arXiv preprint arXiv:1906.11713 (2019)
4. Chopra, S., Rao, M.R.: On the multiway cut polyhedron. Networks **21**(1), 51–89 (1991)
5. Çiçek, Ö., Abdulkadir, A., Lienkamp, S.S., Brox, T., Ronneberger, O.: 3D U-Net: learning dense volumetric segmentation from sparse annotation. In: Ourselin, S., Joskowicz, L., Sabuncu, M.R., Unal, G., Wells, W. (eds.) MICCAI 2016. LNCS, vol. 9901, pp. 424–432. Springer, Cham (2016). https://doi.org/10.1007/978-3-319-46723-8_49
6. Funke, J., Saalfeld, S., Bock, D., Turaga, S., Perlman, E.: Cremi challenge (2016). https://cremi.org. Accessed 15 Nov 2019
7. Funke, J., et al.: Large scale image segmentation with structured loss based deep learning for connectome reconstruction. IEEE Trans. Pattern Anal. Mach. Intell. **41**, 1669–1680 (2018)
8. Gao, N., et al.: SSAP: Single-shot instance segmentation with affinity pyramid. In: The IEEE International Conference on Computer Vision (ICCV), October 2019
9. He, K., Gkioxari, G., Dollár, P., Girshick, R.: Mask R-CNN. In: Proceedings of the IEEE International Conference on Computer Vision, pp. 2961–2969 (2017)
10. Hirsch, P., Mais, L., Kainmueller, D.: PatchPerPix for instance segmentation. In: Proceedings of the European Conference on Computer Vision (2020, in press)
11. Januszewski, M., et al.: High-precision automated reconstruction of neurons with flood-filling networks. Nat. Methods **15**(8), 605 (2018)
12. Kingma, D.P., Welling, M.: Auto-encoding variational Bayes. arXiv preprint arXiv:1312.6114 (2013)
13. Kong, S., Fowlkes, C.C.: Recurrent pixel embedding for instance grouping. In: Proceedings of the IEEE Conference on Computer Vision and Pattern Recognition, pp. 9018–9028 (2018)
14. Lee, K., Lu, R., Luther, K., Seung, H.S.: Learning dense voxel embeddings for 3D neuron reconstruction. arXiv preprint arXiv:1909.09872 (2019)
15. Lee, K., Zung, J., Li, P., Jain, V., Seung, H.S.: Superhuman accuracy on the SNEMI3D connectomics challenge. arXiv preprint arXiv:1706.00120 (2017)
16. Levinkov, E., Kirillov, A., Andres, B.: A comparative study of local search algorithms for correlation clustering. In: Roth, V., Vetter, T. (eds.) GCPR 2017. LNCS, vol. 10496, pp. 103–114. Springer, Cham (2017). https://doi.org/10.1007/978-3-319-66709-6_9
17. Liu, S., Qi, X., Shi, J., Zhang, H., Jia, J.: Multi-scale patch aggregation (MPA) for simultaneous detection and segmentation. In: Proceedings of the IEEE Conference on Computer Vision and Pattern Recognition, pp. 3141–3149 (2016)

18. Liu, Y., Yang, S., Li, B., Zhou, W., Xu, J., Li, H., Lu, Y.: Affinity derivation and graph merge for instance segmentation. In: Ferrari, V., Hebert, M., Sminchisescu, C., Weiss, Y. (eds.) ECCV 2018. LNCS, vol. 11207, pp. 708–724. Springer, Cham (2018). https://doi.org/10.1007/978-3-030-01219-9_42

19. Meirovitch, Y., et al.: A multi-pass approach to large-scale connectomics. arXiv preprint arXiv:1612.02120 (2016)

20. Meirovitch, Y., Mi, L., Saribekyan, H., Matveev, A., Rolnick, D., Shavit, N.: Cross-classification clustering: an efficient multi-object tracking technique for 3-D instance segmentation in connectomics. In: Proceedings of the IEEE Conference on Computer Vision and Pattern Recognition, pp. 8425–8435 (2019)

21. Pape, C., Beier, T., Li, P., Jain, V., Bock, D.D., Kreshuk, A.: Solving large multicut problems for connectomics via domain decomposition. In: Proceedings of the IEEE International Conference on Computer Vision, pp. 1–10 (2017)

22. Parag, T., et al.: Anisotropic EM segmentation by 3D affinity learning and agglomeration. arXiv preprint arXiv:1707.08935 (2017)

23. Porzi, L., Bulo, S.R., Colovic, A., Kontschieder, P.: Seamless scene segmentation. In: Proceedings of the IEEE Conference on Computer Vision and Pattern Recognition, pp. 8277–8286 (2019)

24. Ren, S., He, K., Girshick, R., Sun, J.: Faster R-CNN: towards real-time object detection with region proposal networks. In: Advances in neural information processing systems. pp. 91–99 (2015)

25. Rezende, D.J., Mohamed, S., Wierstra, D.: Stochastic backpropagation and approximate inference in deep generative models. arXiv preprint arXiv:1401.4082 (2014)

26. Ronneberger, O., Fischer, P., Brox, T.: U-Net: convolutional networks for biomedical image segmentation. In: Navab, N., Hornegger, J., Wells, W.M., Frangi, A.F. (eds.) MICCAI 2015. LNCS, vol. 9351, pp. 234–241. Springer, Cham (2015). https://doi.org/10.1007/978-3-319-24574-4_28

27. Schmidt, U., Weigert, M., Broaddus, C., Myers, G.: Cell detection with star-convex polygons. In: Frangi, A.F., Schnabel, J.A., Davatzikos, C., Alberola-López, C., Fichtinger, G. (eds.) MICCAI 2018. LNCS, vol. 11071, pp. 265–273. Springer, Cham (2018). https://doi.org/10.1007/978-3-030-00934-2_30

28. Turaga, S.C., Briggman, K.L., Helmstaedter, M., Denk, W., Seung, H.S.: Maximin affinity Learning of Image Segmentation, pp. 1865–1873 (2009)

29. Welford, B.: Note on a method for calculating corrected sums of squares and products. Technometrics 4(3), 419–420 (1962)

30. Wolf, S., et al.: The mutex watershed and its objective: Efficient, parameter-free image partitioning. arXiv preprint arXiv:1904.12654 (2019)

31. Wolf, S., et al.: The mutex watershed: efficient, parameter-free image partitioning. In: Ferrari, V., Hebert, M., Sminchisescu, C., Weiss, Y. (eds.) ECCV 2018. LNCS, vol. 11208, pp. 571–587. Springer, Cham (2018). https://doi.org/10.1007/978-3-030-01225-0_34

32. Yarkony, J., Ihler, A., Fowlkes, C.C.: Fast planar correlation clustering for image segmentation. In: Fitzgibbon, A., Lazebnik, S., Perona, P., Sato, Y., Schmid, C. (eds.) ECCV 2012. LNCS, vol. 7577, pp. 568–581. Springer, Heidelberg (2012). https://doi.org/10.1007/978-3-642-33783-3_41

33. Zeng, T., Wu, B., Ji, S.: DeepEM3D: approaching human-level performance on 3D anisotropic EM image segmentation. Bioinformatics 33(16), 2555–2562 (2017)

Unsupervised Part Discovery by Unsupervised Disentanglement

Sandro Braun[✉], Patrick Esser[✉], and Björn Ommer[✉]

Heidelberg Collaboratory for Image Processing/IWR, Heidelberg University,
Heidelberg, Germany
{sandro.braun,patrick.esser,bjorn.ommer}@iwr.uni-heidelberg.de

Abstract. We address the problem of discovering part segmentations of articulated objects without supervision. In contrast to keypoints, part segmentations provide information about part localizations on the level of individual pixels. Capturing both locations and semantics, they are an attractive target for supervised learning approaches. However, large annotation costs limit the scalability of supervised algorithms to other object categories than humans. Unsupervised approaches potentially allow to use much more data at a lower cost. Most existing unsupervised approaches focus on learning abstract representations to be refined with supervision into the final representation. Our approach leverages a generative model consisting of two disentangled representations for an object's shape and appearance and a latent variable for the part segmentation. From a single image, the trained model infers a semantic part segmentation map. In experiments, we compare our approach to previous state-of-the-art approaches and observe significant gains in segmentation accuracy and shape consistency (Code available at https://compvis.github.io/unsupervised-part-segmentation). Our work demonstrates the feasibility to discover semantic part segmentations without supervision.

1 Introduction

Instances of articulated objects such as humans, birds and dogs differ in their articulation (different pose) and also show different colors and textures (appearance). Despite those large variations in articulation and appearance, humans are able to establish correspondences between individual parts across instances.

For example, consider two persons wearing different outfits as in Fig. 1a. One is wearing a plain, blue shirt, the other one is wearing a dotted, white T-shirt. In the first case, arms and chest share the same appearance, thus information about appearances cannot be used to identify the parts. In the second case, arms and chest have different appearances, thus information about appearances could be used to identify the parts.

Most previous approaches for learning part segmentations are based on supervised learning. While this can lead to good performance on a narrow set of object

---

Electronic supplementary material The online version of this chapter (https://doi.org/10.1007/978-3-030-71278-5_25) contains supplementary material, which is available to authorized users.

© Springer Nature Switzerland AG 2021
Z. Akata et al. (Eds.): DAGM GCPR 2020, LNCS 12544, pp. 345–359, 2021.
https://doi.org/10.1007/978-3-030-71278-5_25

Instance 1   Instance 2

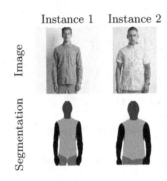

Image

Segmentation

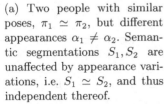

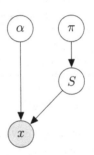

(a) Two people with similar poses, $\pi_1 \simeq \pi_2$, but different appearances $\alpha_1 \neq \alpha_2$. Semantic segmentations S_1, S_2 are unaffected by appearance variations, i.e. $S_1 \simeq S_2$, and thus independent thereof.

(b) In our model, the joint distribution over images x, segmentations S, poses π, and appearances α factorizes into $p(x, S, \pi, \alpha) = p(x|S, \alpha)p(S|\pi)p(\pi)p(\alpha)$. Thus, S is independent of α and dependent on π. While π is a latent representation of pose, S is a semantic segmentation.

Fig. 1. A Probabilistic Model for Unsupervised Part Discovery. As illustrated in a), semantic segmentations are appearance independent, which is reflected in the structure of our probabilistic model shown in b). (Color figure online)

classes, especially that of humans [10], it requires to build a large dataset for each object of interest. To overcome this limitation, we require methods that discover parts and their segmentations solely from observing the data, i.e. we need unsupervised approaches.

Previous works on unsupervised keypoint discovery [13,22,45] produce semantic keypoints which could provide information about parts. However, as we show in our experiments, even when combined with image intensity information to estimate the shape of parts, inferring pixel-wise localizations of parts from keypoints remains ambiguous. An essential ingredient of keypoint-based approaches is the built-in low-dimensional bottleneck which encourages compression and hence learning. The keypoints are represented through heatmaps of spatially normalized activations, which encourages well localized activations i.e. keypoints. In contrast, a segmentation of parts has roughly the same dimensionality as the image itself and allows arbitrary shapes of the segmented parts. Thus we cannot use the segmentation as a built-in bottleneck and must find a different way to enforce the bottleneck.

To learn parts and their segmentation unsupervised, we propose a probabilistic generative model with three hidden variables. We use two low-dimensional, continuous variables, which are independent of each other, to disentangle the instance-specific appearance, from the instance-invariant shape. The third variable is a high-dimensional discrete variable to model the support of parts, hence a segmentation. It is a descendant of the appearance-independent shape variable to ensure independence of instance specific appearance. We show how the mask can be efficiently learned in a variational inference framework assuming suitable priors. Overall, our approach learns to infer a semantic part segmentation map from a single image by learning from a stream of video frames or from pairs of synthetically transformed images.

In experiments on multiple datasets of humans and birds, our method is able to discover parts within the image that are consistent across instances. We compare intersection-over-union metrics (IOU) of our approach to those obtained from previous methods on keypoint learning and observe improvements in two out of three datasets of humans and on the dataset of birds. In addition, the generative nature of our approach enables part-based appearance transfers where it outperforms both pose supervised and keypoint-based unsupervised approaches in terms of shape consistency.

2 Background

Disentangled Representation Learning. To learn more meaningful representations, [6,32] build upon the Variational Autoencoder (VAE) [16,31] to encourage disentanglement. However, non-identifiability issues [12,21] suggest that additional information is required to obtain well-defined factors.

[17] demonstrated a factorization into style and content of digits using a conditional variant of the VAE. Motivated by Generative Adversarial Networks (GANs) [9], [24] adds a discriminator to this architecture. Using videos, [5] uses a classification problem to obtain disentangled representations of the temporally varying factors and its stationary factors. This approach is closely related to estimating and minimizing the Mutual Information of two factors [2] by defining the joint distribution of the two factors through samples from the same video.

Localized Representation Learning. Image segmentation is a well studied problem in computer vision. The seminal work of [27] introduced a variational formulation to approximate images by piecewise constant functions with regularized edge length. Superpixel approaches [30] group nearby pixels according to their similarity and obtain oversegmentations of an image. [1] combines a hierarchy of segmentations with contour detection to improve results. However, these methods rely on low-level image features and cannot account for semantic similarity.

Co-segmentation assumes the availability of a large number of examples showing the object to be segmented. The ability of this paradigm to learn from such a weak source of information resulted in many different approaches [23] ranging from graphical models [38] to deep generative models [34]. But their underlying assumption that the object to be segmented is salient limits them to masks of a single object, whereas our method learns multiple semantic parts, with part-wise correspondences across instances.

Unsupervised Part Discovery. Part based models have been extensively studied [26,33,37,42]. Recent works demonstrated the ability to discover semantic keypoints without supervision. Based on the differentiable score-map to keypoint layer of [43], [35] learns keypoints which are stable under synthetic image transformations by enforcing an equivariance principle. [45] integrates this principle into an autoencoder framework. [13] uses a reconstruction task with two

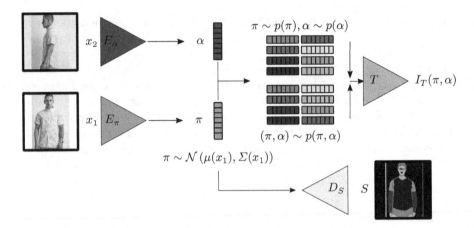

Fig. 2. Learning Appearance Independent Segmentations. To learn segmentations S independent to appearance variation α, we first disentangle a global representation for shape π and appearance α. The disentanglement is achieved through a variational and an adversarial constraint.

images from the same video, instead of synthetic transformations. [22] makes the representation more expressive by considering ellipses instead of circles for keypoints. However, in all cases the intermediate representation of keypoints is crucial, We obtain pixel-accurate part memberships, whereas above approaches can only give a rough heatmap of part localizations. In addition, our approach can handle occlusions robustly which we demonstrate in our experiments.

3   Approach

We have an image x depicting an object o composed of N object parts $o_1, \ldots, o_N$. We would like to build a model that learns about those object parts and assigns each location in the image to its corresponding object part, thus a part segmentation. Without supervision for part segmentations, we rely on a generative approach by looking for the segmentation S^* that explains the image x best. Using Bayes rule, we can rewrite this as follows:

$$S^* = \arg \max_S p(S|x) = \arg \max p(x|S)p(S). \tag{1}$$

The likelihood $p(x|S)$ measures if the segmentation can describe the image well enough and the prior $p(S)$ measures if S is a suitable candidate for a segmentation. We now motivate suitable choices for the priors of S for part learning.

3.1   Appearance Independence of Segmentations

Take two people spontaneously striking the same pose as depicted in Fig. 1a. The two people have the same pose π, but different appearances α_1 and α_2.

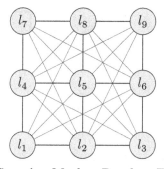

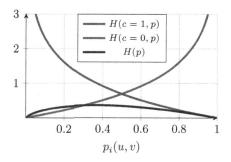

(a) **Gaussian Markov Random Field.** Without any prior assumptions, any image pixel is dependent on any other pixel, (**dense** connectivity). In a GMRF, we only allow adjacent pixels to interact (**sparse** connectivity).

(b) **Entropy Regularization.** To keep the learned segmentation S close to a categorical distribution, we regularize the entropy of part probabilities $p_i(u, v)$.

Fig. 3. Segmentation Priors. We assume two priors for our segmentation model.

This generates two images x_1 and x_2 and we infer corresponding part segmentations S_1 and S_2. Intuitively, the part segmentations S_i are independent of the variation of individual appearances, and as those people share the same poses, clearly the part segmentation of the image will be the same, i.e. $S_1 = S_2$. We argue that we can exploit this independence by modifying the image generation process so that segmentations are a result of pose and shape. We now consider α and π as random variables. In a graphical model sense, the joint distribution over poses, appearances and segmentations should factorize as depicted in Fig. 1: $p(x, S, \pi, \alpha) = p(x|S, \alpha)p(S|\pi)p(\pi)$. Note that this directly reveals the corresponding motivation in (1). If we had access to the underlying shape and appearance variables that generate images x, we know that the part segmentation S must be dependent on shape π. In practice, π and α are hidden variables and we must learn to infer them from observations x.

3.2 Learning Appearance Independent Segmentations

We now explain how we achieve a disentangled representation of shape and appearance. Let $x_i \sim (\alpha_i, \pi_i)$ express that α_i and π_i were the factors generating the image x_i. We then sample $x_1 = (\alpha, \pi_1)$ and $x_2 = (\alpha, \pi_2)$ from the dataset. In practice, this means that we need a pair of images depicting the same object but with varied poses. To infer the latent variables, we use two encoders.

$$E_\alpha : \mathbb{R}^{\dim(x)} \to \mathbb{R}^{\dim(\alpha)}, x \mapsto \alpha, \quad E_\alpha(x_2) = \alpha \qquad (2)$$

$$E_\pi : \mathbb{R}^{\dim(x)} \to \mathbb{R}^{\dim(\pi)}, x \mapsto \pi, \quad E_\pi(x_1) = \pi \qquad (3)$$

Here, α and π are simple low-dimensional latent variables, each represented by a vector. Please refer to the appendix for implementation details. To keep π

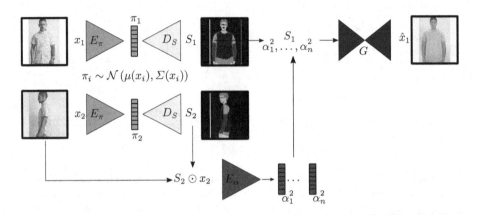

Fig. 4. Complete Method. First, we sample images x_1, x_2 from the dataset and infer their segmentations S_1 and S_2. We extract part based descriptors for appearance $\alpha_1^2, \ldots, \alpha_N^2$ from x_2 by masking out each part using S_2 and mapping it into appearance space using E_α. We then build a likelihood model for x_1 based on S_1 and $\alpha_1^2, \ldots, \alpha_N^2$.

independent of α, we simply keep π close to a standard normal distribution in a variational framework, i.e. $p(\pi) = \mathcal{N}(0, I)$ and $q(\pi|x_1) = \mathcal{N}(\mu(x_1), \Sigma(x_1))$. However, this is not sufficient to guarantee that $p(\pi)$ is factorized into semantically consistent parts. To give an example, the model could also learn to factorize parts based on their average color, i.e. all blue parts and all red parts. To prevent this, we add an additional adversarial constraint that limits the mutual information between shape and appearance $I(\alpha, \pi)$. Following recent works on mutual information estimation [3, 25, 29], we achieve this through an adversary T, which is a simple classifier trained with the following objective

$$\max_T \; \mathrm{E}_{(\pi,\alpha) \sim p(\pi,\alpha)} \; \log\left(\sigma(T(\pi,\alpha))\right) + \tag{4}$$

$$\mathrm{E}_{\pi \sim p(\pi), \alpha \sim p(\alpha)} \; \log\left(1 - \sigma(T(\pi,\alpha))\right) \tag{5}$$

Here, $\sigma(x)$ denotes the sigmoid activation. Intuitively, this means that we sample a batch of B image pairs, $\{x_1^i, x_2^i\}$ $i = 1, \ldots, B$, from the dataset and map them through the encoders, $\alpha = E_\alpha(x_2)$ and $\pi = E_\pi(x_1)$. This gives us a batch of samples from the joint distribution $(\pi, \alpha)_i \sim p(\pi, \alpha)$, $i = 1, \ldots, B$. We then randomly permute the order of $\{\alpha_i\}$ within the batch to obtain a batch of samples from the marginal distribution $\pi \sim p(\pi), \alpha \sim p(\alpha)$.

The procedure is depicted in Fig. 2. Note that the procedure is not a classical image discriminator as used in a GAN [9] training, but rather a neural mutual information estimator [2, 7]. One can derive that in the limit, the adversary converges to an estimate of the mutual information. We thus term $\mathrm{E}_{(\pi,\alpha) \sim p(\pi,\alpha)} T(\pi, \alpha) = I_T(\pi, \alpha) = \hat{I}(\pi, \alpha)$ an estimate for the mutual information of our disentangled representation. This summarizes the objectives used to train the encoders.

$$E_\pi \; : \; \min \; \mathcal{L}_{rec} + \lambda_{\text{variational}} \mathrm{KL}\left(q(\pi|x) \| p(\pi)\right) + \lambda_{\text{adversarial}} I_T(\pi, \alpha) \tag{6}$$

$$E_\alpha \; : \; \min \; \mathcal{L}_{rec} \tag{7}$$

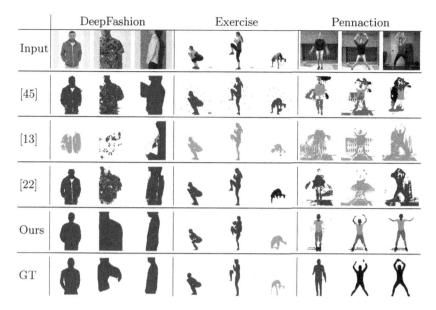

Fig. 5. Qualitative Comparison Against Keypoint Learning. To obtain segmentation masks from keypoint baselines, we use an unsupervised postprocessing based on a conditional random field [18]. We do not apply any postprocessing on our results.

Here $\mathcal{L}_{rec}$ is a reconstruction likelihood, such as a $\mathcal{L}_2$ loss or a perceptual loss between the original and the reconstructed image. $\mathcal{L}_{rec}$ will be explained in more detail in Sect. 3.4. In practice, we rely on the adaptive regularization scheme proposed in [7].

Having a disentangled representation for shape and appearance, we can finally infer segmentations S given shapes π using a simple decoder model D_S. The full procedure of disentanglement and inference for segmentations is depicted in Fig. 2. However, without further prior knowledge, it is in general not clear that D_S will produce what resembles part segmentation under a common prior. We therefore need to formulate suitable priors for S to achieve the desired result.

3.3 Priors for Segmentations

This section motivates suitable priors for the segmentation S. We claim that part segmentations are locally smooth regions within the image, meaning that long-range interactions between pixels are only possible through local connectivity. We illustrate this high-level idea in Fig. 3a. To achieve local smoothness within the image, we interpret S as the output of a per-pixel classifier with probabilities $p_i(u, v), i = 1, \dots, N$. We obtain $p_i(u, v)$ by a softmax normalization of the output of D_S, thus

$$D_S \; : \; \pi \mapsto l, \quad p_i(u, v) = \frac{\exp\left(l_i(u, v)\right)}{\sum_{i=1}^{N} \exp\left(l_i(u, v)\right)}. \tag{8}$$

Table 1. IOU Comparison Against Keypoint Learning. To obtain segmentation masks from keypoint estimates, we use an unsupervised postprocessing based on a conditional random field [18]. See appendix for full details.

Dataset	Method	Arms	Feet	Head	Legs	Torso	Overall
DeepFashion	[45] + CRF	0.194	0.000	0.598	0.293	0.376	0.292
DeepFashion	[13] + CRF	0.0522	0.000	0.118	0.108	0.244	0.104
DeepFashion	[22] + CRF	0.215	0.000	**0.606**	0.309	0.322	0.290
DeepFashion	Ours	**0.508**	0.000	0.530	**0.500**	**0.722**	**0.452**
Exercise	[45] + CRF	0.043	**0.230**	0.096	0.431	0.335	0.227
Exercise	[13] + CRF	0.101	0.190	0.000	0.469	0.357	0.223
Exercise	[22] + CRF	0.212	0.213	**0.366**	**0.445**	0.441	**0.336**
Exercise	Ours	**0.253**	0.104	0.340	0.428	**0.504**	0.326
Pennaction	[45] + CRF	0.066	0.000	**0.327**	0.379	0.442	0.243
Pennaction	[13] + CRF	0.050	**0.122**	0.000	0.316	0.455	0.189
Pennaction	[22] + CRF	0.038	0.000	0.105	0.312	0.402	0.171
Pennaction	Ours	**0.094**	0.101	0.237	**0.371**	**0.484**	**0.257**

In practice, l can be seen as the logits of a classifier. We now assume a Gaussian Markov Random Field prior for l, i.e. $p(l) = \mathcal{N}(0, \nabla)$, where ∇ denotes the spatial gradient operator, which can be approximated using a finite-difference filter. To efficiently train D_S, we use variational inference, meaning that we are looking for a suitable approximate posterior. Using the mean-field approximation we can define $q(l|x) = \prod_i^{\dim(l)} q(l_i|x) = \mathcal{N}(D_S(\pi), I)$. Then, keeping l close to the chosen prior in a KL sense simply results in regularizing the spatial gradient.

$$\mathrm{KL}\left(q\|p\right) = \sum_{i=1}^{N} \sum_{u,v} \|\nabla_{(u,v)} l_i(u,v)\|^2 \tag{9}$$

Unfortunately, this prior is not sufficient. What is still missing is a prior that states that parts are mutually exclusive at every location, i.e. segmentations S are categorical. To enforce this, we have several options: using approximations of categorical distributions [4,14], or add a regularizer that pushes the part segmentations towards a categorical solution, for instance by regularizing the entropy or cross-entropy, as shown in Fig. 3b. In practice, we found that entropy and cross-entropy regularization work best. For simplicity, we restrict us to the entropy regularization.

$$\min\ H(p) = \sum_{u,v} \sum_{i=1}^{N} p_i(u,v) \log p_i(u,v) \tag{10}$$

Here, (u, v) indicate spatial coordinate indices. To summarize, we employ the following objective for D_S.

$$D_S\ :\ \min\ \mathcal{L}_{rec} + \lambda_{\mathrm{GMRF}} \mathrm{KL}\left(q(l|x)\|p(l)\right) + \lambda_{H(p)} H(p) \tag{11}$$

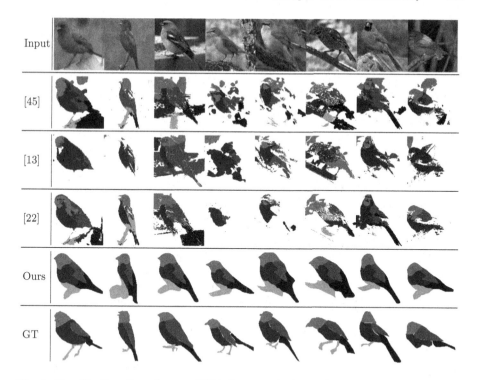

Fig. 6. Qualitative Results on CUB. Despite the lack of multi-view training pairs, we are still able to learn a good part model using our proposed method.

3.4 Part-Based Image Generation

Having introduced all our chosen priors, it remains to specify the likelihood model $p(x|S)$ i.e. how we generate images x from segmentations S. Clearly, S is not sufficient to explain the image because instance-specific appearance details are missing. We therefore would like to build a part-based likelihood model that adds instance-specific part appearances α_i. We employ the following procedure, which is common practice in unsupervised keypoint learning [13,22], as shown in Fig. 4.

1. Sample images x_1, x_2 from the dataset and infer their segmentations S_1 and S_2. As stated in Sect. 3.2, x_1 and x_2 are images of the same instance in different poses.
2. Extract part based descriptors for appearance $\alpha_1^2, \ldots, \alpha_N^2$ from x_2 by masking out each individual part using S_2 and mapping it into appearance space using E_α. The masking out operation is a simple hadamard product of each inferred part segmentation $x_{2,i} = S_{2,i} \odot x_2$ and can be interpreted as a part attention mechanism. We then obtain the part based descriptors using E_α: $\alpha_i^2 = E_\alpha(x_{2,i})$.

Table 2. IOU Comparison Against Keypoint Learning on Birds. To obtain segmentation masks from keypoint baselines, we use an unsupervised postprocessing based on a conditional random field [18]. See appendix for details.

Method	Head	Chest	Wing	Tail	Feet	Overall
[45] + CRF	0.207	0.320	0.318	0.365	0.074	0.257
[13] + CRF	0.000	0.394	0.158	0.189	0.000	0.148
[22] + CRF	0.203	0.477	0.347	0.431	0.068	0.305
Ours	**0.340**	**0.565**	**0.489**	**0.679**	**0.154**	**0.446**

3. We now have a set of vectors representing unlocalized part appearance descriptors and a spatial localization for those descriptors in terms of the segmentation S. To bring back spatial information for the appearance descriptors α_i^2 we calculate the expected appearance descriptor at each pixel which we term S_α: $S_\alpha(u, v) = \sum_{i=1}^{N} \alpha_i^2 \cdot p_i(u, v)$

4. Finally, we reconstruct the image x_1 from S_α using a generator network G. More formally, this gives us the part-based image likelihood.

$$p(x_1|S_1, \alpha_1^2, \dots, \alpha_N^2) = \mathcal{N}\left(G(S_\alpha), I\right)(x_1) \tag{12}$$

With this approach, we make the assumption that part appearances are constant across all poses π for a specific instance. Then, minimizing the negative log-likelihood gives the $\mathcal{L}_{rec}$ objective used in previous sections.

$$\mathcal{L}_{rec} = -\log\left(\mathcal{N}\left(G(S_\alpha), I\right)(x_1)\right) = \|G(S_\alpha) - x_1\|^2 \tag{13}$$

In practice G is a hour-glass style architecture [28] and $\mathcal{L}_{rec}$ is implemented through a perceptual loss [15]. See supplementary for more details.

4   Experiments

Human Object Category. We begin by evaluating our method on datasets of the human object category, namely DeepFashion [19,20], Exercise [40,41] and Pennaction [44]. DeepFashion contains strong variations in viewpoints, poses and appearances but only a simple background. Exercise has strong pose variation but only simple appearances and a simple background. Pennaction introduces the additional challenge of background clutter.

We evaluate the performance of our method using the intersection-over-union (IOU) metric against a ground-truth part annotation. We establish missing ground-truth annotation by using the supervised pretrained model from Dense-pose [10] as a substitute oracle. We calibrate our model on a held-out validation set to match the ground-truth as good as possible. Additional details can be found in the appendix.

We compare against recent work on unsupervised keypoint learning [13,22, 45]. To compare keypoint learning with segmentation learning, we apply a conditional random fields (CRF) [18] postprocessing. This step is a standard technique

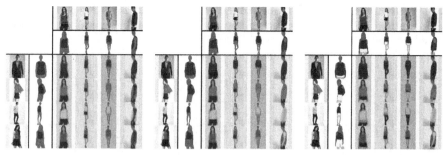

(a) Swapping only chest appearance.

(b) Swapping chest and arm appearance.

(c) Swapping chest, arm, hip and leg appearance.

Fig. 7. Part-Based Appearance Transfer. Parts which are swapped are highlighted in color (active), parts which remain constant are gray (inactive). (a): we transfer appearance of torso parts. (b): we transfer appearance of torso and arm parts. (c): we transfer appearance of torso, arm and leg parts. The transfer succeeds despite strong occlusions and viewpoint variations (Color figure online).

to refine image segmentations [11,36,39]. Note that we *do not* apply any post-processing on top of our proposed method. Additional details can be found in the appendix.

Qualitative results of our method and keypoint learning baselines [13,22,45] are shown in Fig. 5. We observe that keypoint consistency is especially difficult to achieve when dealing with strong viewpoint variations, for instance when switching between frontal and side poses on the DeepFashion and between push-up, squatting position on the Exercise dataset. The results on Pennaction suggest that background clutter is challenging for all methods, especially arm parts in downwards pointing poses. Note that on some images with an extreme amount of part occlusions, even the supervised ground-truth model by [10] fails to segment parts precisely (column 2, Fig. 5).

Finally, we show quantitative results in terms of IOU in Table 1. On Deep-Fashion and Pennaction our method outperforms other methods by a consistent margin in terms of IOU. On Exercise, the method is on par with the state-of-the art keypoint model [22] paired with CRF postprocessing. The quantitative results validate our observation for all the datasets that our method is able to discover semantically consistent parts across instances in form of segmentations.

Other Object Categories. We qualitatively analyze our method on the bird object category in Fig. 6. Note that CUB is a single image dataset, which requires us to use artificial thin-plate-spline transformations (TPS) as an approximation to multi-view pose variations. This approximation is identical to those used in [13,22,35]. We observe that our part discovery method learns local parts and is also able to find appropriate scales for parts for smaller sized birds. To evaluate our method quantitatively, we created a small dataset of bird part annotations

Table 3. Evaluating Shape Consistency. Percentage of Correct Keypoints (PCK) for pose estimation on shape/appearance swapped generations for supervised and unsupervised methods. α is pixel distance divided by image diagonal. Note that [8] serves as upper bound, as it uses the groundtruth shape estimates.

α	PCK@2.5%	PCK@5%	PCK@10%
VU-Net [8]	31.64	54.90	80.83
Lorenz et al. [22]	14.50	37.50	69.63
Ours	**41.56**	**65.76**	**83.12**

as ground-truth information and evaluate against [13,22,45] in terms of IOU in Table 2. The results suggest that our approach can be scaled to other object categories.

4.1 Part-Based Appearance Transfer

We explore the capabilities of part-based appearance transfer between instances in Fig. 7. Parts which are transferred are displayed in color (active), parts which are not transferred are displayed in gray (inactive). The transfer succeeds despite strong occlusions and pose variations. In the most extreme case, occluded appearances can be inferred from partial observations, for instance when transferring from half-body images to full-body images or from frontal to side-ways poses. Note that we do not use any adversarial training, which causes our generated images to look rather smooth and untextured in comparison to state-of-the art image synthesis.

Following [22], we evaluate the resulting pose consistency when transferring parts between instances by calculating the percentage of correct keypoints after swapping the appearance. The results in Table 3 show that our method performs significantly better than [22] and even outperforms the supervised baseline VU-Net [8] by a small margin.

Due to space constraints, we refer the reader to the supplementary materials regarding an ablation study.

5   Conclusion

We have shown that we can build a generative model for part segmentations by a suitable combination of priors. Since the method is generative, it allows learning part segmentations without explicit supervision. Experiments demonstrate the benefits of this approach over models which obtain part masks through keypoints. Overall, this work shows that disentanglement serves as a powerful substitute for supervision and, when combined with appropriate priors, allows to directly discover part segmentations. This is in contrast to most previous works on unsupervised learning, which consider unsupervised learning merely as a pre-training step to be followed by supervised training to obtain the final result.

Acknowledgements. This work has been supported in part by the BW Stiftung project "MULT!nano", the German Research Foundation (DFG) project 421703927, and the German federal ministry BMWi within the project "KI Absicherung".

References

1. Arbelaez, P.: Boundary extraction in natural images using ultrametric contour maps. In: 2006 Conference on Computer Vision and Pattern Recognition Workshop (CVPRW'06), p. 182. IEEE (2006)
2. Belghazi, M.I., et al.: Mine: mutual information neural estimation. ArXiv:180104062 (2018)
3. Belghazi, M.I., et al.: MINE: Mutual Information Neural Estimation. ArXiv:180104062 Cs Stat (2018). http://arxiv.org/abs/1801.04062
4. Bengio, Y., Léonard, N., Courville, A.: Estimating or Propagating Gradients Through Stochastic Neurons for Conditional Computation. ArXiv:13083432 Cs (2013). http://arxiv.org/abs/1308.3432
5. Denton, E.L., et al.: Unsupervised learning of disentangled representations from video. In: Advances in Neural Information Processing Systems, pp. 4414–4423 (2017)
6. Eastwood, C., Williams, C.K.I.: A framework for the quantitative evaluation of disentangled representations. In: International Conference on Learning Representations (2018). https://openreview.net/forum?id=By-7dz-AZ
7. Esser, P., Haux, J., Ommer, B.: Unsupervised robust disentangling of latent characteristics for image synthesis. In: Proceedings of the International Conference on Computer Vision (ICCV) (2019)
8. Esser, P., Sutter, E., Ommer, B.: A variational U-Net for conditional appearance and shape generation. arXiv:1804.04694 (2018)
9. Goodfellow, I., et al.: Generative adversarial nets. In: Advances in Neural Information Processing Systems, pp. 2672–2680 (2014)
10. Güler, R.A., Neverova, N., Kokkinos, I.: Densepose: dense human pose estimation in the wild. In: Proceedings of the IEEE Conference on Computer Vision and Pattern Recognition, pp. 7297–7306 (2018)
11. Hung, W.C., Jampani, V., Liu, S., Molchanov, P., Yang, M.H., Kautz, J.: SCOPS: Self-Supervised Co-Part Segmentation. ArXiv190501298 Cs (2019). http://arxiv.org/abs/1905.01298
12. Hyvärinen, A., Pajunen, P.: Nonlinear independent component analysis: existence and uniqueness results. Neural Netw. **12**(3), 429–439 (1999)
13. Jakab, T., Gupta, A., Bilen, H., Vedaldi, A.: Unsupervised learning of object landmarks through conditional image generation. In: Advances in Neural Information Processing Systems, pp. 4016–4027 (2018)
14. Jang, E., Gu, S., Poole, B.: Categorical Reparameterization with Gumbel-Softmax. ArXiv:161101144 Cs Stat (2016). http://arxiv.org/abs/1611.01144
15. Johnson, J., Alahi, A., Fei-Fei, L.: Perceptual Losses for Real-Time Style Transfer and Super-Resolution. arXiv:1603.08155 (2016)
16. Kingma, D.P., Welling, M.: Auto-encoding variational bayes. ArXiv Prepr. ArXiv:13126114 (2013)
17. Kingma, D.P., Mohamed, S., Rezende, D.J., Welling, M.: Semi-supervised learning with deep generative models. In: Advances in Neural Information Processing Systems, pp. 3581–3589 (2014)

18. Krähenbühl, P., Koltun, V.: Efficient Inference in Fully Connected CRFs with Gaussian Edge Potentials. ArXiv:12105644 Cs (2012). http://arxiv.org/abs/1210.5644
19. Liu, Z., Luo, P., Qiu, S., Wang, X., Tang, X.: Deepfashion: powering robust clothes recognition and retrieval with rich annotations. In: Proceedings of IEEE Conference on Computer Vision and Pattern Recognition (CVPR) (2016)
20. Liu, Z., Yan, S., Luo, P., Wang, X., Tang, X.: Fashion landmark detection in the wild. In: Leibe, B., Matas, J., Sebe, N., Welling, M. (eds.) Computer Vision–ECCV 2016. Lecture Notes in Computer Science, vol. 9906, pp. 229–245. Springer, Cham (2016)
21. Locatello, F., et al.: Challenging common assumptions in the unsupervised learning of disentangled representations. In: International Conference on Machine Learning, pp. 4114–4124 (2019)
22. Lorenz, D., Bereska, L., Milbich, T., Ommer, B.: Unsupervised part-based disentangling of object shape and appearance. In: Proceedings of the IEEE Conference on Computer Vision and Pattern Recognition (CVPR) (2019)
23. Lu, Z., Xu, H., Liu, G.: A survey of object co-segmentation. IEEE Access 7, 62875–62893 (2019). https://doi.org/10.1109/ACCESS.2019.2917152
24. Mathieu, M.F., Zhao, J.J., Zhao, J., Ramesh, A., Sprechmann, P., LeCun, Y.: Disentangling factors of variation in deep representation using adversarial training. In: Advances in Neural Information Processing Systems, pp. 5040–5048 (2016)
25. Mescheder, L., Nowozin, S., Geiger, A.: Adversarial Variational Bayes: Unifying Variational Autoencoders and Generative Adversarial Networks. ArXiv:170104722 Cs (2017). http://arxiv.org/abs/1701.04722
26. Monroy, A., Ommer, B.: Beyond bounding-boxes: learning object shape by model-driven grouping. In: Fitzgibbon, A., Lazebnik, S., Perona, P., Sato, Y., Schmid, C. (eds.) Computer Vision–ECCV 2012. Lecture Notes in Computer Science, vol. 7574, pp. 580–593. Springer, Berlin (2012)
27. Mumford, D.B., Shah, J.: Optimal approximations by piecewise smooth functions and associated variational problems. Commun. Pure Appl. Math. (1989). https://doi.org/10.1002/cpa.3160420503
28. Newell, A., Yang, K., Deng, J.: Stacked Hourglass Networks for Human Pose Estimation. ArXiv:160306937 Cs (2016). http://arxiv.org/abs/1603.06937
29. Poole, B., Ozair, S., van den Oord, A., Alemi, A.A., Tucker, G.: On Variational Bounds of Mutual Information. ArXiv:190506922 Cs Stat (2019). http://arxiv.org/abs/1905.06922
30. Ren, X., Malik, J.: Learning a classification model for segmentation. In: Proceedings Ninth IEEE International Conference on Computer Vision, Nice, France, p. 10. IEEE (2003)
31. Rezende, D.J., Mohamed, S., Wierstra, D.: Stochastic backpropagation and approximate inference in deep generative models. In: Proceedings of the 31st International Conference on International Conference on Machine Learning, vol. 32, pp. II-1278 (2014). JMLR. org
32. Rubenstein, P.K., Schoelkopf, B., Tolstikhin, I.: Learning disentangled representations with wasserstein auto-encoders (2018). https://openreview.net/forum?id=Hy79-UJPM
33. Rubio, J.C., Eigenstetter, A., Ommer, B.: Generative regularization with latent topics for discriminative object recognition. Pattern Recogn. 48(12), 3871–3880 (2015)

34. Singh, K.K., Ojha, U., Lee, Y.J.: FineGAN: unsupervised hierarchical disentanglement for fine-grained object generation and discovery. In: Proceedings of the IEEE Conference on Computer Vision and Pattern Recognition, pp. 6490–6499 (2019)
35. Thewlis, J., Bilen, H., Vedaldi, A.: Unsupervised learning of object landmarks by factorized spatial embeddings. ArXiv:170502193 Cs Stat (2017). http://arxiv.org/abs/1705.02193
36. Tsogkas, S., Kokkinos, I., Papandreou, G., Vedaldi, A.: Deep Learning for Semantic Part Segmentation with High-Level Guidance. ArXiv:150502438 Cs (2015). http://arxiv.org/abs/1505.02438
37. Ufer, N., Ommer, B.: Deep semantic feature matching. In: 2017 IEEE Conference on Computer Vision and Pattern Recognition (CVPR). IEEE, Honolulu, HI (2017)
38. Vicente, S., Rother, C., Kolmogorov, V.: Object cosegmentation. In: CVPR 2011, pp. 2217–2224. IEEE (2011)
39. Wang, P., Shen, X., Lin, Z., Cohen, S., Price, B., Yuille, A.: Joint Object and Part Segmentation using Deep Learned Potentials. ArXiv:150500276 Cs (2015). http://arxiv.org/abs/1505.00276
40. Xue, T., Wu, J., Bouman, K.L., Freeman, W.T.: Visual dynamics: probabilistic future frame synthesis via cross convolutional networks. In: Advances In Neural Information Processing Systems (2016)
41. Xue, T., Wu, J., Bouman, K.L., Freeman, W.T.: Visual dynamics: stochastic future generation via layered cross convolutional networks. IEEE Trans. Pattern Anal. Mach. Intell. (TPAMI) 41(9), 2236–2250 (2019)
42. Yarlagadda, P., Ommer, B.: From meaningful contours to discriminative object shape. In: Fitzgibbon, A., Lazebnik, S., Perona, P., Sato, Y., Schmid, C. (eds.) Computer Vision–ECCV 2012. Lecture Notes in Computer Science, vol. 7572, pp. 766–779. Springer, Berlin Heidelberg (2012). https://doi.org/10.1007/978-3-642-33718-5_55
43. Yi, K.M., Trulls, E., Lepetit, V., Fua, P.: LIFT: learned invariant feature transform. In: Leibe, B., Matas, J., Sebe, N., Welling, M. (eds.) Computer Vision–ECCV 2016. Lecture Notes in Computer Science, vol. 9910, pp. 467–483. Springer, Cham (2016). https://doi.org/10.1007/978-3-319-46466-4_28
44. Zhang, W., Zhu, M., Derpanis, K.G.: From actemes to action: a strongly-supervised representation for detailed action understanding. In: Proceedings of the IEEE International Conference on Computer Vision, pp. 2248–2255 (2013)
45. Zhang, Y., Guo, Y., Jin, Y., Luo, Y., He, Z., Lee, H.: Unsupervised discovery of object landmarks as structural representations. In: Proceedings of the IEEE Conference on Computer Vision and Pattern Recognition, pp. 2694–2703 (2018)

On the Lifted Multicut Polytope for Trees

Jan-Hendrik Lange[1(✉)] and Björn Andres[2]

[1] Max Planck Institute for Informatics, Saarbrücken, Germany
jan-hendrik.lange@mpi-inf.mpg.de
[2] Technical University Dresden, Dresden, Germany

Abstract. We study the lifted multicut problem restricted to trees, which is NP-hard in general and solvable in polynomial time for paths. In particular, we characterize facets of the lifted multicut polytope for trees defined by the inequalities of a canonical relaxation. Moreover, we present an additional class of inequalities associated with paths that are facet-defining. Taken together, our facets yield a complete totally dual integral description of the lifted multicut polytope for paths. This description establishes a connection to the combinatorial properties of alternative formulations such as sequential set partitioning.

1 Introduction

The lifted multicut problem [12] is a graph partitioning model based on real-valued costs attributed to pairs of nodes that are in distinct components. It has been applied successfully to inference tasks in areas such as image segmentation [2,17], object tracking [22] and motion segmentation [16]. In this paper we study the lifted multicut problem and its associated polyhedral geometry under the restriction that the underlying graph is a tree. This constitutes an extreme special case, since it is natural to consider only connected graphs and trees are minimally connected graphs. The resulting *tree partition problem* is equivalently formulated as the minimization of a multi-linear polynomial which exhibits a sparsity pattern that is determined by the tree. Therefore, it is clearly NP-hard, as we show in Sect. 3. In Sect. 4, we study the facial structure of the lifted multicut polytope for trees. We introduce a canonical relaxation in terms of node triplets and characterize under which circumstances the basic inequalities are facet-defining. Furthermore, we present another class of facet-defining inequalities, which we call *intersection inequalities*. Finally, we show that the facets that we present yield a complete totally dual integral description in the case of paths (Sect. 5).

Overall, we contribute to the study of graph partition problems an analysis of the facial structure of the lifted multicut polytope for the extreme case of minimally connected graphs such as trees and paths. Our results establish connections between the lifted multicut problem on trees and pseudo-Boolean optimization as well as sequential set partitioning. Furthermore, our insights can accelerate solvers for the problem based on integer linear programming.

© Springer Nature Switzerland AG 2021
Z. Akata et al. (Eds.): DAGM GCPR 2020, LNCS 12544, pp. 360–372, 2021.
https://doi.org/10.1007/978-3-030-71278-5_26

2   Related Work

Multicut Polytopes and Correlation Clustering. Partitions of a graph into an unconstrained number of components based on similarities or dissimilarities of *neighboring* nodes is referred to as weighted correlation clustering. Weighted correlation clustering has been studied for complete graphs [1] as well as for general graphs [6]. Due to the fact that any partition of a graph into connected components is characterized by the mathematical notion of a multicut, correlation clustering is closely related to the study of multicuts. The combinatorial polytopes associated with the multicuts of a graph have been studied, among others, most notably by [5,7,8,10]. The more general case in which similarities or dissimilarities of *non-neighboring* nodes are taken into account as well has been introduced in [12]. The latter, more expressive formulation has found a number of applications in computer vision [2,16,17,22].

Pseudo-Boolean Optimization. The equivalent formulation of the problem we study in the form of an unconstrained minimization of a multi-linear polynomial connects our work to the field of pseudo-Boolean optimization. The optimization of pseudo-Boolean functions plays an important role in machine learning, for instance, in MAP inference for computer vision. The general problem can be reduced to the quadratic case [3,4], which is responsible for the NP-hardness of the problem. The combinatorial polytope associated with the linearization of quadratic pseudo-Boolean functions was studied extensively by [18]. Recent research also considers the linearization of more general multi-linear forms [19]. Computational approaches to pseudo-Boolean optimization based on the roof duality bound [11,14] have been quite successful in practice [20].

Sequential Set Partitioning. A set partitioning problem where the elements are assumed to adhere to a linear order has been studied by Kernighan [15], who devises a dynamic program to solve the problem in polynomial time. The algorithm essentially solves a shortest path problem in a directed acyclic graph. The corresponding integer linear programming formulation admits a totally unimodular constraint matrix [13]. We derive a complete polyhedral description for the equivalent formulation as a lifted multicut problem on paths.

3   Tree Partition Problem

Let $T = (V, E)$ be a tree. We use the short-hand notation $uv = \{u, v\}$ for any pair of distinct vertices $u, v \in V$, which may or may not correspond to an edge. When convenient, we write $n = |E|$ and $m = |\binom{V}{2}| = n(n-1)/2$ for the number of edges and the total number of vertex pairs, respectively. For any pair of distinct nodes $u, v \in V$, denote by P_{uv} the unique path from u to v in T. Moreover, we denote by $d(u, v)$ the distance of u and v in T, i.e. the length of P_{uv}.

A *multicut* is a set of edges that are between the components of a partition of a graph [7]. The characteristic vector $x \in \{0, 1\}^E$ of a multicut (where $x_e = 1$

if $e \in E$ is cut) is *lifted* to the space $\{0,1\}^{\binom{V}{2}}$ by setting $x_{uv} = 1$ if, and only if u and v are in distinct components of the underlying partition [12].

The lifted multicut problem on trees is to find a partition of a tree that minimizes a sum of costs associated with pairs of nodes that are in distinct components. It is formulated as an integer linear program as follows.

Definition 1 (Lifted Multicut Problem). The *lifted multicut polytope* LMC w.r.t. T is defined as the convex hull of all $x \in \{0,1\}^{\binom{V}{2}}$ that satisfy the *path* and *cut inequalities*:

$$x_{uv} \le \sum_{e \in P_{uv}} x_e \qquad \forall u, v \in V, \, d(u,v) \ge 2, \qquad \text{(path)}$$

$$x_e \le x_{uv} \qquad \forall u, v \in V, \, d(u,v) \ge 2, \, \forall e \in P_{uv}. \qquad \text{(cut)}$$

The *lifted multicut problem* w.r.t. T and $\theta \in \mathbb{R}^{\binom{V}{2}}$ is defined as

$$\min_{x \in \text{LMC}} \sum_{uv \in \binom{V}{2}} \theta_{uv} \, x_{uv}. \qquad \text{(LMP)}$$

Note that any vector $x \in \{0,1\}^E$ is the characteristic vector of a multicut of T as any edge set of a tree defines a partition. The path and cut inequalities ensure that for any distinct pair of nodes $u, v \in V$ it holds that $x_{uv} = 1$ if, and only if the path P_{uv} is cut at any of its edges.

The lifted multicut problem on T can be equivalently formulated as the minimization of a particular multi-linear polynomial over binary inputs, which we refer to as *tree partition problem*.

Definition 2 (Tree Partition Problem). Let $T = (V, E)$ be a tree and $\bar{\theta} \in \mathbb{R}^{\binom{V}{2}}$. The optimization problem

$$\min_{y \in \{0,1\}^E} \sum_{uv \in \binom{V}{2}} \bar{\theta}_{uv} \prod_{e \in P_{uv}} y_e \qquad \text{(TPP)}$$

is called the instance of the *tree partition problem* w.r.t. T and $\bar{\theta}$. If T is a path, then we also refer to (TPP) as the *path partition problem* w.r.t. T and $\bar{\theta}$.

It is straightforward to see, by a change of variables, that the problems (TPP) and (LMP) are equivalent (up to a constant).

Lemma 1. *The vector* $y \in \{0,1\}^E$ *is a solution of problem* (TPP) *w.r.t. the tree* $T = (V, E)$ *and costs* $\bar{\theta} \in \mathbb{R}^{\binom{V}{2}}$ *if, and only if, the unique* $x \in \text{LMC}$ *such that* $x_e = 1 - y_e$ *for all* $e \in E$ *is a solution of problem* (LMP) *w.r.t.* T *and the cost vector* $\theta = -\bar{\theta}$.

Proof. For any distinct pair of nodes $u, v \in V$, we set

$$x_{uv} = 1 - \prod_{e \in P_{uv}} y_e \tag{1}$$

which implies

$$x_{uv} = 0 \quad \Longleftrightarrow \quad \forall e \in P_{uv} : \ y_e = 1 \quad \Longleftrightarrow \quad \forall e \in P_{uv} : \ x_e = 0. \tag{2}$$

Therefore, we can reformulate problem (TPP) in terms of the variables x_{uv} by transforming the objective function according to

$$\bar{\theta}_{uv} \prod_{e \in P_{uv}} y_e = -\bar{\theta}_{uv} \Big(1 - \prod_{e \in P_{uv}} y_e \Big) + \bar{\theta}_{uv} = -\bar{\theta}_{uv}\, x_{uv} + \bar{\theta}_{uv}. \tag{3}$$

This leads to the linear combinatorial optimization problem

$$\min_{x \in \mathsf{LMC}} \sum_{uv \in \binom{V}{2}} \theta_{uv}\, x_{uv} + \bar{\theta}_{uv}, \tag{4}$$

where the definition of LMC captures the relationship (2). $\square$

Apparently, problem (TPP) corresponds to the minimization of a certain class of pseudo-Boolean functions (PBF). More precisely, we call any n-variate PBF *tree-sparse*, if its multi-linear polynomial form can be aligned with a tree such that $n = |E|$ and every non-zero coefficient corresponds to the edge set of a path in the tree. Similarly, we call it *path-sparse* if the tree is a path itself. Tree-sparse PBFs are exactly those PBFs that correspond to tree partition problems (TPP).

Complexity. The tree partition problem, and thus problem (LMP), is NP-hard in general (Lemma 2 below). However, the path partition problem is solvable in polynomial time [15].

Lemma 2. *The tree partition problem is NP-hard.*

Proof. If T is a star (see Fig. 1a for an example), then problem (TPP) is equivalent to the unconstrained binary quadratic program with $|E|$ variables, which is well-known to be NP-hard. $\square$

4   Lifted Multicut Polytope for Trees

In this section we study the facial structure of the lifted multicut polytope LMC. We characterize all trivial facets and offer an outer relaxation of LMC that is tighter than the standard relaxation given by [12]. In Sect. 5, we show that our results yield a complete totally dual integral (TDI) description of the lifted multicut polytope for paths.

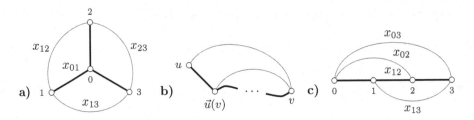

Fig. 1. a) A star with additional (thin, blue) edges between non-neighboring nodes corresponding to non-local variables. **b)** The node $\vec{u}(v)$ is the first internal node on the path P_{uv}. **c)** A path of length at least three gives rise to an *intersection inequality* (10). (Color figure online)

We denote the standard outer relaxation of LMC by

$$P_0 = \Big\{ x \in [0,1]^{\binom{V}{2}} \mid x_{uv} \leq \sum_{e \in P_{uv}} x_e \quad \forall u, v \in V, \ d(u,v) \geq 2,$$
$$x_e \leq x_{uv} \qquad \forall u, v \in V, \ d(u,v) \geq 2, \ \forall e \in P_{uv} \Big\},$$

which is obtained by dropping the integrality constraints from the definition of LMC. In particular, any $x \in P_0$ satisfies all path and cut inequalities. Let $\vec{u}(v)$ be the first node on the path P_{uv} that is different from both u and v (cf. Fig. 1b) and consider the polytope

$$P_1 = \Big\{ x \in [0,1]^{\binom{V}{2}} \mid x_{uv} \leq x_{u,\vec{u}(v)} + x_{\vec{u}(v),v} \quad \forall u, v \in V, \ d(u,v) \geq 2,$$
$$x_{\vec{u}(v),v} \leq x_{uv} \qquad\quad \forall u, v \in V, \ d(u,v) \geq 2 \Big\}.$$

This description is canonical in the sense that it only considers a quadratic number of node triplets, namely those which feature two neighboring nodes and an arbitrary third node. The following lemma states that P_1 is indeed an outer relaxation of LMC that is at least as tight as P_0.

Lemma 3. *It holds that* LMC $\subseteq P_1 \subseteq P_0$.

Proof. We show first that LMC $\subseteq P_1$. For this purpose, let $x \in$ LMC $\cap \, \mathbb{Z}^m$ be a vertex of LMC. If $x_{uv} > x_{u,\vec{u}(v)} + x_{\vec{u}(v),v}$ for some $u, v \in V$, then $x_{uv} = 1$ and $x_{u,\vec{u}(v)} = x_{\vec{u}(v),v} = 0$. This contradicts the fact that x satisfies all cut inequalities w.r.t. $\vec{u}(v), v$ and the path inequality w.r.t. u, v. If $x_{\vec{u}(v),v} > x_{uv}$ for some $u, v \in V$, then $x_{\vec{u}(v),v} = 1$ and $x_{uv} = 0$. This contradicts the fact that x satisfies all cut inequalities w.r.t. u, v and the path inequality w.r.t. $\vec{u}(v), v$. It follows that $x \in P_1$.

Now, we show that $P_1 \subseteq P_0$. Let $x \in P_1$. We need to show that x satisfies all path and cut inequalities. Let $u, v \in V$ with $d(u,v) \geq 2$. We proceed by induction on $d(u,v)$. If $d(u,v) = 2$, then the path and cut inequalities are directly given by the definition of P_1 (for the two possible orderings of u and v). If $d(u,v) > 2$, then the path inequality is obtained from $x_{uv} \leq x_{u,\vec{u}(v)} + x_{\vec{u}(v),v}$ and the induction hypothesis for the pair $\vec{u}(v), v$, since $d(\vec{u}(v), v) = d(u, v) - 1$. Similarly, for any edge e on the path from u to v, we obtain the cut inequality w.r.t. e by using the induction hypothesis and $x_{\vec{u}(v),v} \leq x_{uv}$ such that (w.l.o.g.) e is on the path from $\vec{u}(v)$ to v. It follows that $x \in \mathsf{P}_0$.   □

4.1 Facets

In this section, we show which inequalities in the definition of P_1 define facets of LMC. Moreover, we present another type of inequalities associated with paths in T, which define facets of LMC. We note that further facets can be established by the connection of LMC to the multi-linear polytope and, as a special case, the Boolean quadric polytope [18].

Lemma 4. *The inequality*

$$x_{uv} \leq x_{u,\vec{u}(v)} + x_{\vec{u}(v),v} \tag{5}$$

for some $u, v \in V$ defines a facet of LMC *if, and only if, $d(u, v) = 2$.*

Proof. First, suppose $d(u, v) = 2$. Then P_{uv} is a path of length 2 and thus chordless in the complete graph on V. Hence, the facet-defining property follows directly from [12, Theorem 10]. Now, suppose $d(u, v) > 2$ and let $x \in$ LMC be such that (5) is satisfied with equality. We show that this implies

$$x_{uv} + x_{\vec{u}(v),\vec{v}(u)} = x_{u,\vec{v}(u)} + x_{\vec{u}(v),v}. \tag{6}$$

Then the face of LMC induced by (5) has dimension at most $m - 2$ and hence cannot be a facet. In order to check that (6) holds, we distinguish the following three cases. If $x_{uv} = x_{u,\vec{u}(v)} = x_{\vec{u}(v),v}$, then all terms in (6) vanish. If $x_{uv} = x_{u,\vec{u}(v)} = 1$ and $x_{\vec{u}(v),v} = 0$, then $x_{\vec{u}(v),\vec{v}(u)} = 0$ and $x_{u,\vec{v}(u)} = 1$, so (6) holds. Finally, if $x_{uv} = x_{\vec{u}(v),v} = 1$ and $x_{u,\vec{u}(v)} = 0$, then (6) holds as well, because $x_{\vec{u}(v),\vec{v}(u)} = x_{u,\vec{v}(u)}$ by contraction of the edge $u, \vec{u}(v)$.   □

Lemma 5. *The inequality*

$$x_{\vec{u}(v),v} \leq x_{uv} \tag{7}$$

for some $u, v \in V$ defines a facet of LMC *if, and only if, v is a leaf of T.*

Proof. First, suppose v is not a leaf of T and let $x \in$ LMC be such that (7) is satisfied with equality. Since v is not a leaf, there exists a neighbor $w \in V$ of v such that $P_{\vec{u}(v),v}$ is a subpath of $P_{\vec{u}(v),w}$ We show that x additionally satisfies the equality

$$x_{uw} = x_{\vec{u}(v),w} \tag{8}$$

and thus the face of LMC induced by (7) cannot be a facet. There are two possible cases: Either $x_{uv} = x_{\vec{u}(v),v} = 1$, then $x_{uw} = x_{\vec{u}(v),w} = 1$ as well, or

$x_{uv} = x_{\vec{u}(v),v} = 0$, then $x_{uw} = x_{vw} = x_{\vec{u}(v),w}$ by contraction of the path P_{uv}, so (8) holds.

Now, suppose v is a leaf of T and let Σ be the face of LMC induced by (7). We need to prove that Σ has dimension $m-1$. This can be done explicitly by showing that we can construct $m-1$ distinct indicator vectors $\mathbb{1}_{ww'}$ for $w, w' \in V$ as linear combinations of elements from the set $S = \{x \in \text{LMC} \cap \mathbb{Z}^m \mid x_{uv} = x_{\vec{u}(v),v}\}$.

For any $u, v \in V$, $u \neq v$, let $x^{uv} \in \{0,1\}^m$ be such that $x^{uv}_{ww'} = 0$ for all w, w' on the path P_{uv} and $x^{uv}_{ww'} = 1$ otherwise. Clearly, it holds that $x^{uv} \in \text{LMC}$. Now, for $u, v \in V$ we construct $\mathbb{1}_{uv}$ recursively via the distance $d(u, v)$. If $d(u, v) = 1$, it holds that $\mathbb{1}_{uv} = \mathbb{1} - x^{uv}$. For $d(u, v) > 1$, we have

$$\mathbb{1}_{uv} = \mathbb{1} - x^{uv} - \sum_{\{ww' \neq uv \mid w,w' \text{ on path } P_{uv}\}} \mathbb{1}_{ww'}. \tag{9}$$

Note that, except $\mathbb{1}_{\vec{u}(v),v}$, all indicator vectors are constructed from vectors in S. Thus, we end up with $m-1$ vectors that are linearly independent and constructed as linear combinations of $\mathbb{1} - x^{uv} \in S$ for $u, v \in V$. Hence, the set $\{\mathbb{1}\} \cup \{x^{uv} \mid uv \in \binom{V}{2}\}$ is affine independent and the claim follows. $\square$

Lemma 6. *For any distinct $u, v \in V$, the inequality $x_{uv} \leq 1$ defines a facet of LMC if, and only if, both u and v are leaves of T. Moreover, none of the inequalities $0 \leq x_{uv}$ define facets of LMC.*

Proof. We apply the more general characterization given by [12, Theorem 8] and [12, Theorem 9]. The nodes $u, v \in V$ are a pair of ww'-cut-vertices for some vertices $w, w' \in V$ (with at least one being different from u and v) if, and only if, u or v is not a leaf of V. Thus, the claim follows from [12, Theorem 9]. The second assertion follows from [12, Theorem 8] and the fact that we lift to the complete graph on V. $\square$

Intersection Inequalities. We present another large class of non-trivial facets of LMC. For any $u, v \in V$ with $d(u, v) \geq 3$ consider the inequality

$$x_{uv} + x_{\vec{u}(v),\vec{v}(u)} \leq x_{u,\vec{v}(u)} + x_{\vec{u}(v),v}, \tag{10}$$

which we refer to as *intersection inequality*. As an example consider the graph depicted in Fig. 1c with $u = 0$ and $v = 3$.

Lemma 7. *Any intersection inequality is valid for LMC.*

Proof. Let $x \in \text{LMC} \cap \mathbb{Z}^m$ and suppose that either $x_{u,\vec{v}(u)} = 0$ or $x_{\vec{u}(v),v} = 0$ for some $u, v \in V$ with $d(u, v) \geq 3$. Then, since x satisfies all cut inequalities w.r.t. $u, \vec{v}(u)$, respectively $\vec{u}(v), v$, and the path inequality w.r.t. $\vec{u}(v), \vec{v}(u)$, it must hold that $x_{\vec{u}(v),\vec{v}(u)} = 0$. Moreover, if even $x_{u,\vec{v}(u)} = 0 = x_{\vec{u}(v),v}$, then, by the same reasoning, we have $x_{uv} = 0$ as well. Hence, x satisfies (10). $\square$

Lemma 8. *Any intersection inequality defines a facet of LMC.*

Proof. Let Σ be the face of LMC induced by (10) for some $u, v \in V$ with $d(u, v) \geq 3$. We need to prove that Σ has dimension $m - 1$, which can be done explicitly by showing that we can construct $m - 1$ distinct indicator vectors $\mathbb{1}_{ww'}$ for $w, w' \in V$ as linear combinations of elements from the set $S = \{x \in \mathsf{LMC} \cap \mathbb{Z}^m \mid x_{uv} + x_{\vec{u}(v),\vec{v}(u)} = x_{u,\vec{v}(u)} + x_{\vec{u}(v),v}\}$.

We employ the same construction as used in the proof of Lemma 5. Observe that for any feasible $x \in \mathsf{LMC} \cap \mathbb{Z}^m$ with $x \notin S$, we must have that $x_{\vec{u}(v),\vec{v}(u)} = 0$ and $x_{u,\vec{u}(v)} = x_{\vec{v}(u),v} = 1$. This means that $x^{\vec{u}(v),\vec{v}(u)} \notin S$ in the construction. Thus, we simply omit $\mathbb{1}_{\vec{u}(v),\vec{v}(u)}$ from the construction. By the same reasoning, we conclude that the m constructed vectors $\{\mathbb{1}\} \cup \{x^{uv} \mid uv \in \binom{V}{2} \setminus \{\vec{u}(v)\vec{v}(u)\}\}$ are affine independent, so the claim follows. $\square$

5   Lifted Multicut Polytope for Paths

In this section we show that the facets established in the previous section yield a complete description of LMC when T is a path. To this end, suppose that $V = \{0, \ldots, n\}$ and $E = \{\{i, i + 1\} \mid i \in \{0, \ldots, n - 1\}\}$ are linearly ordered. Therefore, $T = (V, E)$ is path. We consider only paths of length $n \geq 2$, since for $n = 1$, the polytope $\mathsf{LMC} = [0, 1]$ is simply the unit interval. Let $\mathsf{P_{path}}$ be the polytope of all $x \in \mathbb{R}^{\binom{V}{2}}$ that satisfy

$$x_{0n} \leq 1, \tag{11}$$
$$x_{in} \leq x_{i-1,n} \qquad \forall i \in \{1, \ldots, n - 1\}, \tag{12}$$
$$x_{0i} \leq x_{0,i+1} \qquad \forall i \in \{1, \ldots, n - 1\}, \tag{13}$$
$$x_{i-1,i+1} \leq x_{i-1,i} + x_{i,i+1} \qquad \forall i \in \{1, \ldots, n - 1\}, \tag{14}$$
$$x_{j,k} + x_{j+1,k-1} \leq x_{j+1,k} + x_{j,k-1} \qquad \forall j, k \in \{0, \ldots, n\}, j < k - 2. \tag{15}$$

Note that the system of inequalities (11)–(15) consists precisely of those inequalities which we have shown to define facets of LMC in the previous section. We first prove that $\mathsf{P_{path}}$ indeed yields an outer relaxation of LMC.

Lemma 9. *It holds that* $\mathsf{LMC} \subseteq \mathsf{P_{path}} \subseteq \mathsf{P_1}$.

Proof. First, we show that $\mathsf{LMC} \subseteq \mathsf{P_{path}}$. Let $x \in \mathsf{LMC} \cap \mathbb{Z}^m$, then x satisfies (11) and (14) by definition. Suppose x violates (12), then $x_{in} = 1$ and $x_{i-1,n} = 0$. This contradicts the fact that x satisfies all cut inequalities w.r.t. $i - 1, n$ and the path inequality w.r.t. i, n. So, x must satisfy (12) and, by symmetry, also (13). It follows from Lemma 7 that x satisfies (15) as well and thus $x \in \mathsf{P_{path}}$.

Next, we prove that $\mathsf{P_{path}} \subseteq \mathsf{P_1}$. To this end, let $x \in \mathsf{P_{path}}$. We show that x satisfies all inequalities (7). Let $u, v \in V$ with $u < v - 1$. We need to prove that both $x_{u+1,v} \leq x_{uv}$ and $x_{u,v-1} \leq x_{uv}$ hold. For reasons of symmetry, it suffices to show only $x_{u+1,v} \leq x_{uv}$. We proceed by induction on the distance of u from n. If $v = n$, then $x_{u+1,n} \leq x_{un}$ is given by (12). Otherwise, we use (15) for $j = u$ and $k = v + 1$ and the induction hypothesis on $v + 1$:

$$x_{uv} + x_{u+v,v+1} \geq x_{u+1,v} + x_{u,v+1} \tag{16}$$

$$\geq x_{u+1,v} + x_{u+1,v+1} \tag{17}$$

$$\implies x_{uv} \geq x_{u+1,v}. \tag{18}$$

It remains to show that x satisfies all inequalities (5). Let $u, v \in V$ with $u < v-1$. We proceed by induction on $d(u, v) = u - v$. If $d(u, v) = 2$, then (5) is given by (14). If $d(u, v) > 2$, then we use (15) for $j = u$ and $k = v$ as well as the induction hypothesis on $u, v - 1$, which have distance $d(u, v) - 1$:

$$x_{uv} + x_{u+1,v-1} \leq x_{u+1,v} + x_{u,v-1} \tag{19}$$

$$\leq x_{u+1,v} + x_{u,u+1} + x_{u+1,v-1} \tag{20}$$

$$\implies x_{uv} \leq x_{u,u+1} + x_{u+1,v}. \tag{21}$$

Hence, $x \in \mathsf{P}_1$, which concludes the proof. $\qquad\square$

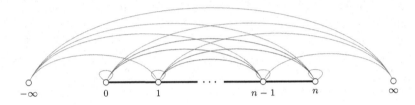

Fig. 2. Illustration of the extension used in the proof of Theorem 1. The additional terms $x_{ii}, x_{-\infty,i}, x_{i,\infty}$ and $x_{-\infty,\infty}$ correspond to the thin green edges. (Color figure online)

As our main result, we prove that $\mathsf{P}_{\mathrm{path}}$ is in fact a complete description of LMC and, moreover, it is *totally dual integral*. For an extensive reference on the subject of total dual integrality we refer the reader to [21].

Definition 3. A system of linear inequalities $Ax \leq b$ with $A \in \mathbb{Q}^{k \times m}$, $b \in \mathbb{Q}^k$ is called *totally dual integral* (TDI) if for any $c \in \mathbb{Z}^m$ such that the linear program $\max\{c^\top x \mid Ax \leq b\}$ is feasible and bounded, there exists an integral optimal dual solution.

Total dual integrality is an important concept in polyhedral geometry as it provides a sufficient condition on the integrality of polyhedra according to the following fact.

Fact 1 ([9]). *If $Ax \leq b$ is totally dual integral and b is integral, then the polytope defined by $Ax \leq b$ is integral.*

Theorem 1. *The system (11)–(15) is totally dual integral.*

Proof. We rewrite system (11)–(15) more compactly in the following way. Introduce two artificial nodes $-\infty$ and ∞ where we associate $-\infty$ to any index less than 0 and ∞ to any index greater than n. Moreover, we introduce variables x_{ii}

for all $0 \le i \le n$ as well as $x_{-\infty,i}$ and $x_{i,\infty}$ for all $1 \le i \le n-1$ and finally $x_{-\infty,\infty}$ (cf. Fig. 2). Now, the system (11)–(15) is equivalent to the system

$$x_{j,k} + x_{j+1,k-1} \le x_{j+1,k} + x_{j,k-1} \qquad \forall j,k \in \{-\infty, 0, \ldots, n, \infty\},\ j \le k-2 \tag{22}$$

given the additional equality constraints

$$x_{ii} = 0 \quad \forall 0 \le i \le n, \tag{23}$$
$$x_{-\infty,i} = 1 \quad \forall 1 \le i \le n-1, \tag{24}$$
$$x_{i,\infty} = 1 \quad \forall 1 \le i \le n-1, \tag{25}$$
$$x_{-\infty,\infty} = 1. \tag{26}$$

Let the system defined by (22)–(26) be represented in matrix form as $Ax \le a$, $Bx = b$. Note that $\mathsf{P}_{\mathrm{path}}$ is non-empty and bounded. Thus, to establish total dual integrality, we need to show that for any $\theta \in \mathbb{Z}^{m+3n}$ the dual program

$$\min\{a^\top y + b^\top z \mid A^\top y + B^\top z = \theta, y \ge 0\} \tag{27}$$

has an integral optimal solution. Here, the y variables, indexed by j,k, correspond to the inequalities (22) and the z variables, indexed by pairs of $i, -\infty$ and ∞, correspond to the equations (23)–(26). Then, the equation system $A^\top y + B^\top z = \theta$ translates to

$$y_{i-1,i+1} + z_{i,i} = \theta_{i,i} \tag{28}$$
$$y_{i-1,i+2} - y_{i-1,i+1} - y_{i,i+2} = \theta_{i,i+1} \tag{29}$$
$$y_{i-1,\ell+1} - y_{i-1,\ell} - y_{i,\ell+1} + y_{i,\ell} = \theta_{i,\ell} \tag{30}$$
$$-y_{-\infty,i+1} + y_{-\infty,i} + z_{-\infty,i} = \theta_{-\infty,i} \tag{31}$$
$$-y_{i-1,\infty} + y_{i,\infty} + z_{i,\infty} = \theta_{i,\infty} \tag{32}$$
$$y_{-\infty,\infty} + z_{-\infty,\infty} = \theta_{-\infty,\infty}, \tag{33}$$

where (28)–(30) hold for all $0 \le i < i+1 < \ell \le n$ and (31), (32) hold for all $1 \le i \le n-1$. Observe that (28) includes only y variables with indices of distance 2, (29) couples y variables of distance 3 with those of distance 2 and (30) couples the remaining y variables of any distance $d > 3$ with those of distance $d-1$ and $d-2$. Hence, any choice of values for the free variables z_{ii} completely determines all y variables. This means we can eliminate y and reformulate the dual program entirely in terms of the z variables, as follows. It holds that

$$0 \le y_{i-1,i+1} = \theta_{i,i} - z_{i,i} \qquad\qquad\qquad\qquad \forall 0 \le i \le n,$$
$$0 \le y_{i-1,i+2} = \theta_{i,i+1} + \theta_{i,i} + \theta_{i+1,i+1} - z_{i,i} - z_{i+1,i+1} \quad \forall 0 \le i < i+1 \le n$$

and thus, by (30),

$$0 \le y_{i-1,\ell+1} = \sum_{i \le j \le k \le \ell} \theta_{j,k} - \sum_{k=i}^{\ell} z_{k,k} \tag{34}$$

for all $0 \leq i \leq \ell \leq n$. Substituting the y variables in (31)–(33) yields the following equivalent formulation of the dual program (27):

$$\min \quad z_{-\infty,\infty} + \sum_{i=1}^{n-1} z_{-\infty,i} + z_{i,\infty} \tag{35}$$

$$\text{s.t.} \quad \sum_{k=i}^{\ell} z_{k,k} \leq \sum_{i \leq j \leq k \leq \ell} \theta_{j,k} \qquad \forall 0 \leq i \leq \ell \leq n \tag{36}$$

$$z_{-\infty,i} + z_{i,i} = \theta_{-\infty,i} + \sum_{0 \leq j \leq i} \theta_{j,i} \qquad \forall 1 \leq i \leq n-1 \tag{37}$$

$$z_{i,\infty} + z_{i,i} = \theta_{i,\infty} + \sum_{i \leq k \leq n} \theta_{i,k} \qquad \forall 1 \leq i \leq n-1 \tag{38}$$

$$z_{-\infty,\infty} - \sum_{k=0}^{n} z_{k,k} = \theta_{-\infty,\infty} - \sum_{0 \leq j \leq k \leq n} \theta_{j,k}. \tag{39}$$

The variables $z_{-\infty,i}$, $z_{i,\infty}$ and $z_{-\infty,\infty}$ occur only in a single equation each. Furthermore, the matrix corresponding to the inequality constraints satisfies the *consecutive-ones* property w.r.t. its rows. Therefore, the constraint matrix of the whole system is totally unimodular, which concludes the proof. $\square$

Remark. The constraint matrix corresponding to the system (11)–(15) is in general *not* totally unimodular. A minimal example is the path of length 4.

Corollary 1. *It holds that* $\mathsf{LMC} = \mathsf{P}_{\text{path}}$.

Proof. This is immediate from Lemma 9, Fact 1 and Theorem 1. $\square$

Remark. The path partition problem admits a more efficient representation as a set partitioning problem as follows. For each $0 \leq i \leq \ell \leq n$, let

$$d_{i,\ell} = \sum_{i \leq j \leq k \leq \ell} \theta_{j,k}, \tag{40}$$

then taking the dual of problem (35) and simplifying yields the problem

$$\min \quad d^\top \lambda \tag{41}$$

$$\text{s.t.} \quad \sum_{0 \leq i \leq k \leq \ell \leq n} \lambda_{i,\ell} = 1, \quad \forall 0 \leq k \leq n$$

$$\lambda \geq 0.$$

Each variable $\lambda_{i,\ell}$ corresponds to the component containing nodes i to ℓ. Problem (41) is precisely the sequential set partitioning formulation of the path partition problem as used by [13]. It admits a quadratic number of variables and a linear number of constraints (opposed to a quadratic number of constraints in the description of LMC). The integrality constraint need not be enforced, since the constraint matrix is totally unimodular.

6   Conclusion

We studied the lifted multicut polytope for the special case of trees lifted to the complete graph. We characterized a number of its facets and provided a tighter relaxation compared to the standard linear relaxation. Our analysis establishes a connection between the lifted multicut problem on trees and pseudo-Boolean optimization. Moreover, the described facets yield a complete totally dual integral description of the lifted multicut polytope for paths. This main results relates the geometry of the path partition problem to the combinatorial properties of the sequential set partitioning problem. Moreover, our insights can accelerate solvers for the tree and path partition problem based on integer linear programming.

References

1. Bansal, N., Blum, A., Chawla, S.: Correlation clustering. Mach. Learn. **56**(1–3), 89–113 (2004). https://doi.org/10.1023/B:MACH.0000033116.57574.95
2. Beier, T., et al.: Multicut brings automated neurite segmentation closer to human performance. Nat. Methods **14**(2), 101–102 (2017). https://doi.org/10.1038/nmeth.4151
3. Boros, E., Gruber, A.: On quadratization of pseudo-Boolean functions. In: International Symposium on Artificial Intelligence and Mathematics, ISAIM 2012, Fort Lauderdale, Florida, USA, 9–11 January 2012 (2012)
4. Boros, E., Hammer, P.L.: Pseudo-Boolean optimization. Discrete Appl. Math. **123**(1–3), 155–225 (2002). https://doi.org/10.1016/S0166-218X(01)00341-9
5. Chopra, S., Rao, M.: The partition problem. Math. Program. **59**(1–3), 87–115 (1993). https://doi.org/10.1007/BF01581239
6. Demaine, E.D., Emanuel, D., Fiat, A., Immorlica, N.: Correlation clustering in general weighted graphs. Theoret. Comput. Sci. **361**(2–3), 172–187 (2006). https://doi.org/10.1016/j.tcs.2006.05.008
7. Deza, M., Grötschel, M., Laurent, M.: Complete descriptions of small multicut polytopes. In: Applied Geometry and Discrete Mathematics, Proceedings of a DIMACS Workshop, pp. 221–252 (1990)
8. Deza, M.M., Laurent, M.: Geometry of Cuts and Metrics. AC, vol. 15. Springer, Heidelberg (1997). https://doi.org/10.1007/978-3-642-04295-9
9. Edmonds, J., Giles, F.R.: A min-max relation for submodular functions on graphs. Ann. Discrete Math. **1**, 185–204 (1977). https://doi.org/10.1016/S0167-5060(08)70734-9
10. Grötschel, M., Wakabayashi, Y.: A cutting plane algorithm for a clustering problem. Math. Program. **45**(1), 59–96 (1989). https://doi.org/10.1007/BF01589097
11. Hammer, P.L., Hansen, P., Simeone, B.: Roof duality, complementation and persistency in quadratic 0–1 optimization. Math. Program. **28**(2), 121–155 (1984). https://doi.org/10.1007/BF02612354
12. Horňáková, A., Lange, J.H., Andres, B.: Analysis and optimization of graph decompositions by lifted multicuts. In: ICML (2017)
13. Joseph, A., Bryson, N.: Partitioning of sequentially ordered systems using linear programming. Comput. Oper. Res. **24**(7), 679–686 (1997). https://doi.org/10.1016/S0305-0548(96)00070-6

14. Kahl, F., Strandmark, P.: Generalized roof duality. Discrete Appl. Math. **160**(16–17), 2419–2434 (2012). https://doi.org/10.1016/j.dam.2012.06.009

15. Kernighan, B.W.: Optimal sequential partitions of graphs. J. ACM **18**(1), 34–40 (1971). https://doi.org/10.1145/321623.321627

16. Keuper, M.: Higher-order minimum cost lifted multicuts for motion segmentation. In: ICCV, pp. 4252–4260 (2017). https://doi.org/10.1109/ICCV.2017.455

17. Keuper, M., Levinkov, E., Bonneel, N., Lavoué, G., Brox, T., Andres, B.: Efficient decomposition of image and mesh graphs by lifted multicuts. In: ICCV (2015). https://doi.org/10.1109/ICCV.2015.204

18. Padberg, M.: The Boolean quadric polytope: some characteristics, facets and relatives. Math. Program. **45**(1), 139–172 (1989). https://doi.org/10.1007/BF01589101

19. Pia, A.D., Khajavirad, A.: A polyhedral study of binary polynomial programs. Math. Oper. Res. **42**(2), 389–410 (2017). https://doi.org/10.1287/moor.2016.0804

20. Rother, C., Kolmogorov, V., Lempitsky, V.S., Szummer, M.: Optimizing binary MRFs via extended roof duality. In: CVPR (2007)

21. Schrijver, A.: Theory of Linear and Integer Programming. John Wiley & Sons Inc., New York (1986)

22. Tang, S., Andriluka, M., Andres, B., Schiele, B.: Multiple people tracking by lifted multicut and person re-identification. In: CVPR (2017). https://doi.org/10.1109/CVPR.2017.394

Conditional Invertible Neural Networks
for Diverse Image-to-Image Translation

Lynton Ardizzone$^{(\boxtimes)}$, Jakob Kruse, Carsten Lüth, Niels Bracher,
Carsten Rother, and Ullrich Köthe

Visual Learning Lab, Heidelberg University, Heidelberg, Germany
lynton.ardizzone@iwr.uni-heidelberg.de

.

Abstract. We introduce a new architecture called a conditional invertible neural network (cINN), and use it to address the task of diverse image-to-image translation for natural images. This is not easily possible with existing INN models due to some fundamental limitations. The cINN combines the purely generative INN model with an unconstrained feed-forward network, which efficiently preprocesses the conditioning image into maximally informative features. All parameters of a cINN are jointly optimized with a stable, maximum likelihood-based training procedure. Even though INN-based models have received far less attention in the literature than GANs, they have been shown to have some remarkable properties absent in GANs, e.g. apparent immunity to mode collapse. We find that our cINNs leverage these properties for image-to-image translation, demonstrated on day to night translation and image colorization. Furthermore, we take advantage of our bidirectional cINN architecture to explore and manipulate emergent properties of the latent space, such as changing the image style in an intuitive way.
Code: github.com/VLL-HD/conditional_INNs.

1 Introduction

INNs occupy a growing niche in the space of generative models. Because they became relevant more recently compared to GANs or VAEs, they have received much less research attention so far. Currently, the task of image generation is still dominated by GAN-based models [4,17,18]. Nevertheless, INNs have some extremely attractive theoretical and practical properties, leading to an increased research interest recently: The training is not adversarial, very stable, and does not require any special tricks. Their loss function is quantitatively meaningful for comparing models, checking overfitting, etc. [32], which is not given with GANs. INNs also do not experience the phenomenon of mode collapse observed in GAN-based models [28]. Compared to VAEs, they are able to generate higher-quality results, because no ELBO approximation or reconstruction loss is needed, which typically leads to modeling errors [1,35]. Furthermore, they allow mapping real

Electronic supplementary material The online version of this chapter (https://doi.org/10.1007/978-3-030-71278-5_27) contains supplementary material, which is available to authorized users.

Z. Akata et al. (Eds.): DAGM GCPR 2020, LNCS 12544, pp. 373–387, 2021.
https://doi.org/10.1007/978-3-030-71278-5_27

images into the latent space for explainability, interactive editing, and concept discovery [15,19]. In addition, they have various connections to information theory, allowing them to be used for lossless compression [12], information-theoretic training schemes [3], and principled out-of-distribution detection [5,24].

In this work, we present a new architecture called a conditional invertible neural network (cINN), and apply it to diverse image-to-image translation. Diverse image-to-image translation is a particular conditional generation task: given a conditioning image Y, the task is to model the conditional probability distribution $p(X|Y)$ over some images X in a different domain. '*Diverse*' implies that the model should generate different X covering the whole distribution, not just a single answer. More specifically, we consider the case of paired training data, meaning matching pairs (x_i, y_i) are given in the training set. Unpaired image-to-image translation in theory is ill-posed, and only possible through inductive bias or explicit regularization.

For this setting, the use of existing INN-based models has so far not been possible in a general way. Some methods for conditional generation using INNs exist, but these are mostly class-conditional, or other cases where the condition directly contains the necessary high-level information [3,26,30]. This is due to the basic limitation that each step in the network must be invertible. Because the condition itself is not part of the invertible transformation, it can therefore not be passed across layers. As a result, it is impossible for an INN to extract useful high-level features from the condition. That would be necessary for effectively performing diverse image-to-image translation, where e.g. the semantic context of the condition is needed.

Our cINN extends standard INNs in three aspects to avoid this shortcoming. Firstly, we use a simple but effective way to inject conditioning into the core building blocks at multiple resolutions in the form of so-called *conditional coupling blocks* (CCBs). Secondly, to provide useful conditions at each resolution level, we couple the INN with a feed-forward *conditioning network*: it produces a feature pyramid C from the condition image Y, that can be injected into the CCBs at each resolution. Lastly, we present a new invertible pooling scheme based on wavelets, that improves the generative capability of the INN model. The entire cINN architecture is visualized in Fig. 2.

Fig. 1. Diverse colorizations, which our network created for the same grayscale image. One of them shows ground truth colors, but which? Solution at the bottom of next page.

The whole cINN can be trained end-to-end with a single maximum likelihood loss function, leading to simple, repeatable, and stable training, without the need for hyperparameter tuning or special tricks. We show that the learned conditioning features C are maximally informative for the task at hand from an information theoretic standpoint. We also show that the cINN will learn the true conditional probability if the networks are powerful enough.

Our contributions are summarized as follows:

- We propose a new architecture called conditional invertible neural network (cINN), which combines an INN with an unconstrained feed-forward network for conditioning. It generates diverse images with high realism, while adding noteworthy and useful properties compared to existing approaches.
- We demonstrate a stable, maximum likelihood training procedure for jointly optimizing the parameters of the INN and the conditioning network. We show that our training causes the conditioning network to extract maximally informative features from the condition, measured by mutual information.
- We take advantage of our bidirectional cINN architecture to explore and manipulate emergent properties of the latent space. We illustrate this for day-to-night image translation and image colorization.

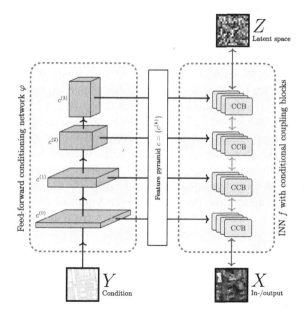

Fig. 2. Illustration of the cINN. It consists of a feed-forward conditioning network (*left half*), and an invertible part (*right half*). **Black arrows:** connections always in the same direction. Green boxes: extracted feature maps $c^{(k)}$. **Purple arrows:** invertible connections, depending on training/testing. Orange arrows: invertible wavelet downsampling. Pink blocks: conditional coupling blocks (CCBs). (Color figure online)

2   Related Work

Image-to-image translation for natural images was first demonstrated with GAN-based models [14]. It was also extended to the unpaired setting by [36]. However, these models are generally not able to produce diverse outputs. Several works attempt to prevent such mode collapse in image-to-image GANs through specialized architectures and regularization [23,25,37]. A hybrid approach between GAN and autoencoder is used in [33] for diversity. While these approaches do lead to visual diversity, there is currently no way to verify if they truly cover the entire distribution, or a lower-dimensional manifold.

Conditional INN models can be divided into methods with a conditional latent space, and methods where the INN itself is conditional. Apart from our cINN, the only example for the second case to our knowledge is [26]: an INN-based model is used to de-modulate mel-spectrograms back into audio waves. While the conditioning scheme is similar to our CCBs, the condition is given externally and directly contains the needed information, instead of being learned. Diversity is also not considered, the model is only used to produce a single output for each condition. For the second category of conditional latent space models, pixel-wise conditioning is in general more difficult to achieve. [19] manipulate latent space after training to generate images with certain global attributes. In [30], a class-conditional latent space is used for training to obtain a class-conditional INN model. A special type of conditional latent space is demonstrated in [2], suitable for non-stochastic inverse problems of small dimensionality. Examples where the approach is extended to spatial conditioning include [31], where two separate INNs define a mapping between medical imaging domains. The model requires an additional loss term with hyperparameters, that has an unknown effect on the output distribution, and diversity is not considered. Closest to our work is [20], where a VAE and INN are trained jointly, to allow a specific form of diverse image-to-image translation. However, the method is only applied for translation between images of the same domain, i.e. generate similar images given a conditioning image. The training scheme requires four losses that have to be balanced with hyperparameters. Our cINN can map between arbitrary domains, is more flexible due to the CCB design instead of a conditional latent space, and only uses a single loss function to train all components jointly.

3   Method

We divide this section into two parts: First, we discuss the architecture itself, split into the invertible components (Fig. 2 right), and the feed-forward conditioning network (Fig. 2 left). Then, we present the training scheme and its effects on each component.

3.1   CINN Architecture

Conditional Coupling Blocks. Our method to inject the conditioning features into the INN is an extension of the affine coupling block architecture established by [8]. There, each network block splits its input u into two parts $[u_1, u_2]$ and applies affine transformations between them that have strictly upper or lower triangular Jacobians:

$$v_1 = u_1 \odot \exp\big(s_1(u_2)\big) + t_1(u_2) , \quad v_2 = u_2 \odot \exp\big(s_2(v_1)\big) + t_2(v_1) . \quad (1)$$

The outputs $[v_1, v_2]$ are concatenated again and passed to the next coupling block. The internal functions s_j and t_j can be represented by arbitrary neural networks, we call these the *subnetworks* of the block. In practice, each $[s_j, t_j]$-pair is jointly modeled by a single subnetwork, instead of separately. Importantly, the subnetworks are only ever evaluated in the forward direction, even when the coupling block is inverted:

$$u_2 = \big(v_2 - t_2(v_1)\big) \oslash \exp\big(s_2(v_1)\big) , \quad u_1 = \big(v_1 - t_1(u_2)\big) \oslash \exp\big(s_1(u_2)\big) . \quad (2)$$

As shown by [8], the logarithm of the Jacobian determinant for such a coupling block is simply the sum of s_1 and s_2 over image dimensions, which we use later.

We adapt the design of Eqs. (1) and (2) to produce a conditional coupling block (CCB): Because the subnetworks s_j and t_j are never inverted, we can concatenate conditioning data c to their inputs without losing the invertibility, replacing $s_1(u_2)$ with $s_1(u_2, c)$ etc. Our CCB design is illustrated in Fig. 3. Multiple coupling

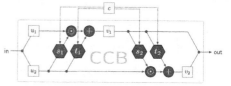

Fig. 3. A single conditional coupling block (CCB).

blocks are then stacked to form the INN-part of the cINN. We denote the entire INN as $f(x; c, \theta)$, with the network parameters θ and the inverse as $g(z; c, \theta)$. Because the resolution does not stay fixed throughout the INN, different sections of the network require different conditions $c^{(k)}$. We then use $c := \{c^{(k)}\}$ to denote the set of all the conditions at once. For any fixed condition c, the invertibility is given as

$$f^{-1}(\,\cdot\,; c, \theta) = g(\,\cdot\,; c, \theta). \quad (3)$$

Haar Wavelet Downsampling. All prior INN architectures use one of two checkerboard patterns for reshaping to lower spatial resolutions ([8] or [16]). Instead, we find it helpful to perform downsampling with Haar wavelets [9], which essentially decompose images into a 2×2 average pool-

Fig. 4. Haar wavelet downsampling reduces spatial dimensions & separates lower frequencies (a) from high (h, v, d).

ing channel as well as vertical, horizontal and diagonal derivatives, see Fig. 4. This results in a more sensible way of distributing the information after downsampling and also contributes to mixing the variables between resolution levels. Similarly, [15] use a single discrete cosine transform as a final transformation in their INN, to replace global average pooling.

Conditioning Network. It is the task of the conditioning network to transform the original condition y into the necessary features $c^{(k)}$ that the INN uses at the different resolution levels k. For this, we simply use a standard feed-forward network, denoted φ, that jointly outputs the different features in the form of the feature pyramid c. The conditioning network can be trained from scratch, jointly with the INN-part, as explained in the next section. It is also possible to use a pretrained model for initialization to speed up the start of training, e.g. a pretrained ResNet [10] or VGG [29].

3.2   Maximum Likelihood Training of cINNs

Training the INN-Part. By prescribing a probability distribution $p_Z(z)$ on latent space z, the INN f assigns any input x a probability, dependent on the conditioning c and the network parameters θ, through the change-of-variables formula:

$$q(x \mid c, \theta) = p_Z \left(f(x; c, \theta) \right) \left| \det \left(\frac{\partial f}{\partial x} \right) \right| . \tag{4}$$

Here, we use the Jacobian matrix $\partial f / \partial x$. We will denote the Jacobian determinant, evaluated at some training sample x_i, as $J_i := \det(\partial f / \partial x |_{x_i})$. With a set of observerd i.i.d. samples $\{(x_i, c_i)\}$, Bayes' theorem gives us the posterior over model parameters as

$$p(\theta \mid \{(x_i, c_i)\}) \propto p_\theta(\theta) \prod_i q(x_i \mid c_i, \theta) \tag{5}$$

This means we can find the most likely model parameters given the known training data by maximizing the right hand side. After taking the logarithm and changing the product to a sum, we get the following loss to minimize: $\mathcal{L} = \mathbb{E}_i \left[- \log \left(q(x_i \mid c_i, \theta) \right) \right]$, which is the same as in classical Bayesian model fitting. Finally, inserting Eq. (4) with a standard normal distribution for $p_Z(z)$, we obtain the *conditional maximum likelihood loss* we use for training:

$$\mathcal{L}_{\mathrm{cML}} = \mathbb{E}_i \left[\frac{\|f(x_i; c_i, \theta)\|_2^2}{2} - \log |J_i| \right] . \tag{6}$$

We can also explicitly include a Gaussian prior over weights $p_\theta = \mathcal{N}(0, \sigma_\theta)$ in Eq. (5), which amounts to the commonly used L2 weight regularization in practice. Training a network with this loss yields an estimate of the maximum likelihood network parameters $\hat{\theta}$. From there, we can perform conditional generation for some c by sampling z and using the inverted network g: $x_{\mathrm{gen}} = g(z; c, \hat{\theta})$, with $z \sim p_Z(z)$.

The maximum likelihood training method makes it virtually impossible for mode collapse to occur: If any mode in the training set has low probability under the current guess $q(x \mid c, \theta)$, the corresponding latent vectors will lie far outside the normal distribution p_Z and receive big loss from the first L2-term in Eq. (6). In contrast, the discriminator of a GAN only supplies a weak signal, proportional to the mode's relative frequency in the training data, so that the generator is not penalized much for ignoring a mode completely.

Jointly Training the Conditioning Network. Next, we consider the result if we also backpropagate the loss through the feature pyramid c, to train the conditioning network φ jointly with the same loss. Intuitively speaking, the more useful the learned features are for the INN's task, the lower the $\mathcal{L}_{\mathrm{cML}}$ loss will become. Therefore, the conditioning network is encouraged to extract useful features.

We can formalize this using the information-theoretical concept of mutual information (MI). MI quantifies the amount of information that two variables share, in other words, how informative one variable is about the other. For any two random variables a and b, It can be written as the KL-divergence between joint and factored distributions: $I(a,b) = D_{\mathrm{KL}}(p(a,b)\|p(a)p(b))$. With this, we can derive the following proposition, details and proof are found in the appendix:

Proposition 1. *Let $\hat{\theta}$ be the INN parameters and $\hat{\varphi}$ the conditioning network that jointly minimize $\mathcal{L}_{\mathrm{cML}}$. Assume that the INN $f(\cdot;\cdot,\theta)$ is optimized over $\mathcal{F}$ defined in Assumption 1 (appendix), and φ over $\mathcal{G}_0$ defined in Assumption 2 (appendix). Then it holds that*

$$I\big(x,\hat{\varphi}(y)\big) = \max_{\varphi \in \mathcal{G}_0} I\big(x,\varphi(y)\big) \tag{7}$$

In other words, the learned features will be the ones that are maximally informative about the generated variable x. Importantly, the assumption about the conditioning networks family $\mathcal{G}_0$ does not say anything about its representational power: the features will be as informative as possible within the limitations of the conditioning network's architecture and number of extracted features.

We can go a step further under the assumption that the power of the conditioning network and number of features in the pyramid are large enough to reach the global minimum of the loss (sufficient condition given by Assumption 3, appendix). In this case, we can also show that the cINN as a whole will learn the true posterior by minimizing the loss (proof in appendix):

Proposition 2. *Assume φ has been optimized over a family $\mathcal{G}_1$ of universal approximators and $\dim(c) \geq \dim(y)$ (Assumption 3, appendix), and the INN is optimized over a family of universal density approximators $\mathcal{F}$ (Assumption 1, appendix). Then the following holds for $(x,y) \in \mathcal{X}$, where $\mathcal{X}$ is the joint domain of the true training distribution $p(x,y)$:*

$$q(x|\hat{\varphi}(y),\hat{\theta}) = p(x|y) \tag{8}$$

4   Experiments

We present results and explore the latent space of our models for two image-to-image generation tasks: day to night image translation, and image colorization. We use the former as a qualitative demonstration, and the latter for a more in-depth analysis and comparison with other methods. MNIST experiments, to purely show the capability of the CCBs without the conditioning network, are given in the appendix.

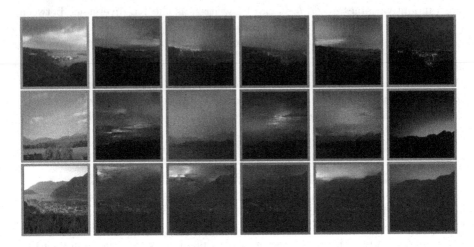

Fig. 5. Examples of conditions y (left), three generated samples (middle), and the original image x (right).

In practice, we use several techniques to improve the network and training. Ablations of the following are included in the appendix.

- We augment the images by adding a small amount of noise, in order to remove the quanitzation into 255 brightness levels. The quantization is known to cause problems in training otherwise [32].
- After each coupling block, we perform a random, fixed permuation of the feature channels. This effectively randomizes the split for the next coupling block.
- We adopt the method from [8], whereby the affine scaling s is parametrized as $\gamma \tanh(r(x))$, where γ is learned directly as a channel-wise parameter, and r is output by the subnetwork. This has exactly the same representational power as directly outputting s, but improves stability, because the $\exp(s)$ term in Eq. (1) does not explode as easily.

4.1 Day to Night Translation

We train on the popular day-to-night dataset [22]. It contains webcam images from approximately 100 different locations, taken at approximately 10–20 times during the day and night each. This results in about 200 combinations of day-night pairs per location. The test set consists of 5 unseen locations. For training, we randomly resize and crop the images to 128×128 pixels. We use the day-images as the condition y, and the night-images as the generated x. For the conditioning network, we use a standard ResNet-18 [10]. We extract the activations after every other layer of the ResNet to form the feature pyramid. As the ResNet contains the usual downsampling operations, the activations already have the correct sizes for the pyramid. We then construct the INN part as

Fig. 6. Conditioning image (top left), and extracted features from different levels of the pyramid. From left to right, top to bottom: 1st level, precise edges and texture; 2nd level, foreground/background; 3rd level, populated area.

described in Sect. 3, with 8 coupling blocks in total, and five wavelet downsampling operations spaced in between. The subnetworks consist of three convolutions, with ReLU activations and batch normalization after the first two convolutions.

We train for 175 000 iterations using the Adam optimizer, with a batch-size of 48, and leave the learning rate fixed at 0.001 throughout. These training parameters are comparable to those of standard feed-forward models.

Despite the relatively small training set, we see little signs of overfitting, and the model generalizes well to the test set. Previously, [31] also found low overfitting and good generalization on small training sets using INNs. Several samples by the model are shown in Fig. 5. The cINN correctly recognizes populated regions and generates lights there, as well as freely synthesizing diverse cloud patterns and weather conditions. At the same time, the edges and structures (e.g. mountains) are correctly aligned with the conditioning image. The features learned by the conditioning network are visualized in Fig. 6. Hereby, independent features were extracted via PCA. The figure shows one example of a feature from the first three levels of the pyramid.

4.2 Diverse Image Colorization

For a more challenging task, we turn to colorization of natural images. The common approach for this task is to represent images in Lab color space and generate color channels a, b by a model conditioned on the luminance channel L. We train on the ImageNet dataset [27]. As the color channels do not require as much resolution as the luminance channel, we condition on 256×256 pixel grayscale images, but generate 64×64 pixel color information. This is in accordance with the majority of existing colorization methods.

For the conditioning network φ, we start with the same VGG-like architecture from [34] and pretrain on the colorization task using their code. We then cut off the network before the second-to-last convolution, resulting in 256 feature maps of size 64×64 from the grayscale image L. To form the feature pyramid, we then add a series of strided convolutions, ReLUs, and batch normaliziation layers on top, to produce the features at each resolution. The ablation study in Fig. 12 confirms that the conditioning network is absolutely necessary to capture semantic information.

The INN-part constist of 22 convolutional CCBs, with three downsampling steps in between. After that, the features are flattened, followed by 8 fully connected CCBs. To conserve memory and computation, we adopt a similar

Grayscale y $z = 0.0 \cdot z^*$ $z = 0.7 \cdot z^*$ $z = 0.9 \cdot z^*$ $z = 1.0 \cdot z^*$ $z = 1.25 \cdot z^*$

Fig. 7. Effects of linearly scaling the latent code z while keeping the condition fixed. Vector z^* is "typical" in the sense that $\|z^*\|^2 = \mathbb{E}[\|z\|^2]$, and results in natural colors. As we move closer to the center of the latent space ($\|z\| < \|z^*\|$), regions with ambiguous colors become desaturated, while less ambiguous regions (e.g. sky, vegetation) revert to their prototypical colors. In the opposite direction ($\|z\| > \|z^*\|$), colors are enhanced to the point of oversaturation. (Color figure online)

Fig. 8. For color transfer, we first compute the latent vectors z for different color images (L, a, b) *(top row)*. We then send the same z vectors through the inverse network with a new grayscale condition L^* *(far left)* to produce transferred colorizations a^*, b^* *(bottom row)*. Differences between reference and output color (e.g. pink rose) can arise from mismatches between the reference colors a, b and the intensity prescribed by the new condition L^*. (Color figure online)

Fig. 10. Failure cases of our method. *Top:* Sampling outliers. *Bottom:* cINN did not recognize an object's semantic class or connectivity.

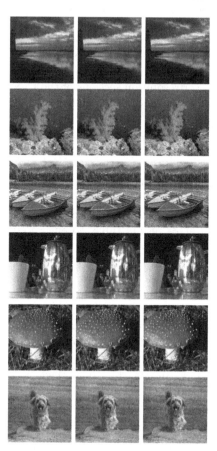

Fig. 9. Diverse colorizations produced by our cINN. (Color figure online)

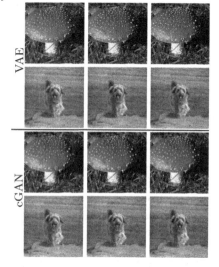

Fig. 11. Other methods have lower diversity or quality, and suffer from inconsistencies in objects, or color blurriness and bleeding (cf. Fig. 9, bottom). (Color figure online)

Fig. 12. In an ablation study, we train a cINN using the grayscale image directly as conditional input, without a conditioning network φ. The resulting colorizations largely ignore semantic content which leads to exaggerated diversity. More ablations are found in the appendix.

splitting- and merging-scheme as in [7]: after each wavelet downsampling step, we split off half the channels. These are not processed any further, but fed into a skip connection and concatenated directly onto the latent output vector. This way, the INN as a whole stays invertible. The reasoning behind this is the following: The high resolution stages have a smaller receptive field and less expressive power, so the channels split off early correspond to local structures and noise. More global information is passed on to the lower resolution sections of the INN and processed further. Overall, the generative performance of the network is not meaningfully impacted, while dramatically reducing the computational cost.

For training, we use the Adam optimizer for faster convergence, and train for roughly 250 000 iterations, and a batch-size of 48. The learning rate is 10^{-3}, decreasing by a factor of 10 at 100 000 and 200 000 iterations. At inference time, we use joint bilateral upsampling [21] to match the resolution of the generated color channels a, b to that of the luminance channel L. This produces visually slightly more pleasing edges than bicubic upsampling, but has little to no impact on the results. It was not used in the quantitative results table, to ensure an unbiased comparison.

Latent space interpolations and color transfer are shown in Figs. 7 and 8, with more experiments in the appendix. In Table 1, a quantitative comparison to existing methods is given. The cINN clearly has the best sample diversity, as summarized by the variance and best-of-8 accuracy. The standard cGAN completely ignores the latent code, and relies only on the condition. As a result, we do not observe any measurable diversity, in line with results from [14]. In terms of FID score, the cGAN performs best, although its results do not appear more realistic to the human eye, cf. Fig. 11. This may be due to the fact that FID is sensitive to outliers, which are unavoidable for a truly diverse method (see Fig. 10), or because the discriminator loss implicitly optimizes for the similarity of deep CNN activations. The VGG classification accuracy of colorized images is decreased for all generative methods equally, because occasional outliers may lead to misclassification.

Table 1. Comparison of conditional generative models for diverse colorization (VAE-MDN: [6]; cGAN: [14]). We additionally compare to a state-of-the-art regression model ('CNN', no diversity, [13]), and the grayscale images alone ('BW'). For each of 5k ImageNet validation images, we compare the best pixel-wise MSE of 8 generated colorization samples, the pixel-wise variance between the 8 samples as an approximation of the diversity, the Fréchet Inception Distance [11] as a measure of realism, and the top 5 accuracy of ImageNet classification performed on the colorized images, to check if semantic content is preserved by the colorization.

	cINN (ours)	VAE-MDN	cGAN	CNN	BW	Ground truth
MSE best of 8	**3.53 ± 0.04**	4.06 ± 0.04	9.75 ± 0.06	6.77 ± 0.05	–	–
Variance	**35.2 ± 0.3**	21.1 ± 0.2	0.0 ± 0.0	–	–	–
FID	25.13 ± 0.30	25.98 ± 0.28	**24.41 ± 0.27**	24.95 ± 0.27	30.91 ± 0.27	14.69 ± 0.18
VGG top 5 acc.	85.00 ± 0.48	85.00 ± 0.48	84.62 ± 0.53	**86.86 ± 0.41**	86.02 ± 0.43	91.66 ± 0.43

5 Conclusion and Outlook

We have proposed a conditional invertible neural network architecture which enables diverse image-to-image translation with high realism. For image colorization, we believe that even better results can be achieved when employing the latest tricks from large-scale GAN frameworks. Especially the non-invertible nature of the conditioning network makes cINNs a suitable method for other computer vision tasks such as diverse semantic segmentation.

Acknowledgements. This work is supported by Deutsche Forschungsgemeinschaft (DFG) under Germany's Excellence Strategy EXC-2181/1 - 390900948 (the Heidelberg STRUCTURES Excellence Cluster). LA received funding by the Federal Ministry of Education and Research of Germany project High Performance Deep Learning Framework (No 01IH17002). JK, CR and UK received financial support from the European Re-search Council (ERC) under the European Unions Horizon2020 research and innovation program (grant agreement No647769). JK received funding by Informatics for Life funded by the Klaus Tschira Foundation.

References

1. Alemi, A., Poole, B., Fischer, I., Dillon, J., Saurous, R.A., Murphy, K.: Fixing a broken elbo. In: International Conference on Machine Learning, pp. 159–168 (2018)
2. Ardizzone, L., Kruse, J., Rother, C., Köthe, U.: Analyzing inverse problems with invertible neural networks. In: International Conference on Learning Representations (2019)
3. Ardizzone, L., Mackowiak, R., Köthe, U., Rother, C.: Exact information bottleneck with invertible neural networks: getting the best of discriminative and generative modeling. arXiv preprint arXiv:2001.06448 (2020)
4. Brock, A., Donahue, J., Simonyan, K.: Large scale GAN training for high fidelity natural image synthesis. In: International Conference on Learning Representations (2019). https://openreview.net/forum?id=B1xsqj09Fm
5. Choi, H., Jang, E., Alemi, A.A.: Waic, but why? generative ensembles for robust anomaly detection. arXiv preprint arXiv:1810.01392 (2018)
6. Deshpande, A., Lu, J., Yeh, M.C., Jin Chong, M., Forsyth, D.: Learning diverse image colorization. In: Conference on Computer Vision and Pattern Recognition (CVPR), pp. 6837–6845 (2017)
7. Dinh, L., Krueger, D., Bengio, Y.: NICE: Non-linear independent components estimation. arXiv:1410.8516 (2014)
8. Dinh, L., Sohl-Dickstein, J., Bengio, S.: Density estimation using Real NVP. arXiv:1605.08803 (2016)
9. Haar, A.: Zur Theorie der orthogonalen Funktionensysteme. Mathematische Annalen **69**(3), 331–371 (1910). https://doi.org/10.1007/BF01456326
10. He, K., Zhang, X., Ren, S., Sun, J.: Deep residual learning for image recognition. In: Proceedings of the IEEE Conference on Computer Vision and Pattern Recognition, pp. 770–778 (2016)
11. Heusel, M., Ramsauer, H., Unterthiner, T., Nessler, B., Hochreiter, S.: GANs trained by a two time-scale update rule converge to a local Nash equilibrium. In: Advances in Neural Information Processing Systems. pp. 6626–6637 (2017)

12. Hoogeboom, E., Peters, J., van den Berg, R., Welling, M.: Integer discrete flows and lossless compression. In: Advances in Neural Information Processing Systems, pp. 12134–12144 (2019)

13. Iizuka, S., Simo-Serra, E., Ishikawa, H.: Let there be color! joint end-to-end learning of global and local image priors for automatic image colorization with simultaneous classification. ACM Trans. Graphics (TOG) 35(4), 110 (2016)

14. Isola, P., Zhu, J.Y., Zhou, T., Efros, A.A.: Image-to-image translation with conditional adversarial networks. In: CVPR 2017, pp. 1125–1134 (2017)

15. Jacobsen, J.H., Behrmann, J., Zemel, R., Bethge, M.: Excessive invariance causes adversarial vulnerability. arXiv preprint arXiv:1811.00401 (2018)

16. Jacobsen, J.H., Smeulders, A.W., Oyallon, E.: i-RevNet: deep invertible networks. In: International Conference on Learning Representations (2018). https://openreview.net/forum?id=HJsjkMb0Z

17. Karras, T., Aila, T., Laine, S., Lehtinen, J.: Progressive growing of GANs for improved quality, stability, and variation. arXiv:1710.10196 (2017)

18. Karras, T., Laine, S., Aila, T.: A style-based generator architecture for generative adversarial networks. In: Proceedings of the IEEE Conference on Computer Vision and Pattern Recognition, pp. 4401–4410 (2019)

19. Kingma, D.P., Dhariwal, P.: Glow: Generative flow with invertible 1x1 convolutions. arXiv:1807.03039 (2018)

20. Kondo, R., Kawano, K., Koide, S., Kutsuna, T.: Flow-based image-to-image translation with feature disentanglement. In: Advances in Neural Information Processing Systems, pp. 4168–4178 (2019)

21. Kopf, J., Cohen, M.F., Lischinski, D., Uyttendaele, M.: Joint bilateral upsampling. In: ACM Transactions on Graphics (ToG), vol. 26, p. 96. ACM (2007)

22. Laffont, P.Y., Ren, Z., Tao, X., Qian, C., Hays, J.: Transient attributes for high-level understanding and editing of outdoor scenes. ACM Trans. graphics (TOG) 33(4), 1–11 (2014)

23. Lee, H.-Y., Tseng, H.-Y., Huang, J.-B., Singh, M., Yang, M.-H.: Diverse image-to-image translation via disentangled representations. In: Ferrari, V., Hebert, M., Sminchisescu, C., Weiss, Y. (eds.) ECCV 2018. LNCS, vol. 11205, pp. 36–52. Springer, Cham (2018). https://doi.org/10.1007/978-3-030-01246-5_3

24. Nalisnick, E., Matsukawa, A., Teh, Y.W., Lakshminarayanan, B.: Detecting out-of-distribution inputs to deep generative models using a test for typicality. arXiv preprint arXiv:1906.02994 5 (2019)

25. Park, T., Liu, M.Y., Wang, T.C., Zhu, J.Y.: Semantic image synthesis with spatially-adaptive normalization. arXiv:1903.07291 (2019)

26. Prenger, R., Valle, R., Catanzaro, B.: Waveglow: a flow-based generative network for speech synthesis. In: ICASSP 2019–2019 IEEE International Conference on Acoustics, Speech and Signal Processing (ICASSP), pp. 3617–3621. IEEE (2019)

27. Russakovsky, O., et al.: ImageNet large scale visual recognition challenge. Int. J. Comput. Vision (IJCV) 115(3), 211–252 (2015). https://doi.org/10.1007/s11263-015-0816-y

28. Salimans, T., Goodfellow, I., Zaremba, W., Cheung, V., Radford, A., Chen, X.: Improved techniques for training gans. In: Advances in Neural Information Processing Systems, pp. 2234–2242 (2016)

29. Simonyan, K., Zisserman, A.: Very deep convolutional networks for large-scale image recognition. CoRR abs/1409.1556 (2014)

30. Sorrenson, P., Rother, C., Köthe, U.: Disentanglement by nonlinear ica with general incompressible-flow networks (gin). arXiv preprint arXiv:2001.04872 (2020)

31. Sun, H., et al.: Dual-glow: Conditional flow-based generative model for modality transfer. In: Proceedings of the IEEE International Conference on Computer Vision, pp. 10611–10620 (2019)

32. Theis, L., Oord, A.V.D., Bethge, M.: A note on the evaluation of generative models. arXiv preprint arXiv:1511.01844 (2015)

33. Ulyanov, D., Vedaldi, A., Lempitsky, V.: It takes (only) two: adversarial generator-encoder networks. In: Thirty-Second AAAI Conference on Artificial Intelligence (2018)

34. Zhang, R., Isola, P., Efros, A.A.: Colorful image colorization. In: Leibe, B., Matas, J., Sebe, N., Welling, M. (eds.) ECCV 2016. LNCS, vol. 9907, pp. 649–666. Springer, Cham (2016). https://doi.org/10.1007/978-3-319-46487-9_40

35. Zhao, S., Song, J., Ermon, S.: Infovae: balancing learning and inference in variational autoencoders. In: Proceedings of the AAAI Conference on Artificial Intelligence, vol. 33, pp. 5885–5892 (2019)

36. Zhu, J.Y., Park, T., Isola, P., Efros, A.A.: Unpaired image-to-image translation using cycle-consistent adversarial networks. In: ICCV 2017, pp. 2223–2232 (2017)

37. Zhu, J.Y., et al.: Toward multimodal image-to-image translation. In: Advances in Neural Information Processing Systems, pp. 465–476 (2017)

Image Inpainting with Learnable Feature Imputation

Håkon Hukkelås$^{(\boxtimes)}$, Frank Lindseth , and Rudolf Mester

Department of Computer Science, Norwegian University of Science and Technology,
Trondheim, Norway
{hakon.hukkelas,rudolf.mester,frankl}@ntnu.no

Abstract. A regular convolution layer applying a filter in the same way over known and unknown areas causes visual artifacts in the inpainted image. Several studies address this issue with feature re-normalization on the output of the convolution. However, these models use a significant amount of learnable parameters for feature re-normalization [41,48], or assume a binary representation of the certainty of an output [11,26].

We propose (layer-wise) feature imputation of the missing input values to a convolution. In contrast to learned feature re-normalization [41,48], our method is efficient and introduces a minimal number of parameters. Furthermore, we propose a revised gradient penalty for image inpainting, and a novel GAN architecture trained exclusively on adversarial loss. Our quantitative evaluation on the FDF dataset reflects that our revised gradient penalty and alternative convolution improves generated image quality significantly. We present comparisons on CelebA-HQ and Places2 to current state-of-the-art to validate our model. (Code is available at: github.com/hukkelas/DeepPrivacy. Supplementary material can be downloaded from: folk.ntnu.no/haakohu/GCPR_supplementary.pdf)

1 Introduction

Image inpainting is the task of filling in missing areas of an image. Use cases for image inpainting are diverse, such as restoring damaged images, removing unwanted objects, or replacing information to preserve the privacy of individuals. Prior to deep learning, image inpainting techniques were generally examplar-based. For example, pattern matching, by searching and replacing with similar patches [4,8,23,29,38,43], or diffusion-based, by smoothly propagating information from the boundary of the missing area [3,5,6].

Convolutional Neural Networks (CNNs) for image inpainting have led to significant progress in the last couple of years [1,24,42]. In spite of this, a standard convolution does not consider if an input pixel is missing or not, making it ill-fitted for the task of image inpainting. Partial Convolution (PConv) [26] propose

---

Electronic supplementary material The online version of this chapter (https://doi.org/10.1007/978-3-030-71278-5_28) contains supplementary material, which is available to authorized users.

Fig. 1. Masked images and corresponding generated images from our proposed single-stage generator.

a modified convolution, where they zero-out invalid (missing) input pixels and re-normalizes the output feature map depending on the number of valid pixels in the receptive field. This is followed by a hand-crafted certainty propagation step, where they assume an output is valid if one or more features in the receptive field are valid. Several proposed improvements replace the hand-crafted components in PConv with fully-learned components [41, 48]. However, these solutions use ∼ 50% of the network parameters to propagate the certainties through the network.

We propose *Imputed Convolution (IConv)*; instead of re-normalizing the output feature map of a convolution, we replace uncertain input values with an estimate from spatially close features (see Fig. 2). IConv assumes that a single spatial location (with multiple features) is associated with a single certainty. In contrast, previous solutions [41, 48] requires a certainty *for each feature* in a spatial location, which allocates half of the network parameters for certainty representation and propagation. Our simple assumption enables certainty representation and propagation to be minimal. In total, replacing all convolution layers with IConv increases the number of parameters by only 1−2%.

We use the DeepPrivacy [15] face inpainter as our baseline and suggest several improvements to stabilize the adversarial training: (1) We propose an improved version of gradient penalties to optimize Wasserstein GANs [2], based on the simple observation that standard gradient penalties causes training instability for image inpainting. (2) We combine the U-Net [35] generator with Multi-Scale-Gradient GAN (MSG-GAN) [19] to enable the discriminator to attend to multiple resolutions simultaneously, ensuring global and local consistency. (3) Finally, we replace the inefficient representation of the pose-information for the FDF dataset [15]. In contrast to the current state-of-the-art, our model requires no post-processing of generated images [16, 25], no refinement network [47, 48], or any additional loss term to stabilize the adversarial training [41, 48]. From our knowledge, our model is the first to be trained exclusively on adversarial loss for image-inpainting.

Our main contributions are the following:

1. We propose IConv which utilize a learnable feature estimator to impute uncertain input values to a convolution. This enables our model to generate visually pleasing images for free-form image inpainting.
2. We revisit the standard gradient penalty used to constrain Wasserstein GANs for image inpainting. Our simple modification significantly improves training stability and generated image quality at no additional computational cost.
3. We propose an improved U-Net architecture, enabling the adversarial training to attend to local and global consistency simultaneously.

2   Related Work

In this section, we discuss related work for generative adversarial networks (GANs), GAN-based image-inpainting, and the recent progress in free-form image-inpainting.

Generative Adversarial Networks. Generative Adversarial Networks [9] is a successful unsupervised training technique for image-based generative models. Since its conception, a range of techniques has improved convergence of GANs. Karras *et al.* [21] propose a *progressive growing* training technique to iteratively increase the network complexity to stabilize training. Karnewar *et al.* [19] replace progressive growing with Multi-Scale Gradient GAN (MSG-GAN), where they use skip connections between the matching resolutions of the generator and discriminator. Furthermore, Karras *et al.* [20] propose a modification of MSG-GAN in combination with residual connections [12]. Similar to [20], we replace progressive growing in the baseline model [15] with a modification of MSG-GAN for image-inpainting.

GAN-Based Image Inpainting. GANs have seen wide adaptation for the image inpainting task, due to its astonishing ability to generate semantically coherent results for missing regions. There exist several studies proposing methods to ensure global and local consistency; using several discriminators to focus on different scales [16,25], specific modules to connect spatially distant features [39,44,45,47], patch-based discriminators [48,49], multi-column generators [40], or progressively inpainting the missing area [11,50]. In contrast to these methods, we ensure consistency over multiple resolutions by connecting different resolutions of the generator with the discriminator. Zheng *et al.* [52] proposes a probabilistic framework to address the issue of mode collapse for image inpainting, and they generate several plausible results for a missing area. Several methods propose combining the input image with auxiliary information, such as user sketches [17], edges [31], or examplar-based inpainting [7]. Hukkelås *et al.* [15] propose a U-Net based generator conditioned on the pose of the face.

GANs are notoriously difficult to optimize reliably [36]. For image inpainting, the adversarial loss is often combined with other objectives to improve training

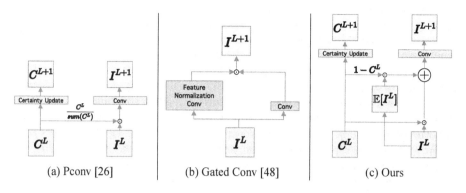

(a) Pconv [26]   (b) Gated Conv [48]   (c) Ours

Fig. 2. Illustration of partial convolution, gated convolution and our proposed solution. $\odot$ is element-wise product and $\oplus$ is addition. Note that C^L is binary for partial convolution.

stability, such as pixel-wise reconstruction [7,16,25,33], perceptual loss [39,51], semantic loss [25], or style loss [41]. In contrast to these methods, we optimize exclusively on the adversarial loss. Furthermore, several studies [17,40,41,47] propose to use Wasserstein GAN [2] with gradient penalties [10]; however, the standard gradient penalty causes training instability for image-inpainting models, as we discuss in Sect. 3.2.

Free-Form Image-Inpainting. Image Inpainting with irregular masks (often referred to as free-form masks) has recently caught more attention. Liu *et al.* [26] propose Partial Convolutions (PConv) to handle irregular masks, where they zero-out input values to a convolution and then perform feature re-normalization based on the number of valid pixels in the receptive field. Gated Convolution [48] modifies PConv by removing the binary-representation constraint, and they combine the mask and feature representation within a single feature map. Xie *et al.* [41] propose a simple modification to PConv, where they reformulate it as "attention" propagation instead of certainty propagation. Both of these PConv adaptations [41,48] doubles the number of parameters in the network when replacing regular convolutions.

3 Method

In this section, we describe a) our modifications to a regular convolution layer, b) our revised gradient penalty suited for image inpainting, and c) our improved U-Net architecture.

3.1 Imputed Convolution (IConv)

Consider the case of a regular convolution applied to a given feature map $I \in \mathbb{R}^N$:

$$f(I) = W_F * I, \tag{1}$$

where $*$ is the convolution and $W_F \in \mathbb{R}^D$ is the filter. To simplify notation, we consider a single filter applied to a single one-dimensional feature map. The generalization to a regular multidimensional convolution layer is straightforward. A convolution applies this filter to all spatial locations of our feature map, which works well for general image recognition tasks. For image inpainting, there exists a set of known and unknown pixels; therefore, a regular convolution applied to all spatial locations is primarily undefined ("unknown" is not the same as 0 or any other fixed value), and naive approaches cause annoying visual artifacts [26].

We propose to replace the missing input values to a convolution with an estimate from spatially close values. To represent known and unknown values, we introduce a certainty C_x for each spatial location x, where $C \in \mathbb{R}^N$, and $0 \leq C_x \leq 1$. Note that this representation enables a single certainty to represent several values in the case of having multiple channels in the input. Furthermore, we define $\tilde{I}_x$ as a random variable with discrete outcomes $\{I_x, h_x\}$, where I_x is the feature at spatial location x, and h_x is an estimate from spatially close features. In this way, we want the output of our convolution to be given by,

$$O = \phi(f(\mathbb{E}[\tilde{I}_x])), \tag{2}$$

where ϕ is the activation function, and O the output feature map. We approximate the probabilities of each outcome using the certainty C_x; that is, $P(\tilde{I}_x = I_x) \approx C_x$ and $P(\tilde{I}_x = h_x) \approx 1 - C_x$, yielding the expected value of $\tilde{I}_x$,

$$\mathbb{E}[\tilde{I}_x] = C_x \cdot I_x + (1 - C_x) \cdot h_x. \tag{3}$$

We assume that a missing value can be approximated from spatially close values. Therefore, we define h_x as a learned certainty-weighted average of the surrounding features:

$$h_x = \frac{\sum_{i=1}^{K} I_{x+i} \cdot C_{x+i} \cdot \omega_i}{\sum_{i=1}^{K} C_{x+i}}, \tag{4}$$

where $\omega \in R^K$ is a learnable parameter. In a sense, our convolutional layer will try to learn the outcome space of $\tilde{I}_x$. Furthermore, h_x is efficient to implement in standard deep learning frameworks, as it can be implemented as a depth-wise separable convolution [37] with a re-normalization factor determined by C.

Propagating Certainties. Each convolutional layer expects a certainty for each spatial location. We handle propagation of certainties as a learned operation,

$$C^{L+1} = \sigma(W_C * C^L), \tag{5}$$

where $*$ is a convolution, $W_C \in \mathbb{R}^D$ is the filter, and σ is the sigmoid function. We constraint W_C to have the same receptive field as f with no bias, and initialize C^0 to 0 for all unknown pixels and 1 else.

The proposed solution is minimal, efficient, and other components of the network remain close to untouched. We use LeakyReLU as the activation function ϕ, and average pooling and pixel normalization [21] after each convolution f.

Replacing all convolutional layers with O_x in our baseline network increases the number of parameters by $\sim 1\%$. This is in contrast to methods based on learned feature re-normalization [41,48], where replacing a convolution with their proposed solution doubles the number of parameters. Similar to partial convolution [26], we use a single scalar to represent the certainty for each spatial location; however, we do not constrain the certainty representation to be binary, and our certainty propagation is fully learned.

U-Net Skip Connection. U-Net [35] skip connection is a method to combine shallow and deep features in encoder-decoder architectures. Generally, the skip connection consists of concatenating shallow and deep features, then followed by a convolution. However, for image inpainting, we only want to propagate certain features.

To find the combined feature map for an input in layer L and $L+l$, we find a weighted average. Assuming features from two layers in the network, (I^L, C^L), (I^{L+l}, C^{L+l}), we define the combined feature map as;

$$I^{L+l+1} = \gamma \cdot I^L + (1 - \gamma) \cdot I^{L+l}, \tag{6}$$

and likewise for C^{L+l+1}. γ is determined by

$$\gamma = \frac{C^L \cdot \beta_1}{C^L \cdot \beta_1 + C^{L+l} \cdot \beta_2}, \tag{7}$$

where $\beta_1, \beta_2 \in \mathbb{R}^+$ are learnable parameters initialized to 1. Our U-Net skip connection is unique compared to previous work and designed for image inpainting. Equation 6 enables the network to only propagate features with a high certainty from shallow layers. Furthermore, we include β_1 and β_2 to give the model the flexibility to learn if it should attend to shallow or deep features.

3.2   Revisiting Gradient Penalties for Image Inpainting

Improved Wasserstein GAN [2,10] is widely used in image inpainting [17,40,41,47]. Given a discriminator D, the objective function for optimizing a Wasserstein GAN with gradient penalties is given by,

$$\mathcal{L}_{total} = \mathcal{L}_{adv} + \lambda \cdot (\|\nabla D(\hat{x})\|_p - 1)^2, \tag{8}$$

where $\mathcal{L}_{adv}$ is the adversarial loss, p is commonly set to 2 (L^2 norm), λ is the gradient penalty weight, and $\hat{x}$ is a randomly sampled point between the real image, x, and a generated image, $\tilde{x}$. Specifically, $\hat{x} = t \cdot x + (1 - t) \cdot \tilde{x}$, where t is sampled from a uniform distribution [10].

Previous methods enforce the gradient penalty only for missing areas [17,40,47]. Given a mask M to indicate areas to be inpainted in the image x, where M is 0 for missing pixels and 1 otherwise (note that $M = C^0$), Yu et al. [47] propose the gradient penalty:

$$\bar{g}(\hat{x}) = (\|\nabla D(\hat{x}) \odot (1 - M)\|_p - 1)^2, \tag{9}$$

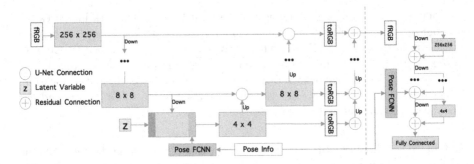

Fig. 3. Illustration of the generator (left of the dashed line) and discriminator architecture. Up and down denotes nearest neighbor upsampling and average pool. The pose information in the discriminator is concatenated to the input of the first convolution layer with 32 × 32 resolution. Note that pose information is only used for the FDF dataset [15].

where $\odot$ is element-wise multiplication. This gradient penalty cause significant training instability, as the gradient sign of $\bar{g}$ shifts depending on the cardinality of M. Furthermore, Eq. 9 impose $||\nabla D(\hat{x})|| \approx 1$, which leads to a lower bound on the Wasserstein distance [18].

Imposing $||\nabla D(\hat{x})|| \leq 1$ will remove the issue of shifting gradients in Eq. 9. Furthermore, imposing the constrain $||\nabla D(\hat{x})|| \leq 1$ is shown to properly estimate the Wasserstein distance [18]. Therefore, we propose the following gradient penalty:

$$g(\hat{x}) = \max(0, ||\nabla D(\hat{x}) \odot (1 - M))||_p - 1) \tag{10}$$

Previous methods enforce the L^2 norm [17,40,47]. Jolicoeur-Martineau *et al.* [18] suggest that replacing the L^2 gradient norm with L^∞ can improve robustness. From empirical experiments (see Appendix 1), we find L^∞ more unstable and sensitive to choice of hyperparameters; therefore, we enforce the L^2 norm (p = 2).

In total, we optimize the following objective function:

$$\mathcal{L}_{total} = \mathcal{L}_{adv} + \lambda \cdot \max(0, ||\nabla D(\hat{x}) \odot (1 - M))||_p - 1) \tag{11}$$

3.3   Model Architecture

We propose several improvements to the baseline U-Net architecture [15]. See Fig. 3 for our final architecture. We replace all convolutions with Eq. 2, average pool layer with a certainty-weighted average and U-Net skip connections with our revised skip connection (see Eq. 6). Furthermore, we replace progressive growing training [21] with Multi-Scale Gradient GAN (MSG-GAN) [19]. For the MSG-GAN, instead of matching different resolutions from the generator with the discriminator, we upsample each resolution and sum up the contribution of the RGB outputs [20]. In the discriminator we use residual connections, similar to [20]. Finally, we improve the representation of pose information in the baseline model (pose information is only used on the FDF dataset [15]).

Representation of Pose Information. The baseline model [15] represents pose information as one-hot encoded images for each resolution in the network, which is extremely memory inefficient and a fragile representation. The pose information, $P \in \mathbb{R}^{K \cdot 2}$, represents K facial keypoints and is used as conditional information for the generator and discriminator. We propose to replace the one-hot encoded representation, and instead pre-process P into a $4 \times 4 \times 32$ feature bank using two fully-connected layers. This feature bank is concatenated with the features from the encoder. Furthermore, after replacing progressive growing with MSG-GAN, we include the same pose pre-processing architecture in the discriminator, and input the pose information as a $32 \times 32 \times 1$ feature map to the discriminator.

4 Experiments

We evaluate our proposed improvements on the Flickr Diverse Faces (FDF) dataset [15], a lower resolution (128×128) face dataset. We present experiments on the CelebA-HQ [21] and Places2 [53] datasets, which reflects that our suggestions generalizes to standard image inpainting. We compare against current state-of-the art [34,41,48,52]. Finally, we present a set of ablation studies to analyze the generator architecture.[1]

Quantitative Metrics. For quantitative evaluations, we report commonly used image inpainting metrics; pixel-wise distance (L1 and L2), peak signal-to-noise ratio (PSNR), and structural similarity (SSIM). Neither of these reconstruction metrics are any good indicators of generated image quality, as there often exist several possible solutions to a missing region, and they do not reflect human nuances [51]. Recently proposed deep feature metrics correlate better with human perception [51]; therefore, we report the Fréchet Inception Distance (FID) [13] (lower is better) and Learned Perceptual Image Patch Similarity (LPIPS) [51] (lower is better). We use LPIPS as the main quantitative evaluation.

4.1   Improving the Baseline

We iteratively add our suggestions to the baseline [15] (Config A-E), and report quantitative results in Table 1. First, we replace the gradient penalty term with Eq. 10, where we use the L^2 norm ($p = 2$), and impose the following constraint (Config B):

$$G_{out} = G(I, C^0) \cdot (1 - C^0) + I \cdot C^0, \tag{12}$$

---

[1]   To prevent ourselves from cherry-picking qualitative examples, we present several images (with corresponding masks) chosen by previous state-of-the-art papers [11, 41,48,52], thus copying their selection. Appendix 5 describes how we selected these samples. The only hand-picked examples in this paper are Fig. 1, Fig. 4, Fig. 6, and Fig. 7. No examples in the Supplementary Material are cherry-picked.

Table 1. Quantitative results on the FDF dataset [15]. We report standard metrics after showing the discriminator 20 million images on the FDF and Places2 validation sets. We report L1, L2, and SSIM in Appendix 3. Note that Config E is trained with MSG-GAN, therefore, we separate it from Config A-D which are trained with progressive growing [21]. * Did not converge. † Same as Config B

Configuration		FDF			Places2		
		LPIPS ↓	PSNR ↑	FID ↓	LPIPS ↓	PSNR ↑	FID ↓
A	Baseline [15]	0.1036	22.52	6.15	–*	–*	–*
B	+ Improved Gradient penalty	0.0757	23.92	1.83	0.1619	20.99	7.96
C	+ Scalar Pose Information	0.0733	24.01	1.76	– †	–†	– †
D	+ Imputed Convolution	0.0739	23.95	1.66	0.1563	21.21	6.81
E	+ No Growing, MSG	**0.0728**	**24.01**	**1.49**	**0.1491**	**21.42**	**5.24**

where C^0 is the binary input certainty and G is the generator. Note that we are not able to converge Config A while imposing G_{out}. We replace the one-hot encoded representation of the pose information with two fully connected layers in the generator (Config C). Furthermore, we replace the input to all convolutional layers with Eq. 3 (Config D). We set the receptive field of h_x to 5×5 ($K = 5$ in Eq. 4). We replace the progressive-growing training technique with MSG-GAN [19], and replace the one-hot encoded pose-information in the discriminator (Config E). These modifications combined improve the LPIPS score by *30.0%*. The authors of [15] report a FID of 1.84 on the FDF dataset with a model consisting of 46M learnable parameters. In comparison, we achieve a FID of 1.49 with 2.94M parameters (config E). For experimental details, see Appendix 2.

4.2 Generalization to Free-Form Image Inpainting

We extend Config E to general image inpainting datasets; CelebA-HQ [21] and Places2 [53]. We increase the number of filters in each convolution by a factor of 2, such that the generator has 11.5M parameters. In comparison, Gated Convolution [48] use 4.1M, LBAM [41] 68.3M, StructureFlow [34] 159M, and PIC [52] use 3.6M parameters. Compared to [48,52], our increase in parameters improves semantic reasoning for larger missing regions. Also, compared to previous solutions, we achieve similar inference time since the majority of the parameters are located at low-resolution layers (8×8 and 16×16). In contrast, [48] has no parameters at a resolution smaller than 64×64. For single-image inference time, our model matches (or outperforms) previous models; on a single NVIDIA 1080 GPU, our network runs at ~89 ms per image on 256×256 resolution, 2× faster than LBAM [41], and PIC [52]. GatedConvolution [48] achieves ~62 ms per image.[2] See Appendix 2.1 for experimental details.

---

[2] We measure runtime for [48,52] with their open-source code, as they do not report inference time for 256×256 resolution in their paper.

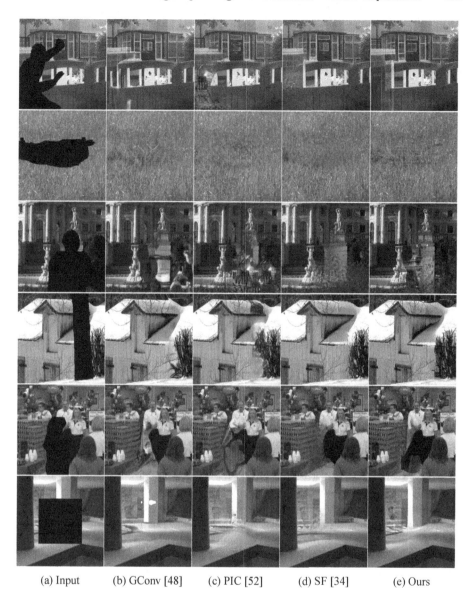

(a) Input   (b) GConv [48]   (c) PIC [52]   (d) SF [34]   (e) Ours

Fig. 4. Qualitative examples on the Places2 validation set with comparisons to Gated Convolution (GConv) [48], StructureFlow (SF) [34], and Pluralistic Image Completion (PIC) [52]. We recommend the reader to zoom-in on missing regions. For non hand-picked qualitative examples, see Appendix 5.

Table 2. Quantitative results on the CelebA-HQ and Places2 datasets. We use the official frameworks to reproduce results from [48,52]. For the (Center) dataset we use a 128 × 128 center mask, and for (Free-Form) we generate free-form masks for each image following the approach in [48]. We report L1, L2, and SSIM in Appendix 3.

Method	Places2 (Center)			Places2 (Free Form)			CelebA-HQ (Center)			CelebA-HQ (Free Form)		
	PSNR	LPIPS	FID	PSNR	LPIPS	FID	PSNR	LPIPS	FID	PSNR	LPIPS	FID
Gated convolutions [48]	21.56	**0.1407**	4.14	**27.59**	**0.0579**	0.90	**25.55**	0.0587	6.05	30.26	0.0366	2.98
Plurastic image inpainting [52]	21.04	0.1584	7.23	26.66	0.0804	2.76	24.59	0.0644	7.50	29.30	0.0394	3.30
Ours	**21.70**	0.1412	**3.99**	27.33	0.0597	0.94	25.29	**0.0522**	**4.43**	**30.32**	**0.0300**	**2.38**

(a) Input   (b) PM [4]   (c) PIC [52]   (d) PC [26]   (e) BA [41]   (f) GC [48]   (g) Ours

Fig. 5. Places2 comparison to PatchMatch (PM) [4], Pluralistic Image Completion (PIC) [52], Partial Convolution (PC) [26], Bidirectional Attention (BA) [41], and Gated Convolution (GC) [48]. Examples selected by authors of [41] (images extracted from their supplementary material). Results of [48,52] generated by using their open-source code and models. We recommend the reader to zoom-in on missing regions.

Quantitative Results. Table 2 shows quantitative results for the CelebA-HQ and Places2 datasets. For CelebA-HQ, we improve LPIPS and FID significantly compared to previous models. For Places2, we achieve comparable results to [48] for free-form and center-crop masks. Furthermore, we compare our model with and without IConv and notice a significant improvement in generated image quality (see Fig. 1 in Appendix 3). See Appendix 5.1 for examples of the center-crop and free-form images.

Qualitative Results. Figure 4 shows a set of hand-picked examples, Fig. 5 shows examples selected by [41], and Appendix 5 includes a large set of examples selected by the authors of [11,41,48,52]. We notice less visual artifacts than models using vanilla convolutions [34,52], and we achieve comparable results to Gated Convolution [48] for free-form image inpainting. For larger missing areas, our model generates more semantically coherent results compared to previous solutions [11,41,48,52].

Fig. 6. Diverse Plausible Results: Images from the FDF validation set [15]. Left column is the input image with the pose information marked in red. Second column and onwards are different plausible generated results. Each image is generated by randomly sampling a latent variable for the generator (except for the second column where the latent variable is set to all 0's). For more results, see Appendix 6.

4.3   Ablation Studies

Pluralistic Image Inpainting. Generating different possible results for the same conditional image (pluralistic inpainting) [52] has remained a problem for conditional GANs [14,54]. Figure 6 illustrates that our proposed model (Config E) generates multiple and diverse results. Even though, for Places2, we observe that our generator suffers from mode collapse early on in training. Therefore, we ask the question; *does a deterministic generator impact the generated image quality for image-inpainting?* To briefly evaluate the impact of this, we train Config D without a latent variable, and observe a 7% degradation in LPIPS score on the FDF dataset. We leave further analysis of this for further work.

Propagation of Certainties. Figure 7 visualizes if the generator attends to shallow or deep features in our encoder-decoder architecture. Our proposed U-Net skip connection enables the network to select features between the encoder and decoder depending on the certainty. Notice that our network attends to deeper features in cases of uncertain features, and shallower feature otherwise.

Fig. 7. U-Net Skip Connections. Visualization of γ from Eq. 6. The left image is the input image, second column and onwards are the values of γ for resolution 8 to 256. Rightmost image is the generated image. Smaller values of γ indicates that the network selects deep features (from the decoder branch).

5   Conclusion

We propose a simple single-stage generator architecture for free-form image inpainting. Our proposed improvements to GAN-based image inpainting significantly stabilizes adversarial training, and from our knowledge, we are the first to produce state-of-the-art results by exclusively optimizing an adversarial objective. Our main contributions are; a revised convolution to properly handle missing values in convolutional neural networks, an improved gradient penalty for image inpainting which substantially improves training stability, and a novel U-Net based GAN architecture to ensure global and local consistency. Our model achieves state-of-the-art results on the CelebA-HQ and Places2 datasets, and our single-stage generator is much more efficient compared to previous solutions.

Acknowledgements. The computations were performed on resources provided by the Tensor-GPU project led by Prof. Anne C. Elster through support from The Department of Computer Science and The Faculty of Information Technology and Electrical Engineering, NTNU. Furthermore, Rudolf Mester acknowledges the support obtained from DNV GL.

References

1. Habibi Aghdam, H., Jahani Heravi, E.: Convolutional neural networks. Guide to Convolutional Neural Networks, pp. 85–130. Springer, Cham (2017). https://doi.org/10.1007/978-3-319-57550-6_3
2. Arjovsky, M., Chintala, S., Bottou, L.: Wasserstein GAN. arXiv preprint arXiv:1701.07875 (2017)
3. Ballester, C., Bertalmio, M., Caselles, V., Sapiro, G., Verdera, J.: Filling-in by joint interpolation of vector fields and gray levels. IEEE Trans. Image Process. **10**(8), 1200–1211 (2001). https://doi.org/10.1109/83.935036
4. Barnes, C., Shechtman, E., Finkelstein, A., Goldman, D.B.: PatchMatch. In: ACM SIGGRAPH 2009 papers on - SIGGRAPH 09. ACM Press (2009). https://doi.org/10.1145/1576246.1531330
5. Bertalmio, M., Sapiro, G., Caselles, V., Ballester, C.: Image inpainting. In: Proceedings of the 27th Annual Conference on Computer Graphics and Interactive Techniques, pp. 417–424 (2000)
6. Criminisi, A., Perez, P., Toyama, K.: Region filling and object removal by exemplar-based image inpainting. IEEE Trans. Image Process. **13**(9), 1200–1212 (2004). https://doi.org/10.1109/tip.2004.833105
7. Dolhansky, B., Ferrer, C.C.: Eye in-painting with exemplar generative adversarial networks. In: 2018 IEEE/CVF Conference on Computer Vision and Pattern Recognition. IEEE, June 2018. https://doi.org/10.1109/cvpr.2018.00824
8. Efros, A.A., Freeman, W.T.: Image quilting for texture synthesis and transfer. In: Proceedings of the 28th Annual Conference on Computer Graphics and Interactive Techniques - SIGGRAPH 2001. ACM Press (2001). https://doi.org/10.1145/383259.383296
9. Goodfellow, I., et al.: Generative adversarial nets. In: Advances in Neural Information Processing Systems, pp. 2672–2680 (2014)
10. Gulrajani, I., Ahmed, F., Arjovsky, M., Dumoulin, V., Courville, A.C.: Improved training of Wasserstein GANs. In: Advances in Neural Information Processing Systems, pp. 5767–5777 (2017)
11. Guo, Z., Chen, Z., Yu, T., Chen, J., Liu, S.: Progressive image inpainting with full-resolution residual network. In: Proceedings of the 27th ACM International Conference on Multimedia. ACM, October 2019. https://doi.org/10.1145/3343031.3351022

12. He, K., Zhang, X., Ren, S., Sun, J.: Deep residual learning for image recognition. In: 2016 IEEE Conference on Computer Vision and Pattern Recognition (CVPR). IEEE, June 2016. https://doi.org/10.1109/cvpr.2016.90
13. Heusel, M., Ramsauer, H., Unterthiner, T., Nessler, B., Hochreiter, S.: GANs trained by a two time-scale update rule converge to a local nash equilibrium. In: Advances in Neural Information Processing Systems, pp. 6626–6637 (2017)
14. Huang, X., Liu, M.-Y., Belongie, S., Kautz, J.: Multimodal unsupervised image-to-image translation. In: Ferrari, V., Hebert, M., Sminchisescu, C., Weiss, Y. (eds.) ECCV 2018. LNCS, vol. 11207, pp. 179–196. Springer, Cham (2018). https://doi.org/10.1007/978-3-030-01219-9_11
15. Hukkelås, H., Mester, R., Lindseth, F.: DeepPrivacy: a generative adversarial network for face anonymization. In: Bebis, G., et al. (eds.) ISVC 2019. LNCS, vol. 11844, pp. 565–578. Springer, Cham (2019). https://doi.org/10.1007/978-3-030-33720-9_44
16. Iizuka, S., Simo-Serra, E., Ishikawa, H.: Globally and locally consistent image completion. ACM Trans. Graph. 36(4), 1–14 (2017). https://doi.org/10.1145/3072959.3073659
17. Jo, Y., Park, J.: SC-FEGAN: face editing generative adversarial network with user's sketch and color. In: The IEEE International Conference on Computer Vision (ICCV), October 2019. https://doi.org/10.1109/ICCV.2019.00183
18. Jolicoeur-Martineau, A., Mitliagkas, I.: Connections between support vector machines, Wasserstein distance and gradient-penalty GANs. arXiv preprint arXiv:1910.06922 (2019)
19. Karnewar, A., Wang, O., Iyengar, R.S.: MSG-GAN: multi-scale gradient GAN for stable image synthesis. In: Proceedings of the IEEE/CVF Conference on Computer Vision and Pattern Recognition, vol. 6 (2019). https://doi.org/10.1109/CVPR42600.2020.00782
20. Karras, T., Laine, S., Aittala, M., Hellsten, J., Lehtinen, J., Aila, T.: Analyzing and improving the image quality of stylegan. In: 2020 IEEE/CVF Conference on Computer Vision and Pattern Recognition (CVPR), pp. 8107–8116 (2020). https://doi.org/10.1109/CVPR42600.2020.00813
21. Karras, T., Aila, T., Laine, S., Lehtinen, J.: Progressive growing of GANs for improved quality, stability, and variation. In: International Conference on Learning Representations (2018)
22. Kingma, D.P., Ba, J.: Adam: a method for stochastic optimization. In: International Conference on Learning Representations (2015)
23. Kwatra, V., Essa, I., Bobick, A., Kwatra, N.: Texture optimization for example-based synthesis. In: ACM SIGGRAPH 2005 Papers on - SIGGRAPH 2005. ACM Press (2005). https://doi.org/10.1145/1186822.1073263
24. Köhler, R., Schuler, C., Schölkopf, B., Harmeling, S.: Mask-specific inpainting with deep neural networks. In: Jiang, X., Hornegger, J., Koch, R. (eds.) GCPR 2014. LNCS, vol. 8753, pp. 523–534. Springer, Cham (2014). https://doi.org/10.1007/978-3-319-11752-2_43
25. Li, Y., Liu, S., Yang, J., Yang, M.H.: Generative face completion. In: 2017 IEEE Conference on Computer Vision and Pattern Recognition (CVPR), pp. 5892–5900. IEEE, July 2017. https://doi.org/10.1109/cvpr.2017.624
26. Liu, G., Reda, F.A., Shih, K.J., Wang, T.-C., Tao, A., Catanzaro, B.: Image inpainting for irregular holes using partial convolutions. In: Ferrari, V., Hebert, M., Sminchisescu, C., Weiss, Y. (eds.) ECCV 2018. LNCS, vol. 11215, pp. 89–105. Springer, Cham (2018). https://doi.org/10.1007/978-3-030-01252-6_6

27. Maas, A.L., Hannun, A.Y., Ng, A.Y.: Rectifier nonlinearities improve neural network acoustic models. In: ICML Workshop on Deep Learning for Audio, Speech and Language Processing (2013)
28. Mescheder, L., Nowozin, S., Geiger, A.: Which training methods for GANs do actually converge? In: International Conference on Machine Learning (ICML) (2018)
29. Meur, O.L., Gautier, J., Guillemot, C.: Examplar-based inpainting based on local geometry. In: 2011 18th IEEE International Conference on Image Processing. IEEE, September 2011. https://doi.org/10.1109/icip.2011.6116441
30. Micikevicius, P., et al.: Mixed precision training. arXiv preprint arXiv:1710.03740 (2017)
31. Nazeri, K., Ng, E., Joseph, T., Qureshi, F.Z., Ebrahimi, M.: EdgeConnect: Generative image inpainting with adversarial edge learning. arXiv preprint arXiv:1901.00212 (2019)
32. Odena, A., Olah, C., Shlens, J.: Conditional image synthesis with auxiliary classifier GANs. In: Proceedings of the 34th International Conference on Machine Learning, vol. 70, pp. 2642–2651. JMLR.org (2017)
33. Pathak, D., Krahenbuhl, P., Donahue, J., Darrell, T., Efros, A.A.: Context encoders: feature learning by inpainting. In: 2016 IEEE Conference on Computer Vision and Pattern Recognition (CVPR). IEEE. June 2016. https://doi.org/10.1109/cvpr.2016.278
34. Ren, Y., Yu, X., Zhang, R., Li, T.H., Liu, S., Li, G.: StructureFlow: image inpainting via structure-aware appearance flow. In: 2019 IEEE/CVF International Conference on Computer Vision (ICCV). IEEE, October 2019. https://doi.org/10.1109/iccv.2019.00027
35. Ronneberger, O., Fischer, P., Brox, T.: U-Net: convolutional networks for biomedical image segmentation. In: Navab, N., Hornegger, J., Wells, W.M., Frangi, A.F. (eds.) MICCAI 2015. LNCS, vol. 9351, pp. 234–241. Springer, Cham (2015). https://doi.org/10.1007/978-3-319-24574-4_28
36. Salimans, T., Goodfellow, I., Zaremba, W., Cheung, V., Radford, A., Chen, X.: Improved techniques for training GANs. In: Advances in Neural Information Processing Systems, pp. 2234–2242 (2016)
37. Sifre, L., Mallat, S.: Rigid-motion scattering for image classification. Ph.D. thesis (2014)
38. Simakov, D., Caspi, Y., Shechtman, E., Irani, M.: Summarizing visual data using bidirectional similarity. In: 2008 IEEE Conference on Computer Vision and Pattern Recognition. IEEE, June 2008. https://doi.org/10.1109/cvpr.2008.4587842
39. Song, Yuhang., et al.: Contextual-based image inpainting: infer, match, and translate. In: Ferrari, Vittorio, Hebert, Martial, Sminchisescu, Cristian, Weiss, Yair (eds.) ECCV 2018. LNCS, vol. 11206, pp. 3–18. Springer, Cham (2018). https://doi.org/10.1007/978-3-030-01216-8_1
40. Wang, Y., Tao, X., Qi, X., Shen, X., Jia, J.: Image inpainting via generative multi-column convolutional neural networks. In: Advances in Neural Information Processing Systems, pp. 331–340 (2018)
41. Xie, C., et al.: Image inpainting with learnable bidirectional attention maps. In: Proceedings of the IEEE International Conference on Computer Vision, pp. 8858–8867 (2019). https://doi.org/10.1109/ICCV.2019.00895
42. Xie, J., Xu, L., Chen, E.: Image denoising and inpainting with deep neural networks. In: Advances in Neural Information Processing Systems, pp. 341–349 (2012)
43. Xu, Z., Sun, J.: Image inpainting by patch propagation using patch sparsity. IEEE Trans. Image Process. **19**(5), 1153–1165 (2010). https://doi.org/10.1109/tip.2010.2042098

44. Yan, Z., Li, X., Li, M., Zuo, W., Shan, S.: Shift-Net: image inpainting via deep feature rearrangement. In: Ferrari, V., Hebert, M., Sminchisescu, C., Weiss, Y. (eds.) Computer Vision – ECCV 2018. LNCS, vol. 11218, pp. 3–19. Springer, Cham (2018). https://doi.org/10.1007/978-3-030-01264-9_1

45. Yang, C., Lu, X., Lin, Z., Shechtman, E., Wang, O., Li, H.: High-resolution image inpainting using multi-scale neural patch synthesis. In: 2017 IEEE Conference on Computer Vision and Pattern Recognition (CVPR). IEEE, July 2017. https://doi.org/10.1109/cvpr.2017.434

46. Yazıcı, Y., Foo, C.S., Winkler, S., Yap, K.H., Piliouras, G., Chandrasekhar, V.: The unusual effectiveness of averaging in GAN training. In: International Conference on Learning Representations (2019)

47. Yu, J., Lin, Z., Yang, J., Shen, X., Lu, X., Huang, T.S.: Generative image inpainting with contextual attention. In: 2018 IEEE/CVF Conference on Computer Vision and Pattern Recognition. IEEE, June 2018. https://doi.org/10.1109/cvpr.2018.00577

48. Yu, J., Lin, Z., Yang, J., Shen, X., Lu, X., Huang, T.S.: Free-form image inpainting with gated convolution. In: Proceedings of the IEEE International Conference on Computer Vision, pp. 4471–4480 (2019). https://doi.org/10.1109/ICCV.2019.00457

49. Zeng, Y., Fu, J., Chao, H., Guo, B.: Learning pyramid-context encoder network for high-quality image inpainting. In: 2019 IEEE/CVF Conference on Computer Vision and Pattern Recognition (CVPR). IEEE, June 2019. https://doi.org/10.1109/cvpr.2019.00158

50. Zhang, H., Hu, Z., Luo, C., Zuo, W., Wang, M.: Semantic image inpainting with progressive generative networks. In: 2018 ACM Multimedia Conference on Multimedia Conference. ACM Press (2018). https://doi.org/10.1145/3240508.3240625

51. Zhang, R., Isola, P., Efros, A.A., Shechtman, E., Wang, O.: The unreasonable effectiveness of deep features as a perceptual metric. In: Proceedings of the IEEE Conference on Computer Vision and Pattern Recognition (2018). https://doi.org/10.1109/CVPR.2018.00068

52. Zheng, C., Cham, T.J., Cai, J.: Pluralistic image completion. In: Proceedings of the IEEE Conference on Computer Vision and Pattern Recognition, pp. 1438–1447 (2019). https://doi.org/10.1109/CVPR.2019.00153

53. Zhou, B., Lapedriza, A., Khosla, A., Oliva, A., Torralba, A.: Places: a 10 million image database for scene recognition. IEEE Trans. Pattern Anal. Mach. Intell. (2017). https://doi.org/10.1109/TPAMI.2017.2723009

54. Zhu, J.Y., et al.: Toward multimodal image-to-image translation. In: Advances in Neural Information Processing Systems, pp. 465–476 (2017)

4Seasons: A Cross-Season Dataset for Multi-Weather SLAM in Autonomous Driving

Patrick Wenzel[1,2(✉)], Rui Wang[1,2], Nan Yang[1,2], Qing Cheng[2],
Qadeer Khan[1,2], Lukas von Stumberg[1,2], Niclas Zeller[2], and Daniel Cremers[1,2]

[1] Technical University of Munich, Munich, Germany
wenzel@cs.tum.edu
[2] Artisense, Munich, Germany

Abstract. We present a novel dataset covering seasonal and challenging perceptual conditions for autonomous driving. Among others, it enables research on visual odometry, global place recognition, and map-based re-localization tracking. The data was collected in different scenarios and under a wide variety of weather conditions and illuminations, including day and night. This resulted in more than 350 km of recordings in nine different environments ranging from multi-level parking garage over urban (including tunnels) to countryside and highway. We provide globally consistent reference poses with up-to centimeter accuracy obtained from the fusion of direct stereo visual-inertial odometry with RTK-GNSS. The full dataset is available at https://www.4seasons-dataset.com.

Keywords: Autonomous driving · Long-term localization · SLAM · Visual learning · Visual odometry

1 Introduction

During the last decade, research on visual odometry (VO) and simultaneous localization and mapping (SLAM) has made tremendous strides [11,12,29,30] particularly in the context of autonomous driving (AD) [9,28,44,46]. One reason for this progress has been the publication of large-scale datasets [6,7,14] tailored for benchmarking these methods. Naturally, the next logical step towards progressing research in the direction of visual SLAM has been to make it robust under dynamically changing and challenging conditions. This includes VO, *e.g.* at night or rain, as well as long-term place recognition and re-localization against a pre-built map. In this regard, the advent of deep learning has exhibited itself have promising potential in complementing the performance of visual SLAM [8,20,22,39]. Therefore, it has become all the more important to have datasets that are commensurate with handling the challenges of any real-world environment while also being capable of discerning the performance of state-of-the-art approaches.

© Springer Nature Switzerland AG 2021
Z. Akata et al. (Eds.): DAGM GCPR 2020, LNCS 12544, pp. 404–417, 2021.
https://doi.org/10.1007/978-3-030-71278-5_29

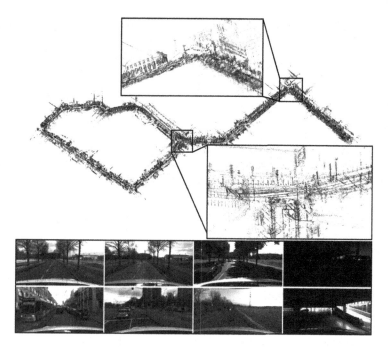

Fig. 1. Dataset overview. Top: overlaid maps recorded at different times and environmental conditions. The points from the reference map (black) align well with the points from the query map (blue), indicating that the reference poses are indeed accurate. Bottom: sample images demonstrating the diversity of our dataset. The first row shows a collection from the same scene across different weather and lighting conditions: snowy, overcast, sunny, and night. The second row depicts the variety of scenarios within the dataset: inner city, suburban, countryside, and a parking garage. (Color figure online)

To accommodate this demand, we present in this paper a versatile cross-season and multi-weather dataset on a large-scale focusing on long-term localization for autonomous driving. By traversing the same stretch under different conditions and over a long-term time horizon, we capture variety in illumination and weather as well as in the appearance of the scenes. Figure 1 visualizes two overlaid 3D maps recorded at different times as well as sample images of the dataset.

In detail this work adds the following contributions to the state-of-the-art:

- A cross-season/multi-weather dataset for long-term visual SLAM in automotive applications containing more than 350 km of recordings.
- Sequences covering nine different kinds of environments ranging from multi-level parking garage over urban (including tunnels) to countryside and highway.

- Global six degrees of freedom (6DoF) reference poses with up-to centimeter accuracy obtained from the fusion of direct stereo visual-inertial odometry (VIO) with RTK-GNSS.
- Accurate cross-seasonal pixel-wise correspondences to train dense feature representations.

2   Related Work

There exists a variety of benchmarks and datasets focusing on VO and SLAM for AD. Here, we divide these datasets into the ones which focus only on the task of VO as well as those covering different weather conditions and therefore aiming towards long-term SLAM.

2.1   Visual Odometry

The most popular benchmark for AD certainly is KITTI [14]. This multi-sensor dataset covers a wide range of tasks including not only VO, but also 3D object detection, and tracking, scene flow estimation as well as semantic scene understanding. The dataset contains diverse scenarios ranging from urban over countryside to highway. Nevertheless, all scenarios are only recorded once and under similar weather conditions. Ground truth is obtained based on a high-end inertial navigation system (INS).

Another dataset containing LiDAR, inertial measurement unit (IMU), and image data at a large-scale is the Málaga Urban dataset [4]. However, in contrast to KITTI, no accurate 6DoF ground truth is provided and therefore it does not allow for a quantitative evaluation based on this dataset.

Other popular datasets for the evaluation of VO and VIO algorithms not related to AD include [40] (handheld RGB-D), [5] (UAV stereo-inertial), [10] (handheld mono), and [36] (handheld stereo-inertial).

2.2   Long-Term SLAM

More related to our work are datasets containing multiple traversals of the same environment over a long period of time. With respect to SLAM for AD the Oxford RobotCar Dataset [27] represents a kind of pioneer work. This dataset consists of large-scale sequences recorded multiple times for the same environment over a period of one year. Hence, it covers large variations in the appearance and structure of the scene. However, the diversity of the scenarios is only limited to an urban environment. Also, the ground truth provided for the dataset is not accurate up-to centimeter-level and therefore, requires additional manual effort to establish accurate cross-sequence correspondences.

The work [34] represents a kind of extension to [27]. This benchmark is based on subsequences from [27] as well as other datasets. The ground truth of the RobotCar Seasons [34] dataset is obtained based on structure from motion (SfM) and LiDAR point cloud alignment. However, due to inaccurate GNSS measurements [27], a globally consistent ground truth up-to centimeter-level can not be guaranteed. Furthermore, this dataset only provides one reference traversal in the overcast condition. In contrast, we provide globally consistent reference models for all traversals covering a wide variety of conditions. Hence, every traversal can be used as a reference model that allows further research, *e.g.* on analyzing suitable reference-query pairs for long-term localization and mapping.

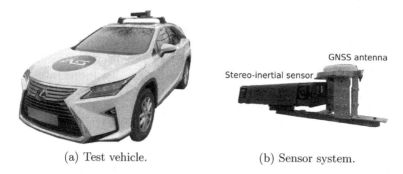

(a) Test vehicle.   (b) Sensor system.

Fig. 2. Recording setup. Test vehicle and sensor system used for dataset recording. The sensor system consists of a custom stereo-inertial sensor with a stereo baseline of 30 cm and a high-end RTK-GNSS receiver from Septentrio.

2.3   Other Datasets

Examples of further multi-purpose AD datasets which also can be used for VO are [6,7,19,45].

As stated in Sect. 1, our proposed dataset differentiates from previous related work in terms of being both large-scale (similar to [14]) as well as having high variations in appearance and conditions (similar to [27]). Furthermore, we are providing accurate reference poses based on the fusion of direct stereo VIO and RTK-GNSS.

3   System Overview

This section presents the sensor setup which is used for data recording (Sect. 3.1). Furthermore, we describe the calibration of the entire sensor suite (Sect. 3.2) as well as our approach to obtain up-to centimeter-accurate global 6DoF reference poses (Sect. 3.3).

3.1   Sensor Setup

The hardware setup consists of a custom stereo-inertial sensor for 6DoF pose estimation as well as a high-end RTK-GNSS receiver for global positioning and global pose refinement. Figure 2 shows our test vehicle equipped with the sensor system used for data recording.

Stereo-Inertial Sensor. The core of the sensor system is our custom stereo-inertial sensor. This sensor consists of a pair of monochrome industrial-grade global shutter cameras (Basler acA2040-35gm) and lenses with a fixed focal length of $f = 3.5$ mm (Stemmer Imaging CVO GMTHR23514MCN). The cameras are mounted on a highly-rigid aluminum rail with a stereo baseline of 30 cm. On the same rail, an IMU (Analog Devices ADIS16465) is mounted. All sensors, cameras, and IMU are triggered over an external clock generated by an field-programmable gate array (FPGA). Here, the trigger accounts for exposure compensations, meaning that the time between the centers of the exposure interval for two consecutive images is always kept constant (1/[frame rate]) independent of the exposure time itself.

Furthermore, based on the FPGA, the IMU is properly synchronized with the cameras. In the dataset, we record stereo sequences with a frame rate of 30 fps. We perform pixel binning with a factor of two and crop the image to a resolution of 800×400. This results in a field of view of approximately 77° horizontally and 43° vertically. The IMU is recorded at a frequency of 2000 Hz. During recording, we run our custom auto-exposure algorithm, which guarantees equal exposure times for all stereo image pairs as well as a smooth exposure transition in highly dynamic lighting conditions, as it is required for visual SLAM. We provide those exposure times for each frame.

GNSS Receiver. For global positioning and to compensate drift in the VIO system we utilize an RTK-GNSS receiver (mosaic-X5) from Septentrio in combination with an Antcom Active G8 GNSS antenna. The GNSS receiver provides a horizontal position accuracy of up-to 6 mm by utilizing RTK corrections. While the high-end GNSS receiver is used for accurate positioning, we use a second receiver connected to the time-synchronization FPGA to achieve synchronization between the GNSS receiver and the stereo-inertial sensor.

3.2   Calibration

Aperture and Focus Adjustment. The lenses used in the stereo-system have both adjustable aperture and focus. Therefore, before performing the geometric calibration of all sensors, we manually adjust both cameras for a matching average brightness and a minimum focus blur [18], across a structured planar target in 10 m distance.

Stereo Camera and IMU. For the intrinsic and extrinsic calibration of the stereo cameras as well as the extrinsic calibration and time-synchronization of the IMU, we use a slightly customized version of *Kalibr*[1] [32]. The stereo cameras are modeled using the Kannala-Brandt model [23], which is a generic camera model

---

[1] https://github.com/ethz-asl/kalibr.

consisting of in total eight parameters. To guarantee an accurate calibration over a long-term period, we perform a feature-based epipolar-line consistency check for each sequence recorded in the dataset and re-calibrate before a recording session if necessary.

GNSS Antenna. Since the GNSS antenna does not have any orientation but has an isotropic reception pattern, only the 3D translation vector between one of the cameras and the antenna within the camera frame has to be known. This vector was measured manually for our sensor setup.

3.3   Ground Truth Generation

Reference poses (*i.e.* ground truth) for VO and SLAM should provide high accuracy in both local relative 6DoF transformations and global positioning. To fulfill the first requirement, we extend the state-of-the-art stereo direct sparse VO [44] by integrating IMU measurements [43], achieving a stereo-inertial SLAM system offering average tracking drift around 0.6% of the traveled distance. To fulfill the second requirement, the poses estimated by our stereo-inertial system are integrated into a global pose graph, each with an additional constraint from the corresponding RTK-GNSS measurement. Our adopted RTK-GNSS system can provide global positioning with up-to centimeter accuracy. The pose graph is optimized globally using the Gauss-Newton method, ending up with 6DoF camera poses with superior accuracy both locally and globally. For the optimization, we make use of the g2o library [25].

One crucial aspect for the dataset is that the reference poses which we provide are actually accurate enough, even though some of the recorded sequences partially contain challenging conditions in GNSS-denied environments. Despite the fact that the stereo-inertial sensor system has an average drift around 0.6%, this cannot be guaranteed for all cases. Hence, for the reference poses in our dataset, we report whether a pose can be considered to be reliable by measuring the distance to the corresponding RTK-GNSS measurement. Only RTK-GNSS measurements with a reported standard deviation of less than 0.01 m are considered as accurate. For all poses, without corresponding RTK-GNSS measurement we do not guarantee a certain accuracy. Nevertheless, due to the highly accurate stereo-inertial odometry system, these poses still can be considered to be accurate in most cases even in GNSS-denied environments, *e.g.* tunnels or areas with tall buildings.

4   Scenarios

This section describes the different scenarios we have collected for the dataset. The scenarios involve different sequences – ranging from urban driving to parking garage and rural areas. We provide complex trajectories, which include partially overlapping routes, and multiple loops within a sequence. For each scenario, we

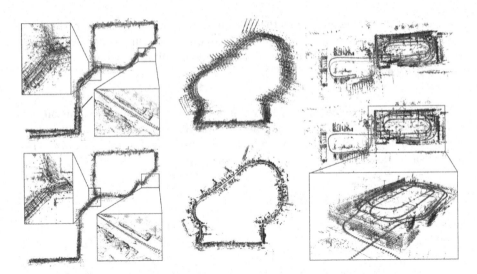

Fig. 3. 3D models of different scenarios contained in the dataset. The figure shows a loop around an industrial area (left), multiple loops around an area with high buildings (middle), and a stretch recorded in a multi-level parking garage (right). The green lines encode the GNSS trajectories, and the red lines encode the VIO trajectories. Top: shows the trajectories before the fusion using pose graph optimization. Bottom: shows the result after the pose graph optimization. Note that after the pose graph optimization the reference trajectory is well aligned. (Color figure online)

have collected multiple traversals covering a large range of variation in environmental appearance and structure due to weather, illumination, dynamic objects, and seasonal effects. In total, our dataset consists of nine different scenarios, *i.e.* industrial area, highway, local neighborhood, ring road, countryside, suburban, inner city, monumental site, and multi-level parking garage.

We provide reference poses and 3D models generated by our ground truth generation pipeline (cf. Fig. 3) along with the corresponding raw image frames and raw IMU measurements. Figure 4 shows another example of the optimized trajectory, which depicts the accuracy of the provided reference poses.

The dataset will challenge current approaches on long-term localization and mapping since it contains data from various seasons and weather conditions as well as from different times of the day as shown in the bottom part of Fig. 1.

4.1   Ground Truth Validation

The top part of Fig. 1 shows two overlaid point clouds from different runs across the same scene. Note that despite the weather and seasonal differences the point clouds align very well. This shows that our reference poses are indeed very accurate. Furthermore, a qualitative assessment of the point-to-point correspondences is shown in Fig. 5. The figure shows a subset of very accurate pixel-wise correspondences across different seasons (*autumn/winter*) in the top

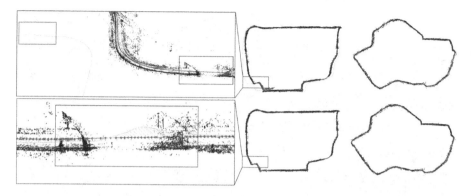

Fig. 4. Reference poses validation. This figure shows two additional 3D models of the scenarios collected. Note that these two sequences are quite large (more than 10 km and 6 km, respectively). Top: before the fusion using pose graph optimization. Bottom: results after optimization. The green lines encode the GNSS trajectories, the red lines show the VIO trajectories (before fusion) and the fused trajectories (after fusion). The left part of the figure shows a zoomed-in view of a tunnel, where the GNSS signal becomes very noisy as highlighted in the red boxes. Besides, due to the large size of the sequence, the accumulated tracking error leads to a significant deviation of the VIO trajectory from the GNSS recordings. Our pose graph optimization, by depending globally on GNSS positions and locally on VIO relative poses, successfully eliminates global VIO drifts and local GNSS positioning flaws. (Color figure online)

and different illumination conditions (*sunny/night*) in the bottom. These point-to-point correspondences are a result of our up-to centimeter-accurate global reference poses and are obtained in a completely self-supervised manner. This makes them suitable as training pairs for learning-based algorithms. Recently, there has been an increasing demand for pixel-wise cross-season correspondences which are needed to learn dense feature descriptors [8,33,38]. However, there is still a lack of datasets to satisfy this demand. The KITTI [14] dataset does not provide cross-seasons data. The Oxford RobotCar Dataset [27] provides cross-seasons data, however, since the ground truth is not accurate enough, the paper does not recommend benchmarking localization and mapping approaches.

Recently, RobotCar Seasons [34] was proposed to overcome the inaccuracy of the provided ground truth. However, similar to the authors of [38], we found that it is still challenging to obtain accurate cross-seasonal pixel-wise matches due to pose inconsistencies. Furthermore, this dataset only provides images captured from three synchronized cameras mounted on a car, pointing to the rear-left, rear, and rear-right, respectively. Moreover, the size of the dataset is quite small and a significant portion of it suffers from strong motion blur and low image quality.

To the best of our knowledge, our dataset is the first that exhibits accurate cross-season reference poses for the AD domain.

Fig. 5. Accurate pixel-wise correspondences, making cross-seasonal training possible. Qualitative assessment of the accuracy of our data collection and geometric reconstruction method for a sample of four different conditions (from top left in clockwise order: *overcast, snowy, night, sunny*) across the same scene. Each same colored point in the four images corresponds to the same geometric point in the world. The cameras corresponding to these images have different poses in the global frame of reference. Please note that the points are not matched but rather a result of our accurate reference poses and geometric reconstruction. This way we are capable of obtaining sub-pixel level accuracy. On average we get more than 1000 of those correspondences per image pair. (Color figure online)

5   Tasks

This section describes the different tasks of the dataset. The provided globally consistent 6DoF reference poses for diverse conditions will be valuable to develop and improve the state-of-the-art for different SLAM related tasks. Here the major tasks are robust VO, global place recognition, and map-based re-localization tracking.

In the following, we will present the different subtasks for our dataset.

5.1   Visual Odometry in Different Weather Conditions

VO aims to accurately estimate the 6DoF pose for every frame relative to a starting position. To benchmark the task of VO there already exist various datasets [10,15,40]. All of these existing datasets consist of sequences recorded at rather homogeneous conditions (indoors, or sunny/overcast outdoor conditions). However, especially methods developed for AD use cases must perform robustly under almost any condition. We believe that the proposed dataset will contribute to improving the performance of VO under diverse weather and lighting conditions in an automotive environment. Therefore, instead of replacing existing

Fig. 6. Challenging scenes for global place recognition. Top: two pictures share the same location with different appearances. Bottom: two pictures have similar appearance but are taken at different locations.

benchmarks and datasets, we aim to provide an extension that is more focusing on challenging conditions in AD. As we provide frame-wise accurate poses for large portions of the sequences, metrics well known from other benchmarks like absolute trajectory error (ATE) or relative pose error (RPE) [15,40] are also applicable to our data.

5.2 Global Place Recognition

Global place recognition refers to the task of retrieving the most similar database image given a query image [26]. In order to improve the searching efficiency and the robustness against different weather conditions, tremendous progress on global descriptors [1,3,13,21] has been seen. For the re-localization pipeline, visual place recognition serves as the initialization step to the downstream local pose refinement by providing the most similar database images as well as the corresponding global poses. Due to the advent of deep neural networks [17,24, 37,41], methods aggregating deep image features are proposed and have shown advantages over classical methods [2,16,31,42].

The proposed dataset is challenging for global place recognition since it contains not only cross-season images that have different appearances but share a similar geographical location but also the intra-season images which share similar appearances but with different locations. Figure 6 depicts example pairs of these scenarios. We suggest to follow the standard metric widely used for global place recognition [2,3,16,35].

5.3   Map-Based Re-localization Tracking

Map-based re-localization tracking [39] refers to the task of locally refining the 6DoF pose between reference images from a pre-built reference map and images from a query sequence. In contrast to wide-baseline stereo matching, for re-localization tracking, it is also possible to utilize the sequential information of the sequence. This allows us to estimate depth values by running a standard VO method. Those depth estimates can then be used to improve the tracking of the individual re-localization candidates.

In this task we assume to know the mapping between reference and query samples. This allows us to evaluate the performance of local feature descriptor methods in isolation. In practice, this mapping can be found using image retrieval techniques like NetVLAD [2] as described in Sect. 5.2 or by aligning the point clouds from the reference and query sequences [34], respectively.

Accurately re-localizing in a pre-built map is a challenging problem, especially if the visual appearance of the query sequence significantly differs from the base map. This makes it extremely difficult especially for vision-based systems since the localization accuracy is often limited by the discriminative power of feature descriptors. Our proposed dataset allows us to evaluate re-localization tracking across multiple types of weather conditions and diverse scenes, ranging from urban to countryside driving. Furthermore, our up to centimeter-accurate ground truth allows us to create diverse and challenging re-localization tracking candidates with an increased level of difficulty. By being able to precisely changing the re-localization distances and the camera orientation between the reference and query samples, we can generate more challenging scenarios. This allows us to determine the limitations and robustness of current state-of-the-art methods.

6   Conclusion

We have presented a cross-season dataset for the purpose of multi-weather SLAM, global visual localization, and local map-based re-localization tracking for AD applications. Compared to other datasets, like KITTI [14] or Oxford RobotCar [27], the presented dataset provides diversity in both multiplicities of scenarios and environmental conditions. Furthermore, based on the fusion of direct stereo VIO and RTK-GNSS we are able to provide up-to centimeter-accurate reference poses as well as highly accurate cross-sequence correspondences. One drawback of the dataset is that the accuracy of the reference poses can only be guaranteed in environments with good GNSS receptions. However, due to the low drift of the stereo VIO system, the obtained reference poses are also very accurate in GNSS-denied environments, e.g. tunnels, garages, or urban canyons.

We believe that this dataset will help the research community to further understand the limitations and challenges of long-term visual SLAM in changing conditions and environments and will contribute to advance the state-of-the-art. To the best of our knowledge, ours is the first large-scale dataset for AD providing cross-seasonal accurate pixel-wise correspondences for diverse scenarios. This will help to vastly increase robustness against environmental changes for deep learning methods. The dataset is made publicly available to facilitate further research.

References

1. Angeli, A., Filliat, D., Doncieux, S., Meyer, J.A.: Fast and incremental method for loop-closure detection using bags of visual words. IEEE Trans. Robot. (T-RO) **24**(5), 1027–1037 (2008)
2. Arandjelovic, R., Gronat, P., Torii, A., Pajdla, T., Sivic, J.: NetVLAD: CNN architecture for weakly supervised place recognition. In: Proceedings of the IEEE Conference on Computer Vision and Pattern Recognition (CVPR), pp. 5297–5307 (2016)
3. Arandjelovic, R., Zisserman, A.: All about VLAD. In: Proceedings of the IEEE Conference on Computer Vision and Pattern Recognition (CVPR), pp. 1578–1585 (2013)
4. Blanco-Claraco, J.L., Ángel Moreno-Dueñas, F., González-Jiménez, J.: The Málaga urban dataset: high-rate stereo and LiDAR in a realistic urban scenario. Int. J. Robot. Res. (IJRR) **33**(2), 207–214 (2014)
5. Burri, M., et al.: The EuRoC micro aerial vehicle datasets. Int. J. Robot. Res. (IJRR) **35**(10), 1157–1163 (2016)
6. Caesar, H., et al.: nuScenes: a multimodal dataset for autonomous driving. In: Proceedings of the IEEE Conference on Computer Vision and Pattern Recognition (CVPR), pp. 11621–11631 (2020)
7. Cordts, M., et al.: The cityscapes dataset for semantic urban scene understanding. In: Proceedings of the IEEE Conference on Computer Vision and Pattern Recognition (CVPR), pp. 3213–3223 (2016)
8. Dusmanu, M., Rocco, I., Pajdla, T., Pollefeys, M., Sivic, J., Torii, A., Sattler, T.: D2-Net: a trainable CNN for joint detection and description of local features. In: Proceedings of the IEEE Conference on Computer Vision and Pattern Recognition (CVPR), pp. 8092–8101 (2019)
9. Engel, J., Stückler, J., Cremers, D.: Large-scale direct SLAM with stereo cameras. In: Proceedings of the IEEE/RSJ Conference on Intelligent Robots and Systems (IROS), pp. 1935–1942 (2015)
10. Engel, J., Usenko, V., Cremers, D.: A photometrically calibrated benchmark for monocular visual odometry. arXiv preprint arXiv:1607.02555 (2016)
11. Engel, J., Koltun, V., Cremers, D.: Direct sparse odometry. IEEE Trans. Pattern Anal. Machine Intell. (PAMI) **40**(3), 611–625 (2017)
12. Engel, J., Schöps, T., Cremers, D.: LSD-SLAM: large-scale direct monocular SLAM. In: Fleet, D., Pajdla, T., Schiele, B., Tuytelaars, T. (eds.) ECCV 2014. LNCS, vol. 8690, pp. 834–849. Springer, Cham (2014). https://doi.org/10.1007/978-3-319-10605-2_54
13. Gálvez-López, D., Tardos, J.D.: Bags of binary words for fast place recognition in image sequences. IEEE Trans. Robot. (T-RO) **28**(5), 1188–1197 (2012)
14. Geiger, A., Lenz, P., Stiller, C., Urtasun, R.: Vision meets robotics: the KITTI dataset. Int. J. Robot. Res. (IJRR) **32**(11), 1231–1237 (2013)
15. Geiger, A., Lenz, P., Urtasun, R.: Are we ready for autonomous driving? the KITTI vision benchmark suite. In: Proceedings of the IEEE Conference on Computer Vision and Pattern Recognition (CVPR), pp. 3354–3361 (2012)

16. Gordo, A., Almazán, J., Revaud, J., Larlus, D.: Deep image retrieval: learning global representations for image search. In: Leibe, B., Matas, J., Sebe, N., Welling, M. (eds.) ECCV 2016. LNCS, vol. 9910, pp. 241–257. Springer, Cham (2016). https://doi.org/10.1007/978-3-319-46466-4_15

17. He, K., Zhang, X., Ren, S., Sun, J.: Deep residual learning for image recognition. In: Proceedings of the IEEE Conference on Computer Vision and Pattern Recognition (CVPR), pp. 770–778 (2016)

18. Hu, H., de Haan, G.: Low cost robust blur estimator. In: Proceedings of the IEEE International Conference on Image Processing (ICIP), pp. 617–620 (2006)

19. Huang, X., et al.: The ApolloScape dataset for autonomous driving. In: Proceedings of the IEEE Conference on Computer Vision and Pattern Recognition Workshops (CVPRW), pp. 954–960 (2018)

20. Jaramillo, C.: Direct multichannel tracking. In: Proceedings of the International Conference on 3D Vision (3DV), pp. 347–355 (2017)

21. Jégou, H., Douze, M., Schmid, C., Pérez, P.: Aggregating local descriptors into a compact image representation. In: Proceedings of the IEEE Conference on Computer Vision and Pattern Recognition (CVPR), pp. 3304–3311 (2010)

22. Jung, E., Yang, N., Cremers, D.: Multi-frame GAN: image enhancement for stereo visual odometry in low light. In: Conference on Robot Learning (CoRL), pp. 651–660 (2019)

23. Kannala, J., Brandt, S.S.: A generic camera model and calibration method for conventional, wide-angle, and fish-eye lenses. IEEE Trans. Pattern Anal. Mach. Intell. (PAMI) **28**(8), 1335–1340 (2006)

24. Krizhevsky, A., Sutskever, I., Hinton, G.E.: ImageNet classification with deep convolutional neural networks. In: Neural Information Processing Systems (NeurIPS), pp. 1097–1105 (2012)

25. Kümmerle, R., Grisetti, G., Strasdat, H., Konolige, K., Burgard, W.: g2o: a general framework for graph optimization. In: Proceedings of the IEEE International Conference on Robotics and Automation (ICRA), pp. 3607–3613 (2011)

26. Lowry, S., et al.: Visual place recognition: a survey. IEEE Trans. Robot. (T-RO) **32**(1), 1–19 (2015)

27. Maddern, W., Pascoe, G., Linegar, C., Newman, P.: 1 year, 1000 km: the oxford robotcar dataset. Int. J. Robot. Res. (IJRR) **36**(1), 3–15 (2017)

28. Mur-Artal, R., Tardós, J.D.: ORB-SLAM2: an open-source SLAM system for monocular, stereo, and RGB-D cameras. IEEE Trans. Robot. (T-RO) **33**(5), 1255–1262 (2017)

29. Mur-Artal, R., Montiel, J.M.M., Tardos, J.D.: ORB-SLAM: a versatile and accurate monocular SLAM system. IEEE Trans. Robot. (T-RO) **31**(5), 1147–1163 (2015)

30. Newcombe, R.A., Lovegrove, S.J., Davison, A.J.: DTAM: dense tracking and mapping in real-time. In: Proceedings of the International Conference on Computer Vision (ICCV), pp. 2320–2327 (2011)

31. Radenović, F., Tolias, G., Chum, O.: Fine-tuning CNN image retrieval with no human annotation. IEEE Trans. Pattern Anal. Mach. Intell. (PAMI) **41**(7), 1655–1668 (2018)

32. Rehder, J., Nikolic, J., Schneider, T., Hinzmann, T., Siegwart, R.: Extending kalibr: calibrating the extrinsics of multiple IMUs and of individual axes. In: Proceedings of the IEEE International Conference on Robotics and Automation (ICRA), pp. 4304–4311 (2016)

33. Revaud, J., Weinzaepfel, P., de Souza, C.R., Humenberger, M.: R2D2: repeatable and reliable detector and descriptor. In: Neural Information Processing Systems (NeurIPS), pp. 12405–12415 (2019)
34. Sattler, T., et al.: Benchmarking 6DOF outdoor visual localization in changing conditions. In: Proceedings of the IEEE Conference on Computer Vision and Pattern Recognition (CVPR), pp. 8601–8610 (2018)
35. Sattler, T., Weyand, T., Leibe, B., Kobbelt, L.: Image retrieval for image-based localization revisited. In: Proceedings of the British Machine Vision Conference (BMVC) (2012)
36. Schubert, D., Goll, T., Demmel, N., Usenko, V., Stückler, J., Cremers, D.: The TUM VI benchmark for evaluating visual-inertial odometry. In: Proceedings of the IEEE/RSJ Conference on Intelligent Robots and Systems (IROS), pp. 1680–1687 (2018)
37. Simonyan, K., Zisserman, A.: Very deep convolutional networks for large-scale image recognition. In: Proceedings of the International Conference on Learning Representations (ICLR) (2015)
38. Spencer, J., Bowden, R., Hadfield, S.: Same features, different day: Weakly supervised feature learning for seasonal invariance. In: Proceedings of the IEEE Conference on Computer Vision and Pattern Recognition (CVPR), pp. 6459–6468 (2020)
39. von Stumberg, L., Wenzel, P., Khan, Q., Cremers, D.: GN-Net: the Gauss-Newton loss for multi-weather relocalization. IEEE Robot. Autom. Lett. (RA-L) **5**(2), 890–897 (2020)
40. Sturm, J., Engelhard, N., Endres, F., Burgard, W., Cremers, D.: A benchmark for the evaluation of RGB-D SLAM systems. In: Proceedings of the IEEE/RSJ Conference on Intelligent Robots and Systems (IROS), pp. 573–580 (2012)
41. Szegedy, C., et al.: Going deeper with convolutions. In: Proceedings of the IEEE Conference on Computer Vision and Pattern Recognition (CVPR), pp. 1–9 (2015)
42. Tolias, G., Sicre, R., Jégou, H.: Particular object retrieval with integral max-pooling of CNN activations. arXiv preprint arXiv:1511.05879 (2015)
43. Von Stumberg, L., Usenko, V., Cremers, D.: Direct sparse visual-inertial odometry using dynamic marginalization. In: Proceedings of the IEEE International Conference on Robotics and Automation (ICRA), pp. 2510–2517 (2018)
44. Wang, R., Schwörer, M., Cremers, D.: Stereo DSO: large-scale direct sparse visual odometry with stereo cameras. In: Proceedings of the International Conference on Computer Vision (ICCV), pp. 3903–3911 (2017)
45. Wang, S., et al.: TorontoCity: seeing the world with a million eyes. In: Proceedings of the International Conference on Computer Vision (ICCV) (2017)
46. Yang, N., Wang, R., Stückler, J., Cremers, D.: Deep virtual stereo odometry: leveraging deep depth prediction for monocular direct sparse odometry. In: Ferrari, V., Hebert, M., Sminchisescu, C., Weiss, Y. (eds.) ECCV 2018. LNCS, vol. 11212, pp. 835–852. Springer, Cham (2018). https://doi.org/10.1007/978-3-030-01237-3_50

Inline Double Layer Depth Estimation with Transparent Materials

Christian Kopf[1(✉)], Thomas Pock[1], Bernhard Blaschitz[2], and Svorad Štolc[3]

[1] Graz University of Technology, Graz, Austria
{christian.kopf,pock}@icg.tugraz.at
[2] AIT Austrian Institute of Technology GmbH, Vienna, Austria
[3] Photoneo s.r.o., Bratislava, Slovakia

Abstract. 3D depth computation from stereo data has been one of the most researched topics in computer vision. While state-of-art approaches have flourished over time, reconstruction of transparent materials is still considered an open problem. Based on 3D light field data we propose a method to obtain smooth and consistent double-layer estimates of scenes with transparent materials. Our novel approach robustly combines estimates from models with different layer hypotheses in a cost volume with subsequent minimization of a joint second order TGV energy on two depth layers. Additionally we showcase the results of our approach on objects from common inspection use-cases in an industrial setting and compare our work to related methods.

1 Introduction

Reconstructing 3D depth information through multi-view stereo methods is a well researched topic of computer vision and has lead to countless variations and approaches over the years. In order to obtain depth from correspondences in different views, the majority of approaches rely on the assumption that scene points reflect light in all directions uniformly, *i.e.* the Lambertian assumption [11]. Unfortunately this assumption is violated for scene points from transparent materials which poses a special case that is rarely considered in state-of-the-art approaches. Moreover little to no surface structure and refractive effects make the task even more difficult. Despite the challenging nature of the problem, computing depth from scenes with transparent materials is a desirable ability for industrial applications. The amount of publications regarding depth estimation for transparent materials is very sparse. Earlier publications explore different approaches to tackle the problem as for example through visible light tomography [20] or polarized light [15]. The trend for recent approaches is towards specialized hardware such as the utilization of Time-of-Flight (ToF) measurements [7,19] from RGB-D cameras and structured light setups [10,16,25]. With the increasing popularity of learning based methods recent approaches show full 3D reconstructions of transparent objects. As an example recent work of Li *et al.* [14] extends the range of capable approaches by computing 3D transparent shapes

© Springer Nature Switzerland AG 2021
Z. Akata et al. (Eds.): DAGM GCPR 2020, LNCS 12544, pp. 418–431, 2021.
https://doi.org/10.1007/978-3-030-71278-5_30

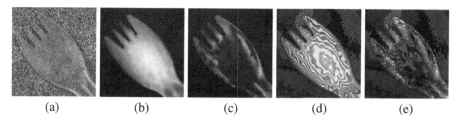

(a)   (b)   (c)   (d)   (e)

Fig. 1. Qualitative double-layer results of our proposed method. (a) shows a view of a transparent fork on top of an opaque background. In (b) and (c) the corresponding front-layer and back-layer depth estimates are shown. (d) and (e) show a high frequency rainbow color mapping of the depth estimates to visualize the consistency of the surfaces.

with a small set of images from a mobile phone camera and a known environment map. Another learning-based publication of recent years is the single-image approach by Chen *et al.* [5] which is able to produce visually convincing transparent shapes from learned refractive flow of transparent objects. Despite the impressive results of these approaches, they are not applicable to our intentions. Our goal is to present a method in a more restricted environment. Firstly, we limit our approach to passive (multi-view) data only. We also assume a front-parallel multi-view alignment along exactly one image axis which is common for stereo data. To allow for applications in an industrial setting, e.g. objects on a conveyor belt, our approach is also required to process scenes which are subjected to linear motion while guaranteeing reasonably fast runtime. The intention is to allow continuous acquisition with parallel processing of the data. The scene structure modeled by our approach consists of a transparent object in front of an opaque background. Thus the goal is to estimate two depth estimates globally across an entire scene, *i.e.* a front layer for non-opaque surfaces and an opaque back-layer for scene points that may be underneath, see Fig. 1 for an example. In the context of passive (multi-view) data, a popular method to describe densely sampled scenes is given by the plenoptic function [2] which enables the description of 3D scene geometry in a light field (LF) [6]. A simplified variant of a light field can be interpreted intuitively through a discrete volume in 3D, comprised of a stack of images from different views. The acquisition of such light fields is commonly subject to epipolar constraints, thus enabling depth estimation through light field analysis. For scene points of transparent materials, multiple depth cues superimpose which results in local multi-orientation structures in light field data. In [22,23] Wanner and Goldlücke show how local structure tensor (ST) analysis [1] in the epipolar domain of light fields can be utilized to solve correspondence problems for Lambertian and non-Lambertian surfaces. The major drawback of this approach is that the depth estimates are bound to their corresponding model hypothesis, which implies that depth estimates are only justified in regions of a scene where the respective hypothesis is valid. Johannsen *et al.* [8,9] present a different approach by solving coding problems with a dictionary of pattern orientations in a local neighborhoods to compute multi-layer depth estimates

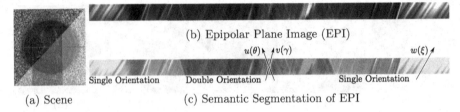

(a) Scene   (c) Semantic Segmentation of EPI

Fig. 2. Depiction of light field and epipolar plane images for the dataset D1 (Coin & Tape) from Table 1. The scene consists of a coin on a flat surface with a strip of clear tape spanned above. Sub-figure (a) partially depicts the central view $V(x, y, s_{\text{ref}})$ and the manually annotated semantic segmentation with respect to single- and double-layer regions. The sub-figures (b) and (c) show EPIs at location $\tilde{x}$ indicated by the red line in (a). (c) shows the semantic segmentation in the EPI domain and the corresponding orientations.

based on 4D light fields. Our experiments have shown that conventional stereo matching and line fitting in the epipolar domain gives bad results when it comes to transparent materials. In comparison we have found that the double orientation structure tensor approach is well suitable for transparent materials since this model is very sensitive to depth cues from non-Lambertian surfaces. In this work we will adopt the basic idea of depth estimation with structure tensors on narrow baseline light field data from an inline setup with the aforementioned requirements. We will start by reviewing our setup and explain the light field structure in Sect. 2. Furthermore we will give a brief introduction to structure tensor analysis in the context of light fields and in Sect. 4 follow up with our proposed method to combine estimates in a cost volume. After that we will explain how we refine our double-layer results through a novel variational method.

2   Setup and Light Fields

Our assumption on the general scene structure is that a transparent object of arbitrary shape is located in front of an opaque second layer as in Fig. 2. This example shows a clear tape spanned over a coin. The objective is to reconstruct the depth of the tape surface as well as the depth of the coin underneath. Acquisition rays that traverse through the transparent object carry composite depth information of at least two surfaces. This leads to some areas for which a single-layer depth hypothesis is valid and some areas for which a double-layer hypothesis is valid. In addition the problem becomes even more complicated due to opaque structures on the transparent object which might occlude the subjacent background. As the final result our goal is to obtain two depth images of size $M \times N$ where the depth estimates among both images coincide in single-layer regions and differ in double-layer regions. The input data is assumed to be 3D light field data which is a mapping of the form

$$\mathcal{L} : (x, y, s) \mapsto \mathbb{R}^3 \,, \tag{1}$$

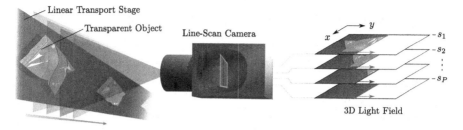

Fig. 3. Illustration of the acquisition system. A line-scan camera is set up in line with a linear transport stage. Through the alignment of the transport direction and the horizontal image axis each scan line acquires a view of the entire scene incrementally while the object on the transport stage passes the camera. The baseline of the resulting light field is determined by the distance between scan lines.

where $x \in \{1, ..., M\}$, $y \in \{1, ..., N\}$ denote the image coordinates and $s \in \{1, ..., P\}$ the view-index. More general 4D light field data can be adapted to this assumption by dropping the dimension corresponding to the movement along the second image axis. To illustrate the content of LF data in this paper we will depict the central view with the view index

$$s_{\text{ref}} = \frac{\lfloor P + 1 \rfloor}{2} \tag{2}$$

as a substitute for the whole light field. All data that is used in this paper is listed in Table 1. To evaluate our proposed method we will use the publicly available "Maria" dataset from [24] and adopt it to our 3D light field setup by discarding all but the 9 central views in the horizontal direction of movement. In addition we acquired multiple light fields of transparent objects with an inline acquisition system based on the principles of [21]. An Illustration and a short Explanation of the setup can be found in Fig. 3.

3   Depth from Light Field Analysis

The depth information of a LF scene can be computed through local orientation analysis in epipolar plane images (EPI)

$$\mathcal{E}(s, y) : (s, y) \mapsto \mathbb{R}^3 . \tag{3}$$

EPIs are images comprised of projections from acquisitions rays sharing common epipolar planes. By choosing a fixed value $x = \tilde{x}$ thus one obtains

$$\mathcal{E}_{\tilde{x}}(s, y) := \mathcal{L}(\tilde{x}, y, s) . \tag{4}$$

An example of such an EPI is depicted in Fig. 2. From this figure it can be seen how different depth layers from the scene impose multi-orientation patterns with certain angles in the epipolar domain of a LF. Note that the angle of these patterns is linked to the depth of the corresponding scene points. To obtain depth estimates we thus apply an orientation analysis through the aforementioned structure tensor approach [22]. Since the orientation analysis is performed locally at each point in the light field ($M \times N \times P$) we obtain a result stack with the same dimensions.

Single Orientation Structure Tensor (SOST): The purpose of the single-orientation model [1] is to compute depth estimate for opaque or occluded scene points, *i.e.* single-layer regions. By finding the local orientation $w(\xi)$ in a local neighborhood Ω through minimization of the least squares energy term

$$\min_{\xi} E_1(\xi) = \int_{\Omega} \left(w(\xi)^{\mathrm{T}} \nabla \mathcal{E}_{\tilde{x}} \right)^2 \mathrm{d}p \ , \tag{5}$$

we obtain an angular estimate ξ, where $p = [s, y]^{\mathrm{T}}$ is the image pixel of the EPI $\mathcal{E}_{\tilde{x}}$ and

$$\nabla \mathcal{E}_{\tilde{x}} = \left[\frac{\partial \mathcal{E}_{\tilde{x}}}{\partial s}, \frac{\partial \mathcal{E}_{\tilde{x}}}{\partial y} \right]^{\mathrm{T}} \tag{6}$$

denotes the gradient vector in the EPI domain. Reformulating Eq. (7) leads to the definition of the single orientation structure tensor $\mathcal{S}$,

$$\min_{\xi} E_1(\xi) = w(\xi)^{\mathrm{T}} \left(\int_{\Omega} \nabla \mathcal{E}_{\tilde{x}} (\nabla \mathcal{E}_{\tilde{x}})^{\mathrm{T}} \mathrm{d}p \right) w(\xi) = w(\xi)^{\mathrm{T}} \mathcal{S} w(\xi) \ . \tag{7}$$

From the eigenvalue analysis on $\mathcal{S}$, $w(\xi)$ is obtained as the eigenvector corresponding to the smaller eigenvalue. The resulting estimate stack for the SOST model will be denoted by $\mathcal{H}_1(x, y, s)$.

Second Order Double Orientation Structure Tensor (SODOST): By extending the single-orientation case to the double-orientation case [1] with an additive composition of patterns in a local neighborhood Ω, the single orientation model can be extended to the double orientation model. In a similar fashion to the SOST model, the optimal solution for a given multi-orientation pattern patch Ω can be computed by minimizing

$$\min_{m(\theta, \gamma)} E_2(m(\theta, \gamma)) = m(\theta, \gamma)^{\mathrm{T}} \left(\int_{\Omega} (D^2 \mathcal{E}_{\tilde{x}})(D^2 \mathcal{E}_{\tilde{x}})^{\mathrm{T}} \mathrm{d}p \right) m(\theta, \gamma)$$
$$= m(\theta, \gamma)^{\mathrm{T}} \mathcal{T} m(\theta, \gamma) \ , \tag{8}$$

where $m(\theta, \gamma)$ denotes the MOP vector [1] and

$$D^2 \mathcal{E}_{\tilde{x}} = \left[\frac{\partial^2 \mathcal{E}_{\tilde{x}}}{\partial s^2}, \frac{\partial^2 \mathcal{E}_{\tilde{x}}}{\partial y \partial s}, \frac{\partial^2 \mathcal{E}_{\tilde{x}}}{\partial y^2} \right]^{\mathrm{T}} \ . \tag{9}$$

Because the structure tensor model $\mathcal{T}$ is comprised of second order derivatives we will refer to this model as the second order double orientation structure tensor (SODOST) model. Through an eigenvalue analysis with a subsequent root solving on this structure tensor we obtain the two orientations $u(\theta)$ and $v(\gamma)$ or the corresponding disparities where it is assumed that $u(\theta)$ denotes the estimate closer to the camera. By performing this local analysis based on this model we thus obtain two further estimate volumes $\mathcal{H}_2(x, y, s)$ (front) and $\mathcal{H}_3(x, y, s)$ (back). The quality of the structure tensor estimates greatly depends on the choice of the inner and outer scale for Gaussian smoothing. This is explained in more detail in [22]. For the "Maria" dataset we chose to use the same parameters as the referenced paper (i.e. $\sigma_i = \sigma_o = 0.8$) and for all other datasets we chose to use a slightly larger outer scale $\sigma_o = 1.2$.

In our experiments we have observed that the structures in EPIs are a lot more prominent on opaque surfaces compared to transparent scene points. Prior to computing the structure tensor estimates we therefore normalize the contrast of the entire light field to enhance the structure imposed by transparent surfaces. Let $I \in \mathbb{R}^{M \times N \times P}$ denote a single channel of the light field, we apply the normalization filtering

$$I_n = \frac{I - k_\sigma * I}{c(k_\sigma * (I - k_\sigma * I)^2)^{\frac{1}{2}} + \epsilon} + \frac{1}{2} , \tag{10}$$

where $*$ is the volumetric convolution and k_σ denotes a Gaussian kernel. The parameter c is used to scale the normalized output to $[0, 1]$. In our implementation we used $c = 6$. We chose k_σ to be a 2D kernel in spatial dimensions with a standard deviation $\sigma = 2$.

4 Double-Layer Cost Volume

The major drawback of the plain structure tensor results are the affiliation to either a single-layer or double-layer hypothesis which is valid in certain regions of the scene. Our approach combines the results from all models and implicitly rules out incorrect or weakly supported estimates. To achieve this in a robust manner we create a cost volume $V \in \mathbb{R}^{M \times N \times D}$ where D is the number of elements in a finite set $\{d_0, d_1, ..., d_{D-1}\}$ of disparity hypotheses. As depicted in Fig. 2 we can observe that the angle of the sloped pattern lines in the epipolar domain of the light field correspond to different depths and subsequently disparities in a light field. Recall that we compute depth estimates at each point in a local neighborhood of the light field, such that we attain the estimate volumes $\mathcal{H}_1$, $\mathcal{H}_2$ and $\mathcal{H}_3$. Because of the locality of the structure tensor operation, we can observe that the estimates along the corresponding sloped line in the light field ideally are constant. If the variance along this sloped line is high we want to penalize this with a high cost. To analyze the estimates along any sloped line we thus formulate a ray which intersects the reference view s_{ref} at the coordinates x, y with an angle α. Since disparities can be translated to angles we can therefore use the set of disparity hypotheses from above to determine the intersected voxel in each view for each hypothesis ray. The relative shift for a certain view s_i with respect to s_{ref} can therefore be computed with

$$\Delta y_{i \to \text{ref}} = \lfloor \tan(\alpha)(s_i - s_{\text{ref}}) \rceil . \tag{11}$$

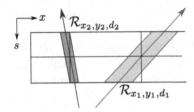

Fig. 4. Rays traversing through an estimate volume. The horizontal dashed line indicates the reference view s_{ref}. The filled area represents the area of influence and defines the set $\mathcal{R}$.

Table 1. List of all datasets used in this paper. The datasets D1–D8 have been acquired by us with the inline line-scan setup shown in Fig. 3. During acquisition, two light sources from opposing directions are strobed alternately at each step to produce 6 channels ($2 \times$ RGB) of data instead of 3.

ID	LF name	Size ($M \times N \times P$)	Channels
-	Maria [24]	$926 \times 926 \times 9$	3
D1	Coin & Tape	$1172 \times 1150 \times 33$	6
D2	Fork	$1172 \times 1150 \times 33$	6
D3	Medical item	$1172 \times 1150 \times 33$	6
D4	Phone case	$1172 \times 1150 \times 33$	6
D5	SMD parts	$1172 \times 1150 \times 33$	6
D6	Syringe	$1172 \times 1150 \times 33$	6
D7	Wallplugs	$1172 \times 1150 \times 33$	6
D8	Wooden balls	$1172 \times 1150 \times 33$	6

In theory each ray thus intersects up to P voxels. We define that all these voxels of a single ray form a finite set $\mathcal{R}_{x,y,d_i}$. In a geometric perspective we can alter the *thickness* of this ray by also selecting neighboring voxels in each view such that we can set its radius of influence resulting in more or fewer voxels in the set. An illustration of this principle is given in Fig. 4. With this in mind the cost for each point of the cost volume can therefore be attained through

$$V(x, y, d_i) = \frac{1}{|\mathcal{R}_{x,y,d_i}|} \sum_{r \in \mathcal{R}_{x,y,d_i}} \min_j h(\mathcal{H}_j(r), d_i) , \qquad (12)$$

where $|\mathcal{R}_{x,y,d_i}|$ denotes the number of elements in the set $\mathcal{R}_{x,y,d_i}$ and r denotes a point in light field coordinates. For the inner cost metric we chose to use a stable absolute distance from the disparity hypothesis d_i

$$h(z, d_i) = \max\{|z - d_i|, \tau\} , \qquad (13)$$

where τ is a hyperparameter. We remark that the structure tensor estimates $\mathcal{H}_j(r)$ from each model may be incorrect and noisy in certain regions. With the given cost measure above a combined robust cost over all three structure tensor models in a local neighbourhood along a hypothesis ray is provided. The cost volume in Fig. 5 shows the unimodal and bimodal distribution along d for any location x, y on the spatial grid. Naturally the distribution can have more than two modes but in our testing we have found that either one or two prominent modes are present in the cost volume. The task now is to find the index d and the corresponding cost for one or two prominent modes on mostly smooth distributions. To isolate the minimal points we use a simple non-maximum-suppression algorithm denoted in Algorithm 1 and define that a local minimum

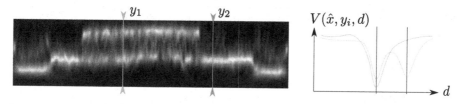

Fig. 5. Example of a cost volume. The left image shows a horizontal cut through the cost volume from the "Coin & Tape" (D1) dataset at $\hat{x}$. The location of the cut is also indicated by the horizontal line in Fig. 2. The structure of the tape spanned over the coin can be seen clearly. On the right hand side the unimodal and bimodal cost distributions at y_1 and y_2 are shown in corresponding colors (visible in the online version of this paper).

is only valid if it is below a certain threshold and no local maximum is within a certain range along d. The algorithm is based upon a local gray value dilation operation with a subsequent comparison to the unchanged cost volume.

For each x, y determining the index of the smallest minimum forms an image which is denoted g_1. For areas with a valid second best minimum we likewise determine the indices which form g_2. For single-layer coordinates we use the corresponding entry from g_1. Since the estimates in both images have no clear affiliation to the front-layer or the back-layer we sort both images such that

$$(g_1)_i \geq (g_2)_i \quad \forall i = 1, ..., MN \ , \tag{14}$$

where $(g_j)_i$ denotes the i-th element of a flattened version of g_j.

5   Joint TGV$^2-\ell_1$ Refinement

In this section, we propose a joint refinement approach to refine noisy depth estimates for the foreground surface u and a background surface v. As a regularizer we make use of the Total Generalized Variation (TGV) regularizer [3] of second order which has been shown to be very suitable for the refinement of depth images [12,17]. The TGV regularizer of second order is defined as

$$\text{TGV}^2(x) = \min_{\tilde{x}} \lambda_0 \left\| \tilde{D}\tilde{x} \right\|_{2,1} + \lambda_1 \left\| Dx - \tilde{x} \right\|_{2,1} \ , \tag{15}$$

where $x \in \mathbb{R}^{M \times N}$ is the image and $\tilde{x}, \in \mathbb{R}^{M \times N \times 2}$ is an auxiliary vector field. $\tilde{D} : \mathbb{R}^{M \times N \times 2} \to \mathbb{R}^{M \times N \times 4}$ and $D : \mathbb{R}^{M \times N} \to \mathbb{R}^{M \times N \times 2}$ are the finite differences approximations of the gradient operator. For robust joint depth refinement we propose to minimize the following variational energy

$$\min_{u,v} \ \text{TGV}^2(u) + \text{TGV}^2(v) + \lambda_u \left\| c_1 \odot (u - g_1) \right\|_1 + \lambda_v \left\| c_2 \odot (v - g_2) \right\|_1 \ s.t. \ u \geq v \ , \tag{16}$$

where the optimal solution for the foreground u is constrained to be at least as close to the camera as the background v. Here $c_1, c_2 \in \mathbb{R}^{M \times N}$ denote the

---

Algorithm 1. Implementation of the non-maximum suppression. The $\odot$ operator denotes the Hadamard product and δ_B is the 1-0 indicator function on condition B.

1: **procedure** NON-MAXIMUM SUPPRESSION($V(x, y, d)$)
2:    $V_d(x, y, d) \leftarrow \text{dilate}(V(x, y, d))$ ▷ Dilate along d with width=3
3:    $V_{\text{NMS}}(x, y, d) \leftarrow V(x, y, d) \odot \delta_{V(x,y,d)=V_d(x,y,d)}$.
4:    **return** $V_{\text{NMS}}(x, y, d)$
5: **end procedure**

---

reciprocal values of the residual cost from the solutions g_1 and g_2 to steer data fidelity locally and the $\odot$ operator denotes the Hadamard product. By transforming Eq. (16) into the corresponding saddle-point notation in the general form of

$$\min_x \max_y \ (Kx)^{\mathrm{T}} y + h(x) - f^*(y) \ , \tag{17}$$

the problem can adequately be solved through the PDHG approach by Chambolle and Pock [4]. To apply the algorithm to our problem, the proximal operators $\text{prox}_{\tau h}$ and $\text{prox}_{\sigma f^*}$ have to be determined. Note that both can be computed element-wise. Since the convex conjugate of $f : \|\cdot\|_{2,1}$ is given by the indicator function on the $2, \infty$-norm ball, the proximal map in dual space is given by

$$\text{prox}_{\sigma f^*}(y)_i = \frac{y_i}{\max\{1, \|y_i\|_2\}} \ . \tag{18}$$

Unfortunately the operator in primal-space is not so straight forward since we need to pay attention to the constraint $u \geq v$. We can formulate the term $h(x)$ from the saddle point notation in the following way

$$h(u, v) = \lambda_u \|c_1 \odot (u - g_1)\|_1 + \lambda_v \|c_2 \odot (v - g_2)\|_1 + \delta_{u \geq v}(u, v) \ , \tag{19}$$

where the last term is the indicator function of the constraint

$$\delta_{u \geq v}(u, v) = \begin{cases} 0 & u \geq v \\ \infty & \text{else} \end{cases} . \tag{20}$$

As a first step we compute the solutions $\hat{u}$ and $\hat{v}$ by the well known shrinkage operator

$$\text{prox}_{\tau h}(x)_i = (g_j)_i + \max\left(0, |x_i - (g_j)_i| - \lambda_x \tau(c_j)_i\right) \text{sign}(x_i - (g_j)_i) \ . \tag{21}$$

In case $\hat{u} < \hat{v}$ the shrinkage operator is invalid. The closest valid solution is given by $u = v$. The joint proximal map in primal space thus becomes

$$\arg\min_u \lambda_u(\|c_1 \odot (u - g_1)\|_1 + \|c_1 \odot (u - g_2)\|_1) + \frac{1}{2\tau}(\|u - \bar{u}\|_2^2 + \|u - \bar{v}\|_2^2) \ , \tag{22}$$

where $\bar{u}$ and $\bar{v}$ are the solutions from the previous step. Since this minimization problem is of the form

$$\arg\min_{u} \sum_{i=1}^{K} w_i |u - g_i| + F(u) , \tag{23}$$

we can solve this through the median formula proposed by Li and Osher [13].

6 Results

To the best of our knowledge there is no dedicated benchmark for transparent depth estimation. Nevertheless we compare our results quantitatively on the publicly available "Maria" dataset [24] and show results on more difficult self acquired datasets.

Quantitative Results: In Fig. 6 the results of our quantitative evaluation are shown. These results show that our procedure is at least on par with comparable methods on the mentioned dataset. However, we use only a small fraction of the data since we deliberately operate in an inline context. As a baseline we consider the results from only the SODOST model with subsequent variational denoising [4,18] of the central view which is closely related to the proposed method in [23]. While the transparent plane is already estimated well, these results lack consistency in single layer regions which is a major requirement of

		MSE	BadX $X=0.2$	$X=0.5$	$X=1.0$
SODOST	Front	1.71	5.73	2.10	0.14
	Back	7.69	12.87	6.09	2.88
	Combined	3.83	8.26	3.51	1.11
SODOST + TV$-\ell_2$	Front	0.45	2.26	0.15	0.03
	Back	4.75	9.96	5.07	1.51
	Combined	1.97	4.98	1.89	0.55
[9]	Front	4.89	-	-	-
	Back	3.20	-	-	-
	Combined	-	-	-	-
[8]	Front	0.79	-	-	-
	Back	1.45	-	-	-
	Combined	-	-	-	-
Ours	Front	1.26	5.21	0.24	0.01
	Back	0.73	3.29	0.06	0.00
	Combined	**1.07**	**4.53**	**0.18**	**0.01**

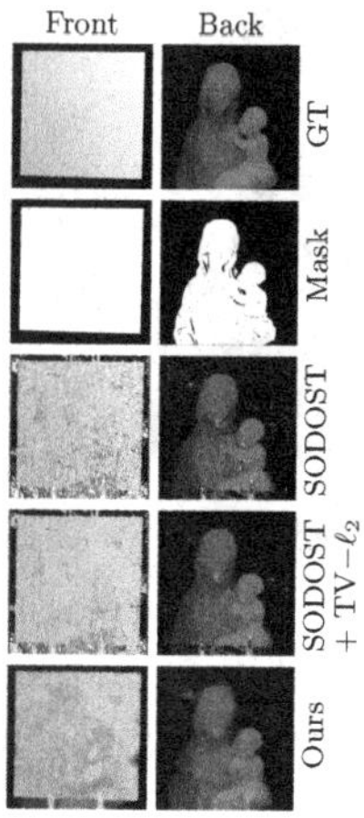

Fig. 6. Quantitative results on the "Maria" dataset [24]. The metrics shown on the left are the mean squared disparity error *MSE* times 100 and *BadX*. The later describes the percentage of pixels with a disparity error greater than a threshold X. On the right the corresponding images including the ground truth are depicted. The stated scores are masked (2nd row on the right) where only white mask pixels are considered. To provide better comparability we also state "Combined" values which are the average masked metrics across both layers (front and back).

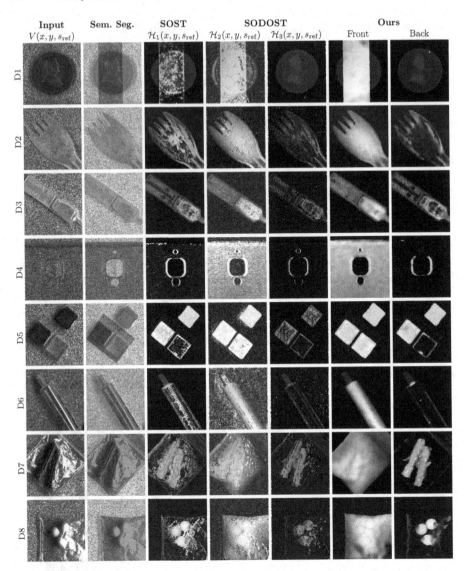

Fig. 7. Results on acquired data. The key information regarding each dataset is stated in Table 1. Since transparent materials in the depicted data may be hard to identify the second column shows the manual semantic segmentation for each scene in green (single-layer) and blue (double-layer). We refer the reader to the online version of this paper for the color coding. The third column shows the result from the single orientation structure tensor model and the 4th and 5th columns show results from the double orientation model. Our final results are shown in the 6th and 7th column.

our application as stated in Sect. 2. [23] solves this problem through consistency checks among results from horizontal and vertical EPIs, which is not applicable in our setup (3D vs. 4D LF). Unfortunately we were not able to provide a proper consistency metric because single orientation regions of the dataset are masked. We therefore leave the interpretation of this aspect to the reader and point to further qualitative results from our acquired datasets. Another crucial point is that we were also able to achieve a low runtime. The runtime for this dataset in our python testing environment is 11.3 seconds. For our larger light fields of size $1150 \times 1172 \times 33$ and 6 channels the average runtime is ≈ 35 s. These times naturally depend on the chosen hyperparameters. Most significantly the size of the cost volume has a major impact on this. For these experiments we chose $D = 80$. These runtime numbers are based on a AMD Ryzen 3900X platform with the support of a NVidia RTX Titan GPU.

Qualitative Results: Based on acquisition with the setup in Fig. 3 we also present qualitative results in Fig. 7 on various use-cases e.g. objects in bags or differently shaped transparent objects which we also consider as a valuable contribution. In case of the "Maria" dataset the transparent object is ideally conditioned for concept studies with a high separation between the front-layer and the back-layer and little to no reflections on a planar structure. However we think that the problem becomes more interesting with the use-cases given by our acquisitions. With the depicted results we show that our approach also works well for more complex shapes of transparent objects in real world data. As an example consider the datasets D7 and D8. It is remarkable that although both examples have a large amount of reflections, the shape of the objects from inside the clear bag can be reconstructed. Also depth of the clear bag itself is estimated well. Similarly the front SODOST estimate of the "Coin & Tape" example (D1) shows heavy reflective effects on the coin and noise at the background. By incorporating single orientation estimates from the single orientation structure tensor, our model is able to circumvent this issue. With our approach it is possible to attain a smooth separation between both layers while maintaining consistency for the rest of the scene. All hyperparameters of our method that have not been stated explicitly have been chosen empirically based on the individual datasets.

7   Conclusion

In this paper we presented a method to obtain smooth double-layer results for a constrained inline setup based on 3D light field data from transparent objects. We covered the preliminaries on various structure tensor models and explained how we combine these estimates robustly in a cost volume. We furthermore presented a refinement procedure based on a joint $\text{TGV}^2 - \ell 1$ regularizer which enables us to handle two layers simultaneously while keeping a global depth ordering. Additionally we presented quantitative numbers and a qualitative evaluation on acquired data to demonstrate the applicability of our approach. To build a basis for future publications, the ability to benchmark depth reconstruction could be improved. Possible future work could therefore include the acquisition of a large high quality dataset with transparent objects which also could enable the development of learned methods. We believe that future approaches can improve the runtime and accuracy further by incorporating learned descriptors for non-Lambertian surfaces.

References

1. Aach, T., Mota, C., Stuke, I., Muhlich, M., Barth, E.: Analysis of superimposed oriented patterns. IEEE Trans. Image Process. **15**(12), 3690–3700 (2006)
2. Adelson, E.H., Bergen, J.R.: The plenoptic function and the elements of early vision. In: Landy, M.S., Movshon, J.A. (eds.) Computational Models of Visual Processing, pp. 3–20. The MIT Press, Cambridge (1991)
3. Bredies, K., Kunisch, K., Pock, T.: Total generalized variation. SIAM J. Imaging Sci. **3**(3), 492–526 (2010). https://doi.org/10.1137/090769521
4. Chambolle, A., Pock, T.: A first-order primal-dual algorithm for convex problems with applications to imaging. J. Math. Imaging Vis. **40**(1), 120–145 (2011). https://doi.org/10.1007/s10851-010-0251-1
5. Chen, G., Han, K., Wong, K.Y.K.: TOM-Net: learning transparent object matting from a single image. In: CVPR (2018)
6. Gortler, S.J., Grzeszczuk, R., Szeliski, R., Cohen, M.F.: The lumigraph. In: Proceedings of the 23rd Annual Conference on Computer Graphics and Interactive Techniques, SIGGRAPH 1996, pp. 43–54. Association for Computing Machinery, New York (1996). https://doi.org/10.1145/237170.237200
7. Ji, Y., Xia, Q., Zhang, Z.: Fusing depth and Silhouette for scanning transparent object with RGB-D sensor. Int. J. Opt. **2017**, 1–11 (2017). https://doi.org/10.1155/2017/9796127
8. Johannsen, O., Sulc, A., Goldlücke, B.: Occlusion-aware depth estimation using sparse light field coding. In: German Conference on Pattern Recognition (Proceedings of the GCPR) (2016)
9. Johannsen, O., Sulc, A., Goldlücke, B.: What sparse light field coding reveals about scene structure. In: IEEE Conference on Computer Vision and Pattern Recognition (CVPR) (2016)
10. Kim, J., Reshetouski, I., Ghosh, A.: Acquiring axially-symmetric transparent objects using single-view transmission imaging. In: Proceedings of the 2017 IEEE Conference on Computer Vision and Pattern Recognition (CVPR), pp. 1484–1492 (2017)
11. Koppal, S.J.: Lambertian reflectance. In: Ikeuchi, K. (ed.) Computer Vision: A Reference Guide, pp. 441–443. Springer, Boston (2014). https://doi.org/10.1007/978-0-387-31439-6_534
12. Kuschk, G., Cremers, D.: Fast and accurate large-scale stereo reconstruction using variational methods. In: Proceedings of the 2013 IEEE International Conference on Computer Vision Workshops, pp. 700–707 (2013)
13. Li, Y., Osher, S.: A new median formula with applications to PDE based denoising. Commun. Math. Sci. **7**(3), 741–753 (2009). https://projecteuclid.org:443/euclid.cms/1256562821
14. Li, Z., Yeh, Y.Y., Chandraker, M.: Through the looking glass: neural 3D reconstruction of transparent shapes. In: The IEEE/CVF Conference on Computer Vision and Pattern Recognition (CVPR) (2020)
15. Miyazaki, D., Ikeuchi, K.: Inverse polarization raytracing: estimating surface shapes of transparent objects. In: Proceedings of the IEEE Computer Society Conference on Computer Vision and Pattern Recognition (CVPR 2005), vol. 2, pp. 910–917 (2005)
16. Qian, Y., Gong, M., Yang, Y.: 3d reconstruction of transparent objects with position-normal consistency. In: Proceedings of the IEEE Conference on Computer Vision and Pattern Recognition (CVPR 2016), pp. 4369–4377 (2016)

17. Ranftl, R., Gehrig, S., Pock, T., Bischof, H.: Pushing the limits of stereo using variational stereo estimation. In: 2012 IEEE Intelligent Vehicles Symposium, pp. 401–407 (2012)
18. Rudin, L.I., Osher, S., Fatemi, E.: Nonlinear total variation based noise removal algorithms. Physica D **60**(1), 259–268 (1992). https://doi.org/10.1016/0167-2789(92)90242-F. http://www.sciencedirect.com/science/article/pii/016727899290242F
19. Tanaka, K., Mukaigawa, Y., Kubo, H., Matsushita, Y., Yagi, Y.: Recovering transparent shape from time-of-flight distortion. In: Proceedings of the IEEE Conference on Computer Vision and Pattern Recognition (CVPR 2016), pp. 4387–4395 (2016)
20. Trifonov, B., Bradley, D., Heidrich, W.: Tomographic reconstruction of transparent objects. In: ACM SIGGRAPH 2006 Sketches, SIGGRAPH 2006, p. 55-es. Association for Computing Machinery, New York (2006). https://doi.org/10.1145/1179849.1179918
21. Štolc, S., Huber-Mörk, R., Holländer, B., Soukup, D.: Depth and all-in-focus images obtained by multi-line-scan light-field approach. In: Electronic Imaging (2014)
22. Wanner, S., Goldlücke, B.: Globally consistent depth labeling of 4D lightfields. In: IEEE Conference on Computer Vision and Pattern Recognition (CVPR) (2012)
23. Wanner, S., Goldlücke, B.: Reconstructing reflective and transparent surfaces from epipolar plane images. In: German Conference on Pattern Recognition (Proceedings of the GCPR, Oral Presentation) (2013)
24. Wanner, S., Meister, S., Goldlücke, B.: Datasets and benchmarks for densely sampled 4D light fields. In: Vision, Modelling and Visualization (VMV) (2013)
25. Wu, B., Zhou, Y., Qian, Y., Cong, M., Huang, H.: Full 3D reconstruction of transparent objects. ACM Trans. Graph. **37**, 1–11 (2018). https://doi.org/10.1145/3197517.3201286

A Differentiable Convolutional Distance Transform Layer for Improved Image Segmentation

Duc Duy Pham[(✉)], Gurbandurdy Dovletov, and Josef Pauli

Intelligent Systems, Faculty of Engineering, University of Duisburg-Essen,
Duisburg, Germany
duc.duy.pham@uni-due.de

Abstract. In this paper we propose using a novel differentiable convolutional distance transform layer for segmentation networks such as U-Net to regularize the training process. In contrast to related work, we do not need to learn the distance transform, but use an approximation, which can be achieved by means of the convolutional operation. Therefore, the distance transform is directly applicable without previous training and it is also differentiable to ensure the gradient flow during backpropagation. First, we present the derivation of the convolutional distance transform by Karam et al. [6]. Then we address the problem of numerical instability for large images by presenting a cascaded procedure with locally restricted convolutional distance transforms. Afterwards, we discuss the issue of non-binary segmentation outputs for the convolutional distance transform and present our solution attempt for the incorporation into deep segmentation networks. We then demonstrate the feasibility of our proposal in an ablation study on the publicly available SegTHOR data set.

1 Introduction

In medical image computing, semantic segmentation of anatomical structures from various imaging modalities is a crucial task to aid in image based diagnostics. Therefore, research on automated segmentation methods is a major topic in the medical computing domain, since manual segmentation is expensive and time consuming. Especially deep learning strategies have become popular approaches to achieve state of the art results. In supervised settings, these usually require a desired ground truth segmentation, used to calculate a loss function, which is minimized during training. Mean squared error, categorical cross entropy and dice loss are common error functions, which directly make use of the ground truth and are applied for segmentation tasks.

These kind of error functions, however, usually employ a pixel-to-pixel comparison and therefore reduce the segmentation task to a pixel-wise classification task. They do not directly leverage information about higher order features, such as shape or texture. Specifically, pixels are considered independent of each other, thus, error correction for one pixel does not influence the error of another

© Springer Nature Switzerland AG 2021
Z. Akata et al. (Eds.): DAGM GCPR 2020, LNCS 12544, pp. 432–444, 2021.
https://doi.org/10.1007/978-3-030-71278-5_31

pixel. Consequently, recent research aims at incorporating shape priors into the segmentation process by means of an additional regularization term within the loss function, that also captures higher order information.

One possible option is to infer shape information by means of a learned latent representation of the ground truth segmentation. Oktay et al. [9] utilize a pre-trained autoencoder for shape preservation. The autoencoder's encoding component is used to regularize the weight adaptation process of a generic segmentation network during training. This is motivated by Girdhar et al.'s work on establishing 3D representations of objects from 2D images [5]. Pham et al. [11] present a 2D end-to-end architecture, in which an autoencoder, trained for shape representation, is imitated in latent space by a separate encoder, to directly leverage the autoencoder's decoder for shape consistent segmentation.

Another strategy is to include a well-established shape representation, particularly the distance transform, into the learning process. Comparing a one-hot encoded segmentation mask with its corresponding distance transform in each channel, the latter contains distance information about the closest object boundary in every pixel, whereas a binary mask only holds binary information of whether the structure of interest is present or not. In Fig. 1 the differences in binary images and in (Manhatten) distance transforms are illustrated on a simple toy example, in which two pixel values are swapped. For the binary representation, one can notice that only the affected pixels yield a difference, whereas in the corresponding distance transforms, the simple swap has a larger impact on the distance transform's landscape.

Rousson et al. [13] leverage the idea of shape representation by means of signed distance transforms, proposed by Paragios et al. [10], to incorporate shape priors into the level set framework. Cremers et al. [3] also base their work on the distance transform's shape representation to enforce shape priors. Naturally, incorporating the distance transform into deep neural networks is a plausible step to model inter-pixel relationships, as also noted in Ma et al.'s work [7]. Dangi et al. [4] apply distance map regression in a multi-task learning setting for cardiac MR image segmentation. They propose a regularization framework by formulating an Euclidean distance map regression objective, that is pursued by a sub-network of their segmentation architecture. In a related fashion Bai and Urtasun [1] incorporate the watershed transform by fundamentally learning the distance transform within image objects for instance segmentation. Bui et al. [2] propose a similar multi-task approach, in which the geodesic distance is approximated as a learning task for neonatal brain segmentation. Similarly, Navarro et al. [8] also include the learning task of distance transform approximation in their multi-task segmentation approach. In these contributions, however, the distance transform needs to be learned, since the implementation of the distance transform is often not differentiable.

In this work we propose using Karam et al.'s [6] derivation of a convolutional distance transform approximation for the application in a deep learning context. To the best of our knowledge, this is the first time an adhoc differentiable convolutional distance transform layer is proposed for deep segmentation networks. In Sect. 2 we will discuss the underlying methods, starting with Karam et al.'s [6]

Convolutional distance Transform, subsequently proposing a cascaded variant for large images, and posing an embedding suggestion into deep learning frameworks. In Sect. 3 we present our experimental setting and discuss our results in Sect. 4 before concluding with Sect. 5.

2 Methods

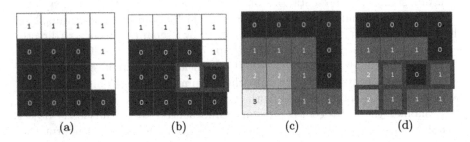

Fig. 1. Differences in binary images (a) and (b) are only visible in the affected pixels, highlighted in (b). For the corresponding (Manhatten) distance transforms (c) and (d), differences propagate to further pixels, emphasized in (d), as these change the foreground shape and thus the distance landscape.

A distance map of a binary image yields the distance of each pixel to its closest foreground boundary pixel. Common applicable distances are the Manhatten and the Euclidean distance. A major advantage of this type of image representation is the provision of information about boundary, shape, and location of the object of interest. We denote the distance between two pixel positions p_i and p_j in an image as $d(p_i, p_j)$. Then a distance transform $D_I : M \times N \to \mathbf{R}_0^+$ for a binary image I of resolution $M \times N$ can be defined pixel-wise as:

$$D_I(p_i) = \min_{p_j : I(p_j)=1} \{d(p_i, p_j)\} \tag{1}$$

To incorporate an adhoc distance transform into the deep learning setting, we follow Karam et al.'s [6] derivation and only consider translation invariant distances, i.e.

$$d(p_i, p_j) = d(p_i + p_k, p_j + p_k) \tag{2}$$

for any image positions $p_i, p_j \in M \times N$ and any translation $p_k \in \mathbf{R} \times \mathbf{R}$. Although most distances are translation invariant, this restriction needs to be mentioned, as there are also counter examples, such as the SNCF distance.

2.1 Convolutional Distance Transform (CDT)

For the computation of the distance transform, Eq. (1) shows, that we need to find the minimal distance to a boundary pixel from a given point. The minimum function can be approximated by a log-sum-exponential. Let $d_1, \ldots, d_n$ denote n distances, then the minimum function can be reformulated as

$$\min\{d_1, \ldots, d_n\} = \lim_{\lambda \to 0} -\lambda \log \left(\sum_{i=1}^{n} \exp \left(-\frac{d_i}{\lambda} \right) \right), \qquad (3)$$

for $1 > \lambda > 0$. The idea is, that the exponential yields very small values, the larger the distances are, as these are artificially increased by λ and negated in the argument. Therefore, larger distances have a significantly smaller impact on the sum than small distances. In the extreme case the exponential of large distances seek zero, leaving only the exponential of the smallest distance in the sum. The subsequent logarithmic function then reverts the exponential operation, leaving an approximation of the minimum function. With Eq. (3), it is possible to reformulate the distance transform of Eq. (1) to

$$D_I(p_i) = \lim_{\lambda \to 0} -\lambda \log \left(\sum_{p_j : I(p_j) = 1} \exp \left(-\frac{d(p_i, p_j)}{\lambda} \right) \right) \qquad (4)$$

Since a translation invariant distance is assumed, the distance between two points can be rewritten to

$$\begin{aligned} d(p_i, p_j) &= d(p_i - p_j, p_j - p_j) \\ &= d(p_i - p_j, 0) \end{aligned} \qquad (5)$$

Therefore, the distance transform can be formulated as

$$D_I(p_i) = \lim_{\lambda \to 0} -\lambda \log \left(\sum_{p_j} I(p_j) \exp \left(-\frac{d(p_i - p_j, 0)}{\lambda} \right) \right), \qquad (6)$$

which is the definition of a convolution. Thus, for a small $\lambda > 0$, the distance transform can be approximated by means of a convolution of the binary image I with a kernel $\exp \left(-\frac{d(\cdot, 0)}{\lambda} \right)$, i.e.:

$$D_I \approx -\lambda \log \left(I * \exp \left(-\frac{d(\cdot, 0)}{\lambda} \right) \right), \qquad (7)$$

where $*$ is the convolutional operator. Since all operations are differentiable, this approximation may be integrated as a differentiable convolutional distance transform layer into current deep learning frameworks. It shall be noted, that Karam et al.'s work [6] also proposes variants of this convolutional distance transform. Initial experiments however showed most promising transforms for the presented formulation.

2.2   Cascaded Distance Transform for Large Images

A major drawback of the convolutional design of the distance transform (in all variants) is that the kernel size theoretically needs to be as large as the diagonal of the input image. This is to ensure that even very sparse binary images can be distance transformed by the proposed method. Otherwise background pixels that are not within the kernel size reach of a foreground pixel would be assigned a distance of 0. This circumstance, however, yields the following two issues:

○ The large kernel size leads to an increased computational complexity for the convolutional operation.
○ For very large distances the exponential term for the kernel design in Eq. (7) may approach zero, decreasing the numeric stability of the logarithmic expression within the convolutional distance transform (CDT). This issue particularly arises for large images with only few foreground pixels. Figure 2(c) shows the CDT of a toy example image (Fig. 2(a)). It is clearly visible, that the CDT was only capable to calculate the distances for a specific range, before becoming unstable, in comparison with a standard Manhatten distance transform implementation in Fig. 2(b).

We address these issues by proposing a cascade of local distance transforms to reduce the computational complexity and overcome the numerical instability. Instead of directly computing the distance transform with a large kernel, we suggest cascading distance transforms with smaller kernels to approximate the actual transform. Since the kernel size determines the maximal distance that can be measured, it is necessary to accumulate the calculated distances to form the final distance transform approximation.

Let k denote the kernel size. Then the maximal distance to a foreground point that can be captured by the CDT is limited to a range of $\lfloor \frac{k}{2} \rfloor$. For all background points that are further away than $\lfloor \frac{k}{2} \rfloor$ from a foreground point, Eq. (7) yields a distance of 0, as within the kernel range, there are only background points. The idea is to iteratively extend the binary input image by the area, for which a distance calculation was possible by the locally restricted CDT, i.e. by all points which fulfill the condition that the calculated distance is greater than zero. This extended binary image can then be used to compute a new locally restricted distance transform by means of the small kernel. The calculated distances can then be utilized with the distances of the previous iterations to form the final distance transform. For the i-th iteration, let $I^{(i)}$ denote the extended binary image, and let $D_I^{(i)}$ denote the local CDT of $I^{(i)}$. For the i-th iteration we assume that the original foreground area has been extended by a margin of $i \cdot \lfloor \frac{k}{2} \rfloor$. Therefore this offset distance is additionally added to the current distances to compensate the lower kernel size. Thus, the cascaded distance transform D_I^* is updated by the current distances by adding $i \cdot \lfloor \frac{k}{2} \rfloor + D_I^{(i)}$, wherever $D_I^{(i)}(p) > 0$ holds. Let $diag$ denote the diagonal of the input image I, then at most $\lceil \frac{diag}{\lfloor \frac{k}{2} \rfloor} \rceil$ of such local distance transforms are necessary to cover the whole image. Algorithm 1 summarizes this suggested procedure. Let w, h denote width and height of the input image and k the kernel size used to compute the CDT. Then in general its computational complexity is given by $\mathcal{O}(w \cdot h \cdot k^2)$ operations. Our proposed procedure can drastically reduce the number of operations from initial $\mathcal{O}(w \cdot h \cdot diag^2)$ (for a naive implementation without using separable kernels)

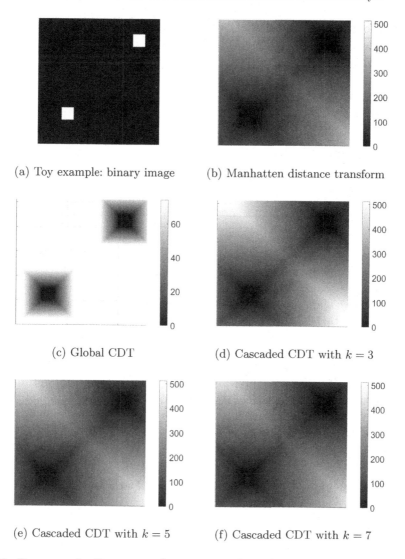

(a) Toy example: binary image   (b) Manhatten distance transform

(c) Global CDT   (d) Cascaded CDT with $k = 3$

(e) Cascaded CDT with $k = 5$ (f) Cascaded CDT with $k = 7$

Fig. 2. Toy example. By means of a toy example with a resolution of 512×512 (a) it becomes apparent, that compared to a standard Manhatten distance transform (b) the original global CDT (c) becomes numerically unstable for background points with large distances. (d)–(f) show the resulting cascaded CDTs with $k = 3, 5, 7$, respectively.

to $\mathcal{O}(w \cdot h \cdot k \cdot diag)$(also for a naive implementation without using separable kernels), if the kernel size is chosen much smaller than the image diagonal, i.e. $k << diag$.

Since the maximally possible measured distance of $d(\cdot, 0)$ in Eq. (7) is restricted by the kernel size, a small kernel size additionally yields a more stable

---

Algorithm 1. Cascaded Convolutional Distance Transform

1: **function** CASCADED_CDT($I, diag, k$)
2:     $s \leftarrow \lceil \frac{diag}{\lfloor \frac{k}{2} \rfloor} \rceil$
3:     $I^{(0)} \leftarrow I$
4:     $D_I^* \leftarrow I \cdot 0$
5:     **for** i=0 **to** s **do**
6:         $D_I^{(i)} \leftarrow \text{CDT}(I^{(i)}, k)$
7:         $I^{(i+1)} \leftarrow I^{(i)}$
8:         **for all** $p : D_I^{(i)}(p) > 0$ **do**
9:             $D_I^*(p) \leftarrow D_I^*(p) + i \cdot \lfloor \frac{k}{2} \rfloor + D_I^{(i)}(p)$
10:            $I^{(i+1)}(p) \leftarrow 1$
11:    **return** D_I^*

---

computation of the logarithmic term as the exponential does not tend to approach zero. Figures 2(d)–(f) show the cascaded CDTs with kernel sizes of 3, 5, 7, respectively. In comparison to the standard Manhatten distance transform in Fig. 2(b), it becomes apparent that the offset assumption after each iteration yields an error that is propagated to points with further distances. This error decreases with increasing kernel size. Thus, with our proposed procedure there is a trade-off between numerical stability by means of smaller kernel sizes and accuracy through larger kernel sizes that needs to be considered. We argue that for the purpose of considering inter-pixel relationships in the weight optimization process of training a convolutional neural network this approximation of the distance transform suffices. Karam et al. [6] also address this issue by multiplexing multiple λ values, however initial experiments showed that the choice of these values heavily influence.

2.3 Convolutional Distance Transform for Deep Learning

The previous section describes an adhoc cascaded convolutional approximation method of the distance transform for binary images. We propose using this approximation to extend common segmentation networks, such as Ronneberger et al.'s U-Net [12], in order to equip the segmentation loss with an additional regression loss, which compares the distance transform of the network's prediction with the distance transform of the ground truth. Since distance transforms are particularly sensible to distortions, the comparison of distance transformed ground truth to the distance transformed prediction may lead to less noisy segmentation results. Figure 3 shows the general idea, of how to extend the U-Net segmentation network with the proposed distance transform layer. In addition to the usual segmentation loss, e.g. the Dice loss, the predicted segmentation and the ground truth segmentation are both passed through the cascaded CDT layer to achieve the distance transforms of prediction and ground truth, respectively. These distance transforms contribute to a regression loss, e.g. the mean squared error, that considers inter-pixel relationships through the distance transforms.

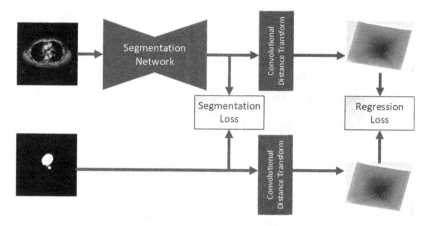

Fig. 3. Network Architecture. The Convolutional Distance Transform Layer can be attached to arbitrary segmentation networks. In addition to the segmentation loss, a regression loss of distance maps is calculated.

However, it needs to be noted that the segmentation's output is usually not binary. Assuming a final softmax or sigmoid layer, the output values for each channel vary between 0 and 1. It was necessary to assume a binary image to be able to restructure Eqs. (4) to (6). A major disadvantage of Eq. (6) is that for gray scale images, $I(p_i)$ may be a lot larger than the exponential for large distances, even if $I(p_i)$ is rather small. Therefore, even small probabilities of the segmentation output are considered in the sum and may be depicted as foreground pixels, distorting the actual distance map, as can be seen in Fig. 4. Here a toy example of a gray scale image is shown (Fig. 4(a)), in which the lower left square is set to a very low intensity of 0.001. The computed CDT (Fig. 4(c)), however, appears nearly identical to the CDT of the corresponding binary toy example (Fig. 4(b)). As can be observed in Fig. 4(d), the only differences of the computed distance transforms occur within the low intensity pixels, whereas the remaining distance transform's landscape does not show any changes. This is very problematic, as in the segmentation prediction even pixels with very low probabilities would then be considered as foreground pixels in the CDT. We address this problem by proposing the following soft-threshold work around. Let C denote the number of classes, and let $y_c(p_i)$ be the prediction for class c at position p_i. Then we can soft-threshold the prediction by

$$\tilde{y}_c(p_i) := ReLU\left(y_c(p_i) - \frac{C-1}{C}\right). \tag{8}$$

This soft-threshold sets any prediction score below $\frac{C-1}{C}$ to zero. Therefore, we enforce strong and correct predictions, as weak correct prediction scores are not registered for the distance transform and negatively impact the regression loss. Figure 4(e) shows the resulting CDT after applying the proposed soft-threshold,

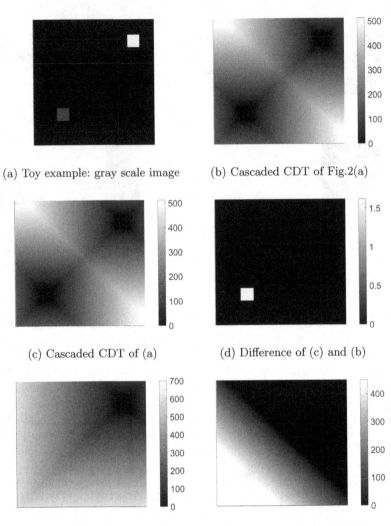

(a) Toy example: gray scale image   (b) Cascaded CDT of Fig.2(a)

(c) Cascaded CDT of (a)   (d) Difference of (c) and (b)

(e) Cascaded CDT of (a) with soft threshold   (f) Difference of (e) and (b)

Fig. 4. Toy example. (a) shows a gray scale image, in which the intensity within the lower left gray square is set to 0.001. (b) shows the cascaded CDT of Fig. 2(a) as reference. Although the intensity of the gray area is very low, the cascaded CDT (c) shows a nearly identical landscape to (b). The only difference can be found within the lower left square (d). Application of the soft-threshold (assuming 2 classes) diminishes the low intensity area (e)–(f).

assuming two classes. Low intensity pixels are considered as background, so that a significant change in the distance transform landscape compared to the reference can be observed in Fig. 4(f).

3   Experiments

3.1   Data

We conducted ablation studies on the example of thoracic segmentation from CT scans of the thorax, using the publicly available SegTHOR CT data set [14] consisting of 40 CT volumes with corresponding ground truths of esophagus, heart, trachea, and aorta, for training and 20 CT volumes for testing. The goal of our ablation study is to investigate the influence of our proposed distance transform layer on the segmentation output. Therefore, we did not aim to outperform optimized ensemble methods within the SegTHOR challenge, but set value on a valid comparison. Thus, we trained a U-Net architecture with and without our proposed layer on the same training and validation set to ensure fair comparability. In a hold-out validation manner, we trained both models on the 40 available training volumes, and submitted the predictions. For evaluation the Dice Similarity Coefficient (DSC) and the Hausdorff Distance (HDD) were considered as evaluation metrics, which were both provided by the challenge's submission platform.

3.2   Implementation Details

We implemented a 2D U-Net and the convolutional distance transform layer in Tensorflow 1.12 with an input size of 256×256. Thus, we resized the CT slices to the corresponding image size. Our U-Net implementation yields 5 image size levels with 2 convolutional layers, batch normalization and a max-pooling layer in each level. Starting with 32 kernels for each convolutional layer in the first size level of each contracting path, we doubled the number of kernels for each size level on the contracting side and halved the number of kernels on the expansive side. We used a kernel size of 3×3 for every convolutional layer and 2×2 max-pooling in each architecture. We used the standard dice loss $\mathcal{L}_{dice}$ as loss function for U-Net and used mean squared error for the regression loss $\mathcal{L}_{dist}$ of the distance transforms. We constructed a total loss function $\mathcal{L}_{total} := \mathcal{L}_{dice} + w_{dist}\mathcal{L}_{dist}$ with weight $w_{dist} := 0.5$ to train the U-Net, equipped with our additional distance transform layer. The optimization was performed with an Adam Optimizer with an initial learning rate of 0.001. The training slices were augmented by means of random translation, rotation and zooming. With a batch size of 4, we trained both models for 200 epochs and chose the model with best validation loss for evaluation. For the distance transform layer, we chose $\lambda := 0.35$, as suggested by Karam et al. [6]. The experiments were conducted on a GTX 1080 TI GPU.

4   Results

It should be noted that the main focus is the ablation study and the image resizing to 256×256 canonically decreases the segmentation quality in comparison to methods that use the full image size. Table 1 shows the achieved DSC scores

Table 1. Achieved DSCs for each organ.

DSC	Esophagus	Heart	Trachea	Aorta
U-Net	0.726312	0.809626	0.770571	**0.853953**
U-Net with distance transform	**0.739028**	**0.822095**	**0.785847**	0.853011

of the trained models for esophagus, heart, trachea, and aorta. It is observable, that with the extension of the convolutional distance transform the scores increase for all organs, except for the aorta. In this case the standard U-Net yields marginally better results. While the DSC score improves by more than 1% for the other organs, for the aorta the additional layer does not seem to bring any benefit. We applied Wilcoxon significance tests with a significance level of 5% on our evaluation results. We found that the improvements in DSCs for Trachea and Esophagus are significant with p-values of 0.0169 and 0.0040, respectively. For the DSC improvement in heart segmentation and the DSC difference in aorta segmentation, we could, however, not find any significance.

The improvements can also be noticed in Table 2, in which the Hausdorff distances are depicted. For esophagus, heart, and trachea the distances decrease with our proposed layer, showing an improvement by approximately 25–30%. However, for the aorta a slightly worse mean distance is observed. This may be due to the fact, that the aorta seems to be a rather simple structure, that U-Net can already easily extract. The improvements in Hausdorff distance are especially noteworthy, as we use a distance based regularization technique to improve the segmentation. Regarding the Wilcoxon significance test we could observe significant improvements for Trachea, Esophagus and Heart with p-values of 0.0072, 0.0008 and 0.0400, respectively. This underlines our assumption that our proposed layer adds significant value to more complex shapes, whereas simple structures as the almost circular aorta and heart slices are already well extracted by a standard U-Net. Figure 5 shows exemplary segmentations of both models on test data slices. The top images indicate better performance for esophagus segmentation with the proposed layer, while the bottom images show superior segmentation results with our layer for the trachea. In both top and bottom row, the segmentations of the aorta do not show much difference.

Table 2. Achieved Hausdorff Distances (HDD) for each organ.

HDD	Esophagus	Heart	Trachea	Aorta
U-Net	1.511865	1.949374	2.137093	**1.900747**
U-Net with distance transform	**1.113825**	**1.533211**	**1.649077**	2.004237

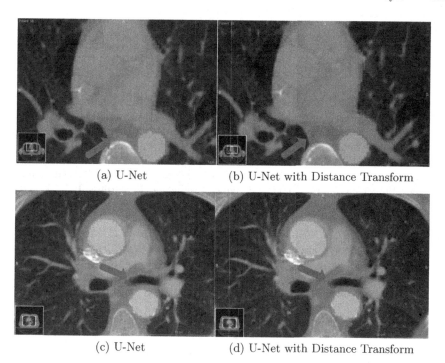

(a) U-Net   (b) U-Net with Distance Transform

(c) U-Net   (d) U-Net with Distance Transform

Fig. 5. Exemplary segmentations on the test set from both models. Top images indicate better performance for esophagus segmentation, bottom images for trachea.

5   Conclusion

In this paper we propose a novel differentiable convolutional distance transform layer for segmentation networks, that can be used adhoc without prior training. We present a cascaded procedure to reduce the computational complexity and to overcome the numerical instability. Additionally we suggest a soft-threshold work around to address the demonstrated issue regarding non-binary segmentation outputs. The proposed layer is used to regularize the training process. We conducted ablation studies on the example of the segmentation of thoracic organs and used the SegTHOR data set for training and evaluation. The experiments show promising results, as compared to an equally trained U-Net our extension yields significant improvements for most organs, particularly regarding Hausdorff distance. We demonstrated on this example, that a combination of proven non-deep learning concepts, such as the distance transform, with deep learning methods may yield great potential. In the future, we aim at extending our proposed layer for 3D segmentation networks.

References

1. Bai, M., Urtasun, R.: Deep watershed transform for instance segmentation. In: Proceedings of the IEEE Conference on Computer Vision and Pattern Recognition, pp. 5221–5229 (2017)
2. Bui, T.D., Wang, L., Chen, J., Lin, W., Li, G., Shen, D.: Multi-task learning for neonatal brain segmentation using 3D dense-Unet with dense attention guided by geodesic distance. In: Wang, Q., et al. (eds.) DART/MIL3ID 2019. LNCS, vol. 11795, pp. 243–251. Springer, Cham (2019). https://doi.org/10.1007/978-3-030-33391-1_28
3. Cremers, D., Sochen, N., Schnörr, C.: Towards recognition-based variational segmentation using shape priors and dynamic labeling. In: Griffin, L.D., Lillholm, M. (eds.) Scale-Space 2003. LNCS, vol. 2695, pp. 388–400. Springer, Heidelberg (2003). https://doi.org/10.1007/3-540-44935-3_27
4. Dangi, S., Linte, C.A., Yaniv, Z.: A distance map regularized CNN for cardiac cine MR image segmentation. Med. Phys. **46**(12), 5637–5651 (2019)
5. Girdhar, R., Fouhey, D.F., Rodriguez, M., Gupta, A.: Learning a predictable and generative vector representation for objects. In: Leibe, B., Matas, J., Sebe, N., Welling, M. (eds.) ECCV 2016. LNCS, vol. 9910, pp. 484–499. Springer, Cham (2016). https://doi.org/10.1007/978-3-319-46466-4_29
6. Karam, C., Sugimoto, K., Hirakawa, K.: Fast convolutional distance transform. IEEE Signal Process. Lett. **26**(6), 853–857 (2019)
7. Ma, J., et al.: How distance transform maps boost segmentation CNNs: an empirical study. In: Medical Imaging with Deep Learning (2020)
8. Navarro, F., et al.: Shape-aware complementary-task learning for multi-organ segmentation. In: Suk, H.-I., Liu, M., Yan, P., Lian, C. (eds.) MLMI 2019. LNCS, vol. 11861, pp. 620–627. Springer, Cham (2019). https://doi.org/10.1007/978-3-030-32692-0_71
9. Oktay, O., Ferrante, E., et al.: Anatomically Constrained Neural Networks (ACNNs): application to cardiac image enhancement and segmentation. IEEE Trans. Med. Imaging **37**(2), 384–395 (2018)
10. Paragios, N., Rousson, M., Ramesh, V.: Matching distance functions: a shape-to-area variational approach for global-to-local registration. In: Heyden, A., Sparr, G., Nielsen, M., Johansen, P. (eds.) ECCV 2002. LNCS, vol. 2351, pp. 775–789. Springer, Heidelberg (2002). https://doi.org/10.1007/3-540-47967-8_52
11. Pham, D.D., Dovletov, G., Warwas, S., Landgraeber, S., Jäger, M., Pauli, J.: Deep learning with anatomical priors: imitating enhanced autoencoders in latent space for improved pelvic bone segmentation in MRI. In: 2019 IEEE 16th International Symposium on Biomedical Imaging (ISBI 2019), pp. 1166–1169. IEEE (2019)
12. Ronneberger, O., Fischer, P., Brox, T.: U-net: convolutional networks for biomedical image segmentation. In: Navab, N., Hornegger, J., Wells, W.M., Frangi, A.F. (eds.) MICCAI 2015. LNCS, vol. 9351, pp. 234–241. Springer, Cham (2015). https://doi.org/10.1007/978-3-319-24574-4_28
13. Rousson, M., Paragios, N.: Shape priors for level set representations. In: Heyden, A., Sparr, G., Nielsen, M., Johansen, P. (eds.) ECCV 2002. LNCS, vol. 2351, pp. 78–92. Springer, Heidelberg (2002). https://doi.org/10.1007/3-540-47967-8_6
14. Trullo, R., Petitjean, C., Dubray, B., Ruan, S.: Multiorgan segmentation using distance-aware adversarial networks. J. Med. Imaging **6**(1), 014001 (2019)

PET-Guided Attention Network for Segmentation of Lung Tumors from PET/CT Images

Varaha Karthik Pattisapu[1(✉)], Imant Daunhawer[1], Thomas Weikert[2], Alexander Sauter[2], Bram Stieltjes[2], and Julia E. Vogt[1]

[1] Department of Computer Science, ETH Zurich, Zurich, Switzerland
karthikp@student.ethz.ch
[2] Clinic of Radiology and Nuclear Medicine,
University Hospital Basel, Basel, Switzerland

Abstract. PET/CT imaging is the gold standard for the diagnosis and staging of lung cancer. However, especially in healthcare systems with limited resources, costly PET/CT images are often not readily available. Conventional machine learning models either process CT or PET/CT images but not both. Models designed for PET/CT images are hence restricted by the number of PET images, such that they are unable to additionally leverage CT-only data. In this work, we apply the concept of visual soft attention to efficiently learn a model for lung cancer segmentation from only a small fraction of PET/CT scans and a larger pool of CT-only scans. We show that our model is capable of jointly processing PET/CT as well as CT-only images, which performs on par with the respective baselines whether or not PET images are available at test time. We then demonstrate that the model learns efficiently from only a few PET/CT scans in a setting where mostly CT-only data is available, unlike conventional models.

1 Introduction

Lung cancer is the second most frequently diagnosed cancer type and the leading cause of cancer-related deaths in men and women alike with high incidence and mortality rates [14]. For the staging of lung cancer, PET/CT imaging is widely used, because it provides complementary information: while the CT component visualizes anatomical properties, the PET component represents the metabolism. This gives additional information on tumor activity and is important for the detection of metastases. Despite its important role, combined PET/CT imaging is often unavailable, due to logistic and economic constraints.

Unfortunately, conventional machine learning models only cater to CT data or PET/CT data, but not both, which poses a significant problem, especially in resource-constrained populations. Prior work [3,6] has highlighted this challenge,

Electronic supplementary material The online version of this chapter (https://doi.org/10.1007/978-3-030-71278-5_32) contains supplementary material, which is available to authorized users.

© Springer Nature Switzerland AG 2021
Z. Akata et al. (Eds.): DAGM GCPR 2020, LNCS 12544, pp. 445–458, 2021.
https://doi.org/10.1007/978-3-030-71278-5_32

and several other approaches have been proposed to deal with it [1,4,19]. They attempt to learn effective joint representations of PET/CT modalities. However, such approaches still assume that a PET image is available for every CT image during training. This assumption greatly reduces the amount of effective training data for a combination of CT-only and PET/CT data. Consequently, while such models might be efficient during inference, they fall short in not being able to learn effective joint representations for a combination of CT-only and PET/CT data. The problem is further compounded by the complexity of the data, which typically includes different types of malignant lesions (e.g., the main tumor, lymph nodes metastases, and distant metastases). As such, conventional models typically cannot make the best use of a combination of CT-only and PET/CT data, a typical scenario in resource-constrained environments.

To solve this problem, we apply the established concept of visual soft attention [12]. The attention mechanism allows us to input PET images when they are available. As such, the model benefits from the additional information contained in PET images but does not mandate them. The model is thus flexible to the availability of PET data. Consequently, it is possible to incorporate two separate models that are trained on unimodal (CT) or bimodal (PET/CT) data, respectively, into one single model. Additionally, since, we do not explicitly enforce the attention mechanism to learn a joint representation of PET/CT modalities, our model can be trained on a mix of CT-only and PET/CT images. Thus, our model has the potential to make efficient use of both CT-only and PET/CT data, unlike conventional models. We present the effectiveness of our model on a large dataset with the goal of segmenting tumorous regions. We acronym our model as PAG, which stands for PET-guided attention gate. To summarize, the three main contributions of the current work are:

i) We propose a novel approach for dealing with a combination of CT-only and PET/CT data based on a visual soft attention mechanism.
ii) Our model combines two discrete functions that deal with unimodal or bimodal data, respectively, in a single model.
iii) We demonstrate a realistic application of the model in scenarios when PET/CT images are scarce relative to CT-only images and show how the model makes efficient use of the combination of CT-only and PET/CT data.

2   Related Work

Segmentation of anatomical structures such as tumors, lesions and lung nodules from PET/CT images is an active and dynamic area of research within medical imaging. [17] implemented the U-Net architecture [13] for the segmentation of nasopharyngeal tumors from dual-modality PET/CT images. [8] learned a probability map of tumorous regions from a CT image and then used a fuzzy variational model that incorporates the probability map as a prior and the PET images to obtain a posterior probability map of tumorous regions. [5] studied different fusion schemes of multi-modal images, all of which fuse the images at the pixel space. [18] refined the segmentation maps obtained separately from CT and PET images, using a graph-cut based co-segmentation model to refine the segmentation maps. [9] used belief functions to fuse the PET and CT images to obtain segmentation masks of the tumors.

The named methods are based on bi-modal inputs i.e. both PET and CT modalities. Such models, typically assume that a complete set of all modalities used during training is available even during inference. Such methods do not have the capacity to incorporate for missing modalities. Accordingly, several other methods have been proposed that deal with missing modalities. [6] proposed the Hemis model to extract representations from multiple modalities—in their case, MR image sequences (such as DWI, T1-weighted, T2-weighted, FLAIR)—and fuse them in a latent space where arithmetic operations such as the first and second moments of the representations can be calculated. This composite representation can then be deconvolved accordingly. The authors tested the applicability of their model to MR image segmentation (on MSGC [15] and BRATS 2015 [10] datasets). They argue that instead of learning all combinations of functions, each dealing with a specific missing modality, one single model can be learned that deals with all such missing modalities. [3] proposed a generic multi-input, multi-output model, which is an improvement over the Hemis model [6] that is equivalently robust to missing modalities. The model, which is based on correlation networks [2], was proposed to tackle the challenge of learning shared representations from multi-modal images. Correlation networks [2] learn effective correlations among individual modality-specific representations in a coordinated representation space. Imposing correlations as such aid in learning a shared (or coordinated) representation space, especially for MR image modalities that are correlated among one another, in the sense that all the tumorous regions show specific distinctive properties from non-tumorous regions, varying only in their intensity patterns. [3] exploited this fact of MR images, by explicitly imposing correlations among representations extracted from individual modalities through the minimization of the Euclidean distance between modalities. However, it is essential to note that for the problem at hand, the PET and CT modality in PET/CT are not as well correlated as MR image modalities are. While tumorous regions show a distinctive glare from non-tumorous ones in a PET image, it is very much plausible that a similar glare can be observed in non-tumorous regions as well. Therefore, enforcing correlations, as done for MR images, may not be the best approach to learn representations of PET/CT scans.

Further, named methods assume that a complete set of modalities is available during training, which may not be a valid assumption. In particular, it is not always possible to compute correlations with incomplete PET/CT data, meaning that a CT scan is available, but no corresponding PET scan. In contrast, the proposed method treats PET representations as an *optional* context vector that is fused with the CT representations through an attention mechanism, which has the capability to amplify the signal in salient and discriminatory regions.

3   Methods

3.1   Objective

Let X^{CT} and X^{PET} represent the domain of CT and PET images respectively. Likewise, let Y represent the domain of segmented tumorous regions. Given is

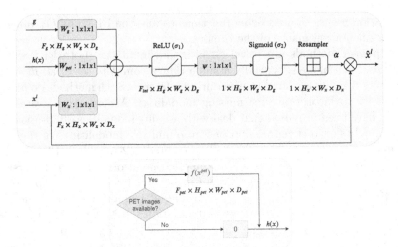

Fig. 1. Schematic of the proposed PET-guided attention gate. The input feature representation x^l is scaled by attention mask α, which is computed by the attention gate. Spatial and contextual information are captured by two gating signals: the encoded feature representation g and the composite function $h(x)$. The composite function is zero when the PET images are missing and the output of the function $f(x^{PET})$ when PET images are available. PET image representation and not the PET image itself is fed to the attention gate.

a dataset consisting of N PET/CT images $\{x_i^{CT}, x_i^{PET}\}_{i=1}^{N}$ and M CT images $\{x_i^{CT}\}_{i=1}^{M}$ ($x_i^{CT} \in X^{CT}$ and $x_i^{PET} \in X^{PET}$) for which corresponding PET images are missing. A typical scenario would be $0 \leq N < M$. For every CT image we have a segmentation mask of tumorous regions i.e. $\{x_j^{CT}, y_j\}_{j=1}^{M}$ where $x_j^{CT} \in X^{CT}$ and $y_j \in Y$.

We are interested in a composite function $H : (X^{CT}, s \cdot X^{PET}) \rightarrow Y$ where s equals 1 if PET images are available and 0 otherwise. The composite function H encompasses two functions: $F : (X^{CT}, X^{PET}) \rightarrow Y$ and $G : X^{CT} \rightarrow Y$. $\hat{y} = H(x^{CT}, s \cdot x^{PET})$ gives the probability map of tumorous regions. The proposed PAG model models the function $H : (X^{CT}, s \cdot X^{PET}) \rightarrow Y$.

3.2   Attention Mechanism

Intuition. The attention gate proposed as part of the current model is built upon the one introduced by [12] for pancreas segmentation on CT images. They based their formulation upon a soft attention mechanism for image classification introduced by [7]. Their attention gate has two inputs: (a) a feature representation and (b) a gating signal. The gating signal filters the input feature representation to select salient regions. The attention gate learns attention masks by attending to parts of the input feature representation. It is enforced by allowing the input feature representation to be compatible with the input gating signal.

[7,12] used the encoded feature representation (output of the encoder) as the gating signal.

The gating signal provides context for the input feature representation to learn salient attention masks. We propose to use PET image feature representation as an additional input along with the encoded feature representation, as shown in Fig. 1. Since PET images can be thought of as a heatmap of tumorous regions in a given CT image, they can help to learn better and discriminatory attention masks by helping the attention masks to focus their attention on regions where PET images show a distinctive glare over their surroundings. Accordingly, the context provided by the encoded feature representation is only enhanced by the input of PET image features whenever they are available. Additionally, this formulation does not mandate the use of PET images features. PET image features can be fed to the model as and when available, making the model flexible to the non-availability of PET images.

Attention Gate. Let g represent an encoded feature representation with $H_g \times W_g \times D_g$ spatial resolution and F_g filter channels respectively for an input CT image x^{CT}. Similarly let x^l represent a feature representation at an intermediate spatial resolution of the encoder branch (skip connection) with $H_x \times W_x \times D_x$ spatial resolution and F_l filter channels respectively for the same input CT image x^{CT}. Likewise, let x^{PET} be an input PET image corresponding to the input CT image x^{CT}.

The attention gate learns attention coefficients $\alpha_i^l \in [0, 1]$ for layer l and voxel position i that identify discriminatory image regions and discard those feature responses to preserve activations that are specific to the appropriate task at hand. The output of the attention gate is an element wise multiplication of the feature representation $x_i^l \in \mathbb{R}^{F_l}$ and attention coefficients α_i^l to obtain the filtered output $\hat{x}_i^l = x_i^l \odot \alpha_i^l$, where $\odot$ denotes the element-wise multiplication. We consider a single attention coefficient for the multi-dimensional vector x_i^l at voxel position i. Also note that $g_i \in \mathbb{R}^{F_g}$ for voxel position i.

Let $\theta_x \in \mathbb{R}^{F_l \times F_{int}}$ and $\theta_g \in \mathbb{R}^{F_g \times F_{int}}$ be linear transformations that are applied to the intermediate feature representation x^l and the encoded feature representation g respectively. Let $f(x^{PET})$ be a function that extracts PET image specific features before they are applied to the attention gate. Define a composite function

$$h(x) = \begin{cases} f(x^{PET}) \text{ when PET images are available} \\ 0 \text{ when PET images are unavailable} \end{cases} \tag{1}$$

Then the attention coefficients are given by

$$q_{att}^l = \psi^T(\sigma_1(\theta_x^T x_i^l + \theta_g^T g_i^l + h(x) + b_g)) + b_\psi \tag{2}$$

$$\alpha_i^l = \sigma_2(q_{att}^l(x_i^l, g_i; \Theta_{att})) \tag{3}$$

where $\sigma_1(x)$ and $\sigma_2(x)$ are ReLU and sigmoid activations respectively. The parameters of attention gate Θ_{att} are given by $\theta_x \in \mathbb{R}^{F_l \times F_{int}}$, $\theta_g \in \mathbb{R}^{F_g \times F_{int}}$, $\psi \in \mathbb{R}^{F_{int} \times 1}$ and bias terms $b_\psi \in \mathbb{R}$, $b_g \in \mathbb{R}^{F_{int}}$. The linear transformations are computed using channel-wise $1 \times 1 \times 1$ convolutions for the input tensors.

The Composite Function. The composite function defined by Eq. 1 represents scenarios when the PET images are either available or missing. In the absence of PET images, the function takes on a value of zero, which boils down to having a simple attention gate akin to the one proposed by [12] on top of the encoder-decoder architecture. However, in the presence of PET images, the function is identical to the PET image feature extractor $f(x^{PET})$. Instead of passing the PET images directly as an input to the attention gate, we pass a higher dimensional feature representation extracted by the function $f(x^{PET})$, which supposedly encompasses a richer spatial and contextual information than the PET images themselves. This function could be any function approximator such as a neural network. A key insight of the proposed model is that, in contrast to previous modality fusion architectures [3,6], there is no fusion of the respective modality-specific embeddings. Such a fusion of embeddings from different modalities can skew the intended embedding space while training the respective models with missing modalities such as in our case. Since there is no fusion of PET and CT embeddings in the proposed PAG model, we do not run the risk of learning skewed embeddings while training the model with a combination of PET/CT and CT images.

The model architecture is an encoder-decoder architecture similar to a U-Net architecture with three skip connections. The three skip connections are filtered through their respective attention gates, with each attention gate having its own set of parameters. More details about the model architecture can be found in the supplementary section.

4 Experiments

We consider four baselines to validate our approach. To make a fair comparison, the PAG model and all baselines use the same backbone architecture [11]. Unimodal and bimodal models process CT-only and PET/CT data respectively. The only difference between unimodal and bimodal models is that PET images are input to bimodal model as an additional channel along with CT images. On the other hand, unimodal+attn and bimodal+attn models are unimodal and bimodal models with the addition of a simple attention gate [12]. Similar to the unimodal and bimodal models, unimodal+attn and bimodal+attn models process CT-only and PET/CT images respectively. Unlike the two discrete unimodal and bimodal models (or unimodal+attn and bimodal+attn models), the PAG model is a single model that handles both unimodal and bimodal scenarios. PAG:ct denotes the PAG model with CT-only inputs during inference, whereas PAG:ct+pet denotes the model with PET/CT inputs respectively.

4.1 Ablation Framework

The ablation study underscores the contribution of the proposed PAG model in contrast to the conventional models. To make things simpler to follow, consider a hypothetical scenario where we have 80 CT and 20 PET/CT scans respectively. So all in all, we have 100 CT scans, of which 20 CT scans have a corresponding PET series. While it is possible to train a unimodal model with 100 CT scans, a bimodal model can only be trained using 20 PET/CT scans. The rest of the 80 CT scans can not be used. On the other hand, since the PAG model is flexible to the availability of PET scans, it is possible to train the model on the 100 CT scans, including the 20 PET scans. Accordingly, the ablation study is designed to examine the performance of the model in scenarios such as these. Concretely, through this ablation study, we examine the performance of the baseline bimodal model and the proposed PAG model as the fraction of the total number of PET series that are made available for training is gradually reduced. With the decrease in the number of PET scans as such, the number of CT scans that can be used for training bimodal model also decreases. However, the PAG model can leverage upon the complete set of CT scans in conjunction with the restricted number of PET scans. It is important to note that since we keep the number of CT scans fixed, the corresponding number of annotated scans (ground truth segmentation masks) is also fixed.

In other words define the ratio $r = n_{pet}/N_{pet}$ where n_{pet} are the number of PET scans available for training and N_{pet} the total number of PET scans in the given dataset. We then decrease the ratio r gradually from 1 to 0 (N_{pet} is fixed). It is expected that with decreasing ratio r, the performance of the bimodal model decreases noticeably. However, we expect the decrease in the performance of the novel PAG model to be less pronounced. At all times, even in the limit of zero PET scans, it should perform at least as good as a unimodal model that is trained on the complete set of CT scans. In the following, we provide details about the dataset and implementation details for the experiments, before we continue with the presentation and discussion of the results.

Evaluation Data. We evaluate our approach on a dataset of 397 PET/CT scans of patients suffering from lung cancer, collected and labeled by the radiology department of the University of Basel, Switzerland. PET/CT images provide complementary information on the regions of interest compared to CT-only data. PET images can be thought of as a heatmap for the corresponding CT images where the tumorous regions show a marked contrast or a distinctive glare between their surroundings. An example of such a pair of CT images and a PET/CT image (PET image superimposed on CT image) is shown in Fig. 2. Note that the tumorous region, which is bound by a red bounding box in the CT image, has a marked contrast over its surroundings in the PET/CT image. This is because of the greater 18F-FDG uptake by the malignant tumors due to higher metabolic activity, which can be detected from PET images.

The dataset contains a rich diversity of primary tumors, lymph node metastases, and other metastases that were independently segmented by two expert

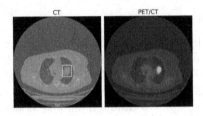

Fig. 2. (Left) An example of a malignant tumor in the right lung. The tumor is surrounded by the bounding box in red. (Right) PET/CT image for the same region. There is a distinctive glare in the region for the corresponding tumorous region.

radiologists. Therefore, the dataset provides a rich data source that is an order of magnitude larger than existing public PET/CT datasets with labelled segmentation maps. More details about the dataset are provided in the supplementary information.

Evaluation Criteria. We use dice coefficient as our metric to evaluate the proposed model. Dice coefficient is one of the most widely used metric to evaluate segmentation algorithms. It measures the degree of overlap between the ground truth and predicted segmentation masks factored by the number of true positives and false positives. It falls within a range of $[0, 1]$ with 0 signifying absolutely no intersection between the two sets while 1 signifying a perfect intersection with no false positives or false negatives, meaning both sets are alike. A correctly predicted segmentation mask has a dice coefficient of 1, whereas a segmentation mask that predicts zeros for all the voxels has a dice coefficient of 0. Therefore we would expect the dice coefficient of a segmentation algorithm to lie in the range of $[0, 1]$ and the higher the dice coefficient, the closer is the predicted segmentation mask to the ground truth segmentation mask.

Training Details. We developed all our models using the PyTorch framework.[1] Each of the models occupies approximately 12 GB of GPU memory for model parameters, forward and backward pass. So with a batch size of 2, the memory requirement is approximately doubled i.e., 24 GB. All models were trained on a server of 8 NVIDIA Tesla V-100-SXM2 32 GB GPUs. We chose a weighted combination of Sorenson-Dice loss and binary cross-entropy loss as our loss function, a default choice for segmentation tasks. All the models were trained using Adam optimizer (default parameters) and group normalization [16]. Initially, the learning rate α was set to 0.0001; the learning rate was then gradually decayed after every training epoch. The model parameters were regularised using L2 regularisation with regularisation parameter β set to 10^{-5}. Augmented data was included for training at every training epoch but with a probability $p_{data-aug} = 0.25$. All models were trained for 75 epochs.

---

[1] https://github.com/pvk95/PAG.

While training the PAG model, it is critical that PET images are randomly excluded at every training step with a non zero probability p. We do this to ensure that the PAG model does not overfit to either of the scenarios when PET images are available or not. We set this probability value $p = 0.5$. (See appendix for a list of hyper-parameters).

From the dataset consisting of 397 PET/CT labeled images, 77 PET/CT images and their corresponding labels were randomly selected and set apart as our test dataset. The remaining 320 samples were used for training and validation. All the baseline and PAG models have been evaluated using four-fold cross-validation experiments. The training and validation dataset is randomly split into four folds. One of the folds was kept out for validation. The remaining three folds were used for training the models. Each of the models was then retrained on the entire 320 PET/CT images before testing the models on the test dataset. When constraining the number of PET images in the ablation study, we randomly sampled the appropriate number of PET images from the samples that were initially earmarked for training and then trained accordingly. It is noteworthy that the respective validation folds across all the models and all the ratios r in the ablation study remain the same. More details about the training are provided in the supplementary information.

4.2 How Well Does the Model Incorporate the Two Scenarios: CT only Images and PET/CT Images?

Figure 3 shows the performance of the individual baseline models and the PAG model when a PET image is available for every CT image while training the models. We thus do not place any restriction on the availability of PET images. We do this primarily to validate whether our model is able to handle the combination of CT and PET/CT images well. We observe that the PAG:ct+pet model performs on par with bimodal and bimodal+attn models. Similarly, PAG:ct performs on par with unimodal and unimodal+attn models.

We incorporated the attention mechanism of [12] to the unimodal and bimodal models, and denote the resulting models with unimodal+attn and bimodal+attn. We expected them to outperform their non-attention counterparts (i.e., the unimodal and bimodal models). However, this is not the case, considering Fig. 3. The reason for this behaviour could be the complexity of our dataset. The attention gate [12] of the unimodal+attn and bimodal+attn models was originally tested on two publicly available datasets for pancreas image segmentation. The pancreas has a definite shape, structure, and morphology. They are found in a single location within the body. However, the tumors of the current dataset exhibit varying shapes, structures, morphologies, and even locations within the body. This could explain why we do not observe a significant performance gain on our dataset, by adding their attention gate to the unimodal and bimodal models. However, this does not imply that the attention mechanism is not at play here, but that the attention masks are not informative enough. However, it becomes clear from Fig. 3 that accommodating PET images as part

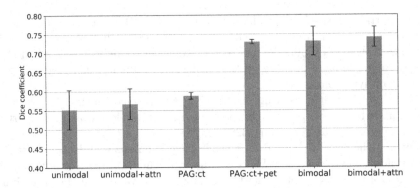

Fig. 3. The Figure shows the performance of the baseline models and the PAG model on the test data set. PAG:ct and PAG:ct+pet are both based on the PAG model. PAG:ct+pet is PAG model when PET images are input to the model in addition to the CT images. Conversely, PAG:ct is PAG model when only CT images are input to the model, but no PET images are used as input to the model. PAG:ct performs at par with the unimodal and unimodal+attn models. Similarly PAG:ct+pet model performs at par with the bimodal and bimodal+attn models.

of the proposed attention gate significantly improves performance of the models, considering the better performance of PAG:ct+pet model (dice coefficient = 0.73) over PAG:ct model (dice coefficient = 0.58). This is not just a consequence of the addition of PET images to the PAG:ct+pet model but because of the addition of PET images to the PAG:ct+pet model in association with the proposed attention gate, the very means of how PET images are fed to the model.

Consequently, we conclude that when PET images are available during inference, PAG:ct+pet performs on par with bimodal and bimodal+attn models, and when they are not available, PAG:ct performs on par with unimodal and unimodal+attn models. This supports the claim that the PAG model successfully encompasses the two discrete models: unimodal and bimodal models. Further, the addition of PET images through the proposed attention gate makes a significant impact on the performance of the PAG model. This validates that the attention gate effectively integrates information from PET images, whenever they are available.

4.3 How Well Does the Model Handle a Combination of CT and PET/CT Images?

Figure 4 shows the result of the ablation study, as described earlier in Sect. 4.1. The performance of the PAG:ct+pet model and the bimodal models is evaluated as the ratio $r = n_{pet}/N_{pet}$ is gradually reduced. The ratio points considered are [1.0, 0.5, 0.3, 0.15, 0.1, 0.05, 0.03]. The majority of examined data points are close to zero, in order to compare and contrast the significance of the PAG model when PET images are very scarce. The unimodal model is illustrated

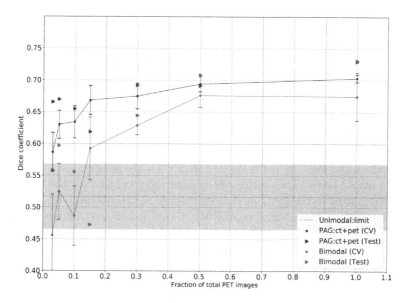

Fig. 4. The Figure shows the dice coefficient for the PAG model and the bimodal model when the fraction of total PET images that are made available for training the models is restricted. Results are shown for the validation (CV) and test (Test) data sets. The green band is the mean and standard deviation of the unimodal model trained on CT images. The degradation in performance of the bimodal model is much more drastic than the PAG:ct+pet model. Note that PAG:ct+pet model always maintains the edge over unimodal model because either of the models were trained on the same number of CT images, with additional PET images for PAG:ct+pet model.

as well with its mean (green dotted line) and standard deviation (green band around the dotted line). It can be seen that for all the values of ratio r, the dice coefficient of PAG:ct+pet is greater than the bimodal model. Consider, for instance a point at $r = 0.15$. This point represents a scenario where one has 36 PET/CT images and 204 CT-only images or 240 CT images in total. The bimodal model was trained on the small set of 36 PET/CT images while the PAG model was trained on 204 CT images and 36 PET/CT images. This shows that the extra 204 CT-only images which would otherwise have been discarded while training the bimodal model could be used for training the PAG model. Clearly, the extra 204 CT-only images make a difference in boosting the dice coefficient of the model. This performance gain becomes more and more extreme as the ratio r approaches values closer to zero.

There is another facet to the PAG:ct+pet model. Irrespective of the ratio r, PAG:ct+pet was trained on the same number of CT images. This implies that even in the limit of zero PET images, the performance of PAG:ct+pet should not degrade below the performance of unimodal model which can be clearly observed for points closer to zero ($[0.03, 0.05, 0.1]$). For example, consider a point $r = 0.03$. This point represents a data set with 7 PET/CT images and 233 CT-only images or 240 CT images in total. The unimodal model was trained on 240 CT images while the PAG model was trained on 240 CT images including 7 PET images. Clearly, the extra number of 7 PET images yielded in significant performance gains (dice coefficient = 0.66) over the unimodal model (dice coefficient = 0.56). Naturally, the improvement in performance becomes more and more obvious with increasing ratios r.

Hence, in the limit of zero PET images, the PAG model is able to successfully leverage upon the extra number of CT-only images. This behaviour is reflected in the higher dice coefficient of PAG:ct+pet model over the bimodal model. In the scenario when the PAG model is trained with CT-only images, the performance boundary would be the unimodal model. Consequently, just with the addition of a few PET images to the PAG model, we observe significant performance gains, considering higher dice coefficient of PAG:ct+pet model over unimodal model. This supports our claim that the model makes efficient use of the combination of CT-only and PET/CT data.

5   Discussion and Conclusion

Although PET/CT imaging is the gold standard for the staging of lung cancer, due to logistic and economic constraints, PET images are often unavailable. This problem is especially prominent in resource-constrained healthcare systems. While conventional methods are unable to handle a combination of CT-only and PET/CT data, we tackled this challenge by adapting an established visual soft attention mechanism to the problem at hand. We demonstrated that our proposed approach performs on par with unimodal and bimodal baselines. We further present that our model is especially useful when the number of PET images is small in comparison to the number of CT images, which is relevant in resource-constrained environments.

It is noteworthy, irrespective of the number of PET/CT images that are available, the model always requires the same number of segmentation masks as the number of total number of CT images. This could be a limitation considering the manual effort in procuring the segmentation masks. In future work, we would like to explore the possibility of reducing the number of segmentation masks by generative models. Thereby, we could extend the resource efficiency of the algorithm to leverage a reduced number of segmented images.

Another interesting direction for future research would be to extend the proposed PAG model to other imaging modalities such as MRIs, as our formulation is not limited to a single additional modality. It would be interesting to investigate further the behaviour of the proposed attention gate with additional modalities.

Acknowledgements. ID is supported by the SNSF grant #200021_188466.

References

1. Cai, L., Wang, Z., Gao, H., Shen, D., Ji, S.: Deep adversarial learning for multi-modality missing data completion. In: Proceedings of the 24th ACM SIGKDD International Conference on Knowledge Discovery and Data Mining, pp. 1158–1166 (2018)
2. Chandar, S., Khapra, M.M., Larochelle, H., Ravindran, B.: Correlational neural networks. Neural Comput. **28**(2), 257–285 (2016)
3. Chartsias, A., Joyce, T., Giuffrida, M.V., Tsaftaris, S.A.: Multimodal MR synthesis via modality-invariant latent representation. IEEE Trans. Med. Imaging **37**(3), 803–814 (2017)
4. Dorent, R., Joutard, S., Modat, M., Ourselin, S., Vercauteren, T.: Hetero-modal variational encoder-decoder for joint modality completion and segmentation. In: Shen, D., et al. (eds.) MICCAI 2019. LNCS, vol. 11765, pp. 74–82. Springer, Cham (2019). https://doi.org/10.1007/978-3-030-32245-8_9
5. Guo, Z., Li, X., Huang, H., Guo, N., Li, Q.: Medical image segmentation based on multi-modal convolutional neural network: study on image fusion schemes. In: 2018 IEEE 15th International Symposium on Biomedical Imaging (ISBI 2018), pp. 903–907. IEEE (2018)
6. Havaei, M., Guizard, N., Chapados, N., Bengio, Y.: HeMIS: hetero-modal image segmentation. In: Ourselin, S., Joskowicz, L., Sabuncu, M.R., Unal, G., Wells, W. (eds.) MICCAI 2016. LNCS, vol. 9901, pp. 469–477. Springer, Cham (2016). https://doi.org/10.1007/978-3-319-46723-8_54
7. Jetley, S., Lord, N.A., Lee, N., Torr, P.H.: Learn to pay attention. arXiv-1804 (2018)
8. Li, L., Zhao, X., Lu, W., Tan, S.: Deep learning for variational multimodality tumor segmentation in PET/CT. Neurocomputing (2019)
9. Lian, C., Ruan, S., Denoeux, T., Li, H., Vera, P.: Joint tumor segmentation in PET-CT images using co-clustering and fusion based on belief functions. IEEE Trans. Image Process. **28**(2), 755–766 (2018)
10. Menze, B.H., et al.: The multimodal brain tumor image segmentation benchmark (BRATS). IEEE Trans. Med. Imaging **34**(10), 1993–2024 (2014)
11. Myronenko, A.: 3D MRI brain tumor segmentation using autoencoder regularization. In: Crimi, A., et al. (eds.) BrainLes 2018. LNCS, vol. 11384, pp. 311–320. Springer, Cham (2019). https://doi.org/10.1007/978-3-030-11726-9_28
12. Oktay, O., et al.: Attention U-Net: learning where to look for the pancreas (2018)
13. Ronneberger, O., Fischer, P., Brox, T.: U-Net: convolutional networks for biomedical image segmentation. In: Navab, N., Hornegger, J., Wells, W.M., Frangi, A.F. (eds.) MICCAI 2015. LNCS, vol. 9351, pp. 234–241. Springer, Cham (2015). https://doi.org/10.1007/978-3-319-24574-4_28
14. American Cancer Society: About lung cancer, March 2020. https://www.cancer.org/cancer/lung-cancer/about.html
15. Styner, M., et al.: 3D segmentation in the clinic: a grand challenge II: MS lesion segmentation. Midas J. **2008**, 1–6 (2008)
16. Wu, Y., He, K.: Group normalization. In: Ferrari, V., Hebert, M., Sminchisescu, C., Weiss, Y. (eds.) ECCV 2018. LNCS, vol. 11217, pp. 3–19. Springer, Cham (2018). https://doi.org/10.1007/978-3-030-01261-8_1
17. Zhao, L., Lu, Z., Jiang, J., Zhou, Y., Wu, Y., Feng, Q.: Automatic nasopharyngeal carcinoma segmentation using fully convolutional networks with auxiliary paths on dual-modality PET-CT images. J. Digit. Imaging **32**(3), 462–470 (2019)

18. Zhong, Z., et al.: 3D fully convolutional networks for co-segmentation of tumors on PET-CT images. In: 2018 IEEE 15th International Symposium on Biomedical Imaging (ISBI 2018), pp. 228–231. IEEE (2018)
19. Zhou, T., Canu, S., Vera, P., Ruan, S.: Brain tumor segmentation with missing modalities via latent multi-source correlation representation. arXiv-2003 (2020)

Self-supervised Disentanglement of Modality-Specific and Shared Factors Improves Multimodal Generative Models

Imant Daunhawer[(✉)], Thomas M. Sutter, Ričards Marcinkevičs,
and Julia E. Vogt

Department of Computer Science, ETH Zurich, Zürich, Switzerland
dimant@inf.ethz.ch

Abstract. Multimodal generative models learn a joint distribution over multiple modalities and thus have the potential to learn richer representations than unimodal models. However, current approaches are either inefficient in dealing with more than two modalities or fail to capture both modality-specific and shared variations. We introduce a new multimodal generative model that integrates both modality-specific and shared factors and aggregates shared information across any subset of modalities efficiently. Our method partitions the latent space into disjoint subspaces for modality-specific and shared factors and learns to disentangle these in a purely self-supervised manner. Empirically, we show improvements in representation learning and generative performance compared to previous methods and showcase the disentanglement capabilities.

1 Introduction

The promise of multimodal generative models lies in their ability to learn rich representations across diverse domains and to generate missing modalities. As an analogy, humans are able to integrate information across senses to make more informed decisions [33], and exhibit cross-modal transfer of perceptual knowledge [41]; for instance, people can visualize objects given only haptic cues [42]. For machine learning, multimodal learning is of interest in any setting where information is integrated across two or more modalities.

Alternatives to multimodal generative models include unimodal models with late fusion or with coordinated representations, as well as conditional models that translate between pairs of modalities [3]. Yet, both alternatives have disadvantages compared to multimodal approaches. While unimodal models cannot handle missing modalities, conditional models only learn a mapping between sources, and neither integrate representations from different modalities into a

Electronic supplementary material The online version of this chapter (https://doi.org/10.1007/978-3-030-71278-5_33) contains supplementary material, which is available to authorized users.

© Springer Nature Switzerland AG 2021
Z. Akata et al. (Eds.): DAGM GCPR 2020, LNCS 12544, pp. 459–473, 2021.
https://doi.org/10.1007/978-3-030-71278-5_33

joint representation. In contrast, multimodal generative models approximate the joint distribution and thus implicitly provide the marginal and conditional distributions. However, learning a joint distribution remains the more challenging task and there still exists a gap in the generative performance compared to unimodal and conditional models.

We bridge this gap by proposing a new self-supervised multimodal generative model that disentangles modality-specific and shared factors. We argue that this disentanglement is crucial for multimodal learning, because it simplifies the aggregation of representations across modalities. For conditional generation, this decomposition allows sampling from modality-specific priors without affecting the shared representation computed across multiple modalities. Further, decomposed representations have been found to be more interpretable [6,17] and more amenable for certain downstream tasks [26].

The main contribution of this work is the development of a new multimodal generative model that learns to disentangle modality-specific and shared factors in a self-supervised manner. We term this new method disentangling multimodal variational autoencoder (DMVAE). It extends the class of multimodal variational autoencoders by modeling modality-specific in addition to shared factors and by disentangling these groups of factors using a self-supervised contrastive objective. In two representative toy experiments, we demonstrate the following advantages compared to previous multimodal generative models:

- Effective disentanglement of modality-specific and shared factors. This allows sampling from modality-specific priors without changing the joint representation computed from multiple modalities.
- Improvements in representation learning over state-of-the-art multimodal generative models. For any subset of modalities, our model aggregates shared information effectively and efficiently.
- Improvements in generative performance over previous work. In a fair comparison, we demonstrate that modeling modality-specific in addition to shared factors significantly improves the conditional generation of missing modalities. For unconditional generation, we demonstrate the effectiveness of using ex-post density estimation [8] to further improve joint generation across all methods, including trained models from previous work.

2   Related Work

Broadly, our work can be categorized as an extension of the class of multimodal generative models that handle more than two modalities (including missing ones) efficiently. Among this class, we present the first method that partitions the latent space into modality-specific and shared subspaces and disentangles these in a self-supervised fashion.

Multimodal Generative Models. Current approaches are mainly based on encoder-decoder architectures which learn the mapping between modalities based on reconstructions or adversarial objectives (for a comprehensive review, see [3]). Among this class, methods can be distinguished by the type of mapping they use to translate between inputs and outputs and by how they handle missing modalities. Early approaches [16,35] try to learn all possible mappings, which in the case of missing modalities results in 2^M encoders for M modalities. A more efficient alternative is proposed by [40] who introduce the multimodal variational autoencoder (MVAE) which uses a joint posterior that is proportional to a product of experts (PoE) [14]. Their method handles missing modalities efficiently, because it has a closed form solution for the aggregation of marginal Gaussian posteriors. However, their derivation of the joint posterior is based on the assumption that all modalities share a common set of factors—an assumption that is often violated in practice, because modalities exhibit a high degree of modality-specific variation. Our model also uses a joint latent space with a product of experts aggregation layer, and thus shares the same theoretical advantages, but it considers modality-specific factors in addition to shared factors. The limitations of the MVAE were shown empirically in [31], where it is stated that the MVAE lacks the abilities of latent factorization and joint generation. With latent factorization the authors refer to the decomposition into modality-specific and shared factors, and by joint generation they mean the semantic coherence of unconditionally generated samples across modalities. They attribute these problems to the joint posterior used by the MVAE and demonstrate empirically that using a mixture of experts, instead of a product, improves generative performance. In contrast, we argue that the product of experts is not a problem per se, but that it is an ill-defined aggregation operation in the presence of modality-specific factors. We resolve this model misspecification by modeling modality-specific factors in addition to shared factors. Compared to the mixture of experts multimodal variational autoencoder (MMVAE) [31], our model has the advantage that it can sample from a modality-specific prior without affecting the shared representation which can still be aggregated efficiently across modalities through the PoE. Especially with more than two modalities, the aggregation of representations, as it is done in our model, shows its benefits compared to the MMVAE (see Sect. 4.2).

Domain Adaption/Translation. The research areas of domain adaption and domain translation are in many regards closely related to multimodal generative models. Approaches that have explored many-to-many mappings between different domains have been based on adversarial methods [7,24], shared autoencoders [36] and cycle-consistency losses [2]. Translation methods have shown remarkable progress on image-to-image style transfer and the conceptual manipulation of images, however, their focus lies on learning conditional mappings, while our method models the joint distribution directly. Further, through the PoE our method aggregates shared representations across any subset of modalities and therefore handles missing modalities efficiently.

Disentanglement. Our goal is not the unsupervised disentanglement of all generative factors, which was shown to be theoretically impossible with a factorizing prior and claimed to be impossible in general [25]. Instead, we are concerned with the disentanglement of modality-specific and shared sets of factors. In the multi-view and multimodal case, there is theoretical evidence for the identifiability of shared factors [9,19,27,37]. Further, the self-supervised disentanglement of shared factors has been previously explored based on grouping information [4], temporal dependencies [23], partly labeled data [18,38,39], and spatial information [5]. We take a first step towards disentanglement given multimodal data with modality-specific factors and an implicit, unknown grouping.

3   Method

In this section, we introduce multimodal generative models and derive the variational approximations and information-theoretic objectives that our method optimizes. All proofs are provided in the appendix.

We consider a generative process with a partition into modality-specific and modality-invariant (i.e., shared) latent factors (Fig. 1). A multimodal sample $\mathbf{x} = (x_1, \ldots, x_M)$ with data from M modalities is assumed to be generated from a set of shared factors c and a set of modality-specific factors s_m. Consequently, samples from different modalities are assumed to be conditionally independent given c. In the following, we denote the set of all modality-specific factors of a multimodal sample as $\mathbf{s} = (s_1, \ldots s_M)$.

Given a dataset $\{\mathbf{x}^{(i)}\}_{i=1}^{N}$ of multimodal samples, our goal is to learn a generative model $p_\theta(\mathbf{x}|c, \mathbf{s})$ with a neural network parameterized by θ. From the above assumptions on the data generating process, it follows the joint distribution

$$p(\mathbf{x}, \mathbf{s}, c) = p(c) \prod_{m=1}^{M} p(s_m) p(x_m \mid c, s_m) \tag{1}$$

which allows to consider only the observed modalities for the computation of the marginal likelihood.

The computation of the exact likelihood is intractable, therefore, we resort to amortized variational inference and instead maximize the evidence lower bound

$$\mathcal{L}_{\text{VAE}}(\mathbf{x}, c) := \sum_{m=1}^{M} \mathbb{E}_{q_\phi(s_m \mid x_m)} \Big[\log p_{\theta_m}(x_m \mid c, s_m) \Big] - D_{\text{KL}} \left(q_\phi(s_m|x_m) \, \| \, p(s_m) \right)$$

which is composed of M log-likelihood terms and KL-divergences between approximate posteriors $q_\phi(s_m \mid \mathbf{x})$ and priors $p(s_m)$. Above objective describes M modality-specific VAEs, each of which takes as input an additional context vector c that encodes shared information (described in Sect. 3.2). We use neural networks for each encoder $q_{\phi_m}(s_m|x_m)$ as well as for each decoder $p_{\theta_m}(x_m|c, s_m)$ and denote the network parameters by the respective subscripts for decoder parameters θ and encoder parameters ϕ. Further, we follow the convention of using an isotropic Gaussian prior and Gaussian variational posteriors parameterized by the estimated means and variances that are the outputs of the encoder.

For each modality-specific VAE, it is possible to control the degree of disentanglement of arbitrary factors with a weight on the respective KL-divergence

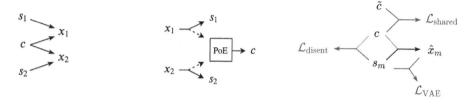

Fig. 1. Graphical model and network architecture for the special case of two modalities. *Left:* A sample x_m from modality m is assumed to be generated by modality-specific factors s_m and modality-invariant factors c. *Center:* Inference network that aggregates shared factors through a product of experts (PoE) layer. Dashed lines represent simulated missing modalities as used during training. *Right:* Decoder network (black) for modality m and loss terms (green). Dotted lines denote paths that are not being backpropagated through. Shared factors are learned by a contrastive objective which takes as input representations c and $\tilde{c}$ computed from different subsets of modalities. Modality-specific factors are inferred by regularizing out shared information from the latent space of the VAE. All loss terms are defined in Subsect. 3.2. (Color figure online)

term, like in the β-VAE [13]. However, there exist theoretical limitations on the feasibility of unsupervised disentanglement of arbitrary factors [25]. In contrast, we focus on the disentanglement of modality-specific and shared factors, for which we use two additional objectives that are introduced in Subsect. 3.2.

3.1 Multimodal Inference Network

A key aspect in the design of multimodal models should be the capability to handle missing modalities efficiently [3]. In our case, only the shared representation depends on all modalities and should ideally be able to cope with any combination of missing inputs, which would require 2^M inference networks in a naive implementation. A more efficient alternative is offered in [40], where a product of experts (PoE) [14] is used to handle missing modalities. Under the assumption of shared factors, previous work [40] has shown that the posterior $p(c \mid \mathbf{x})$ is proportional to a product of unimodal posteriors

$$p(c \mid \mathbf{x}) \propto \frac{1}{p(c)^{M-1}} \prod_{m=1}^{M} p(c \mid x_m) \qquad (2)$$

which—for the special case of Gaussian posteriors—has an efficient closed-form solution (see Appendix A.3). We also assume Gaussian unimodal posteriors $q_{\psi_m}(c \mid x_m)$ where ψ denotes the encoder parameters, part of which can be shared with the encoder parameters ϕ_m of a unimodal VAE. The choice of Gaussian posteriors allows us to employ the PoE as an aggregation layer for shared factors. This allows the model to use M unimodal inference networks to handle all 2^M combinations of missing modalities for the inference of shared factors.

While the PoE is a well defined aggregation operation for shared factors, it is not suitable for modality-specific factors, because it averages over representations from different modalities.[1] Therefore, we partition the latent space into $M + 1$ independent subspaces, one that is specific for each modality (denoted by s_m) and one that has shared content between all modalities (denoted by c), as illustrated in Fig. 1. The PoE is only used for the shared representation, so modality-specific information is not forced through the aggregation layer.

In theory, a partitioned latent space provides the possibility to encode both modality-specific and shared information in separate subspaces; in practice, however, objective $\mathcal{L}_{\mathrm{VAE}}$ does not specify what information (modality-specific or shared) should be encoded in which subspace. For example, the first log-likelihood term $\log p_\theta(x_1 \mid c, s_1)$ can be maximized if *all* information from input x_1 flows through the modality-specific encoder $q_\phi(s_1 \mid x_1)$ and none through the shared encoder. Thus, we posit that the model requires an additional objective for disentangling modality-specific and modality-invariant information. Next, we formalize our notion of disentanglement and introduce suitable contrastive objectives.

3.2   Disentanglement of c and s

We take an information-theoretic perspective on disentanglement and representation learning. Consider multimodal data to be a random variable X and let $h_1(X)$ and $h_2(X)$ be two functions, each of which maps the data to a lower-dimensional encoding. Consider the objective

$$\max_{h_1, h_2 \in \mathcal{H}} I(X; h_1(X)) + I(X; h_2(X)) - I(h_1(X); h_2(X)) \tag{3}$$

where I denotes the mutual information between two random variables and $\mathcal{H}$ is the set of functions that we optimize over, for instance, the parameters of a neural network. Objective (3) is maximized by an encoding that is maximally informative about the data while being maximally independent between $h_1(X)$ and $h_2(X)$. In our case, these two functions should encode modality-specific and shared factors respectively. The proposed model learns such a representation by using suitable estimators for the individual information terms.

The objective optimized by a VAE can be viewed as a lower bound on the mutual information between data and encoding (e.g., see [1,15]). However, on itself a VAE does not suffice to learn a disentangled encoding, because of theoretical limitations on disentanglement in an unsupervised setting [25]. So in addition, we equip the VAE with two contrastive objectives: one that learns an encoding of information shared between modalities, maximizing a lower bound on $I(\mathbf{x}; c)$, and one that infers modality-specific factors by regularizing out shared information from the latent space of a modality-specific VAE. The overall objective that is being maximized is defined as

$$\mathcal{L} = \mathcal{L}_{\mathrm{VAE}} + \gamma \mathcal{L}_{\mathrm{shared}} - \delta \mathcal{L}_{\mathrm{disent}} \tag{4}$$

---

[1] This problem has also been observed in [21] where it is described as "averaging over inseparable individual beliefs".

where $\mathcal{L}_{\mathrm{VAE}}$ is the ELBO optimized by the VAEs, $\mathcal{L}_{\mathrm{shared}}$ learns an encoding of shared factors, $\mathcal{L}_{\mathrm{disent}}$ disentangles shared and modality-specific information, and the hyperparameters γ and δ can be used to control these terms respectively. The proposed objective estimates shared factors directly, while modality-specific factors are inferred indirectly by regularizing out shared information from the encoding of a modality-specific VAE. Further, as in the β-VAE [13], the reconstruction loss and KL-divergence contained in $\mathcal{L}_{\mathrm{VAE}}$ can be traded off to control the quality of reconstructions against the quality of generated samples. Figure 1 shows a schematic of the network including all loss terms that are being optimized. In the following, we define the contrastive objectives used for the approximation of the respective mutual information terms.

To learn shared factors, we use a contrastive objective [10,32] that maximizes a lower bound on the mutual information $I(\mathbf{x};c)$ (see Appendix A for the derivation). We estimate the mutual information with the sample-based InfoNCE estimator [29] adapted to a multimodal setting. The objective is defined as

$$\mathcal{L}_{\mathrm{shared}} := -\mathbb{E}\left[\frac{1}{K}\sum_{i=1}^{K}\log\frac{e^{f(\mathbf{x}^{(i)},\tilde{\mathbf{x}}^{(i)})}}{\frac{1}{K}\sum_{j=1}^{K}e^{f(\mathbf{x}^{(i)},\tilde{\mathbf{x}}^{(j)})}}\right] \tag{5}$$

where the expectation goes over K independent samples $\{\mathbf{x}^{(i)},\tilde{\mathbf{x}}^{(i)}\}_{i=1}^{K}$ from $p(\mathbf{x},\tilde{\mathbf{x}})$ where $\tilde{\mathbf{x}}$ is a subset of modalities $\tilde{\mathbf{x}}\subset\mathbf{x}$ and f is a critic that maps to a real-valued score. In particular, we use an inner product critic $f_\phi(\mathbf{x},\tilde{\mathbf{x}})=\langle c,\tilde{c}\rangle$ where c and $\tilde{c}$ are the representations computed from a full multimodal sample and a subset of modalities respectively. Intuitively, the objective contrasts between a positive pair coming from the same multimodal sample and $K-1$ negative pairs from randomly paired samples [e.g., 11]. By using a large number of negative samples, the bound becomes tighter [29], therefore we use a relatively large batch size of $K = 1024$ such that for every positive, we have 1023 negative samples by permuting the batch. In Appendix A we prove that the contrastive objective is a lower bound on $I(\mathbf{x};c)$ and we further discuss the approximation as well as our choice of critic.

To regularize out shared information from the encoding of a modality-specific VAE, we use a discriminator that minimizes the total correlation $TC(c, s_m)$, a measure of statistical dependence between a group of variables. In the case of two variables, the total correlation is equivalent to the mutual information. We approximate the total correlation using the density-ratio trick [28,34] and refer to the approximation by $\mathcal{L}_{\mathrm{disent}}$ (see Appendix A). This procedure is very similar to the one used by [20] with the important difference that we do not estimate the total correlation between all elements in a single latent representation, but between partitions c, s_m of the latent space, of which c is shared between modalities. In theory, one can use a single discriminator to minimize $TC(c,\mathbf{s})$ jointly, however, we found that in practice one has more control over the disentanglement by using individual terms $\mathcal{L}_{disent} = \delta_m\sum_m\mathcal{L}_{disent}(c, s_m)$ weighted by separate disentanglement coefficients δ_m, instead of a global δ.

4   Experiments

In this section, we compare our method to previous multimodal generative models both qualitatively and quantitatively. In the first experiment, we use a bimodal dataset that has been used in previous studies and compare our method to the MVAE [40] and MMVAE [31], the current state-of-the-art multimodal generative models. In the second experiment, we go beyond two modalities and construct a dataset with 5 simplified modalities that allows us to analyze the aggregation of representations across multiple modalities, which, to the best of our knowledge, has not been done previously.

For the quantitative evaluation, we employ metrics that were used in previous studies. Mainly, we focus on generative coherence [31], which takes a classifier (pretrained on the original data) to classify generated samples and computes the accuracy of predicted labels compared to a ground truth. For unconditional samples, coherence measures how often the generated samples match across all modalities. To measure the quality of generated images, we compute Fréchet Inception Distances (FIDs) [12]. It is important to note that a generative model can have perfect coherence yet very bad sample quality (e.g., blurry images of the correct class, but without any diversity). Analogously, a model can achieve very good FID without producing coherent samples. Therefore, we also propose to compute class-specific conditional FIDs for which the set of input images is restricted to a specific class and the set of conditionally generated images is compared to images of that class only. Hence, class-specific conditional FID provides a measure of both coherence and sample quality. Finally, we evaluate the quality of the learned representations by training a linear classifier on the outputs of the encoders.

4.1   MNIST-SVHN

A popular dataset for the evaluation of multimodal generative models is the MNIST-SVHN dataset [31,38], which consists of digit images from two different domains, hand-written digits from MNIST [22] and street-view house numbers from SVHN [30]. The images are paired by corresponding digit labels, and similar to [31] we use 20 random pairings for each sample in either dataset. The pairing is done for the training and test sets separately and results in a training set of 1,121,360 and test set of 200,000 image pairs. The dataset is convenient for the evaluation of multimodal generative models, because it offers a clear separation between shared semantics (digit labels) and perceptual variations across modalities. This distinctive separation is required for the quantitative evaluation via generative coherence and class-specific conditional FID.

For a fair comparison to previous work, we employ the same architectures, likelihood distributions, and training regimes across all models. The setup is adopted from the official implementation of the MMVAE.[2] For our model we use a 20 dimensional latent space of which 10 dimensions are shared between

---

[2] https://github.com/iffsid/mmvae.

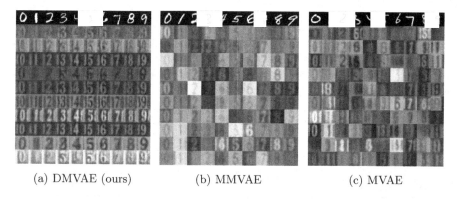

(a) DMVAE (ours)	(b) MMVAE	(c) MVAE

Fig. 2. Comparison of conditionally generated SVHN samples given the respective MNIST digit in the first row. Across a column, we sample from the modality-specific prior (our model) or from the posterior (other models). Only our model keeps consistent styles across rows, as it disentangles modality-specific and shared factors (without supervision).

modalities and 10 dimensions are modality-specific.[3] This does not increase the total number of parameters compared to the MMVAE or MVAE where a 20 dimensional latent space is used respectively. All implementation details are listed in Appendix C.

Qualitative Results. Figure 2 illustrates the conditional generation of SVHN given MNIST. Only our method is capable of keeping consistent styles across rows, because our model allows to draw samples from the modality-specific prior without changing the shared representation computed from the input. For both MVAE and MMVAE, we sample from the posterior to generate diverse images along one column.[4] One can already observe that our model and the MMVAE are both capable of generating images with coherent digit labels, while the MVAE struggles to produce matching digits, as already observed in [31]. The results are similar for the conditional generation of MNIST given SVHN (see Appendix B), demonstrating that our method is effective in disentangling modality-specific and shared factors in a self-supervised manner.

Quantitative Results. Since the setup of this experiment is equivalent to the one used by [31] to evaluate the MMVAE, we report the quantitative results from their paper. However, we decided to implement the MVAE ourselves, because we found that the results reported in [31] were too pessimistic.

---

[3] The size of latent dimensions for modality-specific and shared representations is a hyperparameter of our model. Empirically, we found the effect of changing the dimensionality to be minor, as long as neither latent space is too small.

[4] We further observed that without sampling from the posterior (i.e., reparameterization) both the MVAE and MMVAE tend to generate samples with very little diversity, even if diverse input images are used.

Table 1. Results on MNIST/SVHN, where x_1 corresponds to MNIST and x_2 to SVHN. Numbers denote median values over 5 runs (standard deviations in parentheses). For MMVAE, numbers are based on the original work and standard deviations were computed with the publicly available code. For latent classification, we use linear classifiers and for the DMVAE only the shared representation is used (concatenation further improves the results).

Method	Latent accuracy (in %)			Coherence (in %)		
	x_1	x_2	Aggregated	Joint	$x_1 \to x_2$	$x_2 \to x_1$
MVAE	79.8 (± 3.8)	65.1 (± 4.6)	80.2 (± 3.6)	38.0 (± 1.8)	31.8 (± 1.4)	57.1 (± 3.4)
MMVAE	91.3 (± 0.4)	68.0 (± 0.6)	N/A	42.1 (± 1.9)	86.4 (± 0.5)	69.1 (± 2.5)
DMVAE	**95.0** (± 0.6)	**79.9** (± 1.4)	**92.9** (± 1.8)	**85.9** (± 1.0)	**91.6** (± 0.8)	**76.4** (± 0.4)

Table 2. Comparison of generative quality on MNIST/SVHN, where x_1 corresponds to MNIST and x_2 to SVHN. Numbers represent median FIDs (lower is better) computed across 5 runs with standard deviations in parentheses. For the MMVAE, we computed FIDs based on the publicly available code.

Method	Unconditional FID		Conditional FID		Class-Conditional FID	
	x_1	x_2	$x_1 \to x_2$	$x_2 \to x_1$	$x_1 \to x_2$	$x_2 \to x_1$
MVAE	21.2 (± 1.1)	68.2 (± 1.9)	**65.0** (± 2.2)	19.3 (± 0.4)	**83.8** (± 1.8)	53.6 (± 1.9)
MMVAE	36.6 (± 3.1)	98.9 (± 1.5)	97.0 (± 0.6)	28.6 (± 1.1)	125.3 (± 0.8)	52.6 (± 4.8)
DMVAE	**15.7** (± 0.7)	**57.3** (± 3.6)	67.6 (± 4.0)	**18.7** (± 0.9)	91.9 (± 4.4)	**23.3** (± 1.0)

Table 1 presents linear latent classification accuracies as well as conditional and unconditional coherence results. Across all metrics, our model achieves significant improvements over previous methods. Most strikingly, joint coherence improves from 42.1% to 85.9% as a result of ex-post density estimation. As previously noted, it can be misleading to look only at latent classification and coherence, because these metrics do not capture the diversity of generated samples. Therefore, in Table 2 we also report FIDs for all models. In terms of FIDs, our model shows the best overall performance, with an exception in the conditional generation of SVHN given MNIST, for which the MVAE has slightly lower FIDs. However, looking at the results as a whole, DMVAE demonstrates a notable improvement compared to state-of-the-art multimodal generative models. Ablations across individual loss terms are provided in Appendix B.

Ex-post Density Estimation. [8], which we employ for sampling from the shared space of the DMVAE, proves to be very effective for improving certain metrics (Table 3). In particular, it can be used as an additional step after training, to improve the joint coherence and, partially, unconditional FIDs of already trained models. Note that ex-post density estimation does not influence any other metrics reported in Tables 1 and 2 (i.e., latent classification, conditional coherence, and conditional FID).

Table 3. Comparison of sampling from the prior vs. using ex-post density estimation with a Gaussian mixture model (GMM) with 100 components and full covariance matrix. After training, the GMM is fitted on the embeddings computed from the training data. For FIDs, the first number refers to MNIST, the second to SVHN, respectively. Overall, ex-post density estimation improves most metrics for both MVAE and MMVAE.

Sampling	MVAE		MMVAE		DMVAE	
	FIDs	Coherence	FIDs	Coherence	FIDs	Coherence
Prior	21.2 / 68.2	38.0	36.6 / 98.9	42.1	N/A	N/A
GMM	**13.4** / 73.7	68.5	28.7 / 119.7	80.3	15.7 / **57.3**	**85.9**

4.2 Paired MNIST

To investigate how well the aggregation of shared representations works for more than two modalities, we create a modified version of the MNIST dataset, which consists of M-tuples of images that depict the same digit. We view each image in the tuple $(x_1, \ldots, x_M)$ as coming from a different modality x_m–X_m, even though each instance is drawn from MNIST. Further we perturb each image with a high degree of Gaussian noise, which makes it difficult to infer digit labels from a single image (for an example, see Appendix B), and train the models as denoising variational autoencoders. We use comparable architectures, likelihoods, and training regimes across all methods. All implementation details are provided in Appendix C.

The dataset is generated by repeatedly pairing M images with the same label. We vary $M = 2, \ldots 5$ to investigate how the methods perform with an increasing number of modalities. This pairing is done separately for training and test data and results in 60,000 and 10,000 image M-tuples for the training and test sets respectively. The resulting dataset offers a simple benchmark that requires no modality-specific weights for the likelihood terms, has a clear characterization of shared and modality-specific factors, and allows visual inspection of the results.[5]

The goal of this experiment is to test whether models are able to integrate shared information across multiple modalities and if the aggregated representation improves with more modalities. To the best of our knowledge, experiments evaluating the aggregation with more than two modalities have not been performed before. Unlike the previous experiment, paired MNIST allows measuring how well models generate a missing modality given two or more inputs. To quantify this, we measure the average coherence over leave-one-out mappings $\{x_i\}_{i \neq j} \to x_j$. Further, we compute the average class-specific conditional FID over leave-one-out mappings, which combines both coherence and generative quality in a single metric.

Figure 3 presents the results for an increasing number of input modalities. The left subplot shows that for the MVAE and DMVAE leave-one-out coherence consistently improves with additional modalities, supporting our hypothesis that

---

[5] Note that the weights of likelihood terms have been observed to be important hyperparameters in both [40] and [31].

the PoE is effective in aggregating shared information. Notably, the MMVAE fails to take advantage of more than two modalities, as it does not have a shared representation that aggregates information. The right subplot shows that the DMVAE outperforms the other methods in class-specific conditional FIDs, demonstrating that it can achieve both high sample quality and strong coherence. We provide further metrics and ablations for this experiment in Appendix B.

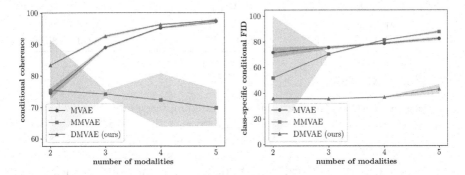

Fig. 3. Results on paired MNIST with varying number of "modalities". Markers denote median values, error-bars standard deviations, computed across 5 runs. *Left:* Leave-one-out conditional coherence (higher is better). *Right:* Class-specific conditional FIDs (lower is better).

5   Conclusion

We have introduced DMVAE, a novel multimodal generative model that learns a joint distribution over multiple modalities and disentangles modality-specific and shared factors completely self-supervised. The disentanglement allows sampling from modality-specific priors and thus facilitates the aggregation of shared information across modalities. We have demonstrated significant improvements in representation learning and generative performance compared to previous methods. Further, we have found that ex-post density estimation, that was used to sample from the shared latent space of the DMVAE, improves certain metrics dramatically when applied to trained models from existing work. This suggests that the latent space learned by multimodal generative models is more expressive than previously expected, which offers exciting opportunities for future work. Moreover, the DMVAE is currently limited to disentangling modality-specific and shared factors and one could extend it to more complex settings, such as graphs of latent factors.

Acknowledgements. Thanks to Mario Wieser for discussions on learning invariant subspaces, to Yuge Shi for providing code, and to Francesco Locatello for sharing his views on disentanglement in a multimodal setting. ID is supported by the SNSF grant #200021_188466.

References

1. Alemi, A.A., Fischer, I., Dillon, J.V., Murphy, K.: Deep variational information bottleneck. In: International Conference on Learning Representations (2017)
2. Almahairi, A., Rajeswar, S., Sordoni, A., Bachman, P., Courville, A.C.: Augmented CycleGAN: learning many-to-many mappings from unpaired data. In: International Conference on Machine Learning (2018)
3. Baltrušaitis, T., Ahuja, C., Morency, L.P.: Multimodal machine learning: a survey and taxonomy. IEEE Trans. Pattern Anal. Mach. Intell. **41**(2), 423–443 (2019)
4. Bouchacourt, D., Tomioka, R., Nowozin, S.: Multi-level variational autoencoder: learning disentangled representations from grouped observations. In: AAAI Conference on Artificial Intelligence (2018)
5. Chartsias, A., et al.: Disentangled representation learning in cardiac image analysis. Med. Image Anal. **58**, 101535 (2019)
6. Chen, X., Duan, Y., Houthooft, R., Schulman, J., Sutskever, I., Abbeel, P.: InfoGAN: interpretable representation learning by information maximizing generative adversarial nets. In: Advances in Neural Information Processing Systems (2016)
7. Choi, Y., Choi, M., Kim, M., Ha, J.W., Kim, S., Choo, J.: StarGAN: unified generative adversarial networks for multi-domain image-to-image translation. In: Conference on Computer Vision and Pattern Recognition (2018)
8. Ghosh, P., Sajjadi, M.S.M., Vergari, A., Black, M., Scholkopf, B.: From variational to deterministic autoencoders. In: International Conference on Learning Representations (2020)
9. Gresele, L., Rubenstein, P.K., Mehrjou, A., Locatello, F., Schölkopf, B.: The incomplete Rosetta Stone problem: identifiability results for multi-view nonlinear ICA. In: Conference on Uncertainty in Artificial Intelligence (2019)
10. Gutmann, M., Hyvärinen, A.: Noise-contrastive estimation: a new estimation principle for unnormalized statistical models. In: International Conference on Artificial Intelligence and Statistics (2010)
11. He, K., Fan, H., Wu, Y., Xie, S., Girshick, R.B.: Momentum contrast for unsupervised visual representation learning. In: Conference on Computer Vision and Pattern Recognition (2020)
12. Heusel, M., Ramsauer, H., Unterthiner, T., Nessler, B., Hochreiter, S.: GANs trained by a two time-scale update rule converge to a local Nash equilibrium. In: Advances in Neural Information Processing Systems (2017)
13. Higgins, I., et al.: beta-VAE: learning basic visual concepts with a constrained variational framework. In: International Conference on Learning Representations (2017)
14. Hinton, G.E.: Training products of experts by minimizing contrastive divergence. Neural Comput. **14**(8), 1771–1800 (2002)
15. Hjelm, R.D., et al.: Learning deep representations by mutual information estimation and maximization. In: International Conference on Learning Representations (2019)
16. Hsu, W.N., Glass, J.: Disentangling by partitioning: a representation learning framework for multimodal sensory data. arXiv preprint arXiv:1805.11264 (2018)
17. Hsu, W.N., Zhang, Y., Glass, J.: Unsupervised learning of disentangled and interpretable representations from sequential data. In: Advances in Neural Information Processing Systems (2017)
18. Ilse, M., Tomczak, J.M., Louizos, C., Welling, M.: DIVA: domain invariant variational autoencoders. arXiv preprint arXiv:1905.10427 (2019)

19. Khemakhem, I., Kingma, D.P., Monti, R.P., Hyvärinen, A.: Variational autoencoders and nonlinear ICA: a unifying framework. In: International Conference on Artificial Intelligence and Statistics (2020)
20. Kim, H., Mnih, A.: Disentangling by factorising. In: International Conference on Machine Learning (2018)
21. Kurle, R., Guennemann, S., van der Smagt, P.: Multi-source neural variational inference. In: AAAI Conference on Artificial Intelligence (2019)
22. LeCun, Y., Bottou, L., Bengio, Y., Haffner, P.: Gradient-based learning applied to document recognition. Proc. IEEE **86**(11), 2278–2324 (1998)
23. Li, Y., Mandt, S.: Disentangled sequential autoencoder. In: International Conference on Machine Learning (2018)
24. Liu, A.H., Liu, Y.C., Yeh, Y.Y., Wang, Y.C.F.: A unified feature disentangler for multi-domain image translation and manipulation. In: Advances in Neural Information Processing Systems (2018)
25. Locatello, F., et al.: Challenging common assumptions in the unsupervised learning of disentangled representations. In: International Conference on Machine Learning (2019)
26. Locatello, F., Abbati, G., Rainforth, T., Bauer, S., Schölkopf, B., Bachem, O.: On the fairness of disentangled representations. In: Advances in Neural Information Processing Systems (2019)
27. Locatello, F., Poole, B., Rätsch, G., Schölkopf, B., Bachem, O., Tschannen, M.: Weakly-supervised disentanglement without compromises. In: International Conference on Machine Learning (2020)
28. Nguyen, X., Wainwright, M.J., Jordan, M.I.: Estimating divergence functionals and the likelihood ratio by convex risk minimization. IEEE Trans. Inf. Theory **56**(11), 5847–5861 (2010)
29. Oord, A.v.d., Li, Y., Vinyals, O.: Representation learning with contrastive predictive coding. arXiv preprint arXiv:1807.03748 (2018)
30. Sermanet, P., Chintala, S., LeCun, Y.: Convolutional neural networks applied to house numbers digit classification. In: International Conference on Pattern Recognition, pp. 3288–3291. IEEE (2012)
31. Shi, Y., Siddharth, N., Paige, B., Torr, P.: Variational mixture-of-experts autoencoders for multi-modal deep generative models. In: Advances in Neural Information Processing Systems (2019)
32. Smith, N.A., Eisner, J.: Contrastive estimation: training log-linear models on unlabeled data. In: Proceedings of the 43rd Annual Meeting on Association for Computational Linguistics, pp. 354–362 (2005)
33. Stein, B.E., Stanford, T.R., Rowland, B.A.: The neural basis of multisensory integration in the midbrain: its organization and maturation. Hear. Res. **258**(1–2), 4–15 (2009)
34. Sugiyama, M., Suzuki, T., Kanamori, T.: Density-ratio matching under the Bregman divergence: a unified framework of density-ratio estimation. Ann. Inst. Stat. Math. **64**(5), 1009–1044 (2012)
35. Suzuki, M., Nakayama, K., Matsuo, Y.: Joint multimodal learning with deep generative models. arXiv preprint arXiv:1611.01891 (2016)
36. Tian, Y., Engel, J.: Latent translation: crossing modalities by bridging generative models. arXiv preprint arXiv:1902.08261 (2019)
37. Träuble, F., et al.: Is independence all you need? On the generalization of representations learned from correlated data. arXiv preprint arXiv:2006.07886 (2020)

38. Tsai, Y.H.H., Liang, P.P., Zadeh, A., Morency, L.P., Salakhutdinov, R.: Learning factorized multimodal representations. In: International Conference on Learning Representations (2019)
39. Wieser, M., Parbhoo, S., Wieczorek, A., Roth, V.: Inverse learning of symmetry transformations. In: Advances in Neural Information Processing Systems (2020)
40. Wu, M., Goodman, N.: Multimodal generative models for scalable weakly-supervised learning. In: Advances in Neural Information Processing Systems (2018)
41. Yildirim, I.: From perception to conception: learning multisensory representations. Ph.D. thesis, University of Rochester (2014)
42. Yildirim, I., Jacobs, R.A.: Transfer of object category knowledge across visual and haptic modalities: experimental and computational studies. Cognition **126**(2), 135–148 (2013)

Multimodal Semantic Forecasting Based on Conditional Generation of Future Features

Kristijan Fugošić[✉], Josip Šarić, and Siniša Šegvić

University of Zagreb Faculty of Electrical Engineering and Computing,
Zagreb, Croatia
kfugosic@gmail.com

Abstract. This paper considers semantic forecasting in road-driving scenes. Most existing approaches address this problem as deterministic regression of future features or future predictions given observed frames. However, such approaches ignore the fact that future can not always be guessed with certainty. For example, when a car is about to turn around a corner, the road which is currently occluded by buildings may turn out to be either free to drive, or occupied by people, other vehicles or roadworks. When a deterministic model confronts such situation, its best guess is to forecast the most likely outcome. However, this is not acceptable since it defeats the purpose of forecasting to improve security. It also throws away valuable training data, since a deterministic model is unable to learn any deviation from the norm. We address this problem by providing more freedom to the model through allowing it to forecast different futures. We propose to formulate multimodal forecasting as sampling of a multimodal generative model conditioned on the observed frames. Experiments on the Cityscapes dataset reveal that our multimodal model outperforms its deterministic counterpart in short-term forecasting while performing slightly worse in the mid-term case.

1 Introduction

Self-driving cars are today's burning topic [27]. With their arrival, the way that we look at passenger and freight traffic will change forever. But in order to solve such a complex task, we must first solve a series of "simpler" problems. One of the most important elements of an autonomous driving system is the ability to recognize and understand the environment [2,7]. It is very important that the system is able to recognize roads, pedestrians moving along or on the pavement,

This work has been funded by Rimac Automobili and supported by the European Regional Development Fund under the grant KK.01.1.1.01.0009 DATACROSS.

Electronic supplementary material The online version of this chapter (https://doi.org/10.1007/978-3-030-71278-5_34) contains supplementary material, which is available to authorized users.

Z. Akata et al. (Eds.): DAGM GCPR 2020, LNCS 12544, pp. 474–487, 2021.
https://doi.org/10.1007/978-3-030-71278-5_34

other cars and all other traffic participants. This makes semantic segmentation a very popular problem [26,30,31].

However, the ability to predict the future is an even more important attribute of intelligent behavior [16,21,23–25]. It is intuitively clear that critical real-time systems such as autonomous driving controllers could immensely benefit from the ability to predict the future by considering the past [4,17,27]. Such systems could make much better decisions than their counterparts which are able to perceive only the current moment. Unfortunately, this turns out to be a very hard problem. Most of the current work in the field approaches it very conservatively, by forecasting only unimodal future [16,22]. However, this approach makes an unrealistic assumption that the future is completely determined by the past, which makes it suitable for guessing only the short-term future. Hence, deterministic forecasting approaches will be prone to allocate most of its forecasts to instances of common large classes such as cars, roads, sky and similar. On the other side, such approaches will often underrepresent smaller objects. When it comes to signs, poles, pedestrians or some other thin objects, it makes more sense for a conservative model to allocate more space to the background than to risk classifying them. Additionally, future locations of dynamic and articulated objects such as pedestrians or domestic animals would also be very hard to forecast by a deterministic approach.

In order to address problems of unimodal forecasting, this work explores how to equip a given forecasting model with somewhat more freedom, by allowing and encouraging prediction of different futures. Another motivation for doing so involves scenarios where previously unseen space is unoccluded. Such scenarios can happen when we are turning around a corner or when another car or some larger vehicle is passing by. Sometimes we can deduce what could be in that new space by observing recent past, and sometimes we simply can't know. In both cases, we would like our model to produce a distribution over all possible outcomes in a stochastic environment [1,4,9,17]. We will address this goal by converting the basic regression model into a conditional generative model based on adversarial learning [18] and moment reconstruction losses [12].

2   Related Work

Dense Semantic Forecasting. Predicting future scene semantics is a prominent way to improve accuracy and reaction speed of autonomous driving systems. Recent work shows that direct semantic forecasting is more effective than RGB forecasting [16]. Further work proposes to forecast features from an FPN pyramid by multiple feature-to-feature (F2F) models [15]. This has recently been improved by single-level F2F forecasting with deformable convolutions [19,20].

Multimodal Forecasting. Future is uncertain and multimodal, especially in long-term forecasting. Hence, forecasting multiple futures is an interesting research goal. An interesting related work forecasts multi-modal pedestrian trajectories [9]. Similar to our work, they also achieve multimodality through a conditional GAN framework. Multi-modality has also been expressed through mixture density networks [17] in order to forecast egocentric localization and emergence prediction. None of these two works consider semantic forecasting.

To the best of our knowledge, there are only a few works in multimodal semantic forecasting, and all these works are either very recent [1] or concurrent [17]. One way to address multimodal semantic forecasting is to express inference within a Bayesian framework [1]. However, Bayesian methods are known for slow inference and poor real-time performance. Multi-modality can also be expressed within a conditional variational framework [4], by modelling interaction between the static scene, moving objects and multiple moving objects. However, the reported performance suggests that the task is far from being solved.

GANs with Moment Reconstruction. GANs [3] and their conditional versions [18] have been used in many tasks [6,11,14,28]. However, these approaches lack output diversity due to mode collapse. Recent work alleviates this problem with moment reconstruction loss [12] which also improves the training stability.

Improving Semantic Segmentation with Adversarial Loss. While most GAN discriminators operate on raw image level, they can also be applied to probabilistic maps. This can be used either as a standalone loss [8] or as a regularizer of the standard cross entropy loss [5].

3   Method

3.1   Conditional MR-GAN

Generative adversarial models [3] are comprised of two neural networks - a generator and a discriminator. Each of them has its own task and separate loss function. The goal of a generator is to produce diverse and realistic samples, while discriminator classifies given sample as either real (drawn from the dataset) or fake (generated). By conditioning the model it is possible to direct the data generation process. Generative adversarial networks can be extended to a conditional model if both the generator and discriminator are conditioned on some additional information [18]. Additional information can be of any kind, in our case it's a blend of features extracted from past frames.

However, both standard GAN and its conditional version are highly unstable to train. To counter the instability, most conditional GANs for image-to-image translation [6] use reconstruction (l1/l2) loss in addition to the GAN loss. While reconstruction loss forces model to generate samples similar to ground-truth, it often results in *mode collapse*. Mode collapse is one of the greatest problems of generative adversarial models. While we desire diverse outputs, mode collapse manifests itself as one-to-one mapping. This problem can be mitigated by replacing the traditional reconstruction loss with *moment reconstruction (MR)* losses which increase training stability and favour multimodal output generation [12].

The main idea of MR-GAN [12] is to use maximum likelihood estimation loss to predict conditional statistics of the real data distribution. Specifically, MR-GAN estimates the central measure and the dispersion of the underlying distribution, which correspond to mean and variance in the Gaussian case.

$$\mathcal{L}_{MLE,Gaussian} = \mathbb{E}_{x,y}\left[\frac{(y-\hat{\mu})^2}{2\hat{\sigma}^2} + \frac{1}{2}\log\hat{\sigma}^2\right], \text{ where } (\hat{\mu},\hat{\sigma}^2) = f_\theta(x) \quad (1)$$

Overall architecture of MR-GAN is similar to conditional GANs, with two important novelties:

1. Generator produces K different samples $\hat{y}_{1:K}$ for each image x by varying random noise $z_{1:K}$.
2. Loss function is applied to the sampled moments (mean and variance) in contrast to the reconstruction loss which is applied directly on the samples.

They estimate the moments of the generated distribution as follows:

$$\tilde{\mu} = \frac{1}{K}\sum_{i=1}^{K}\tilde{y}_i, \ \tilde{\sigma}^2 = \frac{1}{K-1}\sum_{i=1}^{K}(\tilde{y}_i - \tilde{\mu})^2, \text{ where } \tilde{y}_{1:K} = G(x, z_{1:K}). \quad (2)$$

MR loss is calculated by plugging $\tilde{\mu}$ and $\tilde{\sigma}^2$ in Eq. 1. The loss thus obtained is called MR2, while they denote a loss that does not take into account variance with MR1.

For more stable learning, especially at an early stage, the authors suggest a loss called *Proxy Moment Reconstruction (proxy MR)* loss. As It was shown in [12] that MR and proxy MR losses achieve similar results on Pix2Pix [6] problem, we will use simpler MR losses for easier, *end-to-end*, training.

3.2 F2F Forecasting

Most of the previous work in forecasting focuses on predicting raw RGB future frames and subsequent semantic segmentation. Success in that area would be a significant achievement because it would make possible to train on extremely large set of unmarked learning data. However, problems such as autonomous driving require the program to recognize the environment on a semantically meaningful level. In that sense forecasting on RGB level is an unnecessary complication. As many attempts in feature-to-feature forecasting were based on semantic segmentation, in [15] they go a step further and predict the semantic future at the instance level. This step facilitates understanding and prediction of individual objects trajectories. The proposed model shares much of the architecture with the Mask R-CNN, with the addition of predicting future frames. Since the number of objects in the images varies, they do not predict the labels of the objects directly. Instead, they predict convolutional features of fixed dimensions. Those features are then passed through the detection head and upsampling path to get final predictions.

3.3   Single-Level F2F Forecasting

Šarić et al. in their paper [19] proposed a single-level F2F model with deformable convolutions. The proposed model, denoted as DeformF2F, brings few notable changes compared to [15]:

1. Single-level F2F model which performs on last, spatially smallest, resolution
2. Deformable convolutions instead of classic or dilated ones
3. Ability to fine tune two separately trained submodels (F2F and submodel for semantic segmentation).

DeformF2F achieves *state-of-the-art* performance on mid-term (t + 9) prediction, and second best result on short-term (t + 3) prediction.

3.4   Multimodal F2F Forecasting

We use modified single-level F2F forecasting model as a generator, customized PatchGAN [13] as a discriminator, while MR1 and MR2 losses are used in order to achieve diversity in predictions. We denote our model as MM-DeformF2F (Multimodal DeformF2F) (Fig. 1).

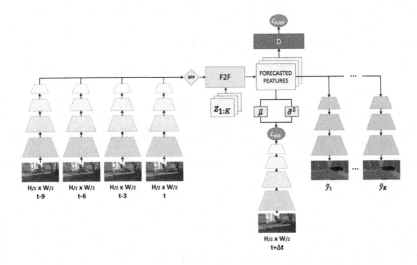

Fig. 1. Structural diagram of our model. Base structure is similar to [19]. It is composed of feature extractor (yellow), upsampling branch (blue), spatial pyramid pooling (SPP) and F2F. Additionally, we introduce random noise z, discriminator D and new loss functions. Our model generates multiple predictions. (Color figure online)

Generator. Generator is based on DeformF2F model. In order to generate diverse predictions, we introduce noise in each forward pass. Gaussian noise tensor has 32 channels and fits the spatial dimensions of the input tensor. Instead of one, we now generate K different predictions with the use of K different noise tensors. Generator is trained with MR and GAN loss applied to those predictions.

Discriminator. According to the proposal from [12], we use PatchGAN as a discriminator. Since input features of our PatchGAN are of significantly smaller spatial dimensions, we use it in a modified form with a smaller number of convolutional layers. Its purpose is still to reduce the features to smaller regions, and then to judge each region as either fake (generated) or real (from dataset). Decisions across all patches are averaged in order to bring final judgment in the form of 0 to 1 scalar. Since the discriminator was too dominant in learning, we introduced dropout in its first convolutional layer. In general, we shut down between 50 and 65% of features.

Dataset. Following the example of [19], we use video sequences from the Cityscapes dataset. The set contains 2975 scenes (video sequences) for learning, 500 for validation and 1525 for testing with labels for 19 classes. Each scene is described with 30 images, with a total duration of 1.8 s. That means that dataset contains a total of 150,000 images with resolution 1024×2048 pixels. Ground-truth semantic segmentation is available for the 20th image of each scene. Since introduction of GAN methods made our model more complicated, all images from the dataset were halved in width and height in order to reduce the number of features and speed up training.

Training Procedure. In last paragraph we described the dataset and the initial processing of the input data. If we denote the current moment as t, then in short-term prediction we use convolutional features at moments $t-9$, $t-6$, $t-3$ in order to predict the semantic segmentation at moment $t+3$, or at moment $t+9$ for mid-term forecasting. Features have spatial dimensions 16×32 and 128 channels. Training can be divided into two parts. First, we jointly train feature extractor and upsampling branch with cross entropy loss [10,19]. All images later used for training are passed through feature extracting branch and the resulting features are stored on SSD drive. We later load those features instead of passing through feature extractor, as that saves us time in successive training and evaluation of the model. In the second part, we train the F2F model in an unsupervised manner. Unlike [19], instead of L2 loss we use MR loss and GAN loss. We give a slightly greater influence to the reconstruction loss ($\lambda_{MR} = 100$) than adversarial ($\lambda_{GAN} = 10$). For both the generator and the discriminator, we use Adam optimizer with a learning rate of $4 \cdot 10^{-4}$ and decay rates 0.9 and 0.99 for the first and second moment estimates, respectively. We reduce the learning rate using cosine annealing without restart to a minimum value of $1 \cdot 10^{-7}$. To balance the generator and the discriminator, we introduce dropout in first convolutional layer of the discriminator. As an example, training short-term forecasting task with MR1 loss without dropout begins to stagnate as early as the fortieth epoch, where mIoU is 1.5 to 2% points less than the best results achieved.

4   Experiments

We show average metrics across 3 trained models and multiple evaluations for each task. In every forward pass we generate 8 predictions. We use mIoU as our main metric for accuracy, while MSE and LPIPS are used to express diversity as explained below.

MSE. Mean Squared Error is our main diversity metric. We measure Euclidean distance on pixel level between every two generated predictions for each scene, and take mean over whole dataset.

LPIPS. Following the example of [12], we also use LPIPS (*Learned Perceptual Image Patch Similarity* [29]) to quantify the diversity of generated images. In [29] they have shown that deep features can be used to describe similarity between two images while outperforming traditional measures like L2 or SSIM. We measure LPIPS between every two generated predictions for each scene, and take mean over whole dataset. Since we don't generate RGB images, but instead predict semantic future which has limited structure, MSE has proved to be sufficient measure for diversity.

4.1   Visual Assessment

In addition to the numerical results, in following subsections we will also show generated predictions, as well as two gray images:

a) Mean logit variance

```
logits.var(dim=0).mean(dim=0)
```

b) Variance of discrete predictions

```
logits.argmax(dim=1).double().var(dim=0)
```

An example of gray images is shown in Fig. 2. The first gray image highlights areas of uncertainty, while on the second image we observe areas that are classified into different classes on different generated samples.

4.2   Experimental Results on Cityscapes Dataset

We conducted experiments on short-term and mid-term forecasting tasks with roughly the same hyperparameters. With MR1 loss we observe mIoU which is on par with baseline model, and slightly greater in the case of short-term forecasting. On the other hand, using MR2 loss resulted in lower mIoU, but predictions are

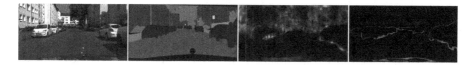

Fig. 2. Shown in the following order: future frame and its ground truth segmentation, mean logit variance and variance of discrete predictions. The first gray image highlights areas of uncertainty, while on the second gray image we see areas that are classified into different classes on different generated samples. The higher the uncertainty, the whiter the area.

a lot more diverse compared to MR1. Although we could get higher mIoU on mid-term forecasting with minimal changes in hyperparameters at the cost of diversity, we do not intervene because mIoU is not the only relevant measure in this task. Accordingly, although MR2 lowers mIoU by 5 or more percentage points, we still use it because of the greater variety. Visually most interesting predictions were obtained by using MR2 on short-term forecasting task. One of those is shown on Fig. 3. Notice that people are visible in the first frame and obscured by the car in the last frame. In the future moment, the car reveals the space behind it, and for the first time our model predicts people in correct place (second row, first prediction). We failed to achieve something like that when using MR1 loss or with baseline model. Such predictions are possible with the MR2 loss, but still rare, as in this particular case model recognized people in the right place in only one of the twelve predictions.

Fig. 3. Short-term forecasting with MR2 loss. Row 1 shows the first and the last input frame, the future frame, and its ground-truth segmentation. Row 2 shows 4 out of 12 model predictions.

Our main results are shown in Tables 1 and 2. Since results in [19] were obtained on images of full resolution, we retrained their model on images with halved height and width. In tables we show average mIoU and mIoU-MO (**M**oving **O**bjects) across five different models. We also show results achieved with Oracle, single-frame model used to train the feature extractor and the upsampling path, which "predicts" future segmentation by observing a future frame. While oracle represents upper limit, *Copy last segmentation* can be seen as lower bound, or as a good difficulty measure for this task. We get a slightly better mIoU if we average the predictions, although this contradicts the original idea of this paper. Like in [1], we also observe slight increase of mIoU when comparing top 5% to averaged predictions. Performance boost is best seen when we look at moving objects accuracy (mIoU-MO) while using MR1 loss, as we show in more detail in supplementary.

Table 1. Short-term prediction results. While our model achieves slightly higher mIoU with MR1 loss, MR2 results in much more diverse predictions.

Method	Loss	mIoU	mIoU-MO	MSE	LPIPS
Oracle	L2	66.12	64.24	/	/
Copy last segmentation	L2	49.87	45.63	/	/
DeformF2F-8 [19]	L2	$58.98^{\pm0.17}$	$56.00^{\pm0.20}$	/	/
MM-DeformF2F-8	MR1	$59.22^{\pm0.05}$	$56.40^{\pm0.08}$	$1.21^{\pm0.05}$	0.0482
MM-DeformF2F-8 avg	MR1	$59.46^{\pm0.06}$	$56.66^{\pm0.09}$	/	/
MM-DeformF2F-8	MR2	$53.85^{\pm0.35}$	$49.52^{\pm0.64}$	$4.38^{\pm0.24}$	0.1519
MM-DeformF2F-8 avg	MR2	$56.81^{\pm0.20}$	$53.38^{\pm0.43}$	/	/

Table 2. Mid-term prediction results. While our model achieves slightly lower mIoU with MR1 loss, MR2 results in much more diverse predictions.

Method	Loss	mIoU	mIoU-MO	MSE	LPIPS
Oracle	L2	66.12	64.24	/	/
Copy last segmentation	L2	37.24	28.31	/	/
DeformF2F-8 [19]	L2	$46.36^{\pm0.44}$	$40.78^{\pm0.99}$	/	/
MM-DeformF2F-8	MR1	$46.23^{\pm0.28}$	$41.07^{\pm0.57}$	$2.81^{\pm0.32}$	0.1049
MM-DeformF2F-8 avg	MR1	$46.96^{\pm0.21}$	$41.90^{\pm0.53}$	/	/
MM-DeformF2F-8	MR2	$37.48^{\pm0.31}$	$29.66^{\pm0.60}$	$7.42^{\pm0.44}$	0.2279
MM-DeformF2F-8 avg	MR2	$40.32^{\pm0.12}$	$32.42^{\pm0.37}$	/	/

4.3   Impact of the Number of Predictions on Performance

Table 3 shows the impact of the number of generated predictions (K) on their diversity and measured mIoU. We can see that larger number of generated predictions contributes to greater diversity, while slightly reducing mIoU. In training, we use $K = 8$ because of the acceptable training time and satisfactory diversity. Training with $K = 16$ would take about twice as long. We discuss memory overhead and evaluation time in supplementary material.

Table 3. Impact of the number of generated samples (K) on mIoU and diversity, measured on mid-term prediction task and MR1 loss. We can see that a larger number of generated samples contributes to greater diversity and slightly reduces mIoU. Testing was performed at an early stage of the paper, and the results are somewhat different from those in the Table 2. Since we evaluate on only one model for each K, due to high variance in both training and evaluation mIoU values don't necessarily represent real situation. Furthermore, in some tasks training with $K = 12$ or $K = 16$ sometimes showed better accuracy. The number of generated samples listed in the table refers to the learning phase, while we measure mIoU, MSE and LPIPS on 8 generated samples in the model exploitation phase.

K	mIoU	MSE	LPIPS
16	45.28	4.105	0.1496
8	45.81	3.004	0.1264
4	46.48	2.078	0.0947
2	46.44	1.424	0.0608
1	46.32	0.014	0.0010

4.4 Importance of GAN Loss

To show that the output diversity is not only due to the use of moment reconstruction losses and random noise, we trained the model without adversarial loss ($\lambda_{GAN} = 0$). In this experiment we were using MR1 loss on short-term forecasting task and with generator producing 8 predictions for each input image. Although some diversity is visible at an early stage (MSE around 0.7), around the 12th epoch the diversity is less and less noticeable (MSE around 0.35), and after the 40th it can barely be seen (MSE around 0.1). The model achieved its best mIoU 59.18 in epoch 160 (although it was trained on 400 epochs), and the MSE measure in that epoch was 0.08. Figure 4a shows the improvement in performance through the epochs, but with gradual weakening of diversity. For this example we chose an image with a lot of void surfaces, due to the fact that greatest diversity is usually seen in those places. On Fig. 4b we show the same scene, but predictions were obtained on a model trained with weights $\lambda_{GAN} = 10$ and $\lambda_{MR} = 100$. It took the model 232 epochs to achieve its best mIoU which is 59.14, but MSE held stable above 1 until the last, 400th, epoch.

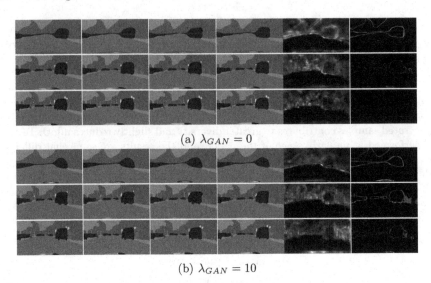

(a) $\lambda_{GAN} = 0$

(b) $\lambda_{GAN} = 10$

Fig. 4. The first four columns show 4 out of 8 generated predictions. The last two columns show mean logit variance and the variance of discrete predictions. Samples in the three rows are generated in epochs 1, 13 and 161, respectively. On image a), in the last row we observe almost indistinguishable predictions with second gray image being entirely black. On image b), we see some diversity in people behind a van, and last image shows some diversity in predicted classes.

4.5   Diversity of Multiple Forecasts

We have seen that averaging generated predictions before grading them increases mIoU - from 0.2 up to 3% points, depending on the task. Therefore, we propose a novel metric for measuring plausibility of multimodal forecasting. The proposed metric measures percentage of pixels that were correctly classified at least once through multiple forecasts. We distinguish three cases by looking at:

1. Every pixel except void class
2. Only pixels of movable objects
3. Only pixels that were correctly classified by Oracle.

We measure at multiple checkpoints (1, 2, 4, ..., 128) and present the obtained results in Fig. 5. On Fig. 5 we compare short term forecasting using MR1 and MR2 losses, while in supplementary material we also show additional line which represents *best so far* prediction.

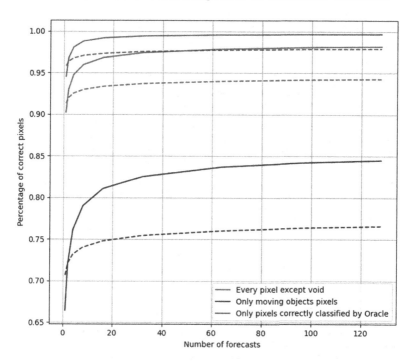

Fig. 5. Number of future pixels that were correctly classified at least once depending on the number of forecasts. We show three different cases with lines of different colors, as described by legend. Full lines represent MR2 short-term model, while dashed lines represent results with MR1 short-term model.

5   Conclusion and Future Work

We have presented a novel approach for multimodal semantic forecasting in road driving scenarios. Our approach achieves multi-modality by injecting random noise into the feature forecasting module. Hence, the feature forecast module becomes a conditional feature-to-feature generator which is trained by minimizing the moment reconstruction loss, and by maximizing the loss of a patch-level discriminator. Both the generator and the discriminator operate on abstract features.

We have also proposed a novel metric for measuring plausibility of multimodal forecasting. The proposed metric measures the number of forecasts required to correctly guess a given proportion of all future pixels. We encourage the metric to reflect forecasting performance by disregarding pixels which are not correctly guessed by the oracle.

The inference speed of our multi-modal model is similar to the uni-modal baseline. Experiments show that the proposed setup is able to achieve considerable diversity in mid-term forecasting. MR2 loss brings more diversity compared to MR1, however it reduces mIoU by around 5% points. Inspection of the generated forecasts, reveals that the model is sometimes still hesitating to replace close and large objects, but it often accepts to take a chance on close and dynamic or small and distant objects, like bikes and pedestrians.

In the future work we shall consider training with *proxy MR1* and *proxy MR2* losses. We should also consider using different discriminator, for example the one with global contextual information. Also, one of the options is to try concatenating features with their spatial pools prior to the discriminator. Other suitable future directions include evaluating performance on the instance segmentation task and experimenting with different generative models.

References

1. Bhattacharyya, A., Fritz, M., Schiele, B.: Bayesian prediction of future street scenes using synthetic likelihoods. In: International Conference on Learning Representations (2019)
2. Chen, L., et al.: Leveraging semi-supervised learning in video sequences for urban scene segmentation. CoRR abs/2005.10266 (2020)
3. Goodfellow, I.J., et al.: Generative adversarial networks (2014)
4. Hu, A., Cotter, F., Mohan, N., Gurau, C., Kendall, A.: Probabilistic future prediction for video scene understanding. In: Vedaldi, A., Bischof, H., Brox, T., Frahm, J.-M. (eds.) ECCV 2020. LNCS, vol. 12361, pp. 767–785. Springer, Cham (2020). https://doi.org/10.1007/978-3-030-58517-4_45
5. Hung, W., Tsai, Y., Liou, Y., Lin, Y., Yang, M.: Adversarial learning for semi-supervised semantic segmentation. In: British Machine Vision Conference 2018, BMVC 2018, Northumbria University, Newcastle, UK, 3–6 September 2018, p. 65 (2018)
6. Isola, P., Zhu, J.Y., Zhou, T., Efros, A.A.: Image-to-image translation with conditional adversarial networks (2016)
7. Kirillov, A., He, K., Girshick, R., Rother, C., Dollár, P.: Panoptic segmentation. arXiv preprint arXiv:1801.00868 (2018)
8. Kohl, S., et al.: Adversarial networks for the detection of aggressive prostate cancer (2017)
9. Kosaraju, V., Sadeghian, A., Martín-Martín, R., Reid, I., Rezatofighi, H., Savarese, S.: Social-BiGAT: multimodal trajectory forecasting using bicycle-GAN and graph attention networks. Adv. Neural Inf. Process. Syst. **32**, 137–146 (2019)
10. Krešo, I., Krapac, J., Šegvić, S.: Efficient ladder-style DenseNets for semantic segmentation of large images. IEEE Trans. Intell. Transp. Syst. (2020)
11. Ledig, C., et al.: Photo-realistic single image super-resolution using a generative adversarial network (2016)
12. Lee, S., Ha, J., Kim, G.: Harmonizing maximum likelihood with GANs for multimodal conditional generation (2019)
13. Li, C., Wand, M.: Precomputed real-time texture synthesis with Markovian generative adversarial networks. In: Leibe, B., Matas, J., Sebe, N., Welling, M. (eds.) ECCV 2016. LNCS, vol. 9907, pp. 702–716. Springer, Cham (2016). https://doi.org/10.1007/978-3-319-46487-9_43
14. Liang, X., Lee, L., Dai, W., Xing, E.P.: Dual motion GAN for future-flow embedded video prediction. In: Proceedings of the IEEE International Conference on Computer Vision. pp. 1744–1752 (2017)

15. Luc, P., Couprie, C., LeCun, Y., Verbeek, J.: Predicting future instance segmentation by forecasting convolutional features. In: Ferrari, V., Hebert, M., Sminchisescu, C., Weiss, Y. (eds.) ECCV 2018. LNCS, vol. 11213, pp. 593–608. Springer, Cham (2018). https://doi.org/10.1007/978-3-030-01240-3_36

16. Luc, P., Neverova, N., Couprie, C., Verbeek, J., LeCun, Y.: Predicting deeper into the future of semantic segmentation. In: Proceedings of the IEEE International Conference on Computer Vision, pp. 648–657 (2017)

17. Makansi, O., Cicek, O., Buchicchio, K., Brox, T.: Multimodal future localization and emergence prediction for objects in egocentric view with a reachability prior. In: The IEEE/CVF Conference on Computer Vision and Pattern Recognition (CVPR), June 2020

18. Mirza, M., Osindero, S.: Conditional generative adversarial nets (2014)

19. Šarić, J., Oršić, M., Antunović, T., Vražić, S., Šegvić, S.: Single level feature-to-feature forecasting with deformable convolutions. In: Fink, G.A., Frintrop, S., Jiang, X. (eds.) DAGM GCPR 2019. LNCS, vol. 11824, pp. 189–202. Springer, Cham (2019). https://doi.org/10.1007/978-3-030-33676-9_13

20. Saric, J., Orsic, M., Antunovic, T., Vrazic, S., Segvic, S.: Warp to the future: joint forecasting of features and feature motion. In: 2020 IEEE/CVF Conference on Computer Vision and Pattern Recognition, CVPR 2020, Seattle, WA, USA, 13–19 June 2020, pp. 10645–10654. IEEE (2020)

21. Su, S., Pyo Hong, J., Shi, J., Soo Park, H.: Predicting behaviors of basketball players from first person videos. In: Proceedings of the IEEE Conference on Computer Vision and Pattern Recognition, pp. 1501–1510 (2017)

22. Sun, J., Xie, J., Hu, J., Lin, Z., Lai, J., Zeng, W., Zheng, W.: Predicting future instance segmentation with contextual pyramid ConvLSTMs. In: ACM MM, pp. 2043–2051 (2019)

23. Terwilliger, A., Brazil, G., Liu, X.: Recurrent flow-guided semantic forecasting. In: 2019 IEEE Winter Conference on Applications of Computer Vision (WACV), pp. 1703–1712. IEEE (2019)

24. Vondrick, C., Pirsiavash, H., Torralba, A.: Anticipating the future by watching unlabeled video. arXiv preprint arXiv:1504.08023 2 (2015)

25. Vukotić, V., Pintea, S.-L., Raymond, C., Gravier, G., van Gemert, J.C.: One-step time-dependent future video frame prediction with a convolutional encoder-decoder neural network. In: Battiato, S., Gallo, G., Schettini, R., Stanco, F. (eds.) ICIAP 2017. LNCS, vol. 10484, pp. 140–151. Springer, Cham (2017). https://doi.org/10.1007/978-3-319-68560-1_13

26. Yang, M., Yu, K., Zhang, C., Li, Z., Yang, K.: DenseASPP for semantic segmentation in street scenes. In: CVPR, pp. 3684–3692 (2018)

27. Yao, Y., Xu, M., Choi, C., Crandall, D.J., Atkins, E.M., Dariush, B.: Egocentric vision-based future vehicle localization for intelligent driving assistance systems. In: ICRA (2019)

28. Yu, J., Lin, Z., Yang, J., Shen, X., Lu, X., Huang, T.S.: Generative image inpainting with contextual attention. In: The IEEE Conference on Computer Vision and Pattern Recognition (CVPR), June 2018

29. Zhang, R., Isola, P., Efros, A.A., Shechtman, E., Wang, O.: The unreasonable effectiveness of deep features as a perceptual metric (2018)

30. Zhao, H., Shi, J., Qi, X., Wang, X., Jia, J.: Pyramid scene parsing network. In: ICCV (2017)

31. Zhen, M., Wang, J., Zhou, L., Fang, T., Quan, L.: Learning fully dense neural networks for image semantic segmentation. In: AAAI (2019)

Author Index